Advances in Experimental Medicine and Biology

Volume 813

Advances in Experimental Medicine and Biology

For further volumes:
http://www.springer.com/series/5584

Helen E. Scharfman • Paul S. Buckmaster
Editors

Issues in Clinical Epileptology: A View from the Bench

Helen E. Scharfman
The Nathan S. Kline Institute for Psychiatric Research,
Orangeburg, NY, USA
Departments of Child & Adolescent Psychiatry,
Physiology & Neuroscience, and Psychiatry,
New York University Langone Medical Center,
New York, NY, USA

Paul S. Buckmaster
Departments of Comparative Medicine
and Neurology & Neurological Sciences,
Stanford University, Stanford, CA, USA

 Springer

Editors
Helen E. Scharfman
The Nathan S. Kline Institute
  for Psychiatric Research
Orangeburg, NY, USA

Departments of Child & Adolescent
  Psychiatry, Physiology &
  Neuroscience, and Psychiatry
New York University Langone
  Medical Center
New York, NY, USA

Paul S. Buckmaster
Departments of Comparative Medicine
  and Neurology & Neurological
  Sciences
Stanford University
Stanford, CA, USA

ISSN 0065-2598   ISSN 2214-8019 (electronic)
ISBN 978-94-017-8913-4   ISBN 978-94-017-8914-1 (eBook)
DOI 10.1007/978-94-017-8914-1
Springer Dordrecht Heidelberg New York London

Library of Congress Control Number: 2014942054

Printed on acid-free paper

Springer is part of Springer Science+Business Media (www.springer.com)

Foreword

This book is a tribute to Phil Schwartzkroin, who, in addition to over four decades of committed original research into fundamental mechanisms of epilepsy, has been a consummate editor, contributing to archival knowledge as well as the synthesis of new information. Phil served as editor-in-chief of *Epilepsia*, the journal of the International League against Epilepsy (ILAE), editor of the *Encyclopedia of Basic Epilepsy Research* [1], and editor of definitive textbooks on animal models of epilepsy [2, 3], brain development and epilepsy [4, 5], and brain plasticity and epilepsy [6]. This book is a fitting acknowledgment of Phil Schwartzkroin's career achievements, as an edited volume that addresses many of the most pressing research issues concerning neuronal mechanisms underlying epilepsy that were, and continue to be, his passion.

Epilepsy is among the most common serious neurological diseases. According to a study by the World Health Organization, epilepsy accounts for 1% of the global burden of disease [7]. This is equivalent to breast cancer in women and lung cancer in men. Among primary disorders of the brain, epilepsy ranks with depression and other affective disorders, Alzheimer's disease and other dementias, and substance abuse [7]. Public attention on epilepsy, however, and the resultant amount of resources devoted to research on epilepsy, is but a small fraction of that for these other medical conditions. The fact that epilepsy has been a stigmatized disease in most cultures since antiquity might be one reason why it has remained in the shadows, but interest in epilepsy also suffers because it is a complicated multifactorial condition with such diverse manifestations that clinical research alone to elucidate comprehensive underlying fundamental neuronal mechanisms is essentially not possible.

Historical Perspective

Epilepsy is an ancient disease, being both common and easy to recognize in antiquity. There is a long history about epilepsy being attributed to many prevailing causes, but it was not until the late nineteenth and early twentieth century that modern concepts of epilepsy as a disease of the brain attributed to excessive neuronal activity was first formulated. Before then, both the clinical phenomenology and some of the brain pathology associated with epilepsy were described, but the advent of electrophysiological recordings from the human brain by Berger and colleagues initiated a new way of studying the disease [8]. It quickly became apparent that during seizures there was a dramatic change in the electroencephalogram (EEG) and that in many patients there was also an alteration in the EEG even between seizures when they were behaviorally "normal" [9, 10]. The interictal EEG could show "spikes" (fast, sharp transients) or spike and slow wave discharges. As the field of basic neurophysiology began to develop around that time, investigators were able to demonstrate, predominantly in animal models of seizures, that some neurons in the cortex fired abnormally during seizures and also during the interictal EEG spikes. From the early 1950s through the early 1970s, single unit studies predominantly were carried out in animal models of either acutely provoked seizures in neocortex [11–13] or hippocampus [14, 15]. Although most of our current concepts about the origin and spread of seizures were developed from this work, opportunities to investigate functional mechanisms at the cellular and subcellular level were limited.

In the early 1970s, remarkable methodological advances occurred in the capacity to understand fundamental physiological and pharmacological properties of mammalian brain function: the development of the brain slice and dissociated cell culture. At that time, Per Andersen in Oslo devised the ability to maintain a slice of mammalian hippocampus in a dish for many hours and to record from the cells with extracellular and intracellular microelectrodes [16]. Within a year or two, Phil was working in Anderson's laboratory to extend his studies on the mechanisms responsible for epileptic seizures and along the way to investigate many other important physiological functions of mammalian cortical neurons. He brought this preparation back to Stanford to collaborate with David Prince and others, and eventually in his own laboratory he continued to make significant contributions to our understanding of mechanisms underlying the development and spread of epileptic seizures.

Over the ensuing four decades, much has been learned about the electrophysiological substrates of epileptic seizure activity from studies with simplified slice preparations and from additional studies utilizing cell cultures of mammalian neocortex and hippocampus. However, two main conceptual problems persisted: the recognition that seizures artificially provoked in an in vitro preparation, or even those induced in a normal animal brain, are not epilepsy, and the mechanisms by which a normal brain can become chronically epileptic were not understood.

A person with epilepsy has an enduring epileptogenic abnormality responsible for the generation of spontaneous seizures, which continues to be present during the interictal state [17]. Although models of chronic focal epilepsy were created in rats and primates in the 1950s and 1960s with topical application of metals such as cobalt, iron, and alumina [2], scars were produced by these metals and made microelectrode recordings at the site of application – the area of most interest – difficult. In the 1970s, kindling became a popular animal model to study mechanisms of chronic epilepsy at the cellular level [18]; however, kindling is a model of secondary epileptogenesis and, as usually performed, kindling results in stimulation-induced seizures, not spontaneously generated seizures. Intensive investigations into mechanisms of chronic epilepsy were facilitated by the introduction of the status epilepticus models of mesial temporal lobe epilepsy (MTLE) with hippocampal sclerosis, the most common, and most pharmacoresistant, form of human epilepsy [19]. Status epilepticus induced by kainic acid, pilocarpine, or electrical stimulation causes a pattern of hippocampal cell loss and neuronal reorganization resembling human hippocampal sclerosis, and eventual spontaneous limbic seizures [20–22]. Opportunities for invasive studies of MTLE in the epilepsy surgery setting made parallel reiterative multidisciplinary animal/human investigations possible [23]. Epileptogenesis, the process by which a normal brain is converted to one that is capable of generating spontaneous seizures, is of increasing interest to epileptologists and, as yet, can only be pursued in animal models. It is now understood that these enduring changes occur in many brain networks, not just in the areas in which the seizures appear to originate, but also in areas into which seizures propagate, and even in more remote brain regions [24].

Epilepsy is a diverse disease, and the extent to which research results obtained from animal models of a few types of epilepsy, such as MTLE, apply to other types of epilepsy remains to be determined. Future research will require the use of a wide variety of animal models of human epilepsy, such as post-traumatic epilepsy, febrile convulsions, neonatal hypoxia, infantile spasms, and genetically engineered models of genetic epilepsies and genetic diseases associated with epilepsy, such as tuberous sclerosis [25, 26]. Creation and validation of experimental animal models of the diverse forms of human epilepsy are now a high priority in order to search for targets not only for antiseizure interventions, but also for antiepileptogenic interventions that can prevent or cure epilepsy [25, 26]. Phil continued to contribute importantly to resolving many of these questions in his later work. The following are brief discussions of the questions he chose for this volume, intended to stimulate the field of epilepsy research today:

The Role of Animal Models

The ideal approach to the study of human epilepsy is to investigate patients with epilepsy; however, ethical concerns, technical constraints, and cost dictate that most of the critical questions concerning fundamental neuronal mechanisms of epilepsy still need to be resolved with experimental animal models. Although attempts are being made to create models of entire epilepsy syndromes and diseases, epilepsy can also be broken down into its component parts, e.g., epileptogenesis, ictogenesis, seizure maintenance, seizure termination, postictal disturbances, and interictal disturbances, each of which may be modeled individually [27].

Is there more to learn about human epilepsy by studying acute seizures in animals? Acute seizures in a normal brain are not the same as chronic epilepsy. The pathophysiological and anatomical substrates of the enduring epileptogenic abnormalities underlying different types of human epilepsy need to be elucidated. Does this mean that we have already learned all we can from studying acute seizures?

Do people with acquired epilepsy have a genetic susceptibility? Although an increasing number of epilepsy genes are being identified as responsible for single-gene epilepsy conditions, these diseases are rare [28]. The genetic bases of inherited diseases *associated with* epilepsy, such as tuberous sclerosis, are being elucidated. Both of these directions provide opportunities to create animal models of specific epilepsy syndromes and diseases, using genetic engineering. More importantly, however, it is now apparent that most genetic epilepsies, formally referred to as idiopathic epilepsies, can result from multiple different genetic mutations, and these represent susceptibility genes rather than epilepsy genes. The distinction between genetic and acquired epilepsies is not absolute, just as the distinction in the 1989 classification of the epilepsies between idiopathic and symptomatic disorders is a false dichotomy [29]. It is likely that some acquired disturbances are necessary for the manifestation of epilepsies primarily due to genetic abnormalities, and that genetic predispositions, susceptibility genes, influence the manifestation of epilepsies with acquired etiologies. Consequently, just as realization that acute seizures induced into a normal animal brain is not the same as epilepsy caused by an enduring epileptogenic abnormality was a paradigm shift, it must now also be realized that artificial introduction of an enduring epileptogenic abnormality into a normal animal brain is not the same as introduction of this abnormality into a brain genetically predisposed to generate specific types of epileptic abnormalities. In order to create more appropriate animal models of human epilepsy, more information is needed regarding specific susceptibility genes.

How relevant are animal models to human epileptic phenomena and how can they be validated? There are many different types of human epilepsy [30], and it is unreasonable to assume that any animal model will completely reproduce all aspects of a human epilepsy disease or syndrome. Rather, models will likely reproduce component parts of human epilepsies, and studies need to be designed to take advantage of the likely similarities while accounting for the differences between any given animal model and the type of human epilepsy that is being modeled. For the rare epilepsies caused by a single gene mutation, these mutations can be introduced into animals to investigate the pathophysiological consequences of their abnormal protein products, even if the phenotype does not resemble the human condition. Reiterative patient/animal investigations utilize clinical data to identify relevant questions make use of relevant animal models to pursue investigations that are not ethically or financially feasible in patients, and then validate results in the clinical population. This has been a valuable paradigm, particularly where invasive EEG recordings can be carried out in an epilepsy surgery setting and tissue is then available for analysis.

Comorbidity: Patients with epilepsy have a high incidence of comorbid conditions that can contribute significantly to disability. Many disturbances, such as depression, anxiety, attention deficit hyperactivity disorder, and autism, have a bidirectional relationship with epilepsy, suggesting shared mechanisms [31]. Because these conditions can precede epilepsy, it is not clear which condition is the comorbid one. Animal models of these conditions in association with epilepsy are necessary to begin investigations into fundamental neuronal mechanisms of epilepsy comorbidity.

Epileptic Activity

Questions persist concerning how to model and investigate the various types of epileptiform activity encountered in patients. In this book, experts in the field address several of the most pressing questions:

Human focal epilepsy is not focal; how can studies in animal models recreate the epileptic network necessary for the manifestation of human epilepsy? There probably is no such thing as a single discretely localized epileptic focus in chronic human epilepsy. Epilepsy manifests as a result of disturbances in distributed networks. The ILAE states: *"Focal epileptic seizures are conceptualized as originating within networks limited to one hemisphere. They may be discretely localized or more widely distributed. Focal seizures may originate in subcortical structures. For each seizure type, ictal onset is consistent from one seizure to another, with preferential propagation patterns that can involve the contralateral hemisphere. In some cases, however, there is more than one network, and more than one seizure type, but each individual seizure type has a consistent site of onset"* [30].

What is generalized epilepsy? The distinction between generalized and focal epilepsies in the 1989 ILAE classification of the epilepsies is a false dichotomy [28]. No epilepsy condition, or epileptic seizure, is truly generalized. The ILAE states: *"Generalized epileptic seizures are conceptualized as originating at some point within, and rapidly engaging, bilaterally distributed networks. Such bilateral networks can include cortical and subcortical structures, but do not necessarily include the entire cortex. Although individual seizure onsets can appear localized, location and lateralization are not consistent from one seizure to another. Generalized seizures can be asymmetric"* [30].

What are interictal EEG spikes and what is their significance? Some interictal EEG spikes may represent exactly the same underlying neuronal mechanisms as an ictal event, as, for instance, is the case with typical absence seizures; the so-called interictal events are too brief to be associated with obvious clinical behavior. In this situation, even with focal seizures, careful investigations can demonstrate behavioral disturbances during the so-called interictal spike [32]. Similarly, generalized paroxysmal fast activity (GPFA) without behavioral correlates, which can be seen in some patients with severe epilepsy, most likely represents the same underlying mechanisms as some low-voltage fast ictal discharges. These, therefore, are fragments of seizures and the terms "interictal spike" or "interictal GPFA" would be oxymorons. There are, however, different types of interictal spikes and not all represent fragments of ictal events. Some may, in fact, reflect seizure-suppressing mechanisms [33].

What are the limitations of studying epileptic phenomena in slice preparations? Epileptic seizures are defined clinically as behavioral events with an electrographic correlate [17]. Electrographic changes that occur in the slice preparation, therefore, cannot be called epileptic seizures, although they may reproduce certain neuronal events similar to those which would underlie behavioral seizures in the intact animal. Disturbances related to ictogenesis at the molecular, cellular, and perhaps microcircuit levels can be studied in slice preparation; however, disconnections from important influences of distant brain areas make it difficult to draw definitive conclusions concerning seizure generation at the level of whole-brain networks.

How is epilepsy mediated by non-neuronal influences? Not all epileptogenic, or homeostatic, mechanisms involve neurons. Glia play an important role in modulating neuronal activity, and other non-neuronal influences, such as hormonal changes, inflammatory and immune-mediated processes, and external toxic substances need to be considered.

Synaptic Plasticity

Concepts of epileptogenesis derived from an understanding of the development of human MTLE with hippocampal sclerosis, and reproduced in the animal laboratory, indicate that an initial epileptogenic insult causes cell loss, which is then followed by synaptic reorganization of surviving neuronal elements. Aberrant excitatory and inhibitory connections ultimately lead to epileptiform hypersynchronization. Epileptogenesis can occur in experimental animals, however, in the absence of obvious cell loss or synaptic reorganization, for instance with classical amygdala kindling, and epilepsy also occurs in patients who have no evidence of cell loss. Cell loss and synaptic reorganization may not, therefore, be a universal mechanism essential for epileptogenesis.

Changes during epileptogenesis can be protective: Neuronal plasticity occurring in response to injury can be responsible for the development of epilepsy, but homeostatic plastic changes also occur, resulting in protective seizure-suppressing influences. Investigations to identify pathophysiologic disturbances following an epileptogenic insult must clearly distinguish epileptogenic from homeostatic protective processes. These homeostatic changes could also be responsible for the appearance of interictal behavioral disturbances.

What is the significance of cell death in acquired epileptogenesis? Cell death is clearly not necessary for all forms of epilepsy, but when it occurs, it can be a cause of the epilepsy, or an affect of epileptic seizures.

Inhibition is not necessarily decreased in human epilepsy, and increased inhibition may be necessary for hypersynchronization: The old concept that epilepsy is due to an increase in excitation and a decrease in inhibition is clearly an oversimplification. In MTLE with hippocampal sclerosis and animal models of this condition, there is an increase in inhibition as well as in excitation [27]. Whereas some increased inhibition may have a protective effect, inhibition is also necessary for hypersynchronization, which is a component of most epileptic seizures. It is the types and location of aberrant excitatory and inhibitory synaptic reorganization that determine the epileptogenic process.

Features of epilepsy in the pediatric population differ considerably from those in patients with more mature brains: Synaptic plasticity leading to epileptogenesis is different in the developing brain than in the mature brain, and epileptic seizures can alter the synaptic plasticity necessary for normal brain development [4].

Research on patients with epilepsy is revealing increasing numbers of genetic aberrations, as well as disturbances in important protein products such as ion channels and neurotransmitter receptors; how do these defects explain the development and maintenance of epileptic phenomena? Plastic changes underlying epileptogenesis involve alterations in expression of genes whose protein products are ion channels, neurotransmitter receptors, and other membrane and intracellular structures that determine excitability. The location of these changes on the cell, and their influence on neuronal interconnections, also determine propensity for hypersynchronization. Although characterization of epileptogenic disturbances at the molecular and cellular levels do not reveal how epilepsy arises at the systems level, this research can help to identify novel targets for antiseizure and antiepileptogenic drugs designed to prevent and cure epilepsy, as well as control ictal events.

Conclusions

The enduring legacy of Phil Schwartzkroin is impossible to summarize here, but reflected well by the discussions in this volume, written by his colleagues, who have watched his contributions evolve over time. These discussions show how complex epilepsy is, that there is much to do to resolve the questions that are associated with epilepsy, and the approaches that have allowed us to make the most advances; using the best neuroscience and clinical epileptology together – an approach Phil mastered, and would want us all to continue.

Acknowledgements

Original research reported by the author was supported in part by NIH Grants P01 NS-02808, U01 NS-15654, R01 NS-33310, and P20 NS-80181, CURE, the Epilepsy Therapy Project, the Epilepsy Foundation, and the Resnick Foundation (JE).

References

1. Schwartzkroin PA (ed) (2009) Encyclopedia of basic epilepsy research. Academic, San Diego
2. Schwartzkroin PA (ed) (2007) Epilepsy: models, mechanisms and concepts. Cambridge University Press, Cambridge
3. Pitkänen A, Schwartzkroin PA, Moshé SL (eds) (2005) Models of seizures and epilepsy. Academic, San Diego
4. Schwartzkroin PA, Moshé SL, Noebels JL, Swann JW (eds) (1995) Brain development and epilepsy. Oxford University Press, New York
5. Schwartzkroin PA, Rho JM (eds) (2002) Epilepsy, infantile spasms, and developmental encephalopathy, vol 49 (International Review of Neurobiology). Academic, San Diego

6. Engel J Jr, Schwartzkroin PA, Moshé SL, Lowenstein DH (eds) Bradley RJ, Harris RA, Jenner P (series eds) (2001) Brain plasticity and epilepsy: a tribute to Frank Morrell, vol 45 (International Review of Neurology). Academic, San Diego

7. Murray CJL, Lopez AD (eds) (1994) Global comparative assessment in the health sector: disease burden, expenditures, and intervention packages. World Health Organization, Geneva

8. Berger H (1929) Uber das Flektrenkephalogram des Menschen. Arch Psychiatr Nervenkr 87:527–570

9. Gibbs FA, Gibbs EL, Lennox WG (1938) Cerebral dysrhythmias of epilepsy. Arch Neurol Psychiatr 39:298–314

10. Blakemore C (1991) The International Brain Research Organization: a brief historical survey. In: IBRO Membership Directory. Pergamon Press, Paris

11. Li CL (1955) Functional properties of cortical neurons with particular reference to synchronization. Electroenceph Clin Neurophysiol 7:475–478

12. Matsumoto H, Ajmone-Marsan C (1964) Cortical cellular phenomena in experimental epilepsy: ictal manifestations. Exp Neurol 9:305–326

13. Matsumoto H, Ajmone-Marsan C (1964) Cortical cellular phenomena in experimental epilepsy: interictal manifestations. Exp Neurol 9:286–304

14. Kandel ER, Spencer WA (1961) The pyramidal cell during hippocampal seizure. Epilepsia 2:63–69

15. Dichter M, Spencer NA (1969) Penicillin-induced interictal discharges from rat hippocampus, I: characteristics and topographical features. J Neurophysiol 32:649–662

16. Andersen P (1975) Organization of hippocampal neurons and their interconnections. In: Isaacson RL, Pribram KH (eds) The hippocampus. Plenum Press, New York

17. Fisher RS, van Emde Boas W, Blume W, Elger C, Engel J Jr, Genton P, Lee P (2005) Epileptic seizures and epilepsy. Definitions proposed by the International League against Epilepsy (ILAE) and the International Bureau for Epilepsy (IBE). Epilepsia 46:470–472

18. Goddard GV (1967) Development of epileptic seizures through brain stimulation at low intensity. Nature 214:1020–1023

19. Engel J Jr (2001) Mesial temporal lobe epilepsy: what have we learned? Neuroscientist 7:340–352

20. Ben-Ari Y, Tremblay E, Riche D, Ghilini G, Naquet R (1981) Electrographic clinical and pathological alterations following systemic administration of kainic acid, bicuculline or pentetrazol: metabolic mapping using the deoxyglucose method with special reference to the pathology of epilepsy. Neuroscience 6:1361–1391

21. Cavalheiro EA, Naffah-Mazzacoratti MG, Mello LE, Leite JP (2006) The pilocarpine model of seizures. In: Pitkänen A, Schwartzkroin PA, Moshé SL (eds) Models of seizures and epilepsy. Elsevier Academic Press, Burlington

22. McIntyre DC, Nathanson D, Edson N (1982) A new model of partial status epilepticus based on kindling. Brain Res 250:53–63

23. Engel J Jr, Ojemann G, Lüders H, Williamson PD (eds) (1987) Fundamental mechanisms of human brain function. Raven Press, New York

24. Engel J Jr, Thompson PM, Stern JM, Staba RJ, Bragin A, Mody I (2013) Connectomics and epilepsy. Curr Opin Neurol 26:186–194

25. Galanopoulou AS, Buckmaster PS, Staley KJ, Moshé SL, Perucca E, Engel J Jr, Löscher W, Noebels JL, Pitkänen A, Stables J, White HS, O'Brien TJ, Simonato M (2012) Identification of new epilepsy treatments: issues in preclinical methodology. Epilepsia 53:571–582

26. Simonato M, Löscher W, Cole AJ, Dudek FE, Engel J Jr, Kaminski RM, Loeb JA, Scharfman H, Staley KJ, Velíšek L, Kitgaard H (2012) WONOEP XI Critical review and invited commentary. Finding a better drug for epilepsy: preclinical screening strategies and experimental trial design. Epilepsia 53:1860–1867

27. Engel J Jr, Dichter MA, Schwartzkroin PA (2008) Basic mechanisms of human epilepsy. In: Engel J Jr, Pedley TA (eds) Epilepsy: a comprehensive textbook, 2nd edn. Lippincott-Raven, Philadelphia

28. Hildebrand MS, Dahl HH, Damiano JA, Smith RJ, Scheffer IE, Berkovic SF (2013) Recent advances in the molecular genetics of epilepsy. J Med Genet 50:271–279

29. Commission on Classification and Terminology of the International League Against Epilepsy (1989) Proposal for revised classification of epilepsies and epileptic syndromes. Epilepsia 30:389–399

30. Berg AT, Berkovic SF, Brodie MJ, Buchhalter J, Cross JH, van Emde Boas W, Engel J Jr, French J, Glauser TA, Mathern GW, Moshé SL, Nordli D Jr, Plouin P, Scheffer IE (2010) Revised terminology and concepts for organization of seizures and epilepsies: report of the ILAE Commission on Classification and Terminology, 2005–2009. Epilepsia 51:676–685

31. Kanner AM (2013) Do psychiatric comorbidities have a negative impact on the course and treatment of seizure disorders? Curr Opin Neurol 26:208–213

32. Shewmon D, Erwin RJ (1989) Transient impairment of visual perception induced by single interictal occipital spikes. J Clin Exp Neuropsych 11:675–691

33. de Curtis M, Avanzini Z (2001) Interictal spikes in focal epileptogenesis. Prog Neurobiol 63:541–567

Los Angeles, CA, USA   Jerome Engel Jr., M.D., Ph.D.
Philadelphia, PA, USA   Marc A. Dichter, M.D., Ph.D.

Preface

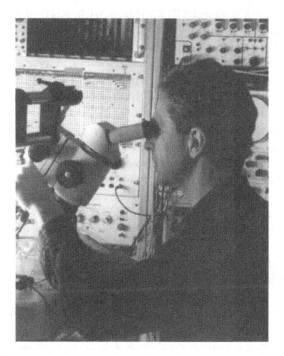

Philip A. Schwartzkroin recently retired after an outstanding, influential career in neuroscience and epilepsy research. He influenced the work of many neuroscientists either by the techniques he developed or his pioneering discoveries of neuronal mechanisms underlying excitability and microcircuitry in health and disease. He personally touched the lives of all those with whom he collaborated and mentored. This volume is dedicated to Phil by many of the numerous colleagues and trainees who respect him both personally and scientifically. As one might expect from Phil's 'hands-on' approach to his work, Phil had a direct influence editing the volume and designing its unusual format, in which key questions in epilepsy research are addressed from both basic science and clinical perspectives. Phil's goal for this volume was to allow experts in the field the opportunity to address critical questions in ways that would stimulate a broad readership. Instead of leaving readers with the sense that they have all the answers, his goal was to encourage them to think about how to address the important questions in epilepsy research today.

Phil began neuroscience research in high school. Through a summer fellowship from the National Science Foundation he investigated mouse behavior at the Jackson Laboratories, where he continued working during several subsequent summer breaks from college. Phil was a National Merit Scholar and graduated *magna cum laude* with highest honors in Psychology from Harvard University. As an undergraduate, Phil investigated neocortical sensory processing in the laboratory of Charles Gross. Even in this early stage of his career Phil was exceptional, publishing articles as first author while still in high school and college.

After graduating from Harvard, Phil moved to Stanford University where he earned a Ph.D. in Neurological Sciences, working in the laboratory of Kao Liang Chow, who had trained with Karl Lashley. Chow served as a role model for Phil, because he was a basic scientist in a clinical department and focused on understanding the relationship between brain structure and function in health and disease. In the coming years, Phil would also serve as a role model in much the same way – for Phil's own trainees.

Phil's dissertation addressed the effects of vestibular stimulation on single cells in cat visual cortex and superior colliculus. Following his dissertation, Phil started to address questions related to epilepsy, which became the major research focus of his career. He started as an Epilepsy Foundation trainee with David Prince for one year before becoming a postdoctoral fellow in Per Anderson's laboratory at the University of Oslo in Norway. Phil returned to the Department of Neurology at Stanford one year later and brought with him a brain slice recording chamber, which had recently been developed. The experiments Phil conducted using brain slices, put him at the forefront of controversy, because there was skepticism that brain slices would be a useful experimental preparation. Nevertheless, he perservered and played a critical role in the ultimate acceptance of the approach. He also was a pioneer; he was the first in the USA (after Yamamoto in Japan) to develop the slice preparation for intracellular recording. He demonstrated the utility of brain slices for studying normal synaptic transmission and synaptic plasticity – and was the first to demonstrate long-term potentiation in the slice preparation. He showed how brain slices could be used to study epileptiform activity, paving the way for decades of epilepsy research based on the slice preparation.

In 1975, Phil was appointed Assistant Professor of Neurology at Stanford, where he was first in the world to carry out intracellular studies in slices of surgically resected human epileptic neocortex and hippocampus. He began what would become a standard structural and functional approach in his laboratory, and characterized numerous cell types in the hippocampus with correlative cellular electrophysiology and intracellular staining techniques. His initial studies began with CA1 pyramidal cells and were followed by some of the most difficult recordings at that time, of GABAergic interneurons.

Phil moved to the Department of Neurological Surgery at the University of Washington in 1978, where he continued to use brain slices to make fundamental discoveries in hippocampal anatomy and physiology. For example, he pioneered the use of the slice preparation for studying brain development. With that approach, he was first to demonstrate depolarizing IPSPs in immature hippocampus. He also was at the forefront of the most challenging electrophysiological techniques, such as the use of simultaneous intracellular recording from two monosynaptically-coupled neurons. Phil was also a leader in applying the slice preparation to questions related to animal models of epilepsy. His laboratory was one of the first to characterize, using both morphology and electrophysiology, transgenic mouse models of epilepsy. Some of this work, such as the studies of the Kv1.1 knockout mouse, were major advances in epilepsy research. In addition, Phil addressed other areas of epilepsy research, including cortical dysplasia. He was an early contributor to studies on the basic mechanisms of the ketogenic diet, and was first to demonstrate that furosemide, a chloride co-transporter antagonist, was anti-epileptic. During this time Phil's productivity was exceptional. For example, over a one-year period in 1988 he was senior author of 14 research articles, six of which appeared in *The Journal of Neuroscience*. Phil earned many awards, including fellowships from the Guggenheim and Klingenstein Foundations, two Jacob Javits Awards from the NIH, and he was one of the first recipients of the American Epilepsy Society/Milken Family Medical Foundation Research Award. In 2001, Phil moved to the Department of Neurological Surgery at the University of California at Davis where he held the Bronte Endowed Chair in Epilepsy Research. He became an emeritus professor in 2013.

In addition to his outstanding contributions to research, Phil was a dedicated member of the epilepsy research community. He served on NIH study sections and on scientific advisory boards for the Epilepsy Foundation and Citizens United for Research in Epilepsy. He led some of the first efforts to address translation, organizing seven workshops and six books that brought together basic and clinical epilepsy researchers. These workshops, and the books that resulted from them, remain some of the most influential in the field. He served as one of the first chairs of Investigators' Workshops for the American Epilepsy Society and was the first basic scientist president of the American Epilepsy Society. He chaired the International League Against Epilepsy (ILAE) Commission on Neurobiology and organized an ILAE Workshop on the Neurobiology of Epilepsy. After these accomplishments, Phil served as co-editor-in-chief of *Epilepsia*, where he strengthened the impact and reputation of the journal.

In addition to his achievements in research and service to the epilepsy research community, Phil trained many students and postdoctoral fellows. Many of these individuals ultimately became independent neuroscientists themselves, including numerous leaders in epilepsy research today. Phil provided his trainees with a great degree of independence. But when help was needed, he was an efficient, "hands-on" trouble-shooter who quickly solved technical problems. Phil trained largely by example. He demonstrated a strong work ethic and began days in the lab at an extremely early hour. During meetings in his office, Phil demonstrated impressive collegiality with colleagues, both near and far, often phoning them in the middle of conversations if he wanted to address a question. He could pick up the phone and call almost anyone in the field, often the original source of information on a given topic. Phil's trainees benefited greatly from exposure to some of these investigators when they visited the laboratory or attended meetings organized by Phil. Despite his considerable accomplishments, Phil was modest and easy to work with, characteristics that helped shape the laboratory environment into one that was truly enjoyable. Writing was an area where Phil's training method was more direct, but just as constructive. Phil routinely transformed manuscripts – often long-hand – with extensive editorial remarks that illustrated how to clearly convey ideas. Remarkably, Phil could do so rapidly and effectively, which left trainees wondering if they could ever master scientific writing and editing as well. In addition, he made it clear that excellent scientific writing was extremely important.

On May 3–5, 2013 a workshop entitled "Issues in Clinical Epileptology: A View from the Bench" was held in honor of Phil. The workshop was supported by several organizations, including the American Epilepsy Society and CURE. It was not possible for all of Phil's colleagues and trainees to attend, but the group that was able to come considered it an excellent meeting – as well as a great opportunity to honor Phil (see review by C. Stafstrom, Epilepsy Curr., 2013). In considering the type of book that would complement this 'festschrift,' Phil provided a great deal of input, as mentioned above. His contribution to this volume shows that – despite his retirement – his influence will be present for years to come.

Issues in Clinical Epileptology: A view from the Bench. A Festschrift in Honor of Philip Schwartzkroin. Pajaro Dunes Resort, Watsonville, California. May 3–5, 2013. First row: Paul Buckmaster, Jong Rho, Jurgen Wenzel, Phil Schwartzkroin, Gerry Chase, Helen Scharfman, Laura Reece, Jean-Claude Lacaille, Mike Haglund, Scott Baraban. Second row: Mareike Wenzel, Catherine Woolley, Carol Robbins, Alan Mueller, Dennis Kunkel, Dennis Turner. Third row: Elsa Rosignol, Daryl Hochman, Robert Fisher. Fourth row: Sloka Iyengar, Jerome (Pete) Engel Jr., James Trimmer, Carl Stafstrom, Damir Janigro, Robert Hunt. Fifth row: Aristea Galanopoulou, Tracy Dixon-Salazar, Solomon (Nico) Moshé, David Prince, Massimo Avoli, Jeffrey Noebels, Robert Wong, Michael Gutnick, Leena Knight. Back row: Satoshi Fujita, Aylin Reid, Charles Behr, Ben Strowbridge, Robert Berman.

Orangeburg, NY, USA   Helen E. Scharfman, Ph.D.
Stanford, CA, USA   Paul S. Buckmaster, Ph.D.

Contents

Part I Seizures, Epileptiform Activities, and Regional Localization

1 How Can We Identify Ictal and Interictal
   Abnormal Activity?... 3
   Robert S. Fisher, Helen E. Scharfman, and Marco deCurtis

2 What Is the Clinical Relevance of *In Vitro*
   Epileptiform Activity? 25
   Uwe Heinemann and Kevin J. Staley

3 What Is the Importance of Abnormal
   "Background" Activity in Seizure Generation? 43
   Richard J. Staba and Gregory A. Worrell

4 What Is a Seizure Focus? 55
   J. Victor Nadler and Dennis D. Spencer

5 What Is a Seizure Network? Long-Range
   Network Consequences of Focal Seizures 63
   Hal Blumenfeld

6 What Is a Seizure Network? Very Fast Oscillations
   at the Interface Between Normal and Epileptic Brain 71
   Roger D. Traub, Mark O. Cunningham,
   and Miles A. Whittington

7 Is There Such a Thing as "Generalized" Epilepsy? 81
   Gilles van Luijtelaar, Charles Behr, and Massimo Avoli

Part II Synaptic Plasticity

8 Are There Really "Epileptogenic" Mechanisms
   or Only Corruptions of "Normal" Plasticity? 95
   Giuliano Avanzini, Patrick A. Forcelli, and Karen Gale

9 When and How Do Seizures Kill Neurons,
   and Is Cell Death Relevant to Epileptogenesis? 109
   Ray Dingledine, Nicholas H. Varvel, and F. Edward Dudek

10   How Is Homeostatic Plasticity Important in Epilepsy?............. 123
     John W. Swann and Jong M. Rho

11   Is Plasticity of GABAergic Mechanisms Relevant
     to Epileptogenesis?.. 133
     Helen E. Scharfman and Amy R. Brooks-Kayal

12   Do Structural Changes in GABA Neurons
     Give Rise to the Epileptic State?................................ 151
     Carolyn R. Houser

13   Does Mossy Fiber Sprouting Give
     Rise to the Epileptic State?.. 161
     Paul S. Buckmaster

14   Does Brain Inflammation Mediate
     Pathological Outcomes in Epilepsy? 169
     Karen S. Wilcox and Annamaria Vezzani

15   Are Changes in Synaptic Function That Underlie
     Hyperexcitability Responsible for Seizure Activity? 185
     John G.R. Jefferys

16   Does Epilepsy Cause a Reversion to Immature Function? 195
     Aristea S. Galanopoulou and Solomon L. Moshé

17   Are Alterations in Transmitter Receptor and Ion
     Channel Expression Responsible for Epilepsies?...................... 211
     Kim L. Powell, Katarzyna Lukasiuk, Terence J. O'Brien,
     and Asla Pitkänen

Part III   Models and Methods

18   How Do We Make Models That Are Useful
     in Understanding Partial Epilepsies?.. 233
     David A. Prince

19   Aligning Animal Models with Clinical
     Epilepsy: Where to Begin?... 243
     Stephen C. Harward and James O. McNamara

20   What Non-neuronal Mechanisms Should
     Be Studied to Understand Epileptic Seizures?.......................... 253
     Damir Janigro and Matthew C. Walker

21   What Epilepsy Comorbidities Are Important
     to Model in the Laboratory? Clinical Perspectives................... 265
     Simon Shorvon

22   Epilepsy Comorbidities: How Can Animal Models Help?........ 273
     Carl E. Stafstrom

23 **What New Modeling Approaches Will Help
    Us Identify Promising Drug Treatments?** 283
    Scott C. Baraban and Wolfgang Löscher

24 **What Are the Arguments For and Against
    Rational Therapy for Epilepsy?** 295
    Melissa Barker-Haliski, Graeme J. Sills,
    and H. Steve White

25 **How Can Advances in Epilepsy Genetics
    Lead to Better Treatments and Cures?** 309
    Renzo Guerrini and Jeffrey Noebels

26 **How Might Novel Technologies Such as Optogenetics
    Lead to Better Treatments in Epilepsy?** 319
    Esther Krook-Magnuson, Marco Ledri,
    Ivan Soltesz, and Merab Kokaia

Index 337

Part I

Seizures, Epileptiform Activities, and Regional Localization

How Can We Identify Ictal and Interictal Abnormal Activity?

Robert S. Fisher, Helen E. Scharfman, and Marco deCurtis

Abstract

The International League Against Epilepsy (ILAE) defined a seizure as "a transient occurrence of signs and/or symptoms due to abnormal excessive or synchronous neuronal activity in the brain." This definition has been used since the era of Hughlings Jackson, and does not take into account subsequent advances made in epilepsy and neuroscience research. The clinical diagnosis of a seizure is empirical, based upon constellations of certain signs and symptoms, while simultaneously ruling out a list of potential imitators of seizures. Seizures should be delimited in time, but the borders of ictal (during a seizure), interictal (between seizures) and postictal (after a seizure) often are indistinct. EEG recording is potentially very helpful for confirmation, classification and localization. About a half-dozen common EEG patterns are encountered during seizures. Clinicians rely on researchers to answer such questions as why seizures start, spread and stop, whether seizures involve increased synchrony, the extent to which extra-cortical structures are involved, and how to identify the seizure network and at what points interventions are likely to be helpful. Basic scientists have different challenges in use of the word 'seizure,' such as distinguishing seizures from normal behavior, which would seem easy but can be very difficult because some rodents have EEG activity during

R.S. Fisher (✉)
Department of Neurology and Neurological Sciences,
Stanford University School of Medicine, Stanford,
CA 94305, USA
e-mail: rfisher@stanford.edu

H.E. Scharfman
The Nathan S. Kline Institute for Psychiatric
Research, Orangeburg, NY, USA

Departments of Child & Adolescent Psychiatry,
Physiology & Neuroscience, and Psychiatry,
New York University Langone Medical Center,
New York, NY, USA
e-mail: hscharfman@nki.rfmh.org

M. deCurtis
Unit of Epileptology and Experimental
Neurophysiology, Fondazione Istituto Nazionale
Neurologico, 20133 Milan, Italy

H.E. Scharfman and P.S. Buckmaster (eds.), *Issues in Clinical Epileptology: A View from the Bench*,
Advances in Experimental Medicine and Biology 813, DOI 10.1007/978-94-017-8914-1_1,
© Springer Science+Business Media Dordrecht 2014

normal behavior that resembles spike-wave discharge or bursts of rhythmic spiking. It is also important to define when a seizure begins and stops so that seizures can be quantified accurately for pre-clinical studies. When asking what causes seizures, the transition to a seizure and differentiating the pre-ictal, ictal and post-ictal state is also important because what occurs before a seizure could be causal and may warrant further investigation for that reason. These and other issues are discussed by three epilepsy researchers with clinical and basic science expertise.

Keywords

Convulsion • Convulsive • Electroencephalogram • Epilepsy • Epileptic • Focal seizure • Epileptiform • Seizure-like • Spike-wave discharge • Theta • Sharp wave • Behavioral arrest • Interictal spike • Ictal • Pre-ictal • Transition to seizure

1.1   Introduction

Seizures are common and important neurological symptoms that may require treatment. Seizures can signal underlying disease. In addition, many research laboratories study mechanisms of seizures. Therefore, a commonly accepted definition of "seizure" is needed for both clinical and research purposes. Some events may obviously be seizures, but others might comprise imitators of seizures [62], epileptiform non-seizure events, or variants of normal laboratory animal behavior.

1.1.1   Clinical Perspective

1.1.1.1 Definition of a Seizure

Webster says that a definition should capture the "essence" of an entity. What then is the essence of a seizure? Table 1.1 highlights definitions from various authorities, dating back to Johns Hughlings Jackson in 1870 [58].

Terms that recur in the various definitions include excessive, disorderly discharge, synchronous, self-limited, abnormal, paroxysmal, neurons, central nervous system (CNS) and cortex. Corresponding symptoms are listed as alteration or loss of consciousness, involuntary movements, sensory, psychic or autonomic disturbances and other clinical manifestations. These terms cover a lot of territory. Delineating

the possible clinical manifestations of seizures is beyond the scope of this chapter, but an overview may be found in [73]. In 2005, a task force of the International League Against Epilepsy [37] provided a parsimonious definition of a seizure as "a transient occurrence of signs and symptoms due to abnormal or synchronous neuronal activity in the brain."

In clinical practice, a clinician rarely sees the abnormal electrical discharge, with the exception of successful video-EEG monitoring, so this discharge is inferred on the basis of a typical constellation of clinical symptoms. Application of the definition also requires ruling out other conditions. For example, abnormal and synchronous firing of thalamic neurons in a patient with Parkinson's disease [17] represents a transient symptom correlated to tremor, but it is not a seizure. Therefore, a definition of seizures must include an implied qualifier: "and not due to other known conditions producing a similar picture."

Some writers use the modifying term "epileptic seizures" to distinguish them from common usage of terms such as heart seizures, psychogenic seizures or other non-epileptic paroxysmal events. However, not all seizures imply epilepsy, particularly for single seizures with low likelihood of recurrence or for provoked seizures. Hence, the phrase "epileptic seizures" tends to be either misleading or redundant.

Table 1.1 Prior Definitions of Seizure

References	Definitions – Note that several say "epilepsy" in place of "seizure"
Jackson [58]	Epilepsy is a symptom… an occasional, an excessive and a disorderly discharge of nerve tissue (in the highest centers)
Penfield and Jasper [75]	An epileptic seizure is a state produced by an abnormal excessive neural discharge within the central nervous system
Aird et al. [3]	Epilepsy may be defined as a paroxysmal disturbance of central nervous system (CNS) function, which is recurrent, stereotyped in character, and associated with excessive neuronal discharge that is synchronous and self-limited
Engel [34]	Epileptic seizures are the clinical manifestations (symptoms and signs) of excessive and/or hypersynchronous, usually self-limited, abnormal activity of neurons in the cerebral cortex… An epileptic seizure may consist of impaired higher mental function or altered consciousness, involuntary movements or cessation of movement, sensory or psychic experiences, or autonomic disturbances
Hauser and Hesdorffer [53]	A seizure can be defined as a paroxysmal disorder of the central nervous system characterized by abnormal cerebral neuronal discharge with or without loss of consciousness
Hopkins et al. [55]	An epileptic seizure is a clinical manifestation presumed to result from an abnormal and excessive discharge of a set of neurons in the brain. The clinical manifestation consists of sudden and transitory abnormal phenomena, which may include alterations of consciousness, motor, sensory, autonomic, or psychic events, perceived by the patient or an observer
Adams et al. [2]	Epilepsy may be defined as an intermittent derangement of the nervous system due presumably to a sudden, excessive, disorderly discharge of cerebral neurons

The seizure definition of excessive neuronal discharges derived from Hughlings Jackson's time, is 144 years old, when awareness of brain electrical activity was new. This mindset has led generations of clinicians and researchers to think of a seizure as an electrical disorder. Abnormal electrical discharges are just one manifestation of seizures, not necessarily more important than metabolic, blood flow, receptor, gene activation, network connectivity and many other changes that are intrinsic to seizures. A contemporary definition of seizures would likely be less electro-centric and focus more on excessive and sustained activation of specific brain networks. The research community should be challenged to invent a better definition for seizures.

1.1.1.2 EEG Manifestations of Seizures

Clinicians rely heavily on electroencephalographic patterns to identify, classify, quantify and localize seizures [7]. Figure 1.1 illustrates common epileptiform EEG patterns. The term epileptiform is used to connote EEG patterns believed to be associated with a relatively high risk for having seizures. Gloor [43] defined spikes as potentials that stand above the background, have a "pointy" shape, duration between 30 and 70–80 ms, asymmetric rise and fall, and followed by a slow wave. The potential should have a sensible field, meaning that it should be reflected in physically adjacent electrodes and perhaps in synaptically linked regions such as the contralateral hemisphere. "Sharp waves" have durations of 70–200 ms. The distinction between spikes and sharp waves is arbitrary in the clinical arena and is discussed further below (see also [28]).

Spikes may be focal or apparently generalized across widespread regions of brain bilaterally. Rhythmic recurrence of spikes followed by slow waves is referred to as spike-waves. Focal spikes tend to be associated with focal seizures with or without secondary generalization. In contrast, generalized spikes tend to be associated with seizures that are nonfocal at their onset. Generalized spike-waves are associated with absence (previously called petit mal) seizures.

The right panel of Fig. 1.1 illustrates the onset of a focal seizure in the top four channels, which are in the left temporal region. The local rhythm can be seen evolving in amplitude,

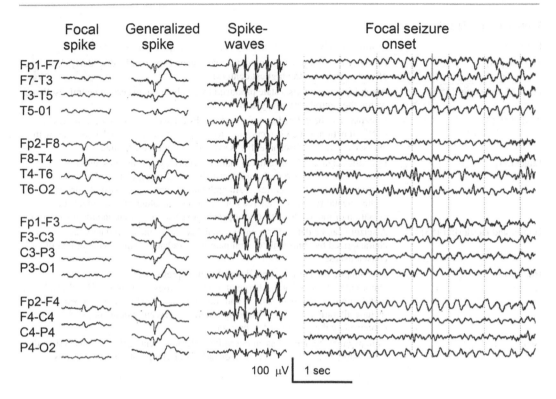

Fig. 1.1 Common epileptiform EEG patterns. Common patterns are shown for individuals with focal spikes, generalized spikes, spike-waves, and a seizure with focal onset (From Fisher, unpublished)

frequency and degree of sharpness. Other channels also reflect the seizure activity, but it is best formed and earliest in the top four channels. Where the potential becomes sharp, there is a phase reversal (down in one channel and up in the next channel) between the top and the second from the top channel. Polarity conventions of the EEG indicate that the electrode common to both these channels is the site of maximum negativity compared to neighbors on either side. Active (discharging) seizure foci are extracellularly negative, since positive ions flow from the extracellular space into the neuron during excitation. Therefore, the phase reversal of a spike or seizure onset can be used to approximately localize the region of seizure origin.

The EEG recorded from the human scalp at the start of the seizure can take at least five different forms, as illustrated in Fig. 1.2. One pattern is rhythmically evolving frequencies in the theta (4–7/s), delta (0–3/s) or alpha (8–12/s) bands.

The rhythmical activity can have varying degrees of sharpness, but spikes and sharp waves are not required to be part of the rhythmical pattern of a focal seizure. An evolution of frequency and amplitude over time is needed to distinguish a seizure from many other normal and abnormal rhythmical events encountered in the EEG. The second pattern of seizure origin is rhythmical spiking. This may be most commonly seen with seizures in hippocampus and neighboring structures. Spike-wave patterns typically occur during generalized absence seizures, but presence of spike-waves cannot be equated with absence epilepsy. Spike-waves also can appear focally during focal seizures or during the course of generalized tonic-clonic seizures. Neocortical seizures often manifest with an electrodecremental pattern, referring to a general flattening of brain rhythms at the start of a seizure. Electrodecremental patterns are commonly seen with tonic, atonic and sometimes tonic-clonic

Rhythmical evolving theta, delta, alpha frequencies	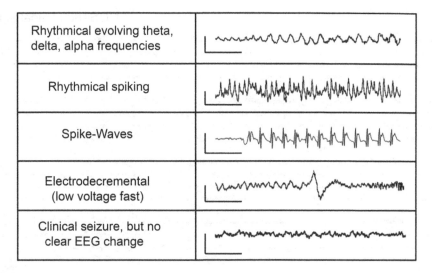
Rhythmical spiking	
Spike-Waves	
Electrodecremental (low voltage fast)	
Clinical seizure, but no clear EEG change	

Fig. 1.2 Common EEG patterns at the start of seizures in patients with epilepsy (From Fisher, unpublished)

seizures [35]. The apparent disappearance of EEG activity is a consequence of the typical 1–70 Hz bandpass filter used to review EEG. In fact, a very low frequency potential heralds the start of such seizures [57, 92] but is largely filtered out by the low frequency filters commonly utilized during scalp EEG revision. Careful examination of the electrodecremental region shows presence of low voltage, high frequency activity [29, 38]. Considerable study has demonstrated importance of frequencies in the beta (13–30 Hz), gamma range (30–100 Hz), ripple (100–250 Hz) and fast ripple (250–1000 Hz) range. Activity in the fast ripple or higher ranges is sometimes referred to as high-frequency oscillations (HFO's) [32, 98, 100]. HFO's can be useful markers for the region of seizure onset. Epilepsy surgery is more successful when regions generating high frequencies are resected [41]. The fifth electrographic pattern of a seizure onset is no change in the scalp EEG. The presumption here is one of sampling error. Two-thirds of cortex is enfolded in sulci and dipole discharges in sulci do not always project to scalp EEG electrodes. Seizures can originate in mesial temporal, orbitofrontal or inter-hemispheric regions far from scalp electrodes. Negative EEG findings therefore do not rule out underlying focal seizures. The EEG

must be correlated with the clinical picture. Of note here is that seizures that begin in the brainstem in experimental animals often lead to convulsions before the forebrain EEG shows any change from normal [42] (personal observations, HES).

1.1.1.3 Ambiguities in EEG Manifestations of Seizures

Electroencephalographers sometimes disagree about whether a particular pattern is epileptiform and representative of associated seizures. Figure 1.3 shows an evolving event over the right mid-temporal region lasting for about 5 s. The EEG technician noted no clinical signs. Such events might be considered too brief to represent a seizure: duration of at least 10 s has occasionally been applied operationally [1], but there is no official minimum time to define a seizure. In animal research, 2–3 s is often used as a minimum time for an electrographic seizure but the length of time that is sufficient to define a seizure is extremely variable [26, 31]. However, discharges accompanied by clinical seizures qualify as electrographic seizures regardless of their duration. In the extreme, a single generalized spike associated with a myoclonic jerk could be considered to be a very brief seizure.

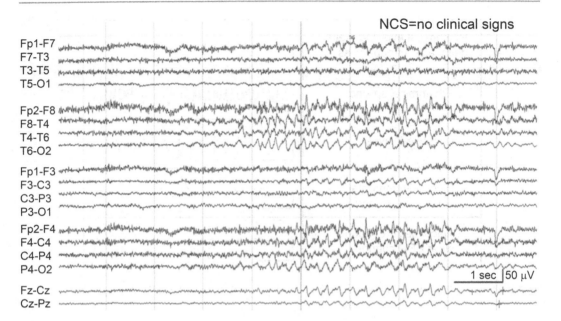

Fig. 1.3 Is this a seizure? Rhythmical brief epileptiform activity, illustrating the ambiguity involved in deciding whether an EEG event corresponds to interictal activity or a seizure (From Fisher, unpublished)

Epileptiform EEG activity has been categorized as ictal, meaning during a seizure, postictal, meaning after a seizure and interictal, meaning between seizures. While ingrained in common usage, these terms may be more confusing than helpful [36]. What sense does it make to designate an interictal spike in cases where there have not been two seizures? Where does the behavioral and EEG pattern of an ictal event merge into the postictal behavioral confusion and EEG slowing? Is postictal slowing always a consequence of the seizure [33]? Delineations between ictal and postictal may not be obvious. Are periodic lateralized epileptiform discharges (PLEDs, Fig. 1.4) interictal, ictal or either depending upon circumstances [76]? When is a burst of generalized spike-waves interictal and when is it ictal? Behavioral manifestations, such as unresponsiveness and automatisms, tend to occur in direct proportion to the duration of spike-wave discharges [77]. Whether a person is noted to have clinical signs such as limited responsiveness depends upon how carefully they are tested. Meticulous studies [4] show that responsive latency and task accuracy declines even during a period of so-called interictal spikes. Research in animals

suggests the same is true for rodents [54], although the assumptions in these studies – that blocking interictal spikes improves behavior and therefore interictal spikes cause behavioral impairment – may not be true. Instead, blocking interictal spikes may only be helpful because of a reduction of other brain abnormalities, not necessarily the spikes *per se*. Clinically, interictal spikes tend to correspond to the zone of origin of a seizure, but not always. Figure 1.5 illustrates interictal spikes from the right temporal region and electrographic seizure onset from the left temporal region in the same patient.

1.1.1.4 Clinical Conclusions

The commonly employed definition of a seizure as a transient occurrence of signs and symptoms due to abnormal or synchronous neuronal activity in the brain is almost a century and a half old, and it does not capture the essential nature of seizures as depicted by modern neuroscience. Seizures are diagnosed clinically, taking into account numerous entities that can imitate seizures, such as syncope, transient ischemic attacks, sleep disorders, confusional migraine, tremor, dystonia, fluctuating delirium and psychological episodes. The scalp EEG is

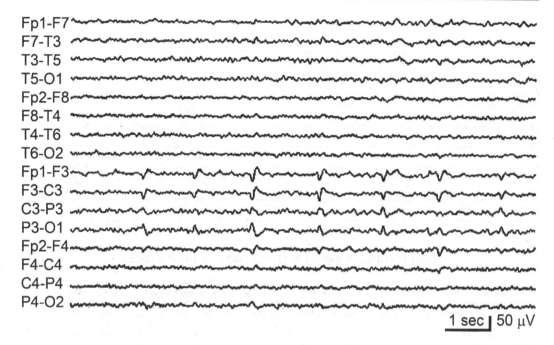

Fig. 1.4 Periodic lateralized epileptiform discharges (PLEDs) – are they ictal or interictal? PLEDs over the left central (C3) region are shown. Some electroencephalog-raphers consider this pattern to be interictal and others ictal, while still others believe it depends upon particular circumstances (From Fisher, unpublished)

a helpful adjunct to diagnosis of seizure disorders, but it is not clear that an EEG pattern should be intrinsic to a definition of seizures. There is no unifying form; instead at least five different EEG patterns can accompany seizures. EEG correlates of high risk for seizures are categorized as ictal (during a seizure), postictal (after seizure) or interictal (between seizures). These distinctions often are unclear and arbitrary, in that the interictal-ictal boundaries are blurred for many seizures. Even so-called interictal spikes can affect behavior.

We need a better understanding of what constitutes the pathophysiological and behavioral essence of a seizure. Numerous questions arise for basic researchers. Need a seizure always involve an excessive discharge and increased synchrony? Have neurons been given excessive primacy in seizures over glia? Do seizures emerge only in cortex or can they develop in subcortical structures as well? Does it make sense to talk about where seizures start, given the involvement of widespread networks? What brain networks are involved in seizures of different types and which behaviors correlate with seizures in these networks? These questions will only be answered with a collaboration between basic researchers and clinicians.

1.2   Defining Seizure Correlates with Intracranial Electrodes in Patients

The advent of intracranial recordings (with grid and strip electrode arrays) and intracerebral recordings (with depth electrodes) during presurgical evaluation in patients with partial epilepsies resistant to pharmacological treatment changed our view of the electrographic correlate of a seizure. During pre-surgical intracranial monitoring, seizures are recorded with electrodes positioned close to the generators of ictal epileptiform discharges. In particular, depth stereo-EEG electrode implants aim at the epileptogenic area. This is done by accurately planning electrode insertion on the basis of the analysis of the

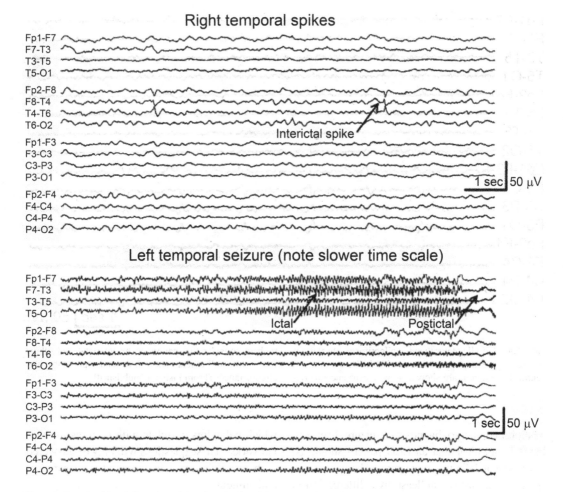

Fig. 1.5 Interictal-ictal disparity with spikes in the right hemisphere and seizures on the left. Interictal-ictal disparity in the same patient as Fig. 1.5, with interictal spikes over the right temporal region, but seizure onset from the left temporal region. Note different time scales for each segment (From Fisher, unpublished)

sequence of localizing clinical features observed during seizures recorded by video monitoring with scalp EEG performed as part of the pre-surgical examination [23, 85]. Intracerebral recordings are finalized to identify the cortical networks activated during a seizure that should be surgically removed to cure the patient. The areas involved in seizure generation are defined as the seizure-onset zone and the epileptogenic zone, which includes the regions of onset and propagation of the ictal epileptiform discharge. Intracranial recordings contribute to outline a larger area, defined as irritative zone, that generates abnormal interictal events/potentials, but is not directly recruited during a seizure discharges.

A large number of pre-surgical studies focused on the functional interactions between the epileptogenic and the irritative zones have been reported in the last 20 years. These studies demonstrate that (i) the irritative area is not coincident and it is usually larger than the epileptogenic/seizure onset zone, (ii) interictal discharges do not show a coherent relationship with seizure discharges, in terms of location and activation patterns, (iii) the rate of interictal discharges can either increase or decrease just ahead of a seizure and (iv) in most cases the electrographic pattern of seizure onset is completely different from the activity recorded during interictal discharges (for review see [28, 29]; Fig. 1.6).

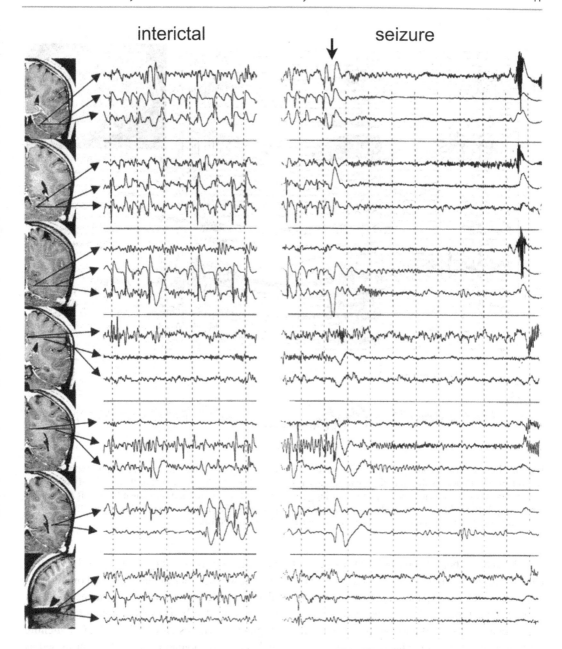

Fig. 1.6 Recordings with intracerebral stereo-EEG electrodes in a patient with focal epilepsy secondary to focal cortical dysplasia. *Far left*: The position of the recording electrodes is illustrated. *Left*: Interictal discharges recorded with intracerebral stereo-EEG electrodes in a patient with focal epilepsy secondary to focal cortical dysplasia. *Right*: Seizure onset is marked by the *arrow*. The *slow spikes* that precede the ictal low-voltage fast activity are different in location and morphology from the interictal spikes (Courtesy of Francione, Tassi and LoRusso of *Claudio Munari* Epilepsy Surgery Center, Niguarda Hospital, Milano)

Intracranial pre-surgical studies revealed that the most consistent pattern observed at the onset of a seizure is characterized by fast activity of low amplitude in the *beta-gamma* range ([5, 38, 48]; for review see [29]) that can be preceded by large amplitude spike potentials. The latter events have often be defined as pre-ictal spikes, but their consistent and reproducible occurrence at the very onset of a seizure include them by definition as integral part of a seizure. Experimental studies in animal

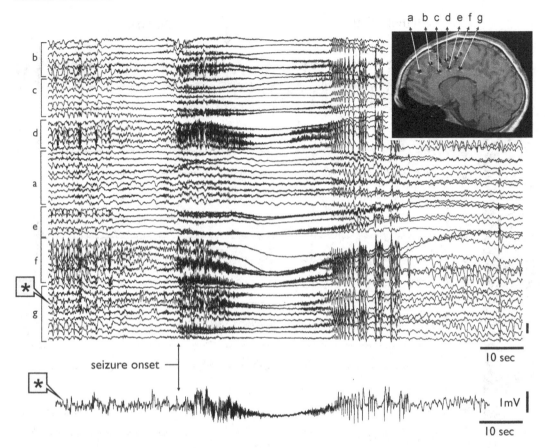

Fig. 1.7 Intracerebral recording of a focal seizure with stereo-EEG electrodes (as shown in the *upper right inset*) in a patient with cryptogenic focal epilepsy during pre-surgical evaluation. Multi-contact electrodes are identified by *letters*. The EEG marked by an *asterisk* is expanded at the *bottom*. When the seizure begins (*seizure onset, arrow*) there is a reduction of background activity, appearance of fast activity, and subsequently there is a very slow potential (From Gnatkovsky, Francione, Tassi and de Curtis, unpublished)

models and in human post-surgical tissue and intracranial stereo-EEG observations demonstrated that these (pre)ictal population spikes are distinct from interictal potentials [21, 44, 56] and are possibly generated by network mechanisms that are different from those sustaining interictal potentials.

More recent studies demonstrated that the low-voltage pattern associated to the initiation of a seizure correlates with the abolition and possibly the desynchronization of background activity. The substitution of background activity with low-voltage fast activity is the intracranial correlate of the electrodecremental pattern defined as EEG "flattening", a phenomenon that is commonly pursued to localize the seizure onset area on the scalp EEG (as discussed above). Low-voltage fast activity is also associated with the appearance

of large amplitude, very slow potentials lasting several seconds that can be identified on intracranial recordings when low EEG frequencies are not filtered out [9, 57]. These three intracranial electrographic features (fast activity, EEG flattening and very slow potentials) have been proposed as biomarkers of seizure-genesis in the epileptogenic zone [45], since a retrospective evaluation demonstrated that their location on stereo-EEG recordings coincides with the area that has been surgically removed to cure the patient (Fig. 1.7).

The above-mentioned triad of electrographic elements defines seizure networks and the epileptogenic zone in the majority of patients selected for stereo-EEG recordings with intracerebral electrodes. The type of epilepsy referred to surgery could be the reason for the homogeneity of seizure

acute seizure in entorhinal cortex in vitro - bicuculline model

chronic seizure in entorhinal cortex (KA model)

Fig. 1.8 Seizures recorded in guinea pig entorhinal cortex. The *upper trace* was recorded in the *in vitro* isolated guinea pig brain after systemic application of 50 μM bicuculline. In the *lower panel* a seizure is shown, which was recorded *in vivo* 3 months after injection of kainic acid in the hippocampus. Both seizures are characterized by fast activity at the onset followed by irregular firing and late periodic bursting (From DeCurtis, unpublished)

pattern reported in the literature. Most of the patients selected for pre-surgical studies, indeed, have pharmacoresistant epilepsies due to either focal cortical dysplasia, low-grade epileptogenic tumors (such as gangliogliomas or dysembryogenetic lesions), or mesial temporal lobe epilepsy with hippocampal sclerosis. Seizures in these types of epilepsy may present with similar EEG features. In mesial temporal lobe epilepsy, seizures that initiate with a hypersynchronous spiking pattern have been reported [8, 93]. Fast activity consistently follows the hypersynchronous discharge, suggesting that this pattern represents a variant of the low-voltage fast activity pattern.

Seizure onset patterns different from low-voltage fast activity have been described during intracranial EEG monitoring, for instance in tuberous sclerosis and in cortical malformations such as polymicrogyria [16, 51, 70, 79]. Whether such patterns are the expression of the epileptogenic network specifically caused by the type of lesion or are due to the failure to implant electrodes precisely in the epileptogenic area, is an open question. Moreover, variable seizure onset patterns have been detected with intracranial and extra-cerebral electrode arrays, such as grid and strips, positioned on the cortical surface in the subdural space. The localizing value of subdural

electrodes has been questioned (e.g., [47, 90]) and, therefore, their ability to define sources and features of ictal patterns is assumed to be less precise than depth electrodes.

Another crucial issue that emerged from intracranial recording studies and can be confirmed by retrospective analysis of earlier reports on seizure patterns, is the demonstration that focal seizures are characterized by a clear sequence of events that starts with a fast activity pattern and ends with highly synchronous, large amplitude bursting. The striking novel finding in this context is the observation that seizures do not initiate with the explosion of sustained, large amplitude, synchronous potentials, as commonly assumed, but feature low amplitude activity and background activity desynchronization that in several occasions last several tenths of seconds. In between seizure onset and seizure termination, a transition from fast, possibly desynchronized activity [59, 82] into an irregular spiking pattern (referred to as "tonic" in several reports) is observed. During the latter phase synchrony of activity builds up and progressively promotes clustering of highly synchronous discharges separated by periods of post-burst depression (Fig. 1.8). The late-seizure bursting (sometimes defined as "clonic phase") precedes seizure termination. Interestingly, if seizure onset is restricted to a spatially limited region, seizure termination characterized by synchronous periodic bursting is usually more diffuse and shows the tendency to involve the entire epileptogenic zone. The mechanism for such a widening of the epileptogenic network during the late seizure is still unclear. A synchronizing influence mediated by the involvement of subcortical structures can be proposed. After the end of a focal seizure, post-ictal depression is evident and can be measured as a reduction of background activity in comparison to the pre-ictal condition. These findings can be reproduced in animal models, as discussed in the next section.

In summary, direct evaluation of seizure-generator networks with intracerebral electrodes in focal human epilepsies demonstrates that specific electrographic patterns with a quite reproducible temporal progression define a seizure (typically a focal seizure). De-synchronization of background activity and the appearance of fast low-voltage rhythms characterize seizure initiation and excessive synchronization correlate with termination of the seizure [59]. Post-ictal depression is typical of focal seizures and should always be verified to identify a seizure.

1.3 Seizures, Seizure-Like Events and Afterdischarges in Animal Models

Based on the intracranial human findings observed in focal epilepsies during pre-surgical monitoring, it is mandatory to re-define the term "seizure" in experimental studies of animal models. We will first address *in vivo* studies performed on animal models of seizures or epilepsy, and then discuss *in vitro* studies carried out on preparations featuring complete or partial preservation of brain networks.

Diverse seizure patterns have been illustrated with *in vivo* intrecerebral recordings in animal models of epilepsy obtained with different methods and protocols. In several studies, seizure-like patterns were defined only with EEG, i.e., without the aid of video monitoring. This approach is

problematic, because the correspondence of EEG patterns with behavioral symptoms should be verified when seizure events are described. The possibility that the reported EEG potentials are interictal events or even physiological patterns, if not artifacts, should be carefully considered (see Sect. 1.4, below). Incidentally, the lack of a precise definition of a normal EEG in different animal species is a serious limitation to the evaluation of pathological patterns in animal models of seizures and epilepsy. These considerations further support the concept that epileptic phenotypes in animal models should always be carefully analyzed with the aid of video-EEG monitoring, to correlate possible seizure patterns to behavioral/motor changes.

Behavioral seizure correlates are not easy to identify in animals, even when careful electro-behavioral evaluation of the video-EEG is performed, because focal seizures may present with minor symptoms that have little, if any, motor sign. This is a major limitation for seizure identification in animal models: we can only be sure of seizures that correlate with enhanced or decreased motor signs, since other critical non-motor symptoms are difficult to detect. Seizures generated in the hippocampus in animal models (and in patients as well), for instance, can occur during immobility ([8, 15, 80]; see Sect. 1.4, below) and are indistinguishable from normal pauses in behavior unless intracerebral EEG recordings are performed in parallel to video monitoring. In this respect, human EEG studies on the definition of electro-clinical seizure patterns are more standardized and detailed than animal reports. The precise electro-clinical correlation of symptoms during seizures performed in humans demonstrates the finer scientific development of clinical epileptology in comparison to experimental epileptology, and sets an example to improve phenotyping in animal models of epilepsy.

In vivo recording of seizures and characterization of seizure patterns have been performed in a relatively small number of studies that describe animal models of epilepsy, largely on temporal lobe epilepsy models developed in rats and mice. Other models in which video-EEG electro-behavioral characterization of focal seizures was analyzed in detail include post-traumatic epilepsy models [25, 65, 66], models of perinatal anoxia-ischemia [61] and infantile spasms [81]. These reports confirmed that EEG correlates of seizures are largely characterized by fast activity at onset, followed by irregular spiking; and periodic bursting that develops with time during seizures (and usually represents the last pattern before seizure termination: [8, 15, 46, 95]). Post-ictal depression ensues and is infrequently characterized in these models.

Other electrographic potentials that supposedly represent the expression of an epileptic brain have been reported and quantified to support the characterization of epilepsy models. The behavioral correlates of these pathological patterns are often not described (and may not be possible to identify), and in some reports the claim is made that a specific pattern that does not respond to the criteria defined above is regarded as seizure. It is frequently assumed that epileptiform discharges that last longer than 2–3 s can be considered as ictal, as mentioned above [26, 31]. The criterion of duration to discriminate between an interictal and ictal discharge is quite subjective and could be misleading when applied to focal epilepsies. Since a consensus on this issue is still missing, more stringent criteria to define a seizure are required and should be identified.

Seizure patterns comparable to those described *in vivo* in animals (and in human focal epilepsies) can be reproduced in preparations of the entire brain or portions of brain tissue maintained *in vitro* in isolation. Obviously, the absence of the peripheral limbs that expresses motor symptoms prevents any definition of seizure in these experimental conditions. Therefore, the identification of interictal and seizure-like patterns on *in vitro* preparations relies exclusively by electrophysiological recordings, and the identification of stringent criteria for seizure definition is quite critical.

Seizure-like events characterized by fast activity at onset, followed by irregular spiking and terminating with periodic bursting discharges are induced by diverse pharmacological manipulations in adult whole guinea pig brain preparation ([44, 89]; Fig. 1.8), in neonatal en-bloc preparation of cortical areas/systems, such as the *in toto* hippocampal-parahippocampal structures [30, 64] and in complex tissue slices, in which connectivity between cortical structures is preserved, such as enthorinal-hippocampal slices ([6, 60]; Fig. 1.8).

In several studies performed on slice preparations, prolonged epileptiform events are described, which are characterized either by repeated spikes or by large paroxysmal depolarizing shifts followed by a depolarizing plateau potential on which decrementing discharges occur (see [28]). These types of discharges are often defined as seizure-like, even though their identification as seizures is questionable: similar events, indeed, are never observed during spontaneous seizures recorded *in vivo*, but can be generated by repeated stimulations, as afterdischarges induced by the kindling procedure. In slice studies, the measurement of the duration of "afterdischarges" is usually reported as a criterion to distinguish between interictal and ictal events. This assumption is based on the idea that the mechanisms that generate interictal and ictal events are similar and differ only by the duration and persistence of repetitive spiking or bursting activity. However, this conclusion may not be correct, based on recent findings demonstrating that seizure-like events in complex preparations are initiated with a prominent activation of inhibitory networks, whereas this may not be true for interictal spikes. The analysis of seizure-like discharges in neocortical and hippocampal slices exposed to different pro-epileptic conditions demonstrate that GABAergic networks are active at the very onset of a seizure [30, 39, 40, 67, 99]. These findings were confirmed in the *in vitro* isolated whole guinea pig brain [29, 44]. In this preparation, pre-ictal (ictal) spikes and fast activity that characterize seizure onset correlate with activation of GABAergic interneurons and with a cessation of neuronal firing in principal excitatory cells that last several seconds. In this model, the progression of seizure activity characterized by the transition to the irregular spiking and periodic bursting phases was sustained by ectopic firing of principal cells driven by changes in extracellular potassium induced by inhibitory network activation at seizure onset [88].

In conclusion, the definition of seizure-like events in *in vitro* preparation should be reconsidered and should rely on the reproduction of seizure patterns observed in humans and in chronic animal models of epilepsy. This "reverse translational" approach might help to focus future *in vitro* studies on the mechanisms of seizure generation that more reliably reproduce human focal epilepsy.

1.4 Defining Seizures in Basic Epilepsy Research: Potential Problems Specific to Rats and Mice

Defining seizures in humans requires consideration of several issues, as discussed above. In basic epilepsy research, conducted mainly in rodents (rat or mouse), there are other issues that are important. In order to quantify seizures for preclinical studies, one would want to be precise about seizure onset and seizure termination. However, not only are seizures hard to define, but the exact time of their onset and termination are also problematic. Other issues are also relevant: if there are brief pauses between seizures, when is the pause sufficient to define the events as two separate seizures? Post-ictal depression is often followed by a series of afterdischarges or spikes that become more and more frequent – when does the repetitive spiking become frequent enough to be called the onset of the next seizure? This issue is not only important in establishing seizure frequency, but it also is important when defining status epilepticus (SE). When examined at high temporal resolution, there are often pauses between seizures during SE. Does this mean it is not SE? If there are no convulsions (non-convulsive SE) how does one determine what is and what is not SE? Similar to humans, defining a seizure in rodents is not as easy as one might think.

1.4.1 Behavioral State

There are several behaviors that make up the vast majority of the lifespan in rats and mice: explora-tion, sleep, grooming, eating and drinking. In addition, there is a state called "quiet immobility," "awake rest" or "behavioral arrest" where rodents stop moving, their eyes are open, and they stare blankly into space. Typically the animal is standing at the time, and has just walked across the cage or explored its surroundings. Unlike humans, this behavioral state can be prolonged (over 10 s). It presents problems for the epilepsy researcher because it appears similar to an absence seizure. Therefore, understanding the normal behavioral states of rodents, and their EEG correlates, is important for epilepsy researchers using these species.

1.4.1.1 Hippocampal EEG Associated with Exploration: Theta Rhythm

Associated with exploration, behavioral arrest, and sleep in rodents are distinct EEG rhythms that can be recorded with chronic electrodes implanted in hippocampus [14]. As shown in Fig. 1.9a and originally described by Green and Arduini [50], EEG oscillations at theta frequency (commonly called theta rhythm) are recorded in hippocampus when an animal explores. Theta oscillations vary in frequency but are typically 6–10 Hz in rats and mice [12, 50, 91].

In animal models of epilepsy, theta rhythm is interesting because epileptic animals are less likely to exhibit seizure activity during exploratory behavior, when theta oscillations occur in hippocampus [69]. This "anticonvulsant" nature of exploration and theta rhythm in hippocampus has been attributed to many potential mechanisms but has not been defined conclusively [22]. It is useful to record theta oscillations *in vivo* because large theta rhythm is found in hippocampus but it is much smaller or not observed elsewhere. Therefore, theta oscillations can be used to confirm the recording is in hippocampus. Theta oscillations are also useful to record because their amplitude can be used to define the specific layer within hippocampus where the recording electrode is located. For example, if a stimulating electrode is used to evoke field EPSPs in area CA1 from the Schaffer collateral input, the field EPSP should be recorded in the layer where theta is relatively small, stratum radiatum.

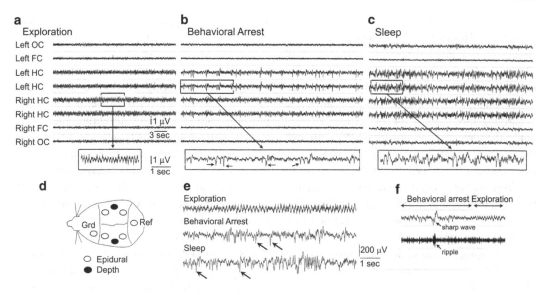

Fig. 1.9 EEG characteristics in the normal adult rat. (**a**) Using 8 electrodes (shown in d), awake behaving rats were recorded in their home cage. During exploration, hippocampal electrodes exhibited theta oscillations. The area outlined by the *box* is expanded at the *bottom*. (**b**) During a spontaneous arrest of behavior, sharp waves (*arrows*) occurred regularly in the hippocampal EEG. (**c**) During sleep, the hippocampal EEG became active. (**d**) The recording arrangement included 4 epidural electrodes and 2 twisted bipolar electrodes in the dorsal hippocampus, one in each hemisphere. *Grd* ground; *Ref* reference. (**e**) A summary of a-c is shown. In three behavioral states there are large differences in the hippocampal EEG with sharp waves (*arrows*) in behavioral arrest and sleep. (**f**) During sharp waves, filtering in the ripple band (100–200 Hz) shows that a ripple occurs at the same time as the sharp wave (From LaFrancois and Scharfman, unpublished)

In contrast, where theta is larger, the adjacent stratum lacunosum-moleculare, the field EPSP evoked by the same stimulus would be small or have a positive polarity. Because the entorhinal cortex is a source of theta rhythm (the other major source originates in the septum; [12]), theta oscillations are very large in stratum lacunosum-moleculare and the outer two-thirds of the molecular layer of the dentate gyrus, were the entorhinal cortical projection (the perforant path) to hippocampus terminates.

1.4.1.2 Hippocampal EEG Associated with Behavioral Arrest: Sharp Wave-Ripples

The hippocampal EEG shown in Fig. 1.9b is taken from a rat that explored and then paused – entering a period of behavioral arrest. As described by Buzsaki originally [10, 11], the hippocampal EEG changes dramatically when an animal stops exploring and pauses in a frozen stance, with eyes still open. Theta oscillations decrease and the EEG becomes irregular. In addition, sharp waves (SPWs) occur intermittently. SPWs are ~100 msec duration spikes that reflect synchronous firing in a subset of area CA3 neurons, which in turn activate area CA1 apical dendrites by the Schaffer collateral axons and the dentate gyrus, most likely by backprojecting axons of CA3 pyramidal cells. Therefore, SPWs can be recorded in many locations within hippocampus [10, 11].

The term SPW is important to discuss in the context of epilepsy, because it is sometimes used interchangeably with the term interictal spikes (IIS). Hippocampal SPWs are distinct from interictal spikes because hippocampal SPWs occur without seizures, i.e., they are not interictal (between ictal events). Hippocampal SPWs are recorded only in hippocampus- if one moves a recording electrode just outside the hippocampus, SPWs are not observed (Pearce and Scharfman, unpublished). IIS in an epileptic rodent can be typically recorded from multiple cortical electrodes simultaneously at many sites in the brain. However, SPWs can be generated by circuits outside hippocampus, i.e., other types of SPWs besides those generated in area CA3. For example, SPWs are generated in entorhinal cortex and piriform cortex [68]. Notably, the underlying mechanisms for an IIS may or may not be the same mechanisms for a SPW, although they do seem related. For example, GABAergic mechanisms may trigger IIS (as discussed in the previous section); GABAergic network oscillations (ripples) are also involved in SPWs. The classic view of the IIS is that it is generated by a giant paroxysmal depolarization shift (see previous section); a synchronous depolarization in pyramidal cells also drives SPWs. Regardless, if SPWs and IIS are terms that are used synonymously, there may be differences in the underlying cellular processes/mechanisms that are overlooked, so it is important to consider the terms carefully.

When recording electrodes are positioned near the CA1 pyramidal cell layer, fast oscillations called ripples [84] can be detected at about the same time as the SPW (Fig. 1.9f). Therefore, the term "SPW-R" (sharp wave-ripple) is now used instead of the original term, sharp wave. Ripples in the hippocampal EEG correspond to synchronous oscillations of pyramidal cells that are caused by rhythmic IPSPs that are initiated by action potentials in a subset of hippocampal GABAergic interneurons that innervate pyramidal cell somata and initial axon segments. As synchronous release of GABA from these peri-somatic targeting interneurons hyperpolarize pyramidal cell somata that are in close proximity, chloride ions enter the pyramidal cells in a repetitive manner and cause a series of extracellular positivities. The positivities wax and wane as the pyramidal cell IPSPs start and stop, leading to an oscillation [19].

1.4.1.3 The Hippocampal EEG Becomes Active During Sleep

The hippocampal EEG becomes extremely active during sleep in the rodent, and is irregular, called large irregular activity (LIA: Fig. 1.9c). The increase in the hippocampal EEG is often

simplified as a type of disinhibitory state that coincides with a 'switch' from sensitivity to sensory input to a state where intrinsic circuitry is active [52]. A similar idea has been proposed for piriform cortex during slow-wave sleep; odor input is reduced in favor of processing between piriform cortex and other forebrain sites [97]. For the epileptologist, it is important to recognize that comparing the hippocampal EEG between animals without considering the behavioral state may make one animal seem normal (if it is exploring) compared to seizure like activity in the other if it is asleep (Fig. 1.9). Compressing the EEG can make this more difficult; for example, if the EEG is compressed it is hard to distinguish a noisy baseline from theta oscillations, so the EEG may look inactive when an animal is exploring. For these reasons, expansion and compression of the EEG should be varied during examination of the EEG for seizures. In addition, the type of electrode and recording system should stay the same for any given set of experiments.

1.4.1.4 When Normal Activity Appears To Be Epileptic

One of the implications of the discussion above for epilepsy research is the possibility that normal EEG activity may be mistaken for seizures. For example, an investigator may think that the animal is freezing because it is having a seizure, but actually exhibiting normal behavioral arrest. This interpretation is based on the limbic seizure stage scale of Racine, who based the scale on behaviors of rats during electrical stimulation of the amygdala during kindling. He suggested that there was initially a period of immobility with small mouth or face movements with small mouth or head movements, and called this a stage 1 or 2 in his scale of limbic seizure severity [78]. The only problem with this idea is that it can be confused with behavioral arrest.

During behavioral arrest, investigators could interpret the irregular activity and repetitive SPW-Rs to be a seizure (Fig. 1.9). Likewise, the transition from behavioral arrest back to exploration may seem like the termination of a seizure, particularly when the EEG is compressed

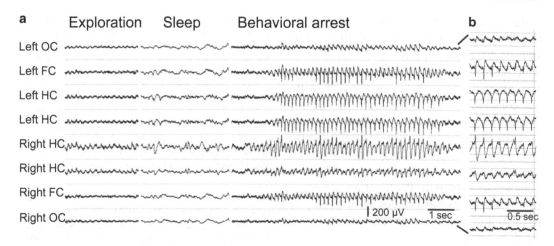

Fig. 1.10 Spike-wave discharges recorded from the normal adult hippocampus of the rat. (**a**) A recording from an adult Sprague-Dawley rat shows typical EEG activity during exploration and behavioral arrest. In behavioral arrest, there were spike-wave discharges. Animals were monitored during the recordings to be sure that artifacts related to grooming or chewing did not occur during spike-wave discharges. (**b**) Recordings in **a** are expanded (From Pearce and Scharfman, unpublished; see also [101])

(Fig. 1.9). In light of these potential problems, describing stage 1 seizures without a hippocampal electrode is problematic. An animal that suddenly stops and appears unresponsive could be interpreted to have a stage 1 seizure when it actually is pausing between episodes of exploration.

Another problem arises in studies of seizure frequency evaluated over time. For example, studies of epileptogenesis often record animals over weeks. There is typically no consideration of behavioral state when the results are quantified. If there is less exploration because an animal is sleeping more, EEG power in the theta band may decrease. EEG power in high frequency bands may increase if there are more SPW-Rs because the animal is pausing more, or sleeping more.

1.4.2 Spike-Wave Discharge

In many strains of rats, the state of behavioral arrest is accompanied by spike-wave discharge in thalamocortical networks [20, 27, 96]. These discharges have been noted in almost every strain of rat, such as Long-Evans [83], where approximately 90 % of female rats exhibited spike-wave discharges spontaneously by 4 months of age. In Wistar rats, Gralewicz [49] reported that 73 % of male rats showed spike-wave discharges by 6 months of age and 93 % of males at 24 months of age. Kelly [63] reported spike-wave discharges in female Fischer 344 rats at 4 and 20 months of age. In rats that are genetic models of absence epilepsy (GAERS, Wag Rij) spike-wave discharges are a characteristic of the strain, and used to gain insight into mechanisms of absence epilepsy [20, 27]. Numerous genetic models of absence epilepsy also exist based on spontaneous mutations in mice (*e.g., lethargic*; [18, 74]). As shown in Fig. 1.10, spike-wave discharges accompany behavioral arrest in naïve Sprague-Dawley rats. These discharges vary according to the sex, age, environment and other factors [13] but are not always observed [96], making control recordings critical to any study of rats in an animal model of epilepsy.

These observations raise several questions: are spike-wave discharges in rodents normal? It has been suggested that they could serve important purposes related to sensory processing [71, 86, 94] or aging and excitability [71].

If this is true in rodents, is human spike-wave discharge normal too? One possibility is that spike-wave discharges and behavioral arrest were present in early stages of evolution and then reduced because behavioral arrests (without complete attention) would be dangerous in the presence of predators – vigilance would be advantageous. In humans, the spike-wave discharges that do arise may be vestiges of rodent circuitry that have not completely been removed by evolution. Photic stimulation can trigger spike-wave discharges in humans [24, 87], and may be a method to trigger these 'vestigial' oscillations.

Another implication of the observations in rats in Fig. 1.10 is relevant to the detection of seizures in hippocampal electrodes in rodent studies of epilepsy. In Fig. 1.10, the hippocampal electrode appears to show rhythmic spiking when spike-wave discharges occur in the frontal and occipital leads. The rhythmic spiking in hippocampus could be volume conducted from thalamus, or it could reflect hippocampal neural activity. In light of the fact that the frontal cortical lead shows spike-wave oscillations, one would know that volume conduction in the hippocampal lead is a possibility. However, if there were only an electrode in hippocampus, which is a common recording arrangement in epilepsy research, the rhythmic activity in hippocampal electrodes might be interpreted to be a seizure generated in hippocampus. Because it is accompanied by a frozen, 'absence' behavior, it could be concluded that there was a Racine stage 1 seizure. Importantly, some of the normal rodents with spike wave discharges also have head nodding or mastications, which could make an investigator more convinced of seizure activity – because these movements were also noted by Racine in his classification of stage 1–2 behaviors. Importantly, most of the spike-wave discharges occur at approximately 7–9 Hz and are stable (in frequency) within a spike-wave episode or across episodes (Fig. 1.10; [13, 20, 27, 96]). Therefore, rhythms at this frequency (e.g., theta rhythm) that occur in hippocampus can be a signal to investigators to interpret their EEG data cautiously.

Acknowledgements RSF: I first met Phil at Stanford in the early 1970's, when he brought the hippocampal slice recording technique to the US. Since then, he has been a pioneer and thought leader in so many ways, and a mentor to generations of epilepsy researchers.

MdC: Phil has been a valued colleague and leader in epilepsy research both in the US and internationally.

HES: The three years I spent in Seattle under Phil's mentorship were some of the most memorable, and important. Long after I left Seattle, his insights and approach to epilepsy research continued to influence my research, and the work of those in my own laboratory.

Other Acknowledgements Section 1.1 was written by RF, Sects. 1.2, 1.3 by MdC and Sect. 1.4 by HES. Supported by The James and Carrie Anderson Fund for Epilepsy Research, The Susan Horngen fund (RF); NIH R01 NS-037562, R01 NS-070173, R21 MH-090606, the Alzheimer's Association, and the New York State Department of Health (HES); Italian Health Ministry grants RC 2011–2013; RF114-2007 and RF151-2010 (MdC).

References

1. Abend NS, Wusthoff CJ (2012) Neonatal seizures and status epilepticus. J Clin Neurophysiol 29:441–448
2. Adams RD, Victor M, Ropper AH (1997) Principles of neurology. McGraw Hill Health Professions Division, New York
3. Aird RB, Masland RL, Woodbury DM (1984) The epilepsies: a critical review. Lippincott Williams & Wilkins, Philadelphia
4. Aldenkamp AP, Beitler J, Arends J, van der Linden I, Diepman L (2005) Acute effects of subclinical epileptiform EEG discharges on cognitive activation. Funct Neurol 20:23–28
5. Allen PJ, Fish DR, Smith SJ (1992) Very high-frequency rhythmic activity during SEEG suppression in frontal lobe epilepsy. Electroencephalogr Clin Neurophysiol 82:155–159
6. Avoli M, D'Antuono M, Louvel J, Kohling R, Biagini G, Pumain R, D'Arcangelo G, Tancredi V (2002) Network and pharmacological mechanisms leading to epileptiform synchronization in the limbic system in vitro. Prog Neurobiol 68:167–207
7. Blume WT, Young GB, Lemieux JF (1984) EEG morphology of partial epileptic seizures. Electroencephalogr Clin Neurophysiol 57:295–302
8. Bragin A, Engel J Jr, Wilson CL, Fried I, Mathern GW (1999) Hippocampal and entorhinal cortex high-frequency oscillations (100–500 hz) in human epileptic brain and in kainic acid–treated rats with chronic seizures. Epilepsia 40:127–137
9. Bragin A, Wilson CL, Fields T, Fried I, Engel J Jr (2005) Analysis of seizure onset on the basis of wideband EEG recordings. Epilepsia 46(Suppl 5):59–63

10. Buzsaki G (1986) Hippocampal sharp waves: their origin and significance. Brain Res 398:242–252

11. Buzsaki G (1989) Two-stage model of memory trace formation: a role for "noisy" brain states. Neuroscience 31:551–570

12. Buzsaki G (2002) Theta oscillations in the hippocampus. Neuron 33:325–340

13. Buzsaki G, Laszlovszky I, Lajtha A, Vadasz C (1990) Spike-and-wave neocortical patterns in rats: genetic and aminergic control. Neuroscience 38:323–333

14. Buzsaki G, Leung LW, Vanderwolf CH (1983) Cellular bases of hippocampal EEG in the behaving rat. Brain Res 287:139–171

15. Carriero G, Arcieri S, Cattalini A, Corsi L, Gnatkovsky V, de Curtis M (2012) A guinea pig model of mesial temporal lobe epilepsy following nonconvulsive status epilepticus induced by unilateral intrahippocampal injection of kainic acid. Epilepsia 53:1917–1927

16. Chassoux F, Landre E, Rodrigo S, Beuvon F, Turak B, Semah F, Devaux B (2008) Intralesional recordings and epileptogenic zone in focal polymicrogyria. Epilepsia 49:51–64

17. Chen H, Zhuang P, Miao SH, Yuan G, Zhang YQ, Li JY, Li YJ (2010) Neuronal firing in the ventrolateral thalamus of patients with Parkinson's disease differs from that with essential tremor. Chin Med J (Engl) 123:695–701

18. Chung WK, Shin M, Jaramillo TC, Leibel RL, LeDuc CA, Fischer SG, Tzilianos E, Gheith AA, Lewis AS, Chetkovich DM (2009) Absence epilepsy in apathetic, a spontaneous mutant mouse lacking the h channel subunit, hcn2. Neurobiol Dis 33:499–508

19. Cobb SR, Buhl EH, Halasy K, Paulsen O, Somogyi P (1995) Synchronization of neuronal activity in hippocampus by individual GABAergic interneurons. Nature 378:75–78

20. Coenen AM, Van Luijtelaar EL (2003) Genetic animal models for absence epilepsy: a review of the WAG/Rij strain of rats. Behav Genet 33:635–655

21. Cohen I, Navarro V, Clemenceau S, Baulac M, Miles R (2002) On the origin of interictal activity in human temporal lobe epilepsy in vitro. Science 298:1418–1421

22. Colom LV, Garcia-Hernandez A, Castaneda MT, Perez-Cordova MG, Garrido-Sanabria ER (2006) Septo-hippocampal networks in chronically epileptic rats: potential antiepileptic effects of theta rhythm generation. J Neurophysiol 95:3645–3653

23. Cossu M, Cardinale F, Castana L, Citterio A, Francione S, Tassi L, Benabid AL, Lo Russo G (2005) Stereoelectroencephalography in the presurgical evaluation of focal epilepsy: a retrospective analysis of 215 procedures. Neurosurgery 57:706–718

24. Covanis A (2005) Photosensitivity in idiopathic generalized epilepsies. Epilepsia 46(Suppl 9):67–72

25. D'Ambrosio R, Fairbanks JP, Fender JS, Born DE, Doyle DL, Miller JW (2004) Post-traumatic epilepsy following fluid percussion injury in the rat. Brain 127:304–314

26. D'Ambrosio R, Miller JW (2010) What is an epileptic seizure? Unifying definitions in clinical practice and animal research to develop novel treatments. Epilepsy Curr 10:61–66

27. Danober L, Deransart C, Depaulis A, Vergnes M, Marescaux C (1998) Pathophysiological mechanisms of genetic absence epilepsy in the rat. Prog Neurobiol 55:27–57

28. de Curtis M, Avanzini G (2001) Interictal spikes in focal epileptogenesis. Prog Neurobiol 63:541–567

29. de Curtis M, Gnatkovsky V (2009) Reevaluating the mechanisms of focal ictogenesis: the role of low-voltage fast activity. Epilepsia 50:2514–2525

30. Derchansky M, Jahromi SS, Mamani M, Shin DS, Sik A, Carlen PL (2008) Transition to seizures in the isolated immature mouse hippocampus: a switch from dominant phasic inhibition to dominant phasic excitation. J Physiol 586:477–494

31. Devinsky O, Kelley K, Porter RJ, Theodore WH (1988) Clinical and electroencephalographic features of simple partial seizures. Neurology 38:1347–1352

32. Engel J, Bragin A, Staba R, Mody I (2009) High-frequency oscillations: what is normal and what is not? Epilepsia 50:598–604

33. Engel J Jr (1984) A practical guide for routine EEG studies in epilepsy. J Clin Neurophysiol 1:109–142

34. Engel JJ (1989) Seizures and epilepsy. F. A. Davis, Philadelphia

35. Fariello RG, Doro JM, Forster FM (1979) Generalized cortical electrodecremental event. Clinical and neurophysiological observations in patients with dystonic seizures. Arch Neurol 36:285–291

36. Fisher RS, Engel JJ Jr (2010) Definition of the postictal state: when does it start and end? Epilepsy Behav 19:100–104

37. Fisher RS, van Emde BW, Blume W, Elger C, Genton P, Lee P, Engel J Jr (2005) Epileptic seizures and epilepsy: definitions proposed by the international league against epilepsy (ILAE) and the international bureau for epilepsy (IBE). Epilepsia 46:470–472

38. Fisher RS, Webber WR, Lesser RP, Arroyo S, Uematsu S (1992) High-frequency EEG activity at the start of seizures. J Clin Neurophysiol 9:441–448

39. Fujiwara-Tsukamoto Y, Isomura Y, Imanishi M, Ninomiya T, Tsukada M, Yanagawa Y, Fukai T, Takada M (2010) Prototypic seizure activity driven by mature hippocampal fast-spiking interneurons. J Neurosci 30:13679–13689

40. Fujiwara-Tsukamoto Y, Isomura Y, Kaneda K, Takada M (2004) Synaptic interactions between pyramidal cells and interneurone subtypes during seizure-like activity in the rat hippocampus. J Physiol 557:961–979

41. Fujiwara H, Greiner HM, Lee KH, Holland-Bouley KD, Seo JH, Arthur T, Mangano FT, Leach JL, Rose DF (2012) Resection of ictal high-frequency oscillations leads to favorable surgical outcome in pediatric epilepsy. Epilepsia 53:1607–1617

42. Gale K (1992) Subcortical structures and pathways involved in convulsive seizure generation. J Clin Neurophysiol 9:264–277

43. Gloor P (1975) Contributions of electroencephalography and electrocorticography to the neurosurgical treatment of the epilepsies. In: Purpura DP, Penry JK, Walter RD (eds) Advances in neurology. Raven, New York, pp 59–105

44. Gnatkovsky V, Librizzi L, Trombin F, de Curtis M (2008) Fast activity at seizure onset is mediated by inhibitory circuits in the entorhinal cortex in vitro. Ann Neurol 64:674–686

45. Gnatkovsky V, Pastori C, Cardinale F, Lo Russo G, Mai R, Nobili L, Sartori I, Tassi L, Francione S, de Curtis M (2014) Biomarkers of epileptogenic zone defined by quantified stereo-EEG analysis. Epilepsia 55(2):296–305

46. Goffin K, Nissinen J, Van Laere K, Pitkanen A (2007) Cyclicity of spontaneous recurrent seizures in pilocarpine model of temporal lobe epilepsy in rat. Exp Neurol 205:501–505

47. Gonzalez-Martinez J, Bulacio J, Alexopoulos A, Jehi L, Bingaman W, Najm I (2013) Stereoelectroencephalography in the "difficult to localize" refractory focal epilepsy: early experience from a North American epilepsy center. Epilepsia 54:323–330

48. Gotman J, Levtova V, Olivier A (1995) Frequency of the electroencephalographic discharge in seizures of focal and widespread onset in intracerebral recordings. Epilepsia 36:697–703

49. Gralewicz S, Wiaderna D, Stetkiewicz J, Tomas T (2000) Spontaneous spike-wave discharges in rat neocortex and their relation to behaviour. Acta Neeurobiol Exp (Wars) 60:323–332

50. Green JD, Arduini AA (1954) Hippocampal electrical activity in arousal. J Neurophysiol 17:533–557

51. Guerrini R, Dravet C, Raybaud C, Roger J, Bureau M, Battaglia A, Livet MO, Gambarelli D, Robain O (1992) Epilepsy and focal gyral anomalies detected by MRI: electroclinico-morphological correlations and follow-up. Dev Med Child Neurol 34:706–718

52. Hasselmo ME (2005) What is the function of hippocampal theta rhythm? – linking behavioral data to phasic properties of field potential and unit recording data. Hippocampus 15:936–949

53. Hauser WA, Hesdorffer DC (1990) Epilepsy: frequency, causes and consequences. Demos Medical, New York

54. Holmes GL, Lenck-Santini PP (2006) Role of interictal epileptiform abnormalities in cognitive impairment. Epilepsy Behav 8:504–515

55. Hopkins A, Shorvon S, Cascino GD (1995) Epilepsy. Chapman & Hall, London

56. Huberfeld G, Menendez de la Prida L, Pallud J, Cohen I, Le Van Quyen M, Adam C, Clemenceau S, Baulac M, Miles R (2011) Glutamatergic pre-ictal discharges emerge at the transition to seizure in human epilepsy. Nat Neurosci 14:627–634

57. Ikeda A, Taki W, Kunieda T, Terada K, Mikuni N, Nagamine T, Yazawa S, Ohara S, Hori T, Kaji R,

Kimura J, Shibasaki H (1999) Focal ictal direct current shifts in human epilepsy as studied by subdural and scalp recording. Brain 122(Pt 5):827–838

58. Jackson JH (1870) A study of convulsions. Trans St Andrews Med Grad Assoc 3:162–204

59. Jiruska P, de Curtis M, Jefferys JG, Schevon CA, Schiff SJ, Schindler K (2013) Synchronization and desynchronization in epilepsy: controversies and hypotheses. J Physiol 591:787–797

60. Jones RS, Lambert JD (1990) The role of excitatory amino acid receptors in the propagation of epileptiform discharges from the entorhinal cortex to the dentate gyrus in vitro. Exp Brain Res 80:310–322

61. Kadam SD, White AM, Staley KJ, Dudek FE (2010) Continuous electroencephalographic monitoring with radio-telemetry in a rat model of perinatal hypoxia-ischemia reveals progressive post-stroke epilepsy. J Neurosci 30:404–415

62. Kaplan PW, Fisher RS (2005) Imitators of epilepsy. Demos Publishing, New York

63. Kelly KM, Shiau DS, Jukkola PI, Miller ER, Mercadante AL, Quigley MM, Nair SP, Sackellares JC (2011) Effects of age and cortical infarction on EEG dynamic changes associated with spike wave discharges in F344 rats. Exp Neurol 232:15–21

64. Khalilov I, Dzhala V, Medina I, Leinekugel X, Melyan Z, Lamsa K, Khazipov R, Ben-Ari Y (1999) Maturation of kainate-induced epileptiform activities in interconnected intact neonatal limbic structures in vitro. Eur J Neurosci 11:3468–3480

65. Kharatishvili I, Nissinen JP, McIntosh TK, Pitkanen A (2006) A model of posttraumatic epilepsy induced by lateral fluid-percussion brain injury in rats. Neuroscience 140:685–697

66. Kharatishvili I, Pitkanen A (2010) Association of the severity of cortical damage with the occurrence of spontaneous seizures and hyperexcitability in an animal model of posttraumatic epilepsy. Epilepsy Res 90:47–59

67. Lopantsev V, Avoli M (1998) Laminar organization of epileptiform discharges in the rat entorhinal cortex in vitro. J Physiol 509(Pt 3):785–796

68. Manabe H, Kusumoto-Yoshida I, Ota M, Mori K (2011) Olfactory cortex generates synchronized top-down inputs to the olfactory bulb during slow-wave sleep. J Neurosci 31:8123–8133

69. Miller JW, Turner GM, Gray BC (1994) Anticonvulsant effects of the experimental induction of hippocampal theta activity. Epilepsy Res 18:195–204

70. Mohamed AR, Bailey CA, Freeman JL, Maixner W, Jackson GD, Harvey AS (2012) Intrinsic epileptogenicity of cortical tubers revealed by intracranial EEG monitoring. Neurology 79:2249–2257

71. Nair SP, Jukkola PI, Quigley M, Wilberger A, Shiau DS, Sackellares JC, Pardalos PM, Kelly KM (2008) Absence seizures as resetting mechanisms of brain dynamics. Cybern Syst Anal 44:664–672

72. Nicolelis MA, Fanselow EE (2002) Dynamic shifting in thalamocortical processing during different behavioural states. Philos Trans R Soc Lond B Biol Sci 357:1753–1758

73. Noachtar S, Peters AS (2009) Semiology of epileptic seizures: a critical review. Epilepsy Behav 15:2–9
74. Noebels J (2006) Spontaneous epileptic mutations in the mouse. In: Pitkanen A, Moshe SL, Schwartzroin PA (eds) Models of seizures and epilepsy. Elsevier, London, pp 222–233
75. Penfield W, Jasper H (1954) Epilepsy and the functional anatomy of the human brain. J. & A. Churchill, Ltd., London
76. Pohlmann-Eden B, Hoch DB, Cochius JI, Chiappa KH (1996) Periodic lateralized epileptiform discharges – a critical review. J Clin Neurophysiol 13:519–530
77. Porter RJ, Penry JK (1973) Responsiveness at the onset of spike-wave bursts. Electroencephalogr Clin Neurophysiol 34:239–245
78. Racine RJ (1972) Modification of seizure activity by electrical stimulation: II. Motor seizure. Electroencephalogr Clin Neurophysiol 32:281–294
79. Ramantani G, Koessler L, Colnat-Coulbois S, Vignal JP, Isnard J, Catenoix H, Jonas J, Zentner J, Schulze-Bonhage A, Maillard LG (2013) Intracranial evaluation of the epileptogenic zone in regional infrasylvian polymicrogyria. Epilepsia 54:296–304
80. Riban V, Bouilleret V, Pham-Le BT, Fritschy JM, Marescaux C, Depaulis A (2002) Evolution of hippocampal epileptic activity during the development of hippocampal sclerosis in a mouse model of temporal lobe epilepsy. Neuroscience 112:101–111
81. Scantlebury MH, Galanopoulou AS, Chudomelova L, Raffo E, Betancourth D, Moshe SL (2010) A model of symptomatic infantile spasms syndrome. Neurobiol Dis 37:604–612
82. Schindler K, Leung H, Elger CE, Lehnertz K (2007) Assessing seizure dynamics by analysing the correlation structure of multichannel intracranial EEG. Brain 130:65–77
83. Shaw FZ (2004) Is spontaneous high-voltage rhythmic spike discharge in Long Evans rats an absence-like seizure activity? J Neurophysiol 91:63–77
84. Sirota A, Buzsaki G (2005) Interaction between neocortical and hippocampal networks via slow oscillations. Thalamus Relat Syst 3:245–259
85. Talairach J, Bancaud J, Szikla G, Bonis A, Geier S, Vedrenne C (1974) New approach to the neurosurgery of epilepsy. Stereotaxic methodology and therapeutic results. 1. Introduction and history. Neurochirurgie 20(Suppl 1):1–240
86. Tort AB, Fontanini A, Kramer MA, Jones-Lush LM, Kopell NJ, Katz DB (2010) Cortical networks produce three distinct 7–12 hz rhythms during single sensory responses in the awake rat. J Neurosci 30:4315–4324
87. Trenite DG (2006) Photosensitivity, visually sensitive seizures and epilepsies. Epilepsy Res 70(Suppl 1):S269–S279
88. Trombin F, Gnatkovsky V, de Curtis M (2011) Changes in action potential features during focal seizure discharges in the entorhinal cortex of the in vitro isolated guinea pig brain. J Neurophysiol 106:1411–1423
89. Uva L, de Curtis M (2005) Polysynaptic olfactory pathway to the ipsi- and contralateral entorhinal cortex mediated via the hippocampus. Neuroscience 130:249–258
90. Vadera S, Mullin J, Bulacio J, Najm I, Bingaman W, Gonzalez-Martinez J (2013) Stereoelectroencephalography following subdural grid placement for difficult to localize epilepsy. Neurosurgery 72:723–729
91. Vanderwolf CH (1969) Hippocampal electrical activity and voluntary movement in the rat. Electroencephalogr Clin Neurophysiol 26:407–418
92. Vanhatalo S, Holmes MD, Tallgren P, Voipio J, Kaila K, Miller JW (2003) Very slow EEG responses lateralize temporal lobe seizures: an evaluation of non-invasive DC-EEG. Neurology 60:1098–1104
93. Wendling F, Bartolomei F, Bellanger JJ, Bourien J, Chauvel P (2003) Epileptic fast intracerebral EEG activity: evidence for spatial decorrelation at seizure onset. Brain 126:1449–1459
94. Wiest MC, Nicolelis MA (2003) Behavioral detection of tactile stimuli during 7–12 hz cortical oscillations in awake rats. Nat Neurosci 6:913–914
95. Williams PA, White AM, Clark S, Ferraro DJ, Swiercz W, Staley KJ, Dudek FE (2009) Development of spontaneous recurrent seizures after kainate-induced status epilepticus. J Neurosci 29:2103–2112
96. Willoughby JO, Mackenzie L (1992) Nonconvulsive electrocorticographic paroxysms (absence epilepsy) in rat strains. Lab Anim Sci 42:551–554
97. Wilson DA, Yan X (2010) Sleep-like states modulate functional connectivity in the rat olfactory system. J Neurophysiol 104:3231–3239
98. Worrell GA, Gardner AB, Stead SM, Hu S, Goerss S, Cascino GJ, Meyer FB, Marsh R, Litt B (2008) High-frequency oscillations in human temporal lobe: simultaneous microwire and clinical macroelectrode recordings. Brain 131:928–937
99. Ziburkus J, Cressman JR, Barreto E, Schiff SJ (2006) Interneuron and pyramidal cell interplay during in vitro seizure-like events. J Neurophysiol 95:3948–3954
100. Zijlmans M, Jacobs J, Zelmann R, Dubeau F, Gotman J (2009) High-frequency oscillations mirror disease activity in patients with epilepsy. Neurology 72:979–986
101. Pearce PS, Friedman D, Lafrancois JJ, Iyengar SS, Fenton AA, Maclusky NJ, Scharfman HE (2014) Spike-wave discharges in adult Sprague-Dawley rats and their implications for animal models of temporal lobe epilepsy. Epilepsy Behav 32:121–131

What Is the Clinical Relevance of *In Vitro* Epileptiform Activity?

Uwe Heinemann and Kevin J. Staley

Abstract

In vitro preparations provide an exceptionally rapid, flexible, and accessible approach to long-standing problems in epilepsy research including icto-genesis, epileptogenesis, and drug resistance. Acute slices suffer from a reduction in network connectivity that has traditionally been compensated through the application of acute convulsants. The utility and limitations of this approach have become clear over time and are discussed here. Other approaches such as organotypic slice preparations demonstrate the full spectrum of spontaneous epileptic activity and more closely mimic human responses to anticonvulsants, including the development of drug resistance. Newly developed transgenic and vector expression systems for fluorophores, optogenetics, and orphan receptors are being coupled with advances in imaging and image analysis. These developments have created the capacity to rapidly explore many new avenues of epilepsy research such as vascular, astrocytic and mitochondrial contributions to epileptogenesis. Rigorous study design as well as close collaboration with in vivo laboratories and clinical investigators will accelerate the translation of the exciting discoveries that will be revealed by these new techniques.

Keywords

Seizure • Epilepsy • Epileptogenisis • In vitro models • Translation • Ictal • Interictal

U. Heinemann
Neuroscience Research Center,
Charité Universitätsmedizin Berlin,
Garystr. 5, D 14195 Berlin, Germany
e-mail: uwe.heinemann@charite.de

K.J. Staley (✉)
Department of Neurology, Massachusetts General Hospital,
114 16th Street, Charlestown, MA 02129, USA
e-mail: kstaley@partners.org

2.1 Current Challenges

From the standpoint of translation, experimental epilepsy research is confronted with two major problems: the first is the discovery of mechanisms underlying drug resistance in epilepsy and development of new agents that would be useful in seizure control in pharmacoresistant patients.

H.E. Scharfman and P.S. Buckmaster (eds.), *Issues in Clinical Epileptology: A View from the Bench*,
Advances in Experimental Medicine and Biology 813, DOI 10.1007/978-94-017-8914-1_2,
© Springer Science+Business Media Dordrecht 2014

The second is discovering the signaling cascades that lead from an initial brain injury to epilepsy later in life, and conversely, discovering signaling cascades that would prevent this process, that is commonly referred to as epileptogenesis. It is important to note that epileptogenesis occurs in only a small percentage of patients who have suffered brain injuries. For example a recent prognostic study reported that only 8.2 % of patients developed epilepsy after stroke [46]. Thus it is important to identify biomarkers that predict which patients will develop epilepsy. It is also important to analyze how these biomarkers illuminate or participate in the process of epileptogenesis. Research into this issue will provide opportunities to identify mechanisms which protect the brain against seizures. To keep these and other research projects in epileptogenesis from being "lost in translation", a number of general guidelines should be kept in mind:

(i) Spectrum: Carefully consider the disease and the entire range of observations during the disease in probing for their potential role. This applies, for example, to the observation that astrocyte activation often precedes epilepsy in many animal models. This raises the possibility that one component of acquired epileptogenesis may involve the influence of astrocytes on synaptic and cellular properties of neurons, microglia, NG2 cells and vascular cells. Thus it may be useful to test whether prevention of astrocyte activation has an antiepileptogenic effect.

(ii) Statistics: matching the experiment to the disease. When experimental groups are too small relative to the variance of the parameters that will be studied, the chance of falsely positive results increases. Many of our current animal models of acquired epilepsy have been developed to ensure that a large percentage of animals develop epilepsy. This hinders development of useful biomarkers, because their predictive value depends on the incidence of the disorder. The incidence of epilepsy after brain injury is much lower in humans, so the predictive value of a biomarker needs to be tested in an experimental population with a similar

incidence. Finally, animals that do not develop epilepsy after brain injury are useful for more than service as controls – we may be overlooking antiepileptogenic characteristics and processes that may provide additional prognostic, mechanistic, and therapeutic insights into epileptogenesis after brain injury.

(iii) Heterogeneity: To maximize the chance that results will extrapolate to humans, preclinical studies should include more than one species and take intra and interspecies inhomogeneity into account. Most animal studies including in vitro studies are done on animals which are rather young and come from genetically homogenous breeding stocks. Hence the genetic inhomogeneity of human species is not taken into account. A second source of inhomogeneity is the injury itself. Common human brain injury mechanisms include trauma, infection, and both global and focal hypoxic- ischemic insults. The severity and anatomical location of each of these injuries varies profoundly from patient to patient. Understanding which circuit elements are altered after both experimental and clinical brain injury will be a necessary step in evaluating the risk and rate of subsequent epileptogenesis.

(iv) Comorbidity: multiple hits are often a critical factor in human disease, but this is usually not considered in experimental work. One approach to correcting this oversight could entail choosing the right animals for study. An example would be using stroke models of epileptogenesis in rodent models of chronic hypertension or type 2 diabetes.

(v) Communication between basic and clinical epileptologists: "Losses in translation" often arise from the different perspectives of experimentalists and clinicians, combined with the barriers to free communication between these groups. Ideally, interactions between clinical and basic investigators should be sufficiently close that experimentalists can contribute to clinical study focus and design. This requires a centralized infrastructure to provide close scientific collaboration as well as institutional mechanisms for patient access, patient monitoring, access to biostatistical resources, and guidance for approved use of patients in research. Perhaps the greatest institutional challenge is provision of protected time for interactions between clinical and basic researchers.

(vi) Pipeline repair: The interruption of classical translational pipelines also leads to losses in translation. Because drug resistant epilepsy is relatively rare, most of the pharmaceutical industry has lost interest in drug development for this type of epilepsy. Filling this gap, including toxicological and pharmacokinetic studies in preclinical and clinical populations, will require the training of clinician scientists who are equipped for drug development in an academic environment. This will require new ways of financing such research, as well as developing processes to provide academic credit for the type of applied research that is essential for the later stages of drug development involving toxicology, compounding, and pharmacokinetics.

2.2 How Can In Vitro Research Help Meet These Challenges?

The advantages of in vitro models have long been recognized in epilepsy research, starting with the pioneering work of P. Schwartzkroin [77]. These advantages include speed, convenience, low cost, the availability of a wide variety of genetically modified animals from which slices can be prepared, and electrophysiological, pharmacological, and optical accessibility. In vitro models make it possible to understand pathophysiology at a high level of electrophysiological, molecular and cell biological resolution.

In vitro models have a number of drawbacks. In vitro preparations usually have no blood brain barrier, and there is no circulation. Rather, drugs are applied in an aqueous solution, and reach their targets by routes that are more relevant to CSF administration than oral or intravenous routes. Brain slice preparation induces massive damage to afferent and efferent circuitries, which

is a particularly significant problem in the investigation of network-level phenomena such as seizures. Experiments are done at non-physiological oxygen and glucose concentrations. Many of the preparations are based on tissue from perinatal animals. For example, organotypic slice cultures, or the intact (whole) in vitro hippocampus preparation are best prepared before the 8th postnatal day (P8). Although acute brain slice preparations can be obtained from animals at any age, there is only a relatively brief period of time when the slice is physiologically stable. This time limit can restrict the types of experimental manipulations that can be performed in vitro, such as those involving viral or expression of exogenous proteins.

Some scientists and clinicians argue that in vitro models are too far removed from human epilepsy, and therefore one should focus on in vivo models. However, in vivo models have the dual problems of complexity and access, such that it is difficult to identify the pathogenic mechanisms in sufficient detail to initiate pharmacological or genetic interference. Moreover, studying acquired epileptogenesis in vivo involves brain injury. Therefore the "3R" strategies of replacement, reduction and refinement (3Rs) in research using animals are relevant.[1] "Replacement" refers to the use of other preparations, such as induced pluripotent stem cells derived from patient fibroblasts. "Refinement" refers to alteration of the experiment to focus the experiment to minimize pain and maximize information return. For example, many conditions leading to epilepsy are associated with activation of astrocytes. Addressing this question specifically might involve replacing a status epilepticus model with a model in which astrocytes are primarily activated [50, 66]. This will – if some investigators are correct – still cause epilepsy but presumably with less damage to the brain. "Reduction" refers to minimization of the number of animals used. Preparation of multiple brain slices per animal can make possible multiple independent tests of the hypothesis for each experimental subject. Using the reactive astrocyte hypothesis as an example, many of the

---

[1] http://www.nc3rs.org.uk/

consequences of astrocyte activation can also be studied in vitro. Some questions may not be feasibly studied in vivo – for example disturbance of potassium homeostasis [42] and/or glutamate homeostasis [29] may not be detectable with currently available in vivo methods. Focusing on the 3 Rs can have beneficial consequences – for example, markers of astrocyte activation might prove to be a biomarker predictive of epileptogenesis.

In the end it is important to recognize "in vivo veritas," i.e., in vitro studies should be complemented by in vivo studies. For example, in vitro studies can be used to rapidly screen drug libraries or target proteins and RNA, and slower, more costly, but more relevant in vivo studies can be used to study the most promising lead compounds. Indeed in vitro studies often underestimate potential side effects. In vivo experiments can determine whether translational relevance is hampered by the unwanted side effects of a given intervention, by toxic effects on organs other than the brain, by long-term loss of efficacy due to development of tolerance, or ineffectiveness due to interference with attention or sleep states of an animal and a patient.

If in vivo studies are used to complement in vitro work, it is important to optimize the in vivo protocols for maximal translational relevance. One improvement in the translational efficacy of in vivo studies could be achieved by completely phenotyping animals undergoing epileptogenesis and experimental therapeutic studies. A critical aspect of thorough and unbiased phenotyping of mice or rats includes continuous seizure surveillance using video EEG monitoring [71]. Wherever possible, experimental approaches should be employed that are based on clinical observations.

One chance to strengthen epilepsy research is also to take advantage of pathophysiological discoveries in other disciplines. For example the abnormalities observed in patients with Alzheimer's disease may not only be relevant for neurodegenerative disease but also for epilepsy because many patients with Alzheimer's disease may also develop a symptomatic form of epilepsy [88]. Mitochondrial disorders are not only

observed in certain forms of Parkinson's disease [13] but also in epilepsy [48]. Elements of the inflammatory response are observed in many brain injuries (for example after trauma, stroke and status epilepticus) which may contribute to epileptogenesis [44, 89]. If useful discoveries and approaches in other areas of applied neuroscience are exploited, the translational gap may be more readily overcome.

2.3 Lessons Learned: In Vitro Techniques for Epilepsy Research

Ictogenesis: The utility of in vitro preparations for epilepsy research was first suggested by a study in which seizure like events were induced by lowering of extracellular Cl^- concentration in the perfusate of acute hippocampal slices [93]. It was therefore surprising that $GABA_A$ receptor antagonists induced only short interictal-like discharges in the hippocampal slice preparation, because the hippocampus was presumed to be the most epileptogenic region [77]. Similar findings were observed in cortical slice preparations. These data suggested that $GABA_A$ receptor blockade was not a sufficient condition for seizure induction in vitro, and that other conditions were necessary for ictogenesis, i.e. the induction of seizure-like events. The first seizure-like events recorded in vitro were generated by conditions that accompany seizures in vivo, such as low concentrations of extracellular Mg^{2+} [90] or Ca^{2+} [43] or elevated concentrations of extracellular K^+ [87]. While low Ca^{2+} and high K^+ induced seizure-like events in hippocampal subregions, low Mg^{2+} and application of 4-aminopyridine, a potassium channels blocker, initiated seizure-like events more reliably in cortical structures than in hippocampal slices [60], unless juvenile tissue was used [31]. These studies suggested that the hippocampus is not as seizure prone as originally thought from the pathological studies of patients with epilepsy, or that seizure generation involves distributed circuits that are lost after slicing-induced deafferentation. In light of this, it is interesting to note that seizures in patients with temporal lobe epilepsy often originate outside the hippocampus, for example in the amygdala and the entorhinal cortex [81]. These cortical structures are now considered to be more seizure prone than the hippocampus or other cortical areas. The ionic manipulations that were used to study ictogenesis also provided early insights as to why and where (cortex vs. hippocampus) Mg administration acts to antagonize eclamptic seizures [24].

The next surprising finding from in vitro studies of epilepsy was that seizure induction was more easily accomplished in control slices than in slices obtained from animals with epilepsy and from specimens of patients with drug resistant epilepsy [30, 98]. This suggests either that ictogenic processes are active only transiently in epilepsy, or that epilepsy also entails protective mechanisms that are more robustly preserved in vitro compared to ictogenic mechanisms. Understanding such protective mechanisms could lead to new antiepileptogenic strategies. A related insight from recordings of tissue from patients with epilepsy was that the transections that accompany slice preparation may be more functionally important in chronically epileptic tissue than normal tissue. One interpretation is that a fundamental and widespread alteration in connectivity occurs in the chronically epileptic brain. On the other hand, many patients with refractory epilepsy know that following a seizure there is usually a seizure- free interval, sometimes of considerable duration. Here, translation is bidirectional: Clinical questions can be "translated" into an experimental approach and experimental observations suggest new possibilities for interfering with epilepsy and epileptogenesis.

2.4 In Vitro Models: The "Nuts and Bolts"

Among the in vitro preparations available for studying ictogenesis, perhaps the most versatile is the slice preparation. Slices can be obtained from any mammalian species including humans following neurosurgical interventions. Slices can also be obtained from animals that are epileptic

as a result of trauma, tumors, status epilepticus, inflammation, etc. However, slices have circuits that have been reduced in size by transection of processes, and therefore more intact preparations are sometimes required to gain insight into epilepsy. There are two acute preparations that address the connectivity issue: one is the isolated, intact hippocampus (also referred to as the whole or in toto hippocampal preparation [61]). The other is the intact isolated brain preparation [19]. The first preparation is only feasible if animals are used at young ages, and the second preparation is feasible only if guinea pigs are used (unless one uses non-mammalian species). Also, in these preparations, many aspects of epilepsy cannot be readily studied. A preparation that can be used for long-term observations is the organotypic slice culture, in which many different aspects of epileptogenesis can be studied "in a dish." Slice cultures represent a model of brain trauma by virtue of the trauma involved in slice preparation, and also of developmentally increased seizure susceptibility because as mentioned above, organotypic slices are most reliably prepared from animals in the first postnatal week. In the following section we will discuss each of these preparations in more detail, and discuss some translational aspects of this research.

2.4.1 Acute In Vitro Brain Slices

2.4.1.1 Interictal Activity and Seizures
Interictal activity refers to paroxysmal epileptic discharges that are much more brief and occur more much more frequently than seizures. Most epileptiform events in acute slices have these two characteristics. Currently, it is not known whether interictal epileptiform activity is proconvulsive or epileptogenic. Because EEGs are not performed routinely in brain injured patients, it is not known whether interictal spikes precede seizures after brain injury. In acute in vitro preparations, brief recurrent epileptiform events can be readily induced by ionic and pharmacological manipulations. Thus many experimentalists have argued that these interictal-like events observed in vitro embody the essential features of ictogenesis.

Indeed preictal spikes often precede seizures in humans, and in in vivo models of epilepsy. During the in vivo spikes, depolarization shifts of the membrane potential occur that are very similar to the membrane potential changes recorded in vitro [45, 78]. On the other hand, questions have been raised about the translational relevance of activity that is induced by acute proconvulsant pharmacological or ionic manipulations in vitro [94].

One example is the interictal-like activity induced by low Mg induced in adult rat hippocampal slices. The pharmacological relevance of this activity is modest, because anticonvulsants have limited effects on this activity. Thus this activity cannot be used in drug screening, and pathophysiological studies based on this activity must be interpreted with caution. In fact, seizure-like events were induced when the GABA$_B$ agonist baclofen was added to the perfusate containing low Mg [83]. One interpretation of the combined effect of baclofen and low Mg is that interictal activity was preventing ictogenesis. Another interpretation is that low Mg induces a state similar to Periodic Lateralize Epileptiform Discharges (PLEDs) rather than interictal spikes, and that seizures can only be observed by reducing the severity of the ictogenic conditions, for example by reducing probability of glutamate release through activation of presynaptic GABA$_B$ receptors with baclofen. Similar results are obtained when elevated K, which induces spontaneous epileptiform burst discharges in adult hippocampal slices, is combined with strontium, which reduces the rate of glutamate release [84].

Consistent with these observations of the interaction between manipulations that alter excitability, induction of recurrent epileptiform burst discharges by tetanic stimulation makes it more difficult to induce seizure-like activity using elevated K$^+$ [59]. Barbarosie and Avoli [7] observed that in the presence of the convulsant 4AP, seizure activity could be initiated following transection between CA1 and entorhinal cortex.

These experiments and many others performed over the last two decades, emphasize that the sum of multiple manipulations in vitro are not predictable and are often difficult to interpret

with respect to human epilepsy. A particularly problematic correlate is that in slices exposed to convulsant conditions, anticonvulsants, even at anesthetic concentrations, while blocking seizure-like events [14, 57, 99], rarely inhibit the interictal-like epileptiform activity induced by convulsants in acute in vitro preparations [94]. This is an important area for future optimization, and is discussed in more detail in the section on organotypic slices.

2.4.1.2 Age, Area, and Astrocytes

Slice preparations can be used to determine age dependence of seizure susceptibility, and to compare different regions of the brain with respect to epileptogenesis. Thus, susceptibility to low Mg or low Ca is much higher in tissue from young animals [31] and treatment with ictogenic agents often results in spreading depression [38]. There are many potential developmental mechanisms to explain these data, including circuit development and or the maturation of astrocytes. At the time when Hablitz and colleagues reported their findings, it had not yet been established that astrocytic properties differ in epileptic vs. control animals [29, 40]. It also was not known that activation of astrocytes can be achieved by many perturbations that are considered minor, such as opening of the blood brain barrier, or exposure to albumin. Conditions leading to chronic astrocyte activation can also acutely reduce seizure threshold. Brain injuries are often accompanied by spreading depression episodes, which in cases of disturbed neurovascular coupling can cause neuronal damage and may therefore exacerbate brain injury in ischemic and hemorrhagic stroke. Spreading depression may also be relevant to epilepsy associated with Alzheimer disease [22, 55, 92].

The regional variation of seizure susceptibility can also be investigated with acute brain slices. Seizures never seem to originate from basal ganglia and cerebellum, perhaps because information is relayed by activation of inhibitory cells through disinhibition of target cells. This brings up the fact that all brain activity involves both excitation and inhibition of both inhibitory neurons and principal cells. Thus pharmacoresistance could involve failure of GABAergic agents (which may exacerbate inhibition of interneurons) or Na channel blocking agents (which may reduce GABA release from strongly inhibited interneurons). Of course there are many other possible explanations for pharmacoresistance, such as alterations in the expression of Na channels.

2.4.1.3 Channelopathies

A good example of a genetic channelopathy that has been benefited from the in vitro approach is the murine Nav1.1 knockout model of epilepsy. These mice exhibit a pattern of seizure activity that is similar to the clinical syndrome with a similar defect in Na channels, Dravet syndrome [74]. One hypothesis for the generation of seizures in these animals is that they arise from preferential expression of Nav1.1 channels in interneurons which – if defective – would result in strongly reduced excitability of interneurons and GABA release [95]. The result could be a pharmacoresistant epilepsy.

Defects in ion channels have long been implicated in the epileptiform discharges induced in vitro by low extracellular calcium concentrations [36, 52]. This idea was based on observations from baboons where seizures induced by stroboscopic stimulation were associated with decreases in extracellular Ca concentration to less than 0.2 mM, and where seizures were accompanied by increases in potassium concentration to near 10 mM [69]. Mimicking this condition did not induce seizure-like events in human or animal cortical structures in vitro. An exception was area CA1 in rat and mouse hippocampus, where the packing density of neurons is higher, promoting ephaptic interactions that are thought to be enhanced by lowered extracellular calcium. On the other hand, cation channels [37] and more recently certain TRP channels are activated by decreasing either Ca or Mg concentration [91]; this is not only important for spreading depolarization [80] but possibly also for generation of seizures and cell death [62]. The regulation of these excitatory channels by Ca and the activity-dependent decrease in extracellular Ca suggests a new mechanism for seizure spread and for modifying seizure generalization.

Similarly recent work has identified KCNQ channels as potential targets for the treatment of seizures. The first drug introduced for treatment is retigabine which affects KCNQ2,3 and 5 channels but not KCNQ 1 channels (KCNQ1 is expressed in the heart; [96]). The distribution of KCNQ channels in principal cells and in interneurons varies, with KCNQ5 channels being expressed on GABAergic cells as well as principal cells [97], which can explain the finding that retigabine can reduce GABAergic inhibition, which may limit its use as an antiseizure medication. More recently however, agents were identified which only affect KCNQ2 [12]. These are preferentially expressed on glutamatergic cells therefore are more suitable as antiseizure drugs. Thus slices prepared from transgenic animal models can make possible the rapid testing of potential anticonvulsant effects; however, side effects are better assessed in vivo.

2.4.1.4 Evoked Seizure-Like Events In Vitro as a Model of Status Epilepticus

In vitro models of ictogenesis such as the high K model, the low Mg model, and the 4-AP model are characterized by epileptiform activity that recurs at short intervals without intervening physiological activity. Clinically this pattern of activity is similar to status epilepticus, which is defined as either a seizure lasting for more than a specified time period or seizures that recur without an intervening period of normal consciousness. There are serious clinical implications of the definition, because after 30 min, status epilepticus can become pharmacoresistant (Kapur and MacDonald). Moreover, prolonged experimental status epilepticus can cause considerable neuronal death [33]. The lack of agreement regarding the duration of seizure activity necessary for status epilepticus is related to our lack of knowledge regarding the time course of the damage to neurons.

In vitro, it turns out that shortly after their initiation, exposure to low Mg, 4-AP, high K and low Ca induce seizure like events that all respond well to standard AEDs. However, if the activity persists for some time, then seizure like events gradually shorten, and ultimately short recurrent discharges occur, which are also unresponsive to standard AEDs [15, 23]. The analysis of such events indicates that the transition from long seizure-like events to short discharges is probably due to reduced GABAergic transmission. A variety of pre and postsynaptic processes may underlie this loss of efficacy, such as internalization of GABA receptors, consumption of GABA for synthesis of ATP, and alteration of the anionic transmembrane gradients that subserve $GABA_A$ receptor-mediated inhibition. The reduction in GABAergic function explains why GABAergic agents that prolong the GABAergic signaling lose efficacy during the course of prolonged seizures. However, in some cases, agents that directly activate GABA receptors are still effective [67]. In other situations GABAergic agents are either minimally ineffective or exacerbate epileptiform activity, a situation that can be improved by agents that improve the transmembrane anionic gradient [26, 27]. This improvement has also been observed in human case studies [47] and is being investigated in human trials. In some circumstances these additional processes that are dependent on the duration of seizure activity prior to drug application have provided the key to resolving seemingly contradictory results [1, 27, 28].

Neurons are depolarized during prolonged seizures, and so are their mitochondria. Brian Meldrum's experimental neuropathological studies of status epilepticus in the baboon focused attention on mitochondrial changes accompanying ictal neuronal cell death [33]. Subsequent studies have provided evidence that at least part of the neuronal damage arising during status epiletpicus seems to be due to mitochondrial depolarization and increased production of free radicals [17, 54]. This suggests that some neuroprotection can also be achieved by free radical scavenging [54, 76]. This is an area that can be profitably studied in vitro, where microscopic imaging during epileptiform activity is more feasible than in vivo. Barbiturates and other anesthetics used to terminate status epilepticus are typically titrated to a burst suppression pattern that is very reminiscent of the periodic population discharges that are observed in acute brain slice preparations exposed to convulsants with anesthetic concentrations of barbiturates [23]. Indeed, recurrent epileptiform discharges cause considerable cell loss due to mitochondrial depolarization and increased free radical production sensitive to neuroprotection by free radical scavengers such as tocopherol. Other anticonvulsant and neuroprotective strategies such as cooling of patients' brain by a few degrees or anticonvulsants which do not involve GABAergic signaling should continue to be investigated [75]. It will important to advance these early results in vitro and then translate the results of these experimental findings into good clinical studies.

2.4.1.5 Increased Seizure Threshold of Epileptic Tissue In Vitro

As mentioned above it is often difficult to induce seizure like events in tissue from animals with epilepsy acquired after drug-induced status epilepticus. This may reflect an endogenous anticonvulsant effect. It has not yet been described for kindled animals, and it depends on the number of seizures an animal has experienced [98]. In chronically epileptic human tissue resected for seizure control, it is even more difficult to evoke seizure like events [30, 39]. In resected hippocampal tissue, seizure-like events can often be induced by elevating potassium concentration in the dentate gyrus and subiculum [30]. In neocortex 4-AP can be employed but it works in only a subset of patient specimens [5]. We have also been able to induce seizure activity with high potassium combined with bicuculline. These observations raise important questions as to the mechanisms underlying relative seizure resistance in epileptic tissue.

Kindling is most effective when a critical interval is included between kindling stimuli. It was first suggested that this may relate to upregulation of opioid receptors [73]. Later it was shown that a single repetitive stimulation of the perforant path from the entorhinal cortex to the dentate gyrus could upregulate the GABA synthesizing enzyme GAD with subsequent co-release of GABA and glutamate form mossy fiber terminals which leads to an elevated seizure threshold [35].

These effects seemed to be transient and fade away with time. Additional evidence for endogenous antiepileptic processes come from slices prepared from kainate treated and also from pilocarpine treated animals, where the convulsant 4-AP was ineffective. This effect was due to up-regulation of the enzyme adenosine deaminase acting on RNA (ADAR2). This causes mRNA editing of AMPA type glutamate receptors as well as Kv1 potassium channels that lose some of their sensitivity to 4-AP [82]. An additional mechanism involves arachidonic acid which is directly blocking K channels [11] and in addition can be metabolized to a number of intrinsic convulsant or proconvulsant derivatives [44]. Activity dependent editing of alpha 3 subunits of glycine receptors has also been described. This editing leads to an increased affinity for gylcine and some of its agonists [63]. Although such processes may decrease seizure susceptibility, we need to keep in mind that there is not sufficient circuitry in a slice for seizure generation. Thus we may need to pay more attention to network preservation when studying network phenomena such as seizures and epilepsy in vitro. Nevertheless, hypothesis driven searches for other anti-ictogenic mechanisms that are active in epileptic tissue comprise a promising route for discovering new treatments of pharmacoresistant epilepsies.

2.4.1.6 Analysis of Proepileptogenic Factors

Another translational opportunity for epilepsy research is the in vitro study of mechanisms of epileptogenesis. Trauma and stroke research led to the important discovery that neuronal circuits reorganize following a brain lesion, and this had important implications for the study of epilepsy. For example, the observation of mossy fiber sprouting, that is sprouting of dentate granule cell axons back into the input layer of the dentate gyrus, has been a central model of the recurrent positive feedback that is a necessary component of any sustained network activity, including seizures [56, 65]. However, some investigators now wonder whether this neurocentric approach to the understanding of epilepsy may have been too narrow. Many conditions which lead to epilepsy are associated with an open blood brain barrier [72, 86]. The immediate effects of blood brain barrier disturbances include vasogenic edema due to extravasation of albumin and other serum proteins into the brain interstitial space. This increases intracranial pressure, potentially reducing microperfusion. Opening of the blood brain barrier also increases extracellular potassium and reduces extracellular Ca and Mg concentrations, because these are lower in serum than in the brain interstitial space [79]. Activity-dependent increases in blood flow might not occur under these conditions. Thus if seizures emerge, relative metabolic deprivation may ensue. Seizures and metabolic deprivation lead to cell swelling, i.e. cytotoxic edema [21]. When the blood brain barrier is opened, albumin is absorbed into perivascular macrophages and astrocytes, perhaps reflecting an attempt to reduce the extracellular colloid pressure. This process is associated with activation of TGFß receptors and subsequent activation of astrocytes, including increased expression of GFAP [40] and down regulation of K_{IR} channels. This results in depolarization of astrocytes and changes in the expression of connexins, resulting in reduced astrocytic electrical coupling [16]. Both effects lead to enhanced accumulation of extracellular potassium and perhaps glutamate in the extracellular space. Under these conditions, seizure threshold is strongly reduced and when seizures develop they rapidly progress to spreading depression [55]. Preliminary evidence suggests also that these alterations in astrocyte properties may be associated with increased release of chemokines and cytokines and potentially also with release of gliotransmitters. Importantly these alterations precede appearance of seizures and if stopped may prevent later epileptogenesis. Probing for an open blood brain barrier may be an important biomarker for epileptogenesis following trauma, stroke and encephalitis and some form of tumors. However, not all tumors are associated with an open blood brain barrier and criteria that take into account constraints on the role of the open blood brain barrier in epileptogenesis have still to be evaluated.

2.4.2 Isolated Hippocampus and Isolated Brain

The isolated intact hippocampus has recently received considerable attention because a number of questions can be addressed that are of potential clinical relevance. One is the induction of a mirror focus by using the two hippocampi interconnected by commissural fibers [49]. Induction of seizure-like events in one hippocampus induced a seizure focus in the contralateral hippocampus without any additional pharmacological treatment. This is potentially important as it could explain why in some cases seizures do not stop when one hippocampus is removed. On the other hand this is an acute finding that may be more closely related to mechanisms underlying rapid kindling than the development of mirror foci in chronic epilepsy. The finding is limited to young age, as maintenance of the intact hippocampus beyond postnatal day 10 is presently not possible. For such studies in older age it may be more feasible to use preparations from turtles or birds. Another aspect of studies in juvenile intact hippocampus is that the evoked seizure like events seem to be resistant to clinically employed drugs [70], perhaps reflecting immature ion transport mechanisms [27]. This may therefore be a preparation in which new agents can be tested which specifically address seizures in babies and young infants.

Another intact in vitro preparation is the intact guinea pig isolated brain preparation [19]. It permits studies on long range interactions within the brain during seizure like events and indicates that seizure generation is based on multisite interaction in wide spread neuronal circuits. However it is apparently difficult to induce epilepsy in guinea pigs and the intact brain is difficult to prepare from aged animals.

2.4.3 Use of Human Tissue In Vitro

About 30 % of patients with epilepsy do not become seizure free with presently available drugs. Thus there remains a pressing need for models of pharmacoresistance that are correlated with data from patients. At present, human tissue resected during epilepsy surgery is primarily used for diagnostic purposes. In past years however the neuropathology field has opened itself to molecular biology aspects concerning expression of peptides, transmitter receptors, ion channels, gene regulation and epigenetics [18]. Human tissue samples can to some extent also be used for determination of changes in interneuronal connectivity and in probing for alterations in astrocyte properties [20]. Moreover in human tissue spontaneous events may be detected that might resemble interictal spikes and fast ripple activity [51]. This may permit the study of mechanisms of fast ripple activity in human tissue. Interestingly slices prepared from human specimens often have a relatively long survival time. This might permit development of slice cultures from human tissue.

It is notoriously difficult to induce seizure like events in human tissue. As discussed above, this is probably due to upregulation of anti- ictogenic mechanisms, in addition to the effects of partial network disassembly. Endogenous protective mechanisms are of interest because studies of these mechanisms could lead to identification of novel anticonvulsant and antiepileptogenic therapies. However it is still possible to induce seizure like events in the hippocampus or temporal neocortex of TLE patients, and in the cortex of patients with developmental disorders. In the hippocampus the most effective method to induce seizures is elevation of potassium concentration. In temporal neocortex seizure-like events can be induced in a subset of preparations by 4-AP, or 4-AP combined with elevated potassium concentration. In our hands the best method for induction of seizure like events in temporal neocortex slices is the use of potassium elevation combined with application of bicuculline (unpublished observation). In studies of epileptiform activity induced by elevated potassium in the hippocampus it was noted that the slices do not respond to CBZ if they come from patients with pharmacoresistant epilepsy but do respond if they come from patients which are not resistant to AEDs such as tumor patients [41]. It is noteworthy that in some instances one slice from a pharmacoresistant patient may not respond to AEDs while the other does. This heterogeneity offers itself for studies on mechanisms underlying pharmacoresistance. Obviously if one is able to induce seizure like events in slices from pharmacoresistant patients this opens the possibility to test for agents which might alleviate the epilepsy in drug resistant patients. One argument against this strategy is that the obtained material is too heterogenous and that in many centers the incidence of epilepsy surgery is too low to permit for rapid information. However monkey studies are often indeed based on many repeated measures in the same subject. The amount of human tissue available is often large and would permit to study effects of a multitude of agents on the same patient material if logistics can be surmounted. For example in analogy to multi center clinical studies, it might be possible to set up multicenter studies on resected material, although this might require new funding mechanisms.

2.4.4   Slice Cultures as a Model of Traumatic Epilepsy

2.4.4.1 Ictogenesis

There are a number of different techniques for preparing organotypic slice cultures from cortex or hippocampus. Their properties depend on the way they are fixed to the substrate material, on the age at preparation and on the media used for maintenance in culture [6, 85]. Most studies related to epileptogenesis are done on organotypic hippocampal slice cultures. Cortical organotypic cultures and hippocampal cultures maintained with B27 artificial media often display spontaneous seizures [2, 10] which can be recorded also while the cultures are in the incubator by different techniques such as MEAs or implanted electrodes. In this preparation, epileptogenesis proceeds at a rapid but predictable time course [25]. Interictal activity precedes the onset of ictal activity by several days. Status epilepticus commencing shortly after the appearance of spontaneous seizures is observable for hours to days [2]. Seizure-induced neuronal death is readily apparent, peaks during status epilepticus [54],

and can be prevented by standard anticonvulsants such as phenytoin [9]. The incidence of epilepsy is nearly 100 % in slice cultures from rats and mice, and in fact a current challenge is developing a culture system with a lower incidence of epilepsy that might make a better predictor of biomarkers and therapeutic agents for human epileptogenesis.

Some investigators prefer to induce seizure like events by lowering Mg thereby activating NMDA receptors, or application of bicuculline thereby reducing inhibition. Application of 4-AP leads to strongly enhanced transmitter release due to the strong expression of 4-AP sensitive Kv1.4 and 1.5 as well as some Kv3 channels on presynaptic terminals. Seizure like events are usually characterized by some initial clonic like discharges, followed by a tonic like and thereafter clonic like period followed by a postictal depression and the recurrence of interictal discharges. During seizure like events, ionic changes occur which mimic those observed during seizures in intact animals. If seizures recur with a high incidence they can convert into late recurrent discharges which are characterized by shorter events with synchronous intracellular depolarizations. Thus slice cultures offer themselves for studies on ictogenesis and factors which facilitate ictogenesis such as reorganization of the neuronal networks under study. Epileptogenesis can also be studied. For example typical epileptic circuitry with recurrent axon collaterals, back projection from CA1 to CA3 or DG can be observed [34, 56]. Slice cultures can be maintained for up to 8 weeks and therefore offer themselves also for long term observations. A drawback is that it is rather difficult to make slice cultures from hippocampal tissue beyond postnatal day 16. There are reports that slice cultures can be made after this date but the chances that these can be maintained for more than 2 weeks are rather slim and therefore very labor intensive [58].

2.4.4.2 Pharmacosensitive vs. Parmacoresistance

Depending on duration of culturing and maintenance conditions, evoked seizures can be sensitive to

AEDs or insensitive. In the same slice culture both conditions can coexist: thus while 4-AP and low Mg induced seizure like events in some conditions are pharmacoresistant the seizure like events induced by repetitive stimulation are not [3, 4]. Therefore slice cultures can be used as a model of pharmacoresistant seizures and drugs can be tested which might be useful for the treatment of epilepsies in patients whose seizures cannot be satisfactorily controlled by present medication.

Spontaneously epileptic slice cultures that are not exposed to convulsants respond to anticonvulsants with suppression of ictal but not interictal activities, as is the case clinically [9]. Interestingly, dependent on culture conditions these cultures become resistant to anticonvulsants after 1–3 weeks of exposure, with recrudescence of seizure activity at anticonvulsant concentrations that completely suppress seizure activity in naïve slices of the same age [3, 9]. Thus the organotypic slice culture is a promising tool for the investigation of the phenomenology and pathophysiology of pharmacoresistance.

2.4.4.3 Mechanisms of Ictogenic Cell Death

Slice cultures offer themselves also for studies on ictogenic cell death. A number of methods are available to monitor cell death. These include the measurements of LDH in the supernatant and also of propidium iodide staining and ethidium bromide staining [9, 54]. Of course it is also possible to test for programmed cell death. One approach is to perform experiments with reduced oxygen supply in slice cultures that are generating stimulation-induced seizure like events. These events develop into spreading depolarization which when oxygen tensions falls to near zero cause cumulative cell death, a situation which is similarly observed also in slices from animals which experienced a stroke [68]. On the other hand with normoxic or hyperoxic perfusion it can be shown that seizure like events are associated with increased free radical production and eventually damage of mitochondria leading to disturbances in the coupling of neuronal and metabolic activity causing cell death because of

lack of sufficient ATP supply. Buffering ROS by different means can be shown to be highly neuroprotective.

Another approach to studying cell death in spontaneously epileptic slice cultures is to assay release of lactate dehydrogenase (LDH) into the culture media, which is changed twice weekly [9, 32]. This is a simple procedure that while not linearly related to cell death, provides a rapid and reliable means to assay cell death in higher-throughput experiments in which toxicity of screened agents and prevention of ictal cell death are important endpoints. More detailed studies of ictal cell death employ either exogenous markers such as propidium iodide, or endogenously expressed fluorescent markers of caspase activation. These studies provide a means to follow cell death over time, and to ask important questions as to the activities and features that precede or predict death in identified neurons.

2.4.4.4 Slice Cultures: A Model of Post-traumatic Epilepsy

When slice cultures are prepared a large number of connectivities are severed leading to some extent to retrograde degeneration but also to transformation of a three dimensional organization into a two dimensional organization. Thus the slice culture can be considered to comprise a model of (pediatric) traumatic brain injury. When spontaneous seizures emerge in these cultures they can be used for long term monitoring of drug effects thus facilitating detection of changes in efficacy of a given drug. This includes also detection of toxic side effects with nervous tissue [8, 9].

2.4.4.5 Long Term Monitoring of Anti Epileptogenic Effects

Slice cultures can also be used to study antiepileptogenic strategies. One example is neovascularisation. The density of blood vessels in human and chronic epileptic rodent tissue is often remarkably increased [72]. This makes it possible to address the question as to whether neoformation of blood vessels can be altered [64]. Surprisingly slice cultures present with many blood vessels which are usually equipped with a tight blood brain barrier [53]. Most of these vessels remain intact unless there is infection in the tissue. Therefore slice cultures can be used to determine effects of microglial activation on vessel density, and also whether factors that prevent revascularization have neurotoxic effects that might interfere with epileptogenesis.

Slice cultures permit study of signaling cascades and of factors that may serve as antiepileptogenic factors [8]. These can be neuroprotective agents, for example blockers of signaling cascades such as the TGFß activated pathways or the mTOR pathways, and agents that interfere with neuronal survival or growth factors. At present most of these strategies are not yet ready for transfer into clinical trials, but this area is a promising area for further in vitro and in vivo study. Slice cultures are most useful to study drugs whose effects require time to produce anti-seizure or anti-epileptogenic effects. Most drug testing assays used in vivo or in vitro test for very acute effects although many treatments in psychiatry and epiletogenesis take time to take full efficacy.

2.4.4.6 Use of Transgenic Models of Epileptic Encephalopathies

Many transgenic mice display seizures. Murine models have been developed for several human mutations that cause severe childhood epileptic encephalopathies. In many instances the transgenic models do not live long enough for research on the precise pathogenic cascade. However slice cultures can be prepared from ages ranging from fetal tissue to P16–18. Preparing slice cultures from such animals offers the possibility to look into the precise pathophysiological cascade and to define intervention points by which the epileptogenesis can be prevented. For these studies, the development of chronic slice cultures that do not become epileptic except in the presence of the targeted gene defect would be very useful. However, in the absence of such a slice preparation, the organotypic slice can still be of exceptional utility. For example, slice cultures prepared from transgenic animals expressing cell-type-specific fluorophores that are activated by particular ions, neurotransmitters, or second

messengers, or by cell-type-specific expression of light-sensitive rhodopsins can be studied with targeted path scanning multiphoton microscopy and activity-dependent fluorophores. This provides the means to precisely interrogate critical network elements that are active during ictogenesis and epileptogenesis.

2.5 Some Conclusions

The above discussion is not intended to be a thorough review of epileptogenesis. We tried to illustrate some of the successes and challenges of in vitro preparations for translational research. In vitro preparations offer a large number of research possibilities to address clinical questions and therapeutic options, including long term observations in slice cultures, detailed cellular analysis, imaging and optogenetic studies, and expression of orphan receptors that permit activation and silencing of select populations. Expression and suppression of specific RNA and proteins can be achieved semi-acutely or chronically. All these technologies can now be employed for studies on ictogenesis, epileptogenesis and aspects of disease such as signaling cascades, development of pharmacoresistance, and neuroprotection. Exploiting the multiple technical possibilities for translational research will be substantially enhanced by improved contact between clinicians and scientists. This would culminate in clinical research executed through coordinated multicenter trials where promising, robust preclinical observations could be readily transformed into clinical proof of principle studies.

Acknowledgement UH is funded by DFG EXC Neurocure

References

1. Achilles K, Okabe A, Ikeda M, Shimizu-Okabe C, Yamada J, Fukuda A, Luhmann HJ, Kilb W (2007) Kinetic properties of Cl uptake mediated by Na+–dependent K+–2Cl cotransport in immature rat neocortical neurons. J Neurosci 27:8616–8627

2. Albus K, Heinemann U, Kovacs R (2013) Network activity in hippocampal slice cultures revealed by long-term in vitro recordings. J Neurosci Methods 217:1–8

3. Albus K, Wahab A, Heinemann U (2008) Standard antiepileptic drugs fail to block epileptiform activity in rat organotypic hippocampal slice cultures. Br J Pharmacol 154:709–724

4. Albus K, Wahab A, Heinemann U (2012) Primary afterdischarge in organotypic hippocampal slice cultures: effects of standard antiepileptic drugs. Epilepsia 53:1928–1936

5. Avoli M, Louvel J, Mattia D, Olivier A, Esposito V, Pumain R, D'Antuono M (2003) Epileptiform synchronization in the human dysplastic cortex. Epileptic Disord 5(Suppl 2):S45–S50

6. Baldino F Jr, Wolfson B, Heinemann U, Gutnick MJ (1986) An N-methyl-D-aspartate (NMDA) receptor antagonist reduces bicuculline-induced depolarization shifts in neocortical explant cultures. Neurosci Lett 70:101–105

7. Barbarosie M, Avoli M (1997) CA3-driven hippocampal-entorhinal loop controls rather than sustains *in vitro* limbic seizures. J Neurosci 17:9308–9314

8. Berdichevsky Y, Dryer AM, Saponjian Y, Mahoney MM, Pimentel CA, Lucini CA, Usenovic M, Staley KJ (2013) PI3K-Akt signaling activates mTOR-mediated epileptogenesis in organotypic hippocampal culture model of post-traumatic epilepsy. J Neurosci 33:9056–9067

9. Berdichevsky Y, Dzhala V, Mail M, Staley KJ (2012) Interictal spikes, seizures and ictal cell death are not necessary for post-traumatic epileptogenesis in vitro. Neurobiol Dis 45:774–785

10. Berdichevsky Y, Sabolek H, Levine JB, Staley KJ, Yarmush ML (2009) Microfluidics and multielectrode array-compatible organotypic slice culture method. J Neurosci Methods 178:59–64

11. Bittner K, Müller W (1999) Oxidative downmodulation of the transient K-current I_A by intracellular arachidonic acid in rat hippocampal neurons. J Neurophysiol 81:508–511

12. Boehlen A, Schwake M, Dost R, Kunert A, Fidzinski P, Heinemann U, Gebhardt C (2013) The new KCNQ2 activator 4-Chlor-N-(6-chlor-pyridin-3-yl)-benzamid displays anticonvulsant potential. Br J Pharmacol 168:1182–1200

13. Bogaerts V, Theuns J, van Broeckhoven C (2008) Genetic findings in Parkinson's disease and translation into treatment: a leading role for mitochondria? Genes Brain Behav 7:129–151

14. Brückner C, Heinemann U (2000) Effects of standard anticonvulsant drugs on different patterns of epileptiform discharges induced by 4-aminopyridine in combined entorhinal cortex-hippocampal slices. Brain Res 859:15–20

15. Brückner C, Stenkamp K, Meierkord H, Heinemann U (1999) Epileptiform discharges induced by combined application of bicucculline and 4-aminopyridine are resistant to standard anticonvulsants in slices of rats. Neurosci Lett 268:163–165

16. Cacheaux LP, Ivens S, David Y, Lakhter AJ, Bar-Klein G, Shapira M, Heinemann U, Friedman A, Kaufer D (2009) Transcriptome profiling reveals TGF-beta signaling involvement in epileptogenesis. J Neurosci 29:8927–8935

17. Cock H (2007) The role of mitochondria in status epilepticus. Epilepsia 48:24–27

18. De Lanerolle NC, Gunel M, Sundaresan S, Shen MY, Brines ML, Spencer DD (1995) Vasoactive intestinal polypeptide and its receptor changes in human temporal lobe epilepsy. Brain Res 686:182–193

19. de Curtis M, Pare D (2004) The rhinal cortices: a wall of inhibition between the neocortex and the hippocampus. Prog Neurobiol 74:101–110

20. DeFelipe J, Sola RG, Marco P (1996) Changes in excitatory and inhibitory synaptic circuits in the human epileptogenic neocortex. In: Conti F, Hicks TP (eds) Excitatory amino acids and the cerebral cortex. MIT Press, Cambridge, MA, pp 299–312

21. Dietzel I, Heinemann U, Hofmeier G, Lux HD (1980) Transient changes in the size of the extracellular space in the sensorimotor cortex of cats in relation to stimulus induced changes in potassium concentration. Exp Brain Res 40:432–439

22. Dreier JP, Major S, Pannek HW, Woitzik J, Scheel M, Wiesenthal D, Martus P, Winkler MK, Hartings JA, Fabricius M, Speckmann EJ, Gorji A (2012) Spreading convulsions, spreading depolarization and epileptogenesis in human cerebral cortex. Brain 135:259–275

23. Dreier JP, Zhang CL, Heinemann U (1998) Phenytoin, phenobarbital, and midazolam fail to stop status epilepticus-like activity induced by low magnesium in rat entorhinal slices, but can prevent its development. Acta Neurol Scand 98:154–160

24. Duley L, Gulmezoglu AM, Henderson-Smart DJ, Chou D (2010) Magnesium sulphate and other anticonvulsants for women with pre-eclampsia. Cochrane Database Syst Rev 11, CD000025

25. Dyhrfjeld-Johnsen J, Berdichevsky Y, Swiercz W, Sabolek H, Staley KJ (2010) Interictal spikes precede ictal discharges in an organotypic hippocampal slice culture model of epileptogenesis. J Clin Neurophysiol 27:418–424

26. Dzhala VI, Brumback AC, Staley KJ (2008) Bumetanide enhances phenobarbital efficacy in a neonatal seizure model. Ann Neurol 63:222–235

27. Dzhala VI, Kuchibhotla KV, Glykys JC, Kahle KT, Swiercz WB, Feng G, Kuner T, Augustine GJ, Bacskai BJ, Staley KJ (2010) Progressive NKCC1-dependent neuronal chloride accumulation during neonatal seizures. J Neurosci 30:11745–11761

28. Dzhala VI, Talos DM, Sdrulla DA, Brumback AC, Mathews GC, Benke TA, Delpire E, Jensen FE, Staley KJ (2005) NKCC1 transporter facilitates seizures in the developing brain. Nat Med 11:1205–1213

29. Eid T, Thomas MJ, Spencer DD, Runden-Pran E, Lai JCK, Malthankar GV, Kim JH, Danbolt NC, Ottersen OP, De Lanerolle NC (2004) Loss of glutamine synthetase in the human epileptogenic hippocampus: possible mechanism for raised extracellular glutamate in mesial temporal lobe epilepsy. Lancet 363:28–37

30. Gabriel S, Njunting M, Pomper JK, Merschhemke M, Sanabria ERG, Eilers A, Kivi A, Zeller M, Meencke HJ, Cavalheiro EA, Heinemann U, Lehmann TN (2004) Stimulus and potassium-induced epileptiform activity in the human dentate gyrus from patients with and without hippocampal sclerosis. J Neurosci 24:10416–10430

31. Gloveli T, Albrecht D, Heinemann U (1995) Properties of low Mg^{2+} induced epileptiform activity in rat hippocampal and entorhinal cortex slices during adolescence. Dev Brain Res 87:145–152

32. Graulich J, Hoffmann U, Maier RF, Ruscher K, Pomper JK, Ko HK, Gabriel S, Obladen M, Heinemann U (2002) Acute neuronal injury after hypoxia is influenced by the reoxygenation mode in juvenile hippocampal slice cultures. Brain Res Dev Brain Res 137:35–42

33. Griffiths T, Evans MC, Meldrum BS (1982) Intracellular sites of early calcium accumulation in the rat hippocampus during status epilepticus. Neurosci Lett 30:329–334

34. Gutierrez R, Heinemann U (1999) Synaptic reorganization in explanted cultures of rat hippocampus. Brain Res 815:304–316

35. Gutierrez R, Heinemann U (2001) Kindling induces transient fast inhibition in the dentate gyrus–CA3 projection. Eur J Neurosci 13:1371–1379

36. Haas HL, Jefferys JGR (1984) Low-calcium field burst discharges of CA1 pyramidal neurones in rat hippocampal slices. J Physiol Lond 354:185–201

37. Hablitz JJ, Heinemann U, Lux HD (1986) Step reductions in extracellular Ca^{2+} activate a transient inward current in chick dorsal root ganglion cells. Biophys J 50:753–757

38. Hablitz JJ, Heinemann U, Weiss DS (1984) Epileptiform activity and spreading depression in the neocortical slice. Epilepsia 25:670

39. Huberfeld G, Menendez dP, Pallud J, Cohen I, Le Van Quyen M, Adam C, Clemenceau S, Baulac M, Miles R (2011) Glutamatergic pre-ictal discharges emerge at the transition to seizure in human epilepsy. Nat Neurosci 14:627–634; 2

40. Ivens S, Kaufer D, Flores LP, Bechmann I, Zumsteg D, Tomkins O, Seiffert E, Heinemann U, Friedman A (2007) TGF-beta receptor-mediated albumin uptake into astrocytes is involved in neocortical epileptogenesis. Brain 130:535–547

41. Jandova K, Pasler D, Antonio LL, Raue C, Ji S, Njunting M, Kann O, Kovacs R, Meencke HJ, Cavalheiro EA, Heinemann U, Gabriel S, Lehmann TN (2006) Carbamazepine-resistance in the epileptic dentate gyrus of human hippocampal slices. Brain 129:3290–3306

42. Jauch R, Windmüller O, Lehmann TN, Heinemann U, Gabriel S (2002) Effects of barium, furosemide, ouabaine and 4,4′-diisothiocyanatostilbene-2,2′-disulfonic acid (DIDS) on ionophoretically-induced changes in extracellular potassium concentration in hippocampal slices from rats and from patients with epilepsy. Brain Res 925:18–27

43. Jefferys JGR, Haas HL (1982) Synchronized bursting of CA1 hippocampal pyramidal cells in the absence of synaptic transmission. Nature 300:448–450

44. Jiang J, Quan Y, Ganesh T, Pouliot WA, Dudek FE, Dingledine R (2013) Inhibition of the prostaglandin receptor EP2 following status epilepticus reduces delayed mortality and brain inflammation. Proc Natl Acad Sci U S A 110:3591–3596

45. Johnston D, Brown TH (1981) Giant synaptic potential hypothesis for epileptiform activity. Science 211:294–297

46. Jungehulsing GJ, Heuschmann PU, Holtkamp M, Schwab S, Kolominsky-Rabas PL (2013) Incidence and predictors of post-stroke epilepsy. Acta Neurol Scand 127:427–430

47. Kahle KT, Barnett SM, Sassower KC, Staley KJ (2009) Decreased seizure activity in a human neonate treated with bumetanide, an inhibitor of the Na(+)-K(+)-2Cl(−) cotransporter NKCC1. J Child Neurol 24:572–576

48. Kann O, Kovacs R, Njunting M, Behrens CJ, Otahal J, Lehmann TN, Gabriel S, Heinemann U (2005) Metabolic dysfunction during neuronal activation in the ex vivo hippocampus from chronic epileptic rats and humans. Brain 128:2396–2407

49. Khalilov I, Holmes GL, Ben-Ari Y (2003) In vitro formation of a secondary epileptogenic mirror focus by interhippocampal propagation of seizures. Nat Neurosci 6:1079–1085

50. Khurgel M, Switzer RC III, Teskey GC, Spiller AE, Racine RJ, Ivy GO (1995) Activation of astrocytes during epileptogenesis in the absence of neuronal degeneration. Neurobiol Dis 2:23–35

51. Köhling R, Lücke A, Straub H, Speckmann E-J, Tuxhorn I, Wolf P, Pannek H, Oppel F (1998) Spontaneous sharp waves in human neocortical slices excised from epileptic patients. Brain 121:1073–1087

52. Konnerth A, Heinemann U, Yaari Y (1984) Slow transmission of neural activity in hippocampal area CA1 in the absence of active chemical synapses. Nature 307:69–71

53. Kovacs R, Heinemann U, Steinhauser C (2012) Mechanisms underlying blood-brain barrier dysfunction in brain pathology and epileptogenesis: role of astroglia. Epilepsia 53(Suppl 6):53–59

54. Kovacs R, Schuchmann S, Gabriel S, Kann O, Kardos J, Heinemann U (2002) Free radical-mediated cell damage after experimental status epilepticus in hippocampal slice cultures. J Neurophysiol 88:2909–2918

55. Lapilover EG, Lippmann K, Salar S, Maslarova A, Dreier JP, Heinemann U, Friedman A (2012) Peri-infarct blood-brain barrier dysfunction facilitates induction of spreading depolarization associated with epileptiform discharges. Neurobiol Dis 48:495–506

56. Lehmann TN, Gabriel S, Eilers A, Njunting M, Kovacs R, Schulze K, Lanksch WR, Heinemann U (2001) Fluorescent tracer in pilocarpine-treated rats shows widespread aberrant hippocampal neuronal connectivity. Eur J Neurosci 14:83–95

57. Leschinger A, Stabel J, Igelmund P, Heinemann U (1993) Pharmacological and electrographic properties of epileptiform activity induced by elevated K⁺ and lowered Ca^{2+} and Mg^{2+} concentration in rat hippocampal slices. Exp Brain Res 96:230–240

58. Leutgeb JK, Frey JU, Behnisch T (2003) LTP in cultured hippocampal-entorhinal cortex slices from young adult (P25-30) rats. J Neurosci Methods 130:19–32

59. Liotta A, Caliskan G, Ul HR, Hollnagel JO, Rosler A, Heinemann U, Behrens CJ (2011) Partial disinhibition is required for transition of stimulus-induced sharp wave-ripple complexes into recurrent epileptiform discharges in rat hippocampal slices. J Neurophysiol 105:172–187

60. Lücke A, Nagao T, Köhling R, Avoli M (1995) Synchronous potentials and elevations in $[K^+]_O$ in the adult rat entorhinal cortex maintained in vitro. Neurosci Lett 185:155–158

61. Luhmann HJ, Dzhala VI, Ben Ari Y (2000) Generation and propagation of 4-AP-induced epileptiform activity in neonatal intact limbic structures in vitro. Eur J Neurosci 12:2757–2768

62. MacDonald JF, Xiong ZG, Jackson MF (2006) Paradox of Ca^{2+} signaling, cell death and stroke. Trends Neurosci 29:75–81

63. Meier JC, Henneberger C, Melnick I, Racca C, Harvey RJ, Heinemann U, Schmieden V, Grantyn R (2005) RNA editing produces glycine receptor alpha3(P185L), resulting in high agonist potency. Nat Neurosci 8:736–744

64. Morin-Brureau M, Rigau V, Lerner-Natoli M (2012) Why and how to target angiogenesis in focal epilepsies. Epilepsia 53(Suppl 6):64–68

65. Okazaki MM, Evenson DA, Nadler JV (1995) Hippocampal mossy fiber sprouting and synapse formation after status epilepticus in rats: Visualization after retrograde transport of biocytin. J Comp Neurol 352:515–534

66. Papageorgiou IE, Gabriel S, Fetani AF, Kann O, Heinemann U (2011) Redistribution of astrocytic glutamine synthetase in the hippocampus of chronic epileptic rats. Glia 59:1706–1718

67. Pfeiffer M, Draguhn A, Meierkord H, Heinemann U (1996) Effects of γ-aminobutyric acid (GABA) agonists and GABA uptake inhibitors on pharmacosensitive and pharmacoresistant epileptiform activity in vitro. Br J Pharmacol 119:569–577

68. Pomper JK, Haack S, Petzold GC, Buchheim K, Gabriel S, Hoffmann U, Heinemann U (2006) Repetitive spreading depression-like events result in cell damage in juvenile hippocampal slice cultures maintained in normoxia. J Neurophysiol 95:355–368

69. Pumain R, Menini C, Heinemann U, Silva-Barrat C, Louvel J (1985) Chemical synaptic transmission is not necessary for epileptic activity to persist in the neocortex of the photosensitive baboon. Exp Neurol 89:250–258

70. Quilichini PP, Diabira D, Chiron C, Milh M, Ben-Ari Y, Gozlan H (2003) Effects of antiepileptic drugs on refractory seizures in the intact immature corticohippocampal formation in vitro. Epilepsia 44:1365–1374

71. Rector DM, Burk P, Harper RM (1993) A data acquisition system for long-term monitoring of physiological and video signals. Electroencephalogr Clin Neurophysiol 87:380–384

72. Rigau V, Morin M, Rousset MC, de Bock F, Lebrun A, Coubes P, Picot MC, Baldy-Moulinier M, Bockaert J, Crespel A, Lerner-Natoli M (2007) Angiogenesis is associated with blood-brain barrier permeability in temporal lobe epilepsy. Brain 130:1942–1956

73. Rocha L, Engel J Jr, Ackermann RF (1991) Effects of chronic naloxone pretreatment on amygdaloid kindling in rats. Epilepsy Res 10:103–110

74. Scheffer IE, Berkovic SF (2003) The genetics of human epilepsy. Trends Pharmacol Sci 24:428–433

75. Schmitt FC, Buchheim K, Meierkord H, Holtkamp M (2006) Anticonvulsant properties of hypothermia in experimental status epilepticus. Neurobiol Dis 23:689–696

76. Schuchmann S, Müller W, Heinemann U (1998) Altered Ca^{2+}-signaling and mitochondrial deficiencies in hippocampal neurons of trisomy 16 mice: a model of Down's syndrome. J Neurosci 18:7216–7231

77. Schwartzkroin PA, Prince DA (1977) Penicillin-induced epileptiform activity in the hippocampal in vitro preparation. Ann Neurol 1:463–469

78. Schwartzkroin PA, Prince DA (1978) Cellular and field potential properties of epileptogenic hippocampal slices. Brain Res 147:117–130

79. Seiffert E, Dreier JP, Ivens S, Bechmann I, Tomkins O, Heinemann U, Friedman A (2004) Lasting blood-brain barrier disruption induces epileptic focus in the rat somatosensory cortex. J Neurosci 24:7829–7836

80. Somjen GG, Kager H, Wadman WJ (2009) Calcium sensitive non-selective cation current promotes seizure-like discharges and spreading depression in a model neuron. J Comput Neurosci 26:139–147

81. Spencer SS, Spencer DD (1994) Entorhinal-hippocampal interactions in medial temporal lobe epilepsy. Epilepsia 35:721–727

82. Streit AK, Derst C, Wegner S, Heinemann U, Zahn RK, Decher N (2011) RNA editing of Kv1.1 channels may account for reduced ictogenic potential of 4-aminopyridine in chronic epileptic rats. Epilepsia 52:645–648

83. Swartzwelder HS, Lewis DV, Anderson WW, Wilson WA (1987) Seizure-like events in brain slices: suppression by interictal activity. Brain Res 410:362–366

84. Swiercz W, Cios K, Hellier J, Yee A, Staley K (2007) Effects of synaptic depression and recovery on synchronous network activity. J Clin Neurophysiol 24:165–174

85. Thompson SM, Gähwiler BH (1989) Activity-dependent disinhibition. II. Effects of extracellular potassium, furosemide, and membrane potential on E_{Cl}^- in hippocampal CA3 neurons. J Neurophysiol 61:512–523

86. Tomkins O, Shelef I, Kaizerman I, Misk A, Afawi Z, Eliushin A, Gidon M, Cohen A, Zumsteg D, Friedman A (2007) Blood-brain barrier disruption in post-traumatic epilepsy. J Neurol Neurosurg Psychiatry 79:774–777

87. Traynelis SF, Dingledine R (1988) Potassium-induced spontaneous electrographic seizures in the rat hippocampal slice. J Neurophysiol 59:259–276

88. Verret L, Mann EO, Hang GB, Barth AM, Cobos I, Ho K, Devidze N, Masliah E, Kreitzer AC, Mody I, Mucke L, Palop JJ (2012) Inhibitory interneuron deficit links altered network activity and cognitive dysfunction in Alzheimer model. Cell 149:708–721

89. Vezzani A (2008) Innate immunity and inflammation in temporal lobe epilepsy: new emphasis on the role of complement activation. Epilepsy Curr 8:75–77

90. Walther H, Lambert JDC, Jones RSG, Heinemann U, Hamon B (1986) Epileptiform activity in combined slices of the hippocampus, subiculum and entorhinal cortex during perfusion with low magnesium medium. Neurosci Lett 69:156–161

91. Wei WL, Sun HS, Olah ME, Sun X, Czerwinska E, Czerwinski W, Mori Y, Orser BA, Xiong ZG, Jackson MF, Tymianski M, MacDonald JF (2007) TRPM7 channels in hippocampal neurons detect levels of extracellular divalent cations. Proc Natl Acad Sci U S A 104:16323–16328

92. Winkler MK, Chassidim Y, Lublinsky S, Revankar GS, Major S, Kang EJ, Oliveira-Ferreira AI, Woitzik J, Sandow N, Scheel M, Friedman A, Dreier JP (2012) Impaired neurovascular coupling to ictal epileptic activity and spreading depolarization in a patient with subarachnoid hemorrhage: possible link to blood-brain barrier dysfunction. Epilepsia 53(Suppl 6):22–30

93. Yamamoto C, Kawai N (1967) Seizure discharge evoked in vitro in thin sections from guinea pig hippocampus. Science 155:341–342

94. Yee AS, Longacher JM, Staley KJ (2003) Convulsant and anticonvulsant effects on spontaneous CA3 population bursts. J Neurophysiol 89:427–441

95. Yu FH, Mantegazza M, Westenbroek RE, Robbins CA, Kalume F, Burton KA, Spain WJ, McKnight GS, Scheuer T, Catterall WA (2006) Reduced sodium current in GABAergic interneurons in a mouse model of severe myoclonic epilepsy in infancy. Nat Neurosci 9:1142–1149

96. Yu W, Jegla T, Wagoner K, Wickenden AD (2000) Retigabine, a novel anti-convulsant, enhances activation of KCNQ27Q3 potassium channels. Mol Pharmacol 58:591–600

97. Yus-Najera E, Munoz A, Salvador N, Jensen BS, Rasmussen HB, DeFelipe J, Villarroel A (2003) Localization of KCNQ5 in the normal and epileptic human temporal neocortex and hippocampal formation. Neuroscience 120:353–364

98. Zahn RK, Tolner EA, Derst C, Gruber C, Veh RW, Heinemann U (2008) Reduced ictogenic potential of 4-aminopyridine in the perirhinal and entorhinal cortex of kainate-treated chronic epileptic rats. Neurobiol Dis 29:186–200

99. Zhang CL, Dreier JP, Heinemann U (1995) Paroxysmal epileptiform discharges in temporal lobe slices after prolonged exposure to low magnesium are resistant to clinically used anticonvulsants. Epilepsy Res 20:105–111

What Is the Importance of Abnormal "Background" Activity in Seizure Generation?

Richard J. Staba and Gregory A. Worrell

Abstract

Investigations of interictal epileptiform spikes and seizures have played a central role in the study of epilepsy. The background EEG activity, however, has received less attention. In this chapter we discuss the characteristic features of the background activity of the brain when individuals are at rest and awake (resting wake) and during sleep. The characteristic rhythms of the background EEG are presented, and the presence of $1/f^{\beta}$ behavior of the EEG power spectral density is discussed and its possible origin and functional significance. The interictal EEG findings of focal epilepsy and the impact of interictal epileptiform spikes on cognition are also discussed.

Keywords

Electroencephalogram • Epilepsy • Local Field Potential Oscillations • High Frequency Oscillations • Sleep

3.1 Introduction

The electrical activity of mammalian brain, defined by the electroencephalogram (EEG), has long been a focus of scientific and clinical brain research [16]. The mechanisms underlying various EEG changes associated with brain maturation, behavioral states, cognition, motor function, and neurological disease represent fundamental discoveries of neuroscience. Epilepsy in particular has benefited from EEG investigations [15]. A disorder characterized by unprovoked recurrent seizures, epilepsy has many underlying pathological causes but is unified by the common clinical expression of seizures and the

R.J. Staba, Ph.D.
Department of Neurology, Reed Neurological
Research Center, David Geffen School
of Medicine at UCLA,
710 Westwood Plaza, RNRC 2-155, Los Angeles,
CA 90095, USA
e-mail: rstaba@mednet.ucla.edu

G.A. Worrell, M.D., Ph.D. (✉)
Mayo Systems Electrophysiology Laboratory,
Divisions of Epilepsy and Clinical Neurophysiology,
Departments of Neurology, Physiology
and Bioengineering, Mayo Clinic,
200 1st Street SW, Rochester, MN 55905, USA
e-mail: worrell.gregory@mayo.edu

H.E. Scharfman and P.S. Buckmaster (eds.), *Issues in Clinical Epileptology: A View from the Bench*,
Advances in Experimental Medicine and Biology 813, DOI 10.1007/978-94-017-8914-1_3,
© Springer Science+Business Media Dordrecht 2014

associated pathological brain electrical activity. Not long after the discovery of the human EEG [7], Berger also reported that epileptic seizures had an abnormal EEG signature, and that between the seizures (interictal) there were also transient epileptiform abnormalities not seen in controls (translated in [8]). Thereafter, the significance of interictal epileptiform spikes (IIS) and abnormal transient oscillatory network activity in the development of epilepsy (epileptogenesis), seizure generation (ictogenesis), and associated functional impairments (e.g., cognition, memory, and reaction times) have been active areas of research.

3.2 Physiological Electrical Activity in the Normal Mammalian Brain

Since the first observation of the occipital alpha rhythm [7] (translated in [8]) the interest in brain oscillations and their physiological and pathological correlates has occupied a central position in human neuroscience. Historically clinical and basic research focused on specific oscillations that are prominent in the EEG intermittently, for example the occipital alpha rhythm (α; 8–12 Hz) recorded at rest with eyes closed, beta (β; 12–30 Hz) and gamma frequency activity (γ; 30–50 Hz) during mental and motor tasks, theta frequency

oscillations (θ; 4–8 Hz) during memory tasks or sleep, and delta frequency activity (δ; 0.5–4 Hz) that characterizes slow wave sleep. Similarly, the EEG activity in traditional frequency bands (δ, θ, α, β, γ) became the focus of EEG research and intensively studied in brain maturation [50], normal function, and disease states. However, it is widely recognized that the brain generates activities well outside these classic EEG bands. In fact, the high amplitude EEG activity below δ (<0.5 Hz) including direct current (DC) changes were some of the earliest electrical activities recorded [2].

While EEG research has largely focused on narrow band EEG oscillations (δ, θ, α, β, γ) it is well recognized that there is a broad spectrum of on-going, arrhythmic, background activity that does not contain a dominant characteristic oscillation, but rather is composed of intermixed spectral frequencies [17, 34]. It is out of this arrhythmic background that the traditional EEG oscillations discussed above, e.g. the posterior dominant alpha rhythm, may be evoked or spontaneously emerge (Fig. 3.1). More recently, this broad spectrum of on-going background activity has been a focus of attention [17, 34, 43].

This composition of background electrical activity in mammalian brain generally follows power-law behavior, i.e. the spectral power scales with frequency as $1/f^{\beta}$ where β is called a scaling exponent (Fig. 3.2). This $1/f^{\beta}$ ("one-over-f")

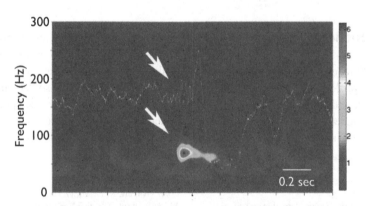

Fig. 3.1 Interictal Background Activity from human hippocampus recorded with intracranial depth electrode. There is an ongoing background activity followed by a paroxysmal gamma frequency oscillation (*bold arrows*), and an interictal epileptiform spike (IIS, *arrow*). The wavelet transform (1–600 Hz, Morlet basis) spectrogram of raw intracranial EEG shows the background theta activity, and the emergent low amplitude gamma oscillation preceding the IIS. The scale bar is normalized units standard deviations from background. Time base is 100 msec per division (Courtesy of Liankun Ren, M.D. unpublished)

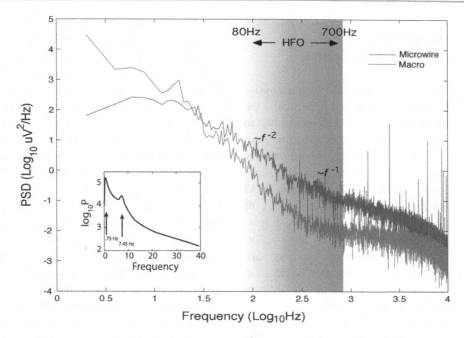

Fig. 3.2 Power spectral density (PSD) from 5 min of human hippocampus during sleep recorded with intracranial EEG (0.05–10,000 Hz, sampled at 32 kHz) using micro- and clinical macro-electrodes. The wide bandwidth recording exhibits $1/f$ β behavior with different scaling regions characterized by different scaling exponents β = 1 and 2. The inset shows an expansion of the PSD in the 0.05–40 Hz range and spectral peaks from a low delta frequency oscillation (~0.75 Hz) and a theta-alpha frequency oscillation (7.45 Hz). The characteristic 0.75 Hz oscillation is persistent throughout the 5 min and modulates the 7.45 Hz intermittent oscillation (Unpublished Data)

spectrum with lower frequency activities exhibiting higher amplitudes than faster frequencies is characterized by the scaling exponent β that can be obtained by plotting log power vs. log frequency (log(Power) v.s. −β log(f)) and ranges over $0 < β < 4$ [31, 34, 52, 80]. The spectral peaks embedded in the $1/f$ β represent ongoing persistent oscillations or organized emergent oscillations that arise out of the ongoing EEG background, such as the traditional EEG rhythms of the human EEG (Fig. 3.2).

The arrhythmic background activity has more recently received attention within the context of the advancing understanding of complex systems. It is recognized that $1/f^β$ behavior in complex systems can be a signature of a self-organized system with scale-free dynamics [31, 34, 55]. It turns out that $1/f^β$ patterns are ubiquitous in nature, from the statistics of earthquakes to stock market dynamics [4]. The origin of $1/f^β$ behavior in EEG and local field potential (LFP) recordings remains unclear [9, 10, 34], but perhaps one of the most intriguing ideas is that it results from hierarchal

nesting of brain activity [19, 34, 73] whereby lower frequency activity modulates higher frequency activity [34]. The modulation of gamma oscillations by theta oscillations is a classic example [6, 14, 18]. At the cellular level, multi-unit activity is correlated with EEG gamma power and phase-locked to the negative-going phase of the delta frequency activity [79]. Synchronization between neuronal assemblies also occurs within arrhythmic brain activity [25, 45, 70].

Maturation of EEG: The continuous maturation of EEG activity through young adulthood reflects brain development, e.g. myelination, and organization [50]. In premature infants (24–27 weeks), the EEG is discontinuous and may alternate between periods containing bursts of high amplitude slow (0.1–1 Hz) activity and intermixed faster rhythms (8–14 Hz). From these earliest electrical rhythms in the infant brain there are long periods of continuous development through late childhood (~12 y.o) when the posterior dominant alpha rhythm reaches ~10 Hz [50].

3.3 Electrical Activity of the Sleeping Brain

There exists substantial evidence for the physiological importance of sleep and in particular the requirement of sleep for normal memory [24]. To better understand how memory benefits from sleep, it would be helpful to first describe briefly the EEG during the two main types of sleep – rapid eye movement (REM) and non-REM sleep – and then how the neurophysiology of sleep might support aspects of memory formation. Since patients with epilepsy often report deficits in sleep and impairment in memory, subsequent sections describe electrophysiological disturbances in the epileptic brain and their likely functional implications for cognition.

EEG of REM sleep: During REM or desynchronized sleep, arising from a background of low-voltage, mixed frequency EEG, are spontaneous synchronous bursts of neuronal activity generated by the pontine tegmentum that spread to the lateral geniculate nucleus and visual cortices in the occipital lobe that are termed "PGO waves". Conspicuous in the EEG of rats and cats and less in humans, PGO waves coincide with rapid eye movements and can become phase-locked with theta oscillations. In rodents, theta oscillations occur with largest amplitude in the hippocampal CA1 area driven by inputs from septum, entorhinal cortex, and CA3. In addition to REM sleep, hippocampal theta can also be observed during awake behaviors in rodents. Theta also occurs in humans during wakefulness, but is more apparent in neocortical areas and less coherent in hippocampal areas.

EEG of non-REM sleep: Non-REM sleep is characterized by high-voltage slow wave activity that includes slow oscillations <1 Hz and delta activity. The slow oscillation persists in isolated neocortical tissue and is abolished if thalamocortical cells are deafferented from cortical inputs, suggesting slow oscillations are generated largely within neocortex [60, 71]. In scalp EEG, the alternating sequence of surface positive (depth negative) and negative (depth positive) waves correspond with periods of neuronal membrane depolarization and hyperpolarization respectively. Periods of membrane depolarization occur within excitatory and inhibitory cells that produces sustained neuronal firing commonly referred to as "UP-states", whereas periods of membrane hyperpolarization are accompanied by neuronal silence denoted as "DOWN-states". The mechanisms generating slow oscillations are not yet clear, although evidence to date suggests UP-states could arise from widespread summation of calcium- and persistent sodium inward current-mediated excitatory postsynaptic potentials in cortical cells, while neuronal disfacilitation associated with DOWN-states could be due to calcium- and sodium-dependent potassium currents, inactivation of persistent sodium currents, and possibly GABA-mediated inhibition.

Slow oscillations strongly modulate two other transient oscillations that occur during non-REM sleep – spindles and sharp wave-ripple complexes – and is another classic example of frequency nesting. Spindle waves are beta frequency oscillations that wax and wane between 10 and 16 Hz and last 0.5–2 s that characterize stage 2 of NREM sleep. Spindles arise from interactions between GABA-containing neurons in the thalamic reticular nucleus as well as thalamocortical cells that facilitate the synchrony and spread of spindles throughout neocortex. Human studies have identified two types of spindles designated slow and fast; however, whether these two types of spindles arise from different neuronal mechanisms or reflect the modulation of a common spindle generator is not known. Slow (10–12 Hz) spindles occur primarily over frontal cortical areas and more frequently during slow wave sleep than stage 2 sleep, and fast (13–15 Hz) spindles appear broadly over central and parietal cortices and are often coincident with increased hippocampal activity.

In hippocampus during non-REM sleep, spontaneous extra-hippocampal impulses drive neuronal firing in CA3 that projects forward via Schaffer collaterals onto dendritic processes of CA1 pyramidal cells and some types of interneurons. This briefly irregular (30–120 milliseconds in duration), increase in neuronal firing registers in the depth EEG as a large amplitude sharp wave with maximum negativity in stratum lucidum,

stratum radiatum, and inner third of stratum moleculare corresponding to input layers of CA3, CA1
and dentate gyrus respectively. In CA1, a similarly
brief high-frequency oscillation (HFO; 80–200 Hz)
termed "ripple" arises from synchronous firing
between pyramidal cells and basket cells that is
largest in amplitude in stratum pyramidale and
superimposed on the sharp wave. During widespread neuronal depolarization associated with the
slow oscillation UP-state, hippocampal ripples can
co-occur with neocortical fast spindles to form
spindle-ripple events with ripples that temporally
coincide with the troughs of spindle waves [64].

*Concept of memory function and putative
neuronal mechanisms:* Memory function generally involves processes of encoding, consolidation, and retrieval. In the awake brain, encoding
occurs when perception of the stimuli produces a
new, yet unstable, memory trace. During subsequent sleep, the labile memory trace becomes
more stable and eventually integrated into brain
networks supporting long-term storage of knowledge in a process termed "consolidation". During
retrieval, the stored memory is accessed and
recalled. A number of theories have been proposed on how sleep supports memory consolidation with some more than others supported by
compelling data from animal and human studies
(for extensive review, see [56]). Central to current
theories (e.g., "active system consolidation") is
the concept of reactivation that involves a sleep-
related replay of neuronal firing patterns corresponding to the neuronal firing patterns that
occurred while encoding, i.e., during prior wakefulness, as well as specific roles for different
types of sleep in memory consolidation.

Considerable research has focused on identifying the neuronal mechanisms that provide
sleep-related benefits on memory formation.
Current models emphasize precisely coordinated
neuronal activity between neocortex and hippocampus for hippocampal-dependent memories
[26]. During non-REM sleep, neocortical slow
oscillation UP-states provide a temporal window
for increased ripple activity and associated
increase in neuronal firing that could reflect
reactivation of hippocampal memories. The coincidence of hippocampal ripples with neocortical

fast spindles (spindle-ripple events) is thought to
promote the transfer and ultimately storage of the
hippocampal memory to neocortex [64], which is
reflected presumably by long-term functional and
structural changes that strengthen synaptic transmission (e.g., long-term potentiation). In addition, evidence suggests REM sleep PGO- and
theta-related neuronal activity could also be
involved with synaptic modifications with theta
possibly playing a role in synaptic downscaling,
which extends the "synaptic homeostasis"
hypothesis that links the regulation of sleep with
mechanisms of synaptic plasticity [72].

*Human single neuron correlates of sleep and
memory:* Microelectrode unit recordings during
natural sleep in humans are few, but available data
indicate hippocampal neuronal firing increases during non-REM sleep and declines during REM sleep
[57, 65–67]. Furthermore, the propensity for burst
discharge is highest during non-REM sleep compared to awake and REM sleep episodes. These
results are similar to the rates and pattern of hippocampal pyramidal cell firing during non-REM
and REM sleep in rodents [64], and are generally
consistent with levels of hippocampal activity that
could be involved with reactivation described in the
preceding paragraphs. Work using the same microelectrode recordings from single neurons in humans
has primarily focused on memory and navigation.
These studies have led to the discovery of place
cells in the human mesial temporal lobe underlying
spatial navigation [26], which resembles the loca-
tion-specific firing patterns of some pyramidal cells
in non-primate hippocampus described in the sections that follow. In addition, studies in humans
have found evidence for neurons that encode
category specific images [40, 41].

3.4 Abnormal Electrical Activity in the Epileptic Brain

In addition to sleep and wake behavioral states of
normal brain, epileptic brain is characterized by
interictal state (between seizures), ictal state
(seizures), and post-ictal states (after the seizure).
It should be noted that while seizures are generally limited to a minute or so, the post-ictal state

as determined by subtle EEG or cognitive and physical changes can be prolonged [30]. In addition to the interictal and post-ictal state, there is emerging evidence for a pre-ictal state that is associated with increase in probability of seizure occurrence [22, 28, 47].

Interictal Epileptiform Discharges: Interictal EEG spikes (IIS) are brief, sharply contoured voltage fluctuations of less than 200 msec that are a signature of epileptic brain. The intracellular correlate of IIS is the paroxysmal depolarizing shift [3] seen in the neuronal membrane potential and is associated paroxysmal burst of neuronal population firing, but also involves a more complex interaction of inhibitory and excitatory neurons [3]. Depth electrode recordings during overnight polysomnographic sleep studies show that in patients with temporal lobe epilepsy (TLE), the highest rates of IIS regularly occur during non-REM stage 3, and in some cases stages 1 and 2, sleep compared to waking and REM sleep [42, 59]. In addition, the spatial distribution of IIS is often broader during non-REM sleep than waking or REM sleep, i.e., IIS appear at electrode recording sites within and remote from where seizures begin [59].

At the level of single neurons, patient studies have not consistently found differences in interictal firing rates and bursting inside versus outside the seizure onset zone (SOZ) during awake episodes [20, 21]. However, during non-REM and REM sleep compared to wakefulness, interictal firing rates, bursting, and synchrony of discharges are significantly higher in mesial temporal lobe (MTL) ipsilateral to the SOZ than contralateral MTL [65–67]. These results provide evidence for sleep-related facilitation of interictal neuronal hyperexcitability within the SOZ of patients with temporal lobe epilepsy.

Pathological HFO: In the epileptic brain, transient abnormally synchronous discharges of principal cells can summate in the extracellular space that give rise to a burst of population spikes commonly termed pathological HFO or pHFO [12, 13]. Chronic animal models of epilepsy and studies in patients with epilepsy indicate hippocampal and neocortical pHFOs are strongly associated with brain areas capable of generating spontaneous seizures [29]. With respect to the wake-sleep cycle, recordings in epileptic rats and patients show the highest rates of hippocampal pHFOs occur during non-REM sleep compared to awake and REM sleep, while equivalent rates can be found during the latter two desynchronized EEG states [65–67]. By contrast, ripples are highest during non-REM sleep, while rates are lower during wakefulness and lowest in REM sleep, which is consistent with their occurrence in the normal non-primate hippocampus [65–67].

Pathological Synchrony: Synchronization of neuronal assemblies is thought to underlie normal brain functions such as perception, learning, and cognition. Alterations in neuronal synchrony are thought to underlie the clinical manifestations of many neurological diseases. Hypersynchrony of pathological neuronal assemblies as the generator of epileptiform activity has been a central theme of epileptic brain electrophysiology [53]. Jasper and Penfield speculated that the local high amplitude interictal epileptiform activity recorded directly from human cortex during surgery was generated by a burst of hypersynchronous neuronal activity [53]. Interestingly, however, many seizures appear to begin with an apparent "asynchronous state" – low amplitude LFP activity that evolves into a hypersynchronous state with high amplitude rhythmic activity [53]. Multiple studies have reported increased local synchrony, i.e. hypersynchrony, within epileptic brain using a range of quantitative measures of synchrony, including spectral coherence [74], magnitude squared coherence [81], and mean phase coherence [61]. In addition, investigations of LFP synchrony during spontaneous human seizures have consistently demonstrated a decrease in local LFP synchrony at seizure onset compared to baseline [48, 62, 78].

Analysis of long records of interictal iEEG from patients with focal epilepsy and control subjects with intractable facial pain found that the spatial distribution of LFP synchronization fell rapidly with the distance between electrodes [77]. Consistent with the hypothesis that the generators of normal and pathological HFOs are more spatially localized than lower frequency oscillations,

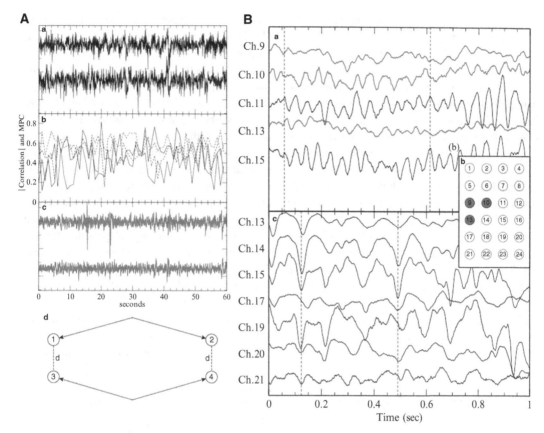

Fig. 3.3 Data and from 2 patients, one with intractable facial pain (*black*) and other with focal epilepsy (*gray*). Data from the patient with intractable facial pain and no history of seizures serves as a control recording for quantitative comparison. (**A**) (**a**) Sample signals from two electrodes of the control brain recording. (**b**) The correlation magnitudes (*solid lines*) and mean phase coherence (*dashed lines*) of the signal pairs in (**a**) and (**c**). Both the correlation and mean phase coherence (MPC) show significant temporal variability over the course of 60 s, with values primarily ranging from (0.2–0.7) (**c**) Sample signals from two electrodes in epileptic cortex outside the seizure onset zone. (**d**) A sample layout of the bipolar reference pair measurement. The dis-tance *d* between one corresponding pair of electrodes 1 and 3 is equal to the distance between the other pair, electrodes 2 and 4, and this is the distance referenced in our bipolar measurements. (**B**) (**a**) Sample interictal iEEG signals from Patient 1 with epilepsy from both inside the seizure onset zone (SOZ), shown in *gray*, and near signals outside the SOZ (*black*). *Dashed line* marks significant phase lag between inside and outside the SOZ. (**b**) Spatial layout of the intracranial electrodes for Patients 1 and A with the SOZ electrodes (9, 10 and 13) of Patient 1 shown in red. (**c**) Sample signals from the control Patient A. The spatial numbering is as shown in (**b**). For clarification, signals are offset vertically (Reproduced from Ref. [77])

the synchrony fall off is frequency dependent [44]. Synchrony in the epileptic brain, however, was shown to be markedly reduced in electrodes bridging connections between the SOZ and surrounding brain (Fig. 3.3). In effect, the SOZ is functionally disconnected and isolated from surrounding brain regions [77].

Focal EEG Slowing: In addition to the IIS and pHFO that have been widely investigated, focal slowing on the EEG is common in the region of epileptic brain. Focal delta frequency slowing was initially described in patients with focal structural abnormalities, such as tumors and strokes [33, 76], but is also common in TLE [11]. When the slowing occurs as intermittent oscillations of monomorphic delta activity in the temporal lobe region it is termed, temporal intermittent rhythmic delta activity and is a signature of focal epilepsy [58, 68]. Focal delta frequency slowing has also recently been shown to be more common than IIS following febrile status epilepticus [51].

Hypsarrhythmia: Hypsarrhythmia is the an EEG pattern characterized by disorganized high-amplitude spikes and spike- and/or sharp-slow wave discharges and commonly observed in infants with West syndrome (Infantile spasms, Hypsarrhythmia, and Developmental regression) (Fig. 3.4). Hypsarrhythmia is modulated by the sleep-wake cycle. During non-REM sleep, there is a tendency for runs of these high-amplitude discharges to become grouped or clustered with a period consistent with non-REM slow wave activity that typically is then followed by episodes of EEG attenuation [32]. By contrast, during REM sleep, there is a significant reduction and in some cases disappearance of hypsarrhythmia that reappears toward the end of the REM sleep episode, and if awakening then hypsarrhythmia continues often with clinical manifestations.

3.5   IIS Impact on Cognition

While epilepsy in general could have detrimental effects on sleep, sleep-related epileptiform activity in particular could contribute to cognitive impairment [27, 39]. Epileptic encephalopathies refer to conditions in some patients with epilepsy who have neurological deterioration associated with frequency or severity of seizures and/or interictal epileptiform discharges and not due to the original cause or etiology [49]. Hypsarrhythmia in West syndrome and Electrical Status Epilepticus During Sleep associated with continuous spike-wave of sleep and Landau-Kleffner syndrome reflect interictal EEG patterns that predominate during non-REM sleep that could support abnormal activity-dependent synaptic plasticity which in turn produce cognitive impairments. Improved control of seizures and interictal EEG in these patients is often associated with improved cognitive performance that suggests ictal and interictal discharges contribute to these deficits [35]. Indeed, studies in adults and children show IIS can be associated with brief episodes of impaired cognitive functioning

referred to as "transient cognitive impairment" [1]. The functional disruption coincides with the location were the IIS occur, e.g., verbal memory task impairment with left-side IIS, spatial task impairment with right-sided interictal epileptiform discharges (IEDs). Other studies observed IIS in occipital cortices disrupted visual stimuli presented in the contralateral visual field [63]. Work using pilocarpine-treated epileptic rats showed that in a hippocampal-dependent memory task, hippocampal IIS occurring during memory retrieval, but not encoding or maintenance, reduced performance [37, 38]. Similar results were observed in presurgical patients during a short-term memory task [38]. In this study, when depth electrode-recorded hippocampal IIS contralateral to SOZ or bilateral occurred during memory retrieval or maintenance, performance was lower. Recent evidence implicates brain network activity underlying memory processing in suppression of the IEDs, however, which raises an important confound of the interplay between cognitive processing on IIS [46].

How IIS disrupt memory processes is not known, but studies in epileptic rats show hippocampal IIS are associated with a decrease in CA1 cell firing [82]. Moreover, a significant reduction in firing was found after compared to before the IIS in interneurons, but not CA1 pyramidal cells that preferentially discharge when the rat is a specific location in the environment ("place cell"). A separate study of CA3 cell firing found significantly lower firing rates in interneurons and pyramidal cells before and after IIS compare to rates during random episodes [82]. These data indicate that neuronal firing surrounding and particularly after IIS could reflect episodes associated with reduced activity-dependent synaptic plasticity. By contrast, spontaneous pHFOs reflect brief episodes of increased neuronal discharges that are two-fold greater or more than the level of activity that occurs during spontaneous hippocampal ripples. Unlike ripples that involves interneuron-mediated regulation of pyramidal cell firing, it appears that the effects of interneurons are diminished during pHFOs. The abnormally high spatial and temporal coincidence of discharges could contribute to pathological synaptic plasticity that is functionally disruptive to memory consolidation. One study of resective sclerotic hippocampal tissue, which is often associated with pHFOs and hyperexcitability in patients with drug-resistant temporal lobe epilepsy, found significantly lower levels of long-term potentiation and its synaptic counterpart long-term depression in sclerotic compared to non-sclerotic tissue [5]. These data suggest morphological alterations associated with hippocampal sclerosis and persistent abnormal interictal activity contribute to diminished capacity for physiological activity-dependent synaptic plasticity.

Disruptions in sleep and sleep-related oscillatory activity could interfere with aspects of memory formation. Patients with epilepsy are two times more likely to complain of sleep disturbances, chiefly excessive daytime sleepiness and insomnia that can have a negative influence on quality of life measures [23, 54]. Studies indicate seizures, comorbidity (e.g., sleep apnea),

and anti-seizure drugs can cause disruptions in the amount and architecture of sleep. During nights with nocturnal seizures, patients often have a greater number of nighttime stage-shifts or awakenings, increased amounts of non-REM stage 1 sleep, and reduced amounts of non-REM stage 3 sleep and REM sleep [72]. Daytime seizures, which themselves could prevent or interfere with learning, also reduce the amount of REM sleep during the subsequent night [72]. Furthermore, older types of anti-seizure drugs, such as barbiturates and benzodiazepines, can reduce the amount of REM sleep and in some cases stage 3 non-REM sleep. Newer drugs have no effect or can even increase the amount of REM sleep, although some reduce amounts of non-REM sleep (e.g., Gabapentin, Lamotrigine). Loss of slow wave-rich non-REM sleep or REM sleep in terms of absolute amounts and relative to when daytime learning occurs could negate the time-dependent benefits of sleep on memory formation.

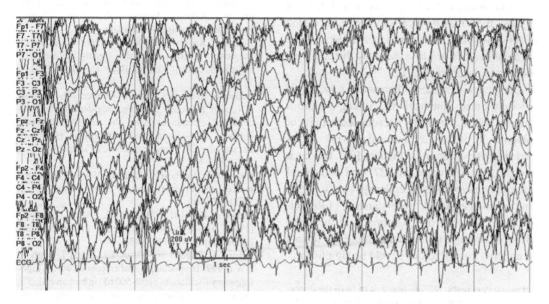

Fig. 3.4 The pattern of hypsarrhythmia is a specific EEG pattern associated with West Syndrome. First described in detail by Gibbs (66) the pattern consists of high voltage, disorganized EEG with multifocal and generalized epileptiform spikes and sharp waves. The characteristic pattern, often described as disorganized or chaotic, is unique in that the normal pattern of spatial synchrony over multiple brain regions is absent. The EEG tracing from each chan-nel (e.g. Fp1 and F7) appear independent of each other despite the anatomical proximity. This is distinct from normal brain activity where there is widespread synchronous activity over multiple brain regions. In the hypsarrhythmia pattern the periods of generalized synchrony are due to generalized epileptiform discharges. The epileptiform spikes fluctuate in time and space, and are various focal, multifocal, and generalized

3.6 Conclusions

Is the "background" activity important? The answer to this question is clearly yes. There is good evidence that the background EEG activity contains important information about brain function and dysfunction in human epilepsy. While the diagnostic importance of IIS and seizures is clear, there is also evidence that the background EEG during sleep and wake is important prognostic tool. In addition, investigations of LFP synchrony and neuronal assemblies are providing mechanistic insights about brain function, cognition, and epilepsy related comorbidities.

Is interictal background really "normal"? In some epilepsy cases there are clear EEG background abnormalities. In West syndrome, Electrical Status Epilepticus in Slow-wave Sleep and other epileptic encephalopathies the EEG background is markedly abnormal. Whether there are more subtle abnormalities in the background in primary generalized epilepsy is an area of active study, but often the EEG on visual review appears normal. In focal epilepsy there may be focal slowing in the region of epileptic brain. In drug resistant epilepsy there are abnormalities in LFP synchrony that are present even in the absence of IIS.

Our cognition and behavior rely upon precisely-timed interactions among neurons forming brain networks by coordinated activity of anatomically distributed neuronal networks mediated via rhythmic brain oscillations [75]. Research into the common cognitive [27] and behavioral [36, 69] comorbidities of epilepsy is only beginning to emerge. As our understanding of the cellular mechanisms of cognition and behavior advance the impact on understanding the impact on epilepsy comorbidities should be significant.

Acknowledgements We are honored to have the opportunity to participate in this book celebrating Philip Schwartzkroin's many contributions to epilepsy research.

Other Acknowledgements This research was supported by NIH R01-NS071048 (RS) and NIH R01-NS63039 (GW).

References

1. Aarts JH, Binnie CD, Smit AM, Wilkins AJ (1984) Selective cognitive impairment during focal and generalized epileptiform EEG activity. Brain 107(Pt 1):293–308
2. Aladjalova NA (1957) Infra-slow rhythmic oscillation of the steady potential of the cerebral cortex. Nature 4567:957–959
3. Ayala GF, Dichter M, Gumnit RJ, Matsumoto H, Spencer WA (1973) Genesis of epileptic interictal spikes. New knowledge of cortical feedback systems suggests a neurophysiological explanation of brief paroxysms. Brain Res 52:1–17
4. Bak P (1996) How nature works: the science of self-organized criticality. Nature 383(6603):772–773
5. Beck H, Goussakov IV, Lie A, Helmstaedter C, Elger CE (2000) Synaptic plasticity in the human dentate gyrus. J Neurosci 20(18):7080–7086
6. Belluscio MA, Mizuseki K, Schmidt R, Kempter R, Buzsáki G (2012) Cross-frequency phase–phase coupling between theta and gamma oscillations in the hippocampus. J Neurosci 32(2):423–435
7. Berger H (1929) Über das elektrenkephalogramm des menschen. I Mitteilung. Arch Psychiatr Nervenkr 87:527–570
8. Berger H, Gloor P (1969) Hans berger on the electroencephalogram of man: the fourteen original reports on the human electroencephalogram. Elsevier Publishing Company, Amsterdam
9. Bédard C, Kröger H, Destexhe A (2006) Does the 1/f frequency scaling of brain signals reflect self-organized critical states? Phys Rev Lett 97(11): 118102
10. Bédard C, Rodrigues S, Roy N, Contreras D, Destexhe A (2010) Evidence for frequency-dependent extracellular impedance from the transfer function between extracellular and intracellular potentials: intracellular-LFP transfer function. J Comput Neurosci 29(3):389–403
11. Blume WT, Borghesi JL, Lemieux JF (1993) Interictal indices of temporal seizure origin. Ann Neurol 34(5):703–709
12. Bragin A, Engel JJ Jr, Wilson CL, Fried I, Buzsaki G (1999) High-frequency oscillations in human brain. Hippocampus 9(2):137–142
13. Bragin A, Engel J Jr, Wilson CL, Fried I, Mathern GW (1999) Hippocampal and entorhinal cortex high-frequency oscillations (100–500 hz) in human epileptic brain and in kainic acid–treated rats with chronic seizures. Epilepsia 40(2):127–137
14. Bragin A, Jando G, Nadasdy Z, Hetke J, Wise K, Buzsaki G (1995) Gamma (40–100 Hz) oscillation in the hippocampus of the behaving rat. J Neurosci 15(1 Pt 1):47–60

15. Brazier MA (1960) The EEG, in epilepsy. A historical note. Epilepsia 1:328–336
16. Brazier MA (1961) A history of the electrical activity of the brain: the first half-century. Pitman Medical Publishing Co. Ltd, London
17. Buzsaki G (2009) Rhythms of the brain. Oxford University Press, London
18. Canolty RT, Edwards E, Dalal SS, Soltani M, Nagarajan SS, Kirsch HE et al (2006) High gamma power is phase-locked to theta oscillations in human neocortex. Science 313(5793):1626–1628
19. Canolty RT, Knight RT (2010) The functional role of cross-frequency coupling. Trends Cogn Sci 14(11): 506–515
20. Colder BW, Frysinger RC, Wilson CL, Harper RM, Engel J Jr (1996) ecreased neuronal burst discharge near site of seizure onset in epileptic human temporal lobes. Epilepsia 37(2):113–121
21. Colder BW, Wilson CL, Frysinger RC, Harper RM, Engel J Jr (1996) Interspike intervals during interictal periods in human temporal lobe epilepsy. Brain Res 719(1–2):96–103
22. Cook MJ, O'Brien TJ, Berkovic SF, Murphy M, Morokoff A, Fabinyi G et al (2013) Prediction of seizure likelihood with a long-term, implanted seizure advisory system in patients with drug-resistant epilepsy: a first-in-man study. Lancet Neurol 12(6):563–571
23. de Weerd A, de Haas S, Otte A, Trenité DK, van Erp G, Cohen A et al (2004) Subjective sleep disturbance in patients with partial epilepsy: a questionnaire-based study on prevalence and impact on quality of life. Epilepsia 45(11):1397–1404
24. Diekelmann S, Born J (2010) The memory function of sleep. Nat Rev Neurosci 11(2):114–126
25. Eckhorn R (1994) Oscillatory and non-oscillatory synchronizations in the visual cortex and their possible roles in associations of visual features. Prog Brain Res 102:405–426
26. Ekstrom AD, Kahana MJ, Caplan JB, Fields TA, Isham EA, Newman EL, Fried I (2003) Cellular networks underlying human spatial navigation. Nature 425(6954):184–188
27. Elger CE, Helmstaedter C, Kurthen M (2004) Chronic epilepsy and cognition. Lancet Neurol 3(11): 663–672
28. Elger CE, Mormann F (2013) Seizure prediction and documentation-two important problems. Lancet Neurol 12(6):531–532
29. Engel J (2011) Biomarkers in epilepsy: introduction. Biomark Med 5(5):537–544
30. Fisher RS, Engel JJ (2010) Definition of the postictal state: when does it start and end? Epilepsy Behav 19(2):100–104
31. Freeman WJ, Holmes MD, Burke BC, Vanhatalo S (2003) Spatial spectra of scalp EEG and EMG from awake humans. Clin Neurophysiol 114(6): 1053–1068
32. Gibbs FA, Gibbs EL (1941) Atlas of electroencephalography. Lew A. Cummings Co., Cambridge, MA
33. Gloor P, Ball G, Schaul N (1977) Brain lesions that produce delta waves in the EEG. Neurology 27(4): 326–329
34. He BJ, Zempel JM, Snyder AZ, Raichle ME (2010) The temporal structures and functional significance of scale-free brain activity. Neuron 66(3):353–369
35. Holmes GL, Lenck-Santini PP (2006) Role of interictal epileptiform abnormalities in cognitive impairment. Epilepsy Behav 8(3):504–515
36. Kanner AM (2011) Anxiety disorders in epilepsy: the forgotten psychiatric comorbidity. Epilepsy Curr 11(3):90–91
37. Kleen JK, Scott RC, Holmes GL, Lenck-Santini PP (2010) Hippocampal interictal spikes disrupt cognition in rats. Ann Neurol 67(2):250–257
38. Kleen JK, Scott RC, Holmes GL, Roberts DW, Rundle MM, Testorf M et al (2013) Hippocampal interictal epileptiform activity disrupts cognition in humans. Neurology 81(1):18–24
39. Kleen JK et al (2012) Cognitive and behavioral co-morbidities of epilepsy. In: Jasper's basic mechanisms of the epilepsies. National Center for Biotechnology Information (US)
40. Kreiman G, Koch C, Fried I (2000) Category-specific visual responses of single neurons in the human medial temporal lobe. Nat Neurosci 3(9):946–953
41. Kreiman G, Koch C, Fried I (2000) Imagery neurons in the human brain. Nature 408(6810):357–361
42. Lieb JP, Joseph JP, Engel J Jr, Walker J, Crandall PH (1980) Sleep state and seizure foci related to depth spike activity in patients with temporal lobe epilepsy. Electroencephalogr Clin Neurophysiol 49(5–6): 538–557
43. Linkenkaer-Hansen K, Nikouline VV, Palva JM, Ilmoniemi RJ (2001) Long-range temporal correlations and scaling behavior in human brain oscillations. J Neurosci 21(4):1370–1377
44. Logothetis NK, Kayser C, Oeltermann A (2007) In vivo measurement of cortical impedance spectrum in monkeys: Implications for signal propagation. Neuron 55(5):809–823
45. Manning JR, Jacobs J, Fried I, Kahana MJ (2009) Broadband shifts in local field potential power spectra are correlated with single-neuron spiking in humans. J Neurosci 29(43):13613–13620
46. Matsumoto JY, Stead M, Kucewicz MT, Matsumoto AJ, Peters PA, Brinkmann BH et al (2013) Network oscillations modulate interictal epileptiform spike rate during human memory. Brain 136(Pt 8):2444–2456
47. Mormann F, Andrzejak RG, Elger CE, Lehnertz K (2007) Seizure prediction: the long and winding road. Brain 130(Pt 2):314–333
48. Mormann F, Lehnertz K, David P, Elger CE (2000) Mean phase coherence as a measure for phase synchronization and its application to the EEG of epilepsy patients. Physica D: Nonlinear Phenom 144(3):358–369
49. Nabbout R, Dulac O (2003) Epileptic encephalopathies: a brief overview. J Clin Neurophysiol 20(6): 393–397

50. Neidermeyer E, Da Silva FL (2005) Electroencephalography: basic principals, clinical applications, and related fields. Lippincott and Wilkins, Philadelphia

51. Nordli DR, Moshé SL, Shinnar S, Hesdorffer DC, Sogawa Y, Pellock JM et al (2012) Acute EEG findings in children with febrile status epilepticus results of the FEBSTAT study. Neurology 79(22): 2180–2186

52. Parish LM, Worrell GA, Cranstoun SD, Stead SM, Pennell P, Litt B (2004) Long-range temporal correlations in epileptogenic and non-epileptogenic human hippocampus. Neuroscience 125(4):1069–1076

53. Penfield J (1954) Epilepsy and the functional anatomy of the human brain. Little Brown, Boston

54. Piperidou C, Karlovasitou A, Triantafyllou N, Terzoudi A, Constantinidis T, Vadikolias K et al (2008) Influence of sleep disturbance on quality of life of patients with epilepsy. Seizure 17(7):588–594

55. Plenz D, Thiagarajan TC (2007) The organizing principles of neuronal avalanches: cell assemblies in the cortex? Trends Neurosci 30(3):101–110

56. Rasch B, Born J (2013) About sleep's role in memory. Physiol Rev 93(2):681–766

57. Ravagnati L, Halgren E, Babb TL, Crandall PH (1979) Activity of human hippocampal formation and amygdala neurons during sleep. Sleep 2(2):161–173

58. Reiher J, Beaudry M, Leduc CP (1989) Temporal intermittent rhythmic delta activity (TIRDA) in the diagnosis of complex partial epilepsy: sensitivity, specificity and predictive value. Can J Neurol Sci 16(4):398–401

59. Sammaritano M, Gigli GL, Gotman J (1991) Interictal spiking during wakefulness and sleep and the localization of foci in temporal lobe epilepsy. Neurology 41(2 (Pt 1)):290–297

60. Sanchez-Vives MV, McCormick DA (2000) Cellular and network mechanisms of rhythmic recurrent activity in neocortex. Nat Neurosci 3(10):1027–1034

61. Schevon CA, Cappell J, Emerson R, Isler J, Grieve P, Goodman R et al (2007) Cortical abnormalities in epilepsy revealed by local EEG synchrony. Neuroimage 35(1):140–148

62. Schindler K, Leung H, Elger CE, Lehnertz K (2007) Assessing seizure dynamics by analysing the correlation structure of multichannel intracranial EEG. Brain 130(Pt 1):65–77

63. Shewmon DA, Erwin RJ (1988) Focal spike-induced cerebral dysfunction is related to the after-coming slow wave. Ann Neurol 23(2):131–137

64. Siapas AG, Wilson MA (1998) Coordinated interactions between hippocampal ripples and cortical spindles during slow-wave sleep. Neuron 21(5): 1123–1128

65. Staba RJ, Wilson CL, Bragin A, Fried I, Engel J (2002) Quantitative analysis of high-frequency oscillations (80–500 hz) recorded in human epileptic hippocampus and entorhinal cortex. J Neurophysiol 88(4):1743–1752

66. Staba RJ, Wilson CL, Bragin A, Fried I, Engel J Jr (2002) Sleep states differentiate single neuron activity recorded from human epileptic hippocampus, entorhinal cortex, and subiculum. J Neurosci 22(13):5694–5704

67. Staba RJ, Wilson CL, Fried I, Engel J Jr (2002) Single neuron burst firing in the human hippocampus during sleep. Hippocampus 12(6):724–734

68. Tao JX, Chen XJ, Baldwin M, Yung I, Rose S, Frim D et al (2011) Interictal regional delta slowing is an EEG marker of epileptic network in temporal lobe epilepsy. Epilepsia 52(3):467–476

69. Téllez-Zenteno JF, Dhar R, Hernandez-Ronquillo L, Wiebe S (2007) Long-term outcomes in epilepsy surgery: antiepileptic drugs, mortality, cognitive and psychosocial aspects. Brain 130(Pt 2):334–345

70. Thivierge JP, Cisek P (2008) Nonperiodic synchronization in heterogeneous networks of spiking neurons. J Neurosci 28(32):7968–7978

71. Timofeev I, Grenier F, Bazhenov M, Sejnowski TJ, Steriade M (2000) Origin of slow cortical oscillations in deafferented cortical slabs. Cereb Cortex 10(12): 1185–1199

72. Tononi G, Cirelli C (2003) Sleep and synaptic homeostasis: a hypothesis. Brain Res Bull 62(2):143–150

73. Tort AB, Komorowski R, Eichenbaum H, Kopell N (2010) Measuring phase-amplitude coupling between neuronal oscillations of different frequencies. J Neurophysiol 104(2):1195–1210

74. Towle VL, Carder RK, Khorasani L, Lindberg D (1999) Electrocorticographic coherence patterns. J Clin Neurophysiol 16(6):528–547

75. Uhlhaas PJ, Singer W (2010) Abnormal neural oscillations and synchrony in schizophrenia. Nat Rev Neurosci 11(2):100–113

76. Walter G (1936) The location of cerebral tumors by electroencephalography. Lancet 2:305–308

77. Warren CP, Hu S, Stead M, Brinkmann BH, Bower MR, Worrell GA (2010) Synchrony in normal and focal epileptic brain: the seizure onset zone is functionally disconnected. J Neurophysiol 104(6):3530–3539

78. Wendling F, Bartolomei F, Bellanger JJ, Bourien J, Chauvel P (2003) Epileptic fast intracerebral EEG activity: evidence for spatial decorrelation at seizure onset. Brain 126(Pt 6):1449–1459

79. Whittingstall K, Logothetis NK (2009) Frequency-band coupling in surface EEG reflects spiking activity in monkey visual cortex. Neuron 64(2):281–289

80. Worrell GA, Cranstoun SD, Echauz J, Litt B (2002) Evidence for self-organized criticality in human epileptic hippocampus. Neuroreport 13(16):2017–2021

81. Zaveri HP, Pincus SM, Goncharova II, Duckrow RB, Spencer DD, Spencer SS (2009) Localization-related epilepsy exhibits significant connectivity away from the seizure-onset area. Neuroreport 20(9):891–895

82. Zhou JL, Lenck-Santini PP, Zhao Q, Holmes GL (2007) Effect of interictal spikes on single-cell firing patterns in the hippocampus. Epilepsia 48(4):720–731

What Is a Seizure Focus?

J. Victor Nadler and Dennis D. Spencer

Abstract

The seizure focus is the site in the brain from which the seizure originated and is most likely equivalent to the epileptogenic zone, defined as the area of cerebral cortex indispensable for the generation of clinical seizures. The boundaries of this region cannot be defined at present by any diagnostic test. Imaging and EEG recording can define regions of functional deficit during the interictal period, regions that generate interictal spikes, regions responsible for the ictal symptoms, regions from which the seizure is triggered, and regions of structural damage. However, these regions define the epileptogenic zone only when they are spatially concordant. The frequent discrepancies suggest the essential involvement of synaptically connected regions, that is a distributive focus, in the origination of most seizures. Here we review supporting evidence from animal studies and studies of persons undergoing surgical resection for medically-intractable epilepsy. We conclude that very few of the common seizures are truly local, but rather depend on nodal interactions that permit spontaneous network excitability and behavioral expression. Recognition of the distributive focus underlying most seizures has motivated many surgical programs to upgrade their intracranial studies to capture activity in as much of the network as possible.

Keywords

Epilepsy • Seizure focus • Epileptogenic zone • Neural network • Simple (elementary) partial seizure • Complex partial seizure • Primary generalized seizure • Epilepsy surgery

J.V. Nadler (✉)
Department of Pharmacology and Cancer Biology,
Duke University Medical Center,
Box 3813, Durham, NC 27710, USA
e-mail: nadle002@acpub.duke.edu

D.D. Spencer
Department of Neurosurgery, Yale University
School of Medicine, Box 208082, New Haven,
CT 06520, USA

H.E. Scharfman and P.S. Buckmaster (eds.), *Issues in Clinical Epileptology: A View from the Bench*,
Advances in Experimental Medicine and Biology 813, DOI 10.1007/978-94-017-8914-1_4,
© Springer Science+Business Media Dordrecht 2014

4.1   Seizure Focus: Relation to Focal Neocortical Abnormalities

The seizure focus is usually defined as the site in the brain from which the seizure originated or, in the case of focal seizure discharge, the totality of the tissue involved. Localizing the site of seizure onset is critical to the understanding of seizure mechanisms and to the probability of "curing" the epilepsy through surgical resection. The goal of resective surgery is to remove or disconnect the epileptogenic zone, defined as the area of cortex indispensable for the generation of clinical seizures [33]. Should the epileptogenic zone then be considered equivalent to the seizure focus? Most likely yes, but unfortunately the boundaries of this region cannot be defined at present by any diagnostic test. Instead, imaging and EEG recording methods define cortical regions related to, but not necessarily contiguous with, the epileptogenic zone. These regions include the symptomatogenic region (the region that when activated by epileptiform discharge produces the ictal symptoms), the irritative region (the region(s) that generate(s) interictal spikes), the ictal onset region (the region from which the seizure is triggered), the region of functional deficit (the region that is functionally abnormal during the interictal period), and the epileptogenic lesion (structural damage that may be causally related to the seizures). When these regions are spatially concordant, they define the location of the epileptogenic zone (or focus). Frequently, however, there are discrepancies. This finding suggests the essential involvement of synaptically interconnected regions in the origination of most seizures. In these instances, the epileptogenic zone may be composed of the pacemaker or ictal onset region and one or more relay areas required to produce the ictal symptoms [13, 37, 44]. These are qualities more often attributed to generalized seizures. If not only generalized seizures but even the generation of many focal seizures requires activity within a spatially distributed network, then our concept of a seizure focus requires some reassessment.

The focus is related to, but is not synonymous with, the ictal onset region. Normal brain function requires the correct balance between excitatory and inhibitory processes at every moment in time, and any disruption in this balance that favors excitation over inhibition promotes synchronous discharge of principal neurons. Thus seizures can be evoked in normal brain under conditions that promote such disruption. In focal epilepsy, the excitatory/inhibitory balance is disrupted chronically in some region or regions of the cerebral cortex, such that synchronous discharges arise under appropriate conditions. If these discharges are sufficient to provoke clinical symptoms, the region of chronic imbalance may be regarded as the ictal onset region. The ictal onset region is normally silent, but infrequently generates synchronous action potentials that can evoke afterdischarges. Its location can be approximated by EEG recording or SPECT imaging, but cannot be precisely defined. A great deal of research on animal models of epilepsy in the last 40 years has identified numerous abnormalities that under certain conditions can support the generation of episodic afterdischarges. Pathologic mechanisms found to promote hyperexcitability and seizure generation in animal models include loss or dysfunction of inhibitory neurons [1, 5, 21], creation or expansion of recurrent excitatory circuits [26], dysfunctional Na^+ and/or K^+ channels [12, 14, 31], enhanced intrinsic bursting [34, 35, 43], abnormal expression of HCN channels [29, 32], and altered glial regulation of extracellular fluid composition [9, 15, 39]. Typically, the process of epileptogenesis causes multiple functional and usually also anatomical changes in some region of brain that then becomes a locus for ictal onset. The relative importance of the various changes reported is unclear and is currently an area of active investigation. It is also possible that a single abnormality might trigger episodic seizure discharge when the ictal onset region is stimulated strongly enough or stimulated at an appropriate frequency. Synchronous firing of principal neurons in the ictal onset region provokes afterdischarge in the epileptogenic zone. The epileptogenic zone may be larger or smaller than the ictal onset region. It may include more than one potential ictal onset region differentiated by threshold. The most readily activated region will normally trigger all the seizures, but regions of higher threshold may become evident if the low threshold region is resected or inhibited selectively. Conversely, resection of the entire epileptogenic zone would eliminate any clinical seizures even if a residual ictal onset region remains intact.

The epileptogenic zone or focus may also be distinguished from any structural lesion detected by MRI or histopathology, as well as from the symptomatogenic zone. Not all cortical lesions detected in a patient or animal having epileptic seizures are themselves epileptogenic. Additional testing is necessary to determine which, if any, is essential for the generation of seizures, and this is not often done. Furthermore, the epileptogenic zone may be larger or smaller than the anatomic lesion. When the epileptogenic zone includes only part of an anatomic lesion, the remaining lesion may not be capable of generating a seizure or may be capable only when driven by the portion of the lesion having a lower afterdischarge threshold. The epileptogenic zone may also extend beyond the anatomic lesions. For example, seizures may arise not because of the lesion itself, but rather from changes in surrounding cortical tissue induced by the lesion. In particular, tumors and vascular malformations induce foci of this type. Alternatively, MRI and even histopathology may not be sensitive enough to detect epileptogenic microlesions that extend some distance from the visible lesion.

Clinical symptoms will arise from the epileptogenic zone only if that zone includes a region of "eloquent cortex," that is, cortex related to a specific function. Most of the human cortex is symptomatically silent, implying that seizures arise from activation of an epileptogenic zone primarily when epileptiform discharge propagates to a region of eloquent cortex with sufficient strength to elicit clinical symptoms. Thus cortical lesions and ictal symptomatology may, but usually do not, define the seizure focus.

To this point, the discussion of seizure focus has been limited to simple (or elementary) partial seizures, defined as seizures that originate from a limited region of the cerebral cortex and do not impair consciousness. The seizure may be provoked by hyperactivity that occurs spontaneously within a highly localized region or may require activation of a distributed network. The concept of a distributive onset for many simple partial seizures links circuitry and mechanisms of these seizures with other seizure types suggested to arise from network activity.

4.2 Focus of Complex Partial Seizures

Complex partial seizures, like simple or elementary partial seizures, originate from a limited region of the cerebral cortex (which includes the limbic system), but cause an impairment of consciousness. These seizures typically involve the temporal lobe, but some persons with epilepsy experience extratemporal complex partial seizures. Although attempts to associate epilepsy with a pathological substrate date from antiquity, Bouchet and Cazauvieilh [10] were the first to describe a "palpable induration of the temporal lobes" in a group of persons with epilepsy. Hughlings Jackson [18] later stimulated decades of searching for a single pathologic source when he replaced the term "psychomotor epilepsy" with temporal lobe epilepsy in the case of Dr. Z who died after having experienced seizures for many years. At autopsy, Hughlings Jackson suspected a lesion in the "taste regions of Ferrier," and softening of the left uncinate gyrus was indeed discovered. Thus a mechanistic approach to ascribing epileptogenic causality to "the organ of the mind" was born.

Even earlier, Sommer [36] had concluded that hippocampal pathology (neuronal loss and gliosis; hippocampal sclerosis) is an important etiological factor in the subsequent development of seizures. The relationship between hippocampal damage and complex partial seizures has been debated ever since. To what extent does hippocampal pathology lead to epileptic attacks and to what extent do repeated seizures occurring over a period of years result in this pathology [25]? Finally, with the development of EEG, paroxysmal changes were found in the temporal cortex of patients with psychomotor seizures and Jasper et al. [20] used the term "temporal lobe epilepsy" to define these regional electrographic abnormalities.

Interictal spikes in the temporal region prompted some groups to resect portions of the temporal lobe in patients with medically-intractable complex partial seizures. Epileptogenic causality of hippocampal sclerosis remained controversial, however, until Margerison and Corsellis [22] combined EEG, clinical, and autopsy findings in patients institutionalized for epilepsy, and found hippocampal sclerosis in 30 of 34 patients in whom premortem EEG had revealed anterior temporal spikes. The uniformity of neuronal cell loss and gliosis, particularly in hippocampal area CA1, in about 70 % of patients who were subjected to temporal lobectomy because depth electrode recordings had indicated medial temporal ictal onset and the 75 % control of seizures in those patients appeared to corroborate the growing sense that hippocampal sclerosis was the cause of temporal lobe epilepsy. In the mid-1980s to the early 1990s, MRI revolutionized the diagnosis of epilepsy, and unilateral hippocampal atrophy associated with interictal or ictal onset from one temporal lobe replaced most intracranial studies in diagnosing medial temporal lobe epilepsy. As MRI began to reveal an assortment of pathologies associated with suspected focal epileptogenesis, most, if not all, cases of temporal lobe epilepsy were ascribed to well-defined substrate pathologies of mesial temporal sclerosis, neoplasms, vascular lesions, developmental abnormalities, or gliosis from trauma, inflammation, etc. These findings coupled with the intrinsic excitability of the sclerotic hippocampus seen in depth electrode studies and human slice electrophysiology led many in the field to anticipate increasingly better surgical outcomes over time. However, surgical outcomes have not changed dramatically. In fact, several clinical and research observations have emphasized that epileptogenesis and ictal behavior is very likely a network phenomenon of aberrant nodes (review: [42]).

Resection of the anterior temporal lobe on the side of seizure onset, particularly removal of the hippocampus and amygdala, usually leads to cessation or at least reduction in frequency of temporal complex partial seizures. It is therefore often assumed that the epileptogenic zone must be confined to these regions. Indeed hippocampal area CA3, the entorhinal cortex, and especially the basolateral amygdala exhibit a low threshold for seizure initiation when challenged with electrical stimulation or certain chemoconvulsants [6, 23, 45]. However, the anterior temporal lobe is a rather large block of tissue that includes distinct, but interconnected, brain regions. The epileptogenic zone has not been localized precisely and indeed a single point of onset may not exist. Seizure onset may arise from multiple sites within the temporal lobe [37, 38]. Wherever the onset, the remaining limbic regions are recruited rapidly, resulting in the same behavior regardless. Surgical outcomes also argue for a network as epileptogenic zone, rather than a localized focus. Bitemporal lobe epilepsy, proved by intracranial study, can be cured 50–60 % of the time by removal of only the more dysfunctional temporal lobe, indicating that the spontaneous ictal events in the contralateral lobe depended on network activation. In unilateral temporal lobe epilepsy, seizures are well-controlled after anterior temporal lobectomy in 75 % of patients, but control drops to 50 % when patients are followed for 10 years and falls even lower if antiseizure drugs are not administered. These results suggest that the entire epileptogenic zone is not being resected at least half the time. When one hippocampus is clearly responsible for ictal onset, a restricted mesial temporal resection (hippocampus + amygdala) yields poorer control (50–60 %) than a standard anterior temporal lobectomy. Also, in the most clearly lateralized cases, resection may stop the typical complex partial seizure, but the aura (sensation related to the ictal onset region) persists 15–20 % of the time. Finally, with the exception of some tumors and vascular lesions, patients almost always demonstrate distributed cognitive deficits on neuropsychological testing, again indicative of diseased network not a single diseased region.

Seizure onset may arise from multiple sites within the temporal lobe in animal models of epilepsy as well [7, 11]. In the kindling model, electrical stimulation of many sites within the limbic system evokes the same seizure type, suggesting that activation of the network *per se* is more important than the precise point at which the network is activated [16]. In the kainic acid model, seizures can originate from either the amygdala or hippocampus at different times in the same rat [30]. These and other observations led to the proposal of a distributive focus for temporal complex partial seizures, which includes the hippocampus, amygdala, entorhinal cortex, anterior and midline thalamic nuclei, and pyriform cortex [8]. This hypothesis is further supported by findings that, in animal models of epilepsy, neurons in all these regions exhibit functional changes expected to promote excitability, creating potential ictal onset regions. In addition, these same brain regions are sites of tissue damage. Although hippocampal sclerosis is the characteristic form of histopathologic damage demonstrable in most patients who have undergone anterior temporal lobectomy, neuronal loss, atrophy, and gliosis have also been reported in the amygdala, entorhinal cortex, and thalamus [3, 28, 46]. Similarly widespread damage to the limbic system is found in commonly used animal models [27, 40]. Pyriform cortex, which is not regularly examined in human tissue specimens, is typically also damaged in animals. Pro-epileptogenic changes documented in animal models include degeneration of inhibitory neurons with subsequent axon sprouting by surviving inhibitory neurons, formation of recurrent excitatory connections by principal neurons, enhanced synaptically-evoked and intrinsic bursting, and altered expression and function of diverse ion channels and neurotransmitter receptors. In animals induced to become epileptic by provoking status epilepticus these changes clearly arise as a consequence of acute seizures. One or more of these changes may also precede the development of epilepsy in genetic models and in models of post-traumatic epilepsy. Regardless of how they were brought about, however, they all probably contribute to the spontaneous seizures, changes in circuit function perhaps being required for the origination of the seizures and histopathologic changes perhaps enhancing their frequency and intensity [17, 47]. The existence of a distributed pathological sub-strate implies that seizures can arise at multiple points in the limbic circuit and that a certain minimum percentage of the circuit must be activated for the electrical activity to alter perception or behavior.

If temporal complex partial seizures can originate from any of several limbic regions, some of which lie outside the tissue normally resected for medically-intractable seizures, the distributive focus may explain, in part, the somewhat consistent percentage of temporal lobe surgeries (~30 %) that fail to achieve adequate seizure control [41]. Conversely, the success of many such surgeries may be attributable not so much to eliminating the seizure focus, but rather to removing enough of the limbic circuitry that synchronous firing in the remaining seizure onset regions fails to activate an epileptogenic zone.

4.3   Focus of Primary Generalized Seizures

Primary generalized seizures appear to begin simultaneously in both cerebral hemispheres when recorded by scalp EEG, but are probably driven, at least in part, by hyperactivity of subcortical structures. Simultaneous activation of both cerebral hemispheres causes behavioral and perceptual signs and symptoms to be manifested bilaterally and there is always some impairment of consciousness. Although primary generalized seizures are not generally thought of as arising from a distinct focus or epileptogenic zone, the concept of a distributive focus appears applicable. This is perhaps best illustrated by the mechanisms underlying absence seizures. The spike-wave discharges of absence seizures require circuit interactions between the thalamus and neocortex [2, 19, 24]. Interruption of this circuit, such as by cutting the reciprocal pathways that connect the two regions or by inactivating either region alone, abolishes the seizures. Neocortex supplies the excitatory drive that is organized into ictal discharge by bursting thalamic relay neurons. Thus thalamus and neocortex together can be said to constitute an epileptogenic zone with excitatory thalamic nuclei serving as the ictal onset region. Increasing evidence suggests the involvement of subcortical ictal onset regions in other forms of generalized epilepsy as well [4].

4.4   Conclusion

Evolution in the thinking about concordance or non-concordance of the seizure-related cortical regions, has revealed discrepancies between the pathology and physiology of seizures. Definition of the epileptogenic zone remains complex and elusive, and the outcomes of surgical approaches have plateaued. Increasing evidence suggests that epileptogenesis is distributed among multiple foci, usurping known anatomical and functional networks. Very few of the most common seizures may be truly local. Rather, they appear to depend on nodal interactions that permit spontaneous network excitability and behavioral expression. The epilepsy community has not yet succeeded in creating a new paradigm that combines the critical derangements of electrophysi-

ology, pathology, metabolism, genetics, and network communication. New approaches must include better correlation of human data with animal models, wherein the hyperexcitable networks can be more intensively studied and manipulated.

Given the difficulty of defining the epileptogenic zone, many surgical programs have upgraded their intracranial studies to improve analysis of the distributed network. The Yale program, for example, has utilized advanced imaging and navigation systems to increase the number of electrode contacts per patient from <100 in 1991 to 200–250 in 2006 and to even greater numbers since then. Utilization of these electrode arrays has demonstrated many examples of distributive foci, such as those shown in Fig. 4.1. In these instances, the foci are located in functional networks revealed by fMRI – the cognitive control network between lateral parietal and frontal lobes (Fig. 4.1a) and a portion of what has been labeled the "default network," observed reproducibly when subjects are at rest and not engaged in a task (Fig. 4.1b). Determinations of regional

a   b

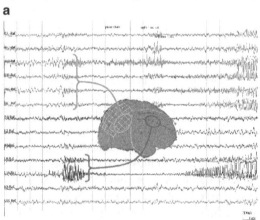
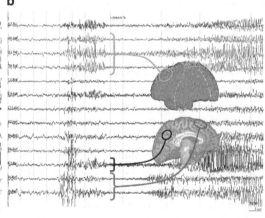
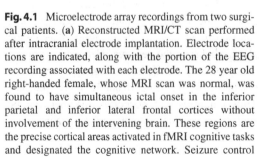

Fig. 4.1 Microelectrode array recordings from two surgical patients. (**a**) Reconstructed MRI/CT scan performed after intracranial electrode implantation. Electrode locations are indicated, along with the portion of the EEG recording associated with each electrode. The 28 year old right-handed female, whose MRI scan was normal, was found to have simultaneous ictal onset in the inferior parietal and inferior lateral frontal cortices without involvement of the intervening brain. These regions are the precise cortical areas activated in fMRI cognitive tasks and designated the cognitive network. Seizure control

was effected by resection of both ictal onset regions. (**b**) Similar superposition of MRI/CT scan, electrode locations, and EEG recording in a second patient. The 30 year old right-handed male, whose MRI scan was also normal, was found to have independent ictal onset in the posterior, medial frontal, and media parietal lobes, with the medial frontal cortex initiating the same behavioral seizure more frequently. Initial treatment with a neurostimulator little affected the behavioral seizures. Subsequent resection of the medial frontal node alone was sufficient to control the seizures extracellular glutamate concentration and metabolic/energetic studies that utilize 7 T MRS have further supported the concept of a distributive focus. At present, unless patients have a clear tumor or cavernoma on MRI or concordance of all data (electrophysiology, mesial temporal sclerosis on MRI, and neuropsychological studies) indicating unilateral temporal lobe epilepsy, intracranial studies are always performed and directed by anatomic MRI, dynamic imaging (FMR, ictal SPECT, MRS), AVEEG, and seizure semiology for distributed electrode placement. It is only by this persistent adaptation of newly developed technology that we can hope to one day understand and properly treat human epilepsy.

Acknowledgment We wish to acknowledge the unparalleled leadership provided by Phil Schwartzkroin in support of basic epilepsy research. His work on the relationship between seizures and synaptic inhibition strongly influenced the search for mechanisms of epileptogenesis in the animal models that emerged during the same period and the subsequent studies pursued in his laboratory taught us a great deal about both the normal and pathological functions of the hippocampus. We have profited enormously from Phil's promotion of basic research through the American Epilepsy Society, including a term as president, his editorship of *Epilepsia*, his organizing numerous symposia, conferences, and books, and his exemplary training of outstanding young basic and translational neuroscientists.

References

1. Avoli M, de Curtis M (2011) GABAergic synchronization in the limbic system and its role in the generation of epileptiform activity. Prog Neurobiol 95:104–132
2. Avoli M, Gloor P (1982) Interaction of cortex and thalamus in spike and wave discharges of feline generalized penicillin epilepsy. Exp Neurol 76:196–217
3. Babb T, Brown W (1987) Pathological findings in epilepsy. In: Engel J (ed) Surgical treatment in epilepsy. Raven, New York, pp 511–540
4. Badawy RAB, Lai A, Vogrin SJ, Cook MJ (2013) Subcortical epilepsy? Neurology 80:1901–1907
5. Ben-Ari Y (2006) Seizures beget seizures: the quest for GABA as a key player. Crit Rev Neurobiol 18:135–144
6. Ben-Ari Y, Tremblay E, Riche D, Ghilini G, Naquet R (1981) Electrographic, clinical and pathological alterations following systemic administration of kainic acid, bicuculline or pentetrazole: metabolic mapping using the deoxyglucose method with special reference to the pathology of epilepsy. Neuroscience 6:1361–1391
7. Bertram EH (1997) The functional anatomy of spontaneous seizures in a rat model of chronic limbic epilepsy. Epilepsia 38:95–105
8. Bertram EH (2009) Temporal lobe epilepsy: where do the seizures really begin? Epilepsy Behav 14(Suppl 1):32–37
9. Boison D (2008) The adenosine kinase hypothesis of epileptogenesis. Prog Neurobiol 84:249–262
10. Bouchet C, Cazauvieilh JB (1825) De l'épilepsie considéré dans ses rapports avec l'aliénation mentale. Arch Gen Med 9:510–542
11. Bragin A, Engel J, Wilson CL, Vizentin E, Mathern GW (1999) Electrophysiologic analysis of a chronic seizure model after unilateral hippocanpal KA injection. Epilepsia 40:1210–1221
12. Catterall WA (2012) Sodium channel mutations in epilepsy. In: Noebels JL, Avoli M, Rogawski MA, Olsen RW, Delgado-Escueta AV (eds) Jasper's basic mechanisms of the epilepsies, 4th edn. Oxford, New York, pp 675–687
13. Chauvel P, Buser P, Badier JM, Liegois-Chauvel C, Marquis P, Bancaud J (1987) The "epileptogenic zone" in humans: representation of intercritical events by spatio-temporal maps. Rev Neurol (Paris) 143:443–450
14. Cooper EC (2012) Potassium channels (including KCNQ) and epilepsy. In: Noebels JL, Avoli M, Rogawski MA, Olsen RW, Delgado-Escueta AV (eds) Jasper's basic mechanisms of the epilepsies, 4th edn. Oxford, New York, pp 55–65
15. de Lanerolle NC, Lee TS, Spencer DD (2010) Astrocytes and epilepsy. Neurotherapeutics 7:424–438
16. Goddard GV, McIntyre DC, Leech CK (1969) A permanent change in brain function resulting from daily electrical stimulation. Exp Neurol 25:295–330
17. Gorter JA, van Vliet EA, Aronica E, Lopes da Silva FH (2001) Progression of spontaneous seizures after status epilepticus is associated with mossy fibre sprouting and extensive bilateral loss of hilar parvalbumin and somatostatin-immunoreactive neurons. Eur J Neurosci 3:657–669
18. Hughlings Jackson J, Colman WS (1898) Case of epilepsy with tasting movements and 'dreamy state' – very small patch of softening in the left uncinate gyrus. Brain 21:580–590
19. Huguenard JR (1999) Neuronal circuitry of thalamocortical epilepsy and mechanisms of antiabsence drug action. Adv Neurol 9:991–999
20. Jasper H, Pertuisset B, Flanigin H (1951) EEG and cortical electrograms in patients with temporal lobe seizures. Arch Neurol Psychiatr 65:272–290
21. Maglóczky Z, Freund TF (2005) Impaired and repaired inhibitory circuits in the epileptic human hippocampus. Trends Neurosci 28:334–340

22. Margerison JH, Corsellis JAN (1966) Epilepsy and the temporal lobes. Brain 89:499–530

23. McNamara JO, Byrne MC, Dasheiff RM, Fitz JG (1980) The kindling model of epilepsy: a review. Prog Neurobiol 15:139–159

24. Meeren HK, Pijn JP, Van Luijtelaar EL, Coenen AM, Lopes da Silva FH (2002) Cortical focus drives widespread corticothalamic networks during spontaneous absence seizures in rats. J Neurosci 22:1480–1495

25. Nadler JV (1989) Seizures and neuronal cell death in epilepsy. In: Chan-Palay V, Köhler C (eds) The hippocampus – new vistas. Liss, New York, pp 463–481

26. Nadler JV (2009) Axon sprouting in epilepsy. In: Schwartzkroin PA (ed) Encyclopedia of epilepsy research, vol 3. Academic, Oxford, pp 1143–1148

27. Nadler JV, Perry BW, Gentry C, Cotman CW (1980) Degeneration of hippocampal CA3 pyramidal cells induced by intraventricular kainic acid. J Comp Neurol 192:333–359

28. Natsume J, Bernasconi N, Andermann F, Bernasconi A (2003) MRI volumetry of the thalamus in temporal, extratemporal, and idiopathic generalized epilepsy. Neurology 60:1296–1300

29. Noam Y, Bernard C, Baram TZ (2011) Towards an integrated view of HCN channel role in epilepsy. Curr Opin Neurobiol 21:873–879

30. Okazaki MM, Nadler JV (1988) Protective effects of mossy fiber lesions against kainic acid-induced seizures and neuronal degeneration. Neuroscience 26:763–781

31. Oliva M, Berkovic SF, Petrou S (2012) Sodium channels and the neurobiology of epilepsy. Epilepsia 53:1849–1859

32. Reid CA, Phillips AM, Petrou S (2012) HCN channelopathies: pathophysiology in genetic epilepsy and therapeutic implications. Br J Pharmacol 165:49–56

33. Rosenow F, Lüders H (2001) Presurgical evaluation of epilepsy. Brain 124:1683–1700

34. Sanabria ER, da Silva AV, Spreafico R, Cavalheiro EA (2002) Damage, reorganization, and abnormal cortical excitability in the pilocarpine model of temporal lobe epilepsy. Epilepsia 43(Suppl 5):96–106

35. Sanabria ER, Su H, Yaari Y (2001) Initiation of network bursts by Ca^{2+}-dependent intrinsic bursting in the rat pilocarpine model of temporal lobe epilepsy. J Physiol 532:205–216

36. Sommer W (1880) Erkrankung des Ammon's horn als aetiologis ches moment der epilepsien. Arch Psychiatr Nurs 10:631–675

37. Spencer SS (2002) Neural networks in human epilepsy: evidence of and implications for treatment. Epilepsia 43:219–227

38. Spenser SS, Spencer DD (1994) Entorhinal-hippocampal interaction in medial temporal epilepsy. Epilepsia 35:721–727

39. Steinhäuser C, Seifert G, Bedner P (2012) Astrocyte dysfunction in temporal lobe epilepsy: K^+ channels and gap junction coupling. Glia 60:1192–1202

40. Turski WA, Cavalheiro EA, Schwarz M, Czuczwar SJ, Kleinrok Z, Tursky L (1983) Limbic seizures produced by pilocarpine in rats: behavioural, electroencephalographic and neuropathological study. Behav Brain Res 9:315–335

41. Vale FL, Pollock G, Benbadis SR (2012) Failed epilepsy surgery for mesial temporal lobe sclerosis: a review of the pathophysiology. Neurosurg Focus 32:(E9)1–6

42. van Diessen E, Diederen SJH, Braun KPJ, Jansen FE, Stam CJ (2013) Functional and structural brain networks in epilepsy: what have we learned? Epilepsia 54:1855–1865

43. Wellmer J, Su H, Beck H, Yaari Y (2002) Long-lasting modification of intrinsic discharge properties in subicular neurons following status epilepticus. Eur J Neurosci 16:259–266

44. Wendling F, Chauvel P, Biraben A, Bartolomei F (2010) From intracerebral EEG signals to brain connectivity: identification of epileptogenic networks in partial epilepsy. Front Syst Neurosci 4:(A154)1–13

45. Westbrook GL, Lothman EW (1983) Cellular and synaptic basis of kainic acid-induced hippocampal epileptiform activity. Brain Res 273:97–109

46. Yilmazer-Hanke DM, Wolf HK, Schramm J, Elger CE, Wiestler OD, Blümcke I (2000) Subregional pathology of the amygdala complex and entorhinal region in surgical specimens from patients with pharmacoresistant temporal lobe epilepsy. J Neuropathol Exp Neurol 59:907–920

47. Zhang X, Cui S-S, Wallace AE, Hannesson DK, Schmued LC, Saucier DM, Honer WG, Corcoran ME (2002) Relations between brain pathology and temporal lobe epilepsy. J Neurosci 22:6052–6061

What Is a Seizure Network? Long-Range Network Consequences of Focal Seizures

5

Hal Blumenfeld

Abstract

What defines the spatial and temporal boundaries of seizure activity in brain networks? To fully answer this question a precise and quantitative definition of seizures is needed, which unfortunately remains elusive. Nevertheless, it is possible to ask under conditions where clearly divergent patterns of activity occur in large-scale brain networks whether certain activity patterns are part of the seizure while others are not. Here we examine brain network activity during focal limbic seizures, including diverse regions such as the hippocampus, subcortical arousal systems and fronto-parietal association cortex. Based on work from patients and from animal models we describe a characteristic pattern of intense increases in neuronal firing, cerebral blood flow, cerebral blood volume, blood oxygen level dependent functional magnetic resonance imaging (BOLD fMRI) signals and cerebral metabolic rate of oxygen consumption in the hippocampus during focal limbic seizures. Similar increases are seen in certain closely linked subcortical structures such as the lateral septal nuclei and anterior hypothalamus, which contain inhibitory neurons. In marked contrast, decreases in all of these parameters are seen in the subcortical arousal systems of the upper brainstem and intralaminar thalamus, as well as in the fronto-parietal association cortex. We propose that the seizure proper can be defined as regions showing intense increases, while those areas showing opposite changes are inhibited by the seizure network and constitute long-range network consequences beyond the seizure itself. Importantly, the fronto-parietal cortex shows sleep-like slow wave activity and depressed metabolism under these conditions, associated with

H. Blumenfeld (✉)
Department of Neurology, Neurobiology
and Neurosurgery, Yale University School
of Medicine, 333 Cedar Street, New Haven,
CT 06520, USA
e-mail: hal.blumenfeld@yale.edu

H.E. Scharfman and P.S. Buckmaster (eds.), *Issues in Clinical Epileptology: A View from the Bench*,
Advances in Experimental Medicine and Biology 813, DOI 10.1007/978-94-017-8914-1_5,
© Springer Science+Business Media Dordrecht 2014

impaired consciousness. Understanding which brain networks are directly involved in seizures versus which sustain secondary consequences can provide new insights into the mechanisms of brain dysfunction in epilepsy, hopefully leading to innovative treatment approaches.

Keywords

Epilepsy • Consciousness • Slow waves • Cortex • Thalamus • Sleep • Hippocampus • Pedunculopontine tegmental nucleus • Acetylcholine • Brainstem • Arousal

5.1   Introduction

Seizures are usually defined as an abnormal pattern of neuronal activity which includes excessive synchrony and high frequency firing of neurons. As in most definitions, the obvious cases are easy to recognize. However, in reality there are no distinct boundaries for precisely when neuronal activity become sufficiently synchronous or intense to be considered a seizure. The situation is complicated further by the fact that seizures occur in neuronal networks, which have both local and long-range effects. Network interactions give rise to abnormal activity in local circuits, but in some cases can also influence remote brain regions. Are these remote network changes part of the seizure proper, or are they "side effects" caused by the seizure but not directly involved in the seizure network? To answer this question it is necessary to identify characteristic features that are seen in seizure activity, and to then determine if these same features are present in the remote network regions. If similar characteristic features are present, then the remote regions are likely to be involved in propagation of the seizure itself. If the activity in the remote regions differs drastically from seizure activity, and instead resembles other well-known patterns of non-seizure brain activity (such as coma or sleep), then the activity in the remote region could be considered outside the seizure network, although influenced by it.

Temporal lobe seizures provide a concrete example of these local and long-range network phenomena. Locally, temporal lobe seizures produce high frequency rhythmic discharges. At the same time remote regions of the fronto-parietal association cortex exhibit 1–3 Hz slow wave activity resembling coma, sleep or encephalopathy [1–3]. Is this slow wave activity part of the seizure, or is it a distinct state of brain activity caused by the seizure? Here we will examine the detailed characteristics of these remote changes in neocortical networks during focal limbic seizures in both patients and in animal models, and also potential mechanisms for these phenomena. We conclude that these remote effects on neocortical networks are best considered outside the seizure network but strongly influenced by it. Analogous to post-ictal depression, which is closely related to and caused by the seizure itself but occurs at a different time, neocortical slow wave activity is closely related to and caused by focal limbic seizures but occurs in a different space.

5.2   Clinical Data

Intracranial recordings from patients with temporal lobe epilepsy show characteristic low voltage fast activity evolving into rhythmic polyspike-and-wave discharges in the medial temporal lobe limbic circuits, often extending into the lateral temporal cortex (Fig. 5.1c). Simultaneously, remote regions of the frontal and parietal association cortex often show 1–3 Hz slow wave activity (Fig. 5.1d). This ictal neocortical slow wave activity has been interpreted as a propagation pattern in temporal lobe epilepsy [1]. However, several features of the fronto-parietal slow wave activity make it likely that this is a distinct,

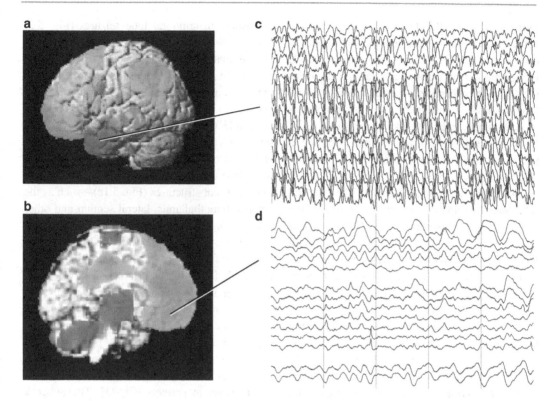

Fig. 5.1 Local and long-range network effects in temporal lobe complex partial seizures. (**a**, **b**) Group analysis of SPECT ictal-interictal difference imaging during temporal lobe seizures. CBF increases (*red*) are present in the temporal lobe (**a**) and in the medial thalamus (**b**). Decreases (*green*) are seen in the lateral frontoparietal association cortex (**a**) and in the interhemispheric frontoparietal regions (**b**). (**c**, **d**) Intracranial EEG recordings from a patient during a temporal lobe seizure. High frequency polyspike-and-wave seizure activity is seen in the temporal lobe (**c**). The orbital and medial frontal cortex (and other regions, EEG not shown) do not show polyspike activity, but instead large-amplitude, irregular slow rhythms resembling coma or sleep (**d**). *Vertical lines* in (**c**) and (**d**) denote 1-s intervals. Note that the EEG and SPECT data were from similar patients, but were not simultaneous, and are shown together here for illustrative purposes only ((**a**, **b**) Modified from Blumenfeld et al. [2] with permission. (**c**, **d**) Modified from Englot et al. [3] with permission)

remote network effect rather than simply seizure propagation, as we discuss below.

Recent work with multiunit recordings in human intracranial EEG has raised new questions about the definition of seizure activity vs. associated changes in surrounding regions. Schevon and colleagues showed that high frequency firing of neurons is highly localized in human seizures [4]. Accompanying local field potential changes measured by conventional intracranial EEG extend over a greater region, but may represent mainly synaptic activity without major changes in local firing of neurons [4]. Whether recording neuronal firing or local field potentials, at least these changes in the vicinity of seizure onset show

high frequency poly-spike activity characteristic of seizure physiology. In contrast, the slow wave activity occurring in distant fronto-parietal regions during temporal lobe seizures occur at a very different frequency (1–3 Hz) from ictal temporal lobe polyspike discharges (broad band >8 Hz) (Fig. 5.1c, d) [2, 3]. Seizure activity on intracranial EEG can be defined as high frequency discharges. Although scalp EEG often exhibits rhythmic theta or delta-frequency slow waves during local seizures [5] direct recording of seizure activity with intracranial electrodes inevitably shows high frequency discharges in these same regions. Therefore, when only slow wave activity is seen in a region *without* high frequency discharges on intracranial EEG, this likely does not represent seizures. As we discuss in the next section, detailed physiological studies from animal models further support this claim. Such slow wave activity seen in the fronto-parietal cortex during temporal lobe seizures is similar to cortical slow waves in deep sleep, coma or encephalopathy [6, 7]. In these states, cortical function and information processing is depressed, leading to impaired level of consciousness [8].

How does focal seizure activity in the temporal lobe lead to remote slow wave activity in the fronto-parietal association cortex? The anatomy and physiology of these changes differs from local "surround inhibition" described for focal cortical seizures [9, 10]. To affect distant lobes, long-range network interactions are required. Some initial clues for the mechanisms of these network changes have come from human cerebral blood flow (CBF) imaging with single photon computed tomography (SPECT) which, unlike fMRI, can be done successfully despite patient movement during seizures. As expected, ictal SPECT in temporal lobe seizures is associated with CBF increases in the temporal lobe (Fig. 5.1a). In addition, *decreases* are seen in frontal and parietal association cortex in the same regions which exhibit slow wave activity (Fig. 5.1a, b) [11–13]. Subcortical networks are also involved in temporal lobe seizures and SPECT imaging shows increases in the medial thalamus and midbrain (Fig. 5.1b) [13–16]. We found that the SPECT increases in the medial thalamus are correlated with the decrease in bilateral fronto-parietal cortex [13], suggesting a mechanistic link between subcortical changes and depressed cortical function in temporal lobe seizures. These long-range network changes in cortical and subcortical function are seen specifically in temporal lobe seizures with impaired consciousness [3, 13, 14, 17, 18]. In contrast, temporal lobe seizures without impaired consciousness are associated with localized seizure activity in the temporal lobe, without these long-range network changes [3, 13].

Based on these findings from patients, we proposed the *network inhibition hypothesis* to explain cortical dysfunction and impaired con-sciousness in temporal lobe seizures (Fig. 5.2) [19, 20]. Normal cortical function and consciousness is maintained by interactions between the cortex and subcortical arousal systems including the thalamus, brainstem and basal forebrain (Fig. 5.1a). Focal temporal lobe seizure activity in simple partial seizures does not have long-range network impact effects, so cortical function and consciousness are spared (Fig. 5.1b). In temporal lobe complex partial seizures, propagation to subcortical structures (Fig. 5.1c)—such as the anterior hypothalamus, lateral septum and other regions—inhibits subcortical arousal systems (Fig. 5.1d). This in turn removes cortical arousal leading to fronto-parietal slow wave activity and impaired level of consciousness. Note that according to this hypothesis, the cortical slow wave activity is not part of the seizure itself, but instead is a long-range network consequence of depressed subcortical arousal.

Further support for the network inhibition hypothesis has come from recent behavioral observations in patients [21–23]. The network inhibition hypothesis predicts that when focal seizures propagate to subcortical structures, this will cause severe and widespread cortical dysfunction. Therefore focal seizures are expected to usually be associated with either marked impairment of many cognitive functions due to depressed level of consciousness, or alternatively to spare most cognitive functions. In support of this hypothesis, we recently found that behavioral deficits in a wide range of verbal and non-verbal test items during partial seizures are bimodally distributed, such that most seizures either globally impair or spare cognition [21–24].

While human studies have provided clinically relevant correlations between physiology and behavioral changes, and suggest that ictal neocortical slow wave activity is distinct from direct seizure involvement, fundamental mechanistic studies are best performed in animal models. An experimental animal model could enable direct physiological measurements to determine if ictal neocortical slow wave activity is indeed distinct from seizure activity, and would allow further investigation of the mechanisms for this phenomenon.

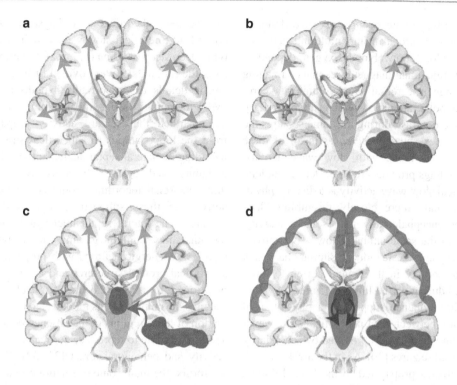

Fig. 5.2 Network inhibition hypothesis. (**a**) Under normal conditions, the upper-brainstem and diencephalic activating systems interact with the cerebral cortex to maintain normal consciousness. (**b**) A focal seizure involving the mesial temporal lobe. If the seizure remains localized, a simple partial seizure will occur without impairment of consciousness. (**c**) Seizure activity often spreads from the temporal lobe to midline subcortical structures and propagation often extends to the contralateral mesial temporal lobe (not shown). (**d**) Inhibition of subcortical arousal systems leads to depressed activity in bilateral frontoparietal association cortex and to loss of consciousness (Modified from Englot et al. [3] with permission)

5.3   Insights from an Experimental Animal Model

Rodent models of limbic seizures replicate many of the behavioral and physiological characteristics of human temporal lobe epilepsy [25–29]. We found that spontaneous focal limbic seizures in awake chronically epileptic rats following pilocarpine status epilepticus exhibited frontal neocortical 1–2 Hz slow wave activity and behavioral arrest similar to human complex partial temporal lobe seizures [30]. Ictal neocortical slow wave activity in this model resembled slow wave activity during natural slow wave sleep in the same animals. In contrast when limbic seizures secondarily generalized, recordings from the frontal cortex showed 9–12 Hz polyspike discharges characteristic of ictal activity, instead of slow waves.

Additional physiological and neuroimaging experiments were performed in an acute lightly anesthetized rat model in which seizures could be induced under controlled conditions [30]. Seizures were induced by brief 2 s stimulus trains at 60 Hz to the hippocampus under ketamine/ xylazine anesthesia reduced to a stage where the cortex showed physiology near to the waking state. Under these conditions, induced partial limbic seizures produced frontal cortical slow wave activity similar to that seen in awake chronically epileptic rats. This acute model enabled detailed physiological measurements to distinguish ictal neocortical slow waves from seizure activity. Measurements from the hippocampus during partial limbic seizures revealed dramatic increases in neuronal firing (multiunit activity), cerebral blood flow, blood oxygen dependent (BOLD) functional magnetic resonance imaging (fMRI) signals, cerebral blood volume, and cerebral metabolic rate of oxygen consumption [30]. In marked contrast, during the same seizures the frontal cortex showed *decreases* in all of these measurements along with slow wave activity. These findings provide strong evidence that ictal neocortical slow wave activity is a distinct physiological state, more closely resembling deep sleep or encephalopathy than seizure activity. Indeed, in the same animals slow wave activity under deep anesthesia induced similar changes in neuronal activity in the frontal cortex to those observed during partial limbic seizures.

Further evidence supporting a physiological distinction between ictal neocortical slow waves and seizure activity was provided by secondarily generalized seizures [30]. As in the awake model, when seizures propagated to the frontal cortex, instead of slow waves the frontal cortex showed high frequency polyspike discharges. Unlike the physiological decreases seen during slow wave activity, during secondary generalized seizures the frontal cortex showed marked *increases* in neuronal firing, cerebral blood flow, BOLD fMRI signals, cerebral blood volume, and cerebral metabolic rate of oxygen consumption.

In summary, direct measurements and neuroimaging during focal limbic seizures revealed very distinct physiology for hippocampal or cortical seizure activity which generally showed marked increases in all neurometabolic functions, contrasting markedly with ictal neurocortical slow activity which showed opposite changes, with decreases in all markers of neurometabolic function. These finding support the hypothesis that ictal neocortical slow wave activity is not part of the seizure itself, but instead is a consequence arising from long-range network effects producing altered physiology in regions remote from the seizure focus.

The next step has been to identify the network mechanisms by which seizure activity in the hippocampus may produce slow wave activity in the neocortex. As we have already discussed, data from patients suggest that focal hippocampal seizures

may depress subcortical arousal systems, which could lead to cortical slow wave activity resembling deep sleep or coma (Fig. 5.2). Experiments from the rat model have provided further mechanistic details to support this hypothesis [31]. fMRI mapping during focal limbic seizures demonstrated that seizure activity propagates from the hippocampus to subcortical structures including the lateral septal nuclei, anterior hypothalamus, and medial thalamus. Subsequent direct neuronal recordings confirmed increased activity in these subcortical regions during seizures. The lateral septal nuclei and anterior hypothalamus contain gamma amino butyric acid (GABA)-ergic neurons with projections to subcortical arousal structures and are thus well positioned to inhibit cortical arousal during seizures. In support of this model, electrical stimulation of these regions in the absence of seizure activity was able to reproduce cortical slow wave activity and behavioral arrest [31, 32]. Cutting the fornix, the main route of seizure propagation from hippocampus to these subcortical structures, prevented cortical slow wave activity and behavioral arrest during seizures.

Additional studies have confirmed decreased subcortical arousal during focal limbic seizures, specifically in the cholinergic arousal systems [32]. fMRI mapping during focal limbic seizures have shown decreased signals in the midbrain reticular formation, thalamic intralaminar nuclei and possibly the basal forebrain. Juxtacellular recordings from the pedunculopontine tegmental nucleus in the brainstem demonstrated decreased firing of identified cholinergic neurons during frontal cortical slow wave activity in focal limbic seizures [32]. In addition, amperometric measurements of choline signals as a surrogate marker of cholinergic neurotransmission showed decreases in both frontal cortex and intralaminar thalamus during focal limbic seizures, but not during secondarily generalized seizures. While it is likely that in addition to cholinergic arousal other subcortical arousal systems are also involved, these findings provide strong evidence that a well characterized subcortical arousal system is depressed during focal limbic seizures, resembling the decreased function seen in slow wave sleep.

5.4   Conclusions and Future Directions

Here we have examined the activity patterns in focal limbic seizures to ask the question: What is a seizure network? In this case, more specifically—which changes in activity during limbic seizures represent the seizure itself and which can be considered long-range network effects arising from, but physiologically distinct from seizure activity? Based on multi-modal measurements including direct recordings of neuronal activity, cerebral blood flow, and neuroimaging-based evaluation of neuroenergetics, we conclude that limbic seizure networks involve intense increases in activity in structures such as the hippocampus and subcortical regions including the lateral septum and anterior hypothalamus. As a long-range network consequence of this abnormal increased activity, there is also abnormal decreased activity in subcortical arousal systems including the brainstem, intralaminar thalamus and basal forebrain which causes the cortex to enter a state resembling deep sleep. These subcortical and cortical decreases in activity are not part of the seizure *per se* since they differ drastically from the increases typically associated with seizures. However, they are an important effect of the seizure network on other parts of the brain, and have a major clinical impact including impaired consciousness.

Important unanswered questions remain about these seizure networks. For example, although the presence of GABAergic neurons in structures involved in seizures (such as the lateral septum or anterior hypothalamus) suggests these may inhibit subcortical arousal systems, direct demonstration of subcortical inhibition has not yet been confirmed. Additional experiments including local infusion of GABAergic agonists and antagonists will be crucial. In addition, while cholinergic arousal was found to be depressed during limbic seizures, the possible involvement of other neurotransmitter systems should be investigated further. Another important direction for future investigation is the development of treatments to prevent long-range network impairment. Although ideally the sei-zures themselves should be stopped, in some patients this is not possible. In these medically and surgically refractory cases, treatments aimed at preventing the impaired consciousness which accompanies depressed cortical function would be highly beneficial. Possible treatments based on the findings above would include deep brain stimulation targeted at arousal regions such as the thalamic intralaminar region [33, 34] or pharmacological treatments such as modafinil [35] aimed at increasing alertness in the ictal and post-ictal periods. Hopefully, further investigation of the interactions between local seizures and long-range network interactions will make such treatments possible, improving the lives of people with epilepsy.

Acknowledgements I would like to especially thank Phil Schwartzkroin for his many leadership roles in the field of epilepsy including original scientific research, organizational leadership, publishing and education. He serves as an exemplary role model for colleagues in the field.

Other Acknowledgements This work was supported by NIH R01NS055829, R01NS066974, R01MH67528, P30NS052519, U01NS045911, and the Betsy and Jonathan Blattmachr Family.

References

1. Lieb JP, Dasheiff RB, Engel J Jr (1991) Role of the frontal lobes in the propagation of mesial temporal lobe seizures. Epilepsia 32(6):822–837
2. Blumenfeld H, Rivera M, McNally KA, Davis K, Spencer DD, Spencer SS (2004) Ictal neocortical slowing in temporal lobe epilepsy. Neurology 63:1015–1021
3. Englot DJ, Yang L, Hamid H, Danielson N, Bai X, Marfeo A et al (2010) Impaired consciousness in temporal lobe seizures: role of cortical slow activity. Brain 133(Pt 12):3764–3777
4. Schevon CA, Weiss SA, McKhann G Jr, Goodman RR, Yuste R, Emerson RG et al (2012) Evidence of an inhibitory restraint of seizure activity in humans. Nat Commun 3:1060
5. Ebersole JS, Pedley TA (2003) Current practice of clinical electroencephalography, 3rd edn. Lippincott Williams & Wilkins, Philadelphia
6. Steriade M, Contreras D, Curro Dossi R, Nunez A (1993) The slow (<1 Hz) oscillation in reticular thalamic and thalamocortical neurons: scenario of sleep rhythm generation in interacting thalamic and neocortical networks. J Neurosci 13(8):3284–3299

7. Haider B, Duque A, Hasenstaub AR, McCormick DA (2006) Neocortical network activity in vivo is generated through a dynamic balance of excitation and inhibition. J Neurosci 26(17):4535–4545

8. Laureys S, Schiff ND (2009) Disorders of consciousness, Annals of the New York academy of sciences. Wiley-Blackwell, New York

9. Schwartz TH, Bonhoeffer T (2001) In vivo optical mapping of epileptic foci and surround inhibition in ferret cerebral cortex. Nat Med 7(9):1063–1067

10. Prince DA, Wilder BJ (1967) Control mechanisms in cortical epileptogenic foci. "Surround" inhibition. Arch Neurol 16(2):194–202

11. Chassagnon S, Namer IJ, Armspach JP, Nehlig A, Kahane P, Kehrli P et al (2009) SPM analysis of ictal-interictal SPECT in mesial temporal lobe epilepsy: relationships between ictal semiology and perfusion changes. Epilepsy Res 85(2–3):252–260

12. Van Paesschen W, Dupont P, Van Driel G, Van Billoen H, Maes A (2003) SPECT perfusion changes during complex partial seizures in patients with hippocampal sclerosis. Brain 126(5):1103–1111

13. Blumenfeld H, McNally KA, Vanderhill SD, Paige AL, Chung R, Davis K et al (2004) Positive and negative network correlations in temporal lobe epilepsy. Cereb Cortex 14(8):892–902

14. Lee KH, Meador KJ, Park YD, King DW, Murro AM, Pillai JJ et al (2002) Pathophysiology of altered consciousness during seizures: subtraction SPECT study [comment]. Neurology 59(6):841–846

15. Hogan RE, Kaiboriboon K, Bertrand ME, Rao V, Acharya J (2006) Composite SISCOM perfusion patterns in right and left temporal seizures. Arch Neurol 63(10):1419–1426

16. Tae WS, Joo EY, Kim JH, Han SJ, Suh Y-L, Kim BT, Hong SC, Hong SB (2005) Cerebral perfusion changes in mesial temporal lobe epilepsy: SPM analysis of ictal and interictal SPECT. Neuroimage 24:101–110

17. Arthuis M, Valton L, Regis J, Chauvel P, Wendling F, Naccache L et al (2009) Impaired consciousness during temporal lobe seizures is related to increased long-distance cortical-subcortical synchronization. Brain 132(Pt 8):2091–2101

18. Guye M, Regis J, Tamura M, Wendling F, McGonigal A, Chauvel P et al (2006) The role of corticothalamic coupling in human temporal lobe epilepsy. Brain 129:1917–1928

19. Norden AD, Blumenfeld H (2002) The role of subcortical structures in human epilepsy. Epilepsy Behav 3(3):219–231

20. Blumenfeld H (2012) Impaired consciousness in epilepsy. Lancet Neurol 11:814–826

21. Yang L, Shklyar I, Lee HW, Ezeani CC, Anaya J, Balakirsky S et al (2012) Impaired consciousness in epilepsy investigated by a prospective respon-siveness in epilepsy scale (RES). Epilepsia 53(3):437–447

22. Bauerschmidt A, Koshkelashvili N, Ezeani CC, Yoo J, Zhang Y, Manganas LN et al (2012) Prospective evaluation of ictal behavior using the revised Responsiveness in Epilepsy Scale (RES II). Soc Neurosci Abstr, Online at http://websfnorg/

23. Cunningham C, Chen WC, Shorten A, McClurkin M, Choezom T, Schmidt CC et al (2014) Impaired consciousness in partial seizures is bimodally distributed. Neurology 82:1736–1744

24. Blumenfeld H, Jackson GD (2013) Should consciousness be considered in the classification of focal (partial) seizures? Epilepsia 54(6):1125–1130

25. Sloviter RS (2008) Hippocampal epileptogenesis in animal models of mesial temporal lobe epilepsy with hippocampal sclerosis: the importance of the "latent period" and other concepts. Epilepsia 49(Suppl 9):85–92

26. Curia G, Longo D, Biagini G, Jones RS, Avoli M (2008) The pilocarpine model of temporal lobe epilepsy. J Neurosci Method 172(2):143–157

27. McIntyre DC, Gilby KL (2008) Mapping seizure pathways in the temporal lobe. Epilepsia 49(Suppl 3):23–30

28. Coulter DA, McIntyre DC, Loscher W (2002) Animal models of limbic epilepsies: what can they tell us? Brain Pathol 12(2):240–256

29. Scharfman HE (2007) The neurobiology of epilepsy. Curr Neurol Neurosci Rep 7(4):348–354

30. Englot DJ, Mishra AM, Mansuripur PK, Herman P, Hyder F, Blumenfeld H (2008) Remote effects of focal hippocampal seizures on the rat neocortex. J Neurosci 28(36):9066–9081

31. Englot DJ, Modi B, Mishra AM, DeSalvo M, Hyder F, Blumenfeld H (2009) Cortical deactivation induced by subcortical network dysfunction in limbic seizures. J Neurosci 29(41):13006–13018

32. Motelow JE, Gummadavelli A, Zayyad Z, Mishra AM, Sachdev RNS, Sanganahalli BG et al (2012) Brainstem cholinergic and thalamic dysfunction during limbic seizures: possible mechanism for cortical slow oscillations and impaired consciousness. Soc Neurosci Abstr, Online at http://am2012.sfn.org/am2012/

33. Schiff ND (2012) Moving toward a generalizable application of central thalamic deep brain stimulation for support of forebrain arousal regulation in the severely injured brain. Ann N Y Acad Sci 1265:56–68

34. Schiff ND, Giacino JT, Kalmar K, Victor JD, Baker K, Gerber M et al (2007) Behavioural improvements with thalamic stimulation after severe traumatic brain injury [see comment][erratum appears in Nature. 2008; 452(7183):120 Note: Biondi, T [added]]. Nature 448(7153):600–603

35. Ballon JS, Feifel D (2006) A systematic review of modafinil: potential clinical uses and mechanisms of action. J Clin Psychiatry 67(4):554–566

What Is a Seizure Network? Very Fast Oscillations at the Interface Between Normal and Epileptic Brain

Roger D. Traub, Mark O. Cunningham, and Miles A. Whittington

Abstract

Although there is a great multiplicity of normal brain electrical activities, one can observe defined, relatively abrupt, transitions between apparently normal rhythms and clearly abnormal, higher amplitude, "epileptic" signals; transitions occur over tens of ms to many seconds. Transitional activity typically consists of low-amplitude very fast oscillations (VFO). Examination of this VFO provides insight into system parameters that differentiate the "normal" from the "epileptic." Remarkably, VFO *in vitro* is generated by principal neuron gap junctions, and occurs readily when chemical synapses are suppressed, tissue pH is elevated, and $[Ca^{2+}]_o$ is low. Because VFO originates in principal cell axons that fire at high frequencies, excitatory synapses may experience short-term plasticity. If the latter takes the form of potentiation of recurrent synapses on principal cells, and depression of these on inhibitory interneurons, then the stage is set for synchronized bursting – if $[Ca^{2+}]_o$ recovers sufficiently. Our hypothesis can be tested (in part) in patients, once it is possible to measure brain tissue parameters (pH, $[Ca^{2+}]_o$) simultaneously with ECoG.

Keywords

pH • Extracellular calcium • Gap junctions • Potentiation • Oscillation • Electrocorticogram

R.D. Traub (✉)
IBM T.J. Watson Research Center,
Yorktown Heights, NY 10598, USA

Department of Neurology, Columbia University,
New York, NY 10032, USA
e-mail: rtraub@us.ibm.com

M.O. Cunningham
Institute of Neuroscience, The Medical School,
Newcastle University, Newcastle upon Tyne, UK
e-mail: mark.cunningham@newcastle.ac.uk

M.A. Whittington
Hull York Medical School, The University of York,
Heslington YO10 5DD, UK
e-mail: miles.whittington@hyms.ac.uk

H.E. Scharfman and P.S. Buckmaster (eds.), *Issues in Clinical Epileptology: A View from the Bench*,
Advances in Experimental Medicine and Biology 813, DOI 10.1007/978-94-017-8914-1_6
© Springer Science+Business Media Dordrecht 2014

Abbreviations

ACSF   Artificial cerebrospinal fluid
DHPG   (S)-3,5-dihydroxyphenylglycine
ECoG   Electrocorticography
TMA   Trimethylamine
VFO   Very fast oscillations (>70 Hz)

6.1   Introduction

The task of defining – or identifying – a seizure network is conceptually very complex and can be approached in a number of different ways. One could, for example, determine which brain regions (and which cell types) are the first to discharge in a "non-normal" fashion that leads to aberrant EEG patterns. This approach has been the conventional one, and has led to the concept of the epileptic "focus" or epileptic "zone." These concepts have been problematic since it is now clear that – at least in the chronic human epileptic brain – cells in rather widespread brain regions are often linked in their aberrant discharge patterns as seizure are initiated (e.g., Worrell's work). A related but more recent approach has been to identify those brain regions that generate high frequency oscillations at the onset of seizure activity. The use of such oscillations as a biomarker for "epileptic brain" has received much attention, and seems to provide a useful guideline for surgical intervention (i.e., removal results in "cure"). With this latter approach, it would appear that the "seizure network" is defined as that group of cells that generate these abnormally high frequency EEG patterns. And thus an understanding of these generators would provide a useful handle on defining a seizure network – and for asking such questions as whether such networks are dynamic, are reflective on tissue pathology, are exclusive to networks in epileptic brain (i.e., do not come into play in normal brain when seizures are exogenously generated), etc. We have therefore approached the question of "epileptic networks" via our interest in very fast oscillations (VFO).

The data discussed below suggest that the transition from normal brain rhythms to seizure is brought about (at least in an immediate sense) by alterations in brain tissue, *in the extracellular environment* rather than by neuronal activities per se – an idea that has been central to the epilepsy scientific endeavor for many years. As shown by many other authors (and also ourselves), very fast oscillations (VFO) – a striking and (we believe) fundamental sort of neuronal activity – are frequently observed prior to and during seizures. What we bring to the table that is new is this: VFO occurs in just those ionic and pH conditions expected to occur after brain activation, and which might in themselves promote seizures. Furthermore, VFO itself could induce synaptic habituation (specifically of pyramidal/interneuron synapses) that would also favor seizures. This emphasis on the extracellular environment, and on the mechanisms for transition from "normal" electrical activity to "seizure" activity, provides perhaps a new view of what might profitably be explored as a defining feature of epileptic networks.

6.2   Very Fast Oscillations in Normal and Epileptic Brain

During epileptic burst complexes (both interictal and ictal), there coexist large field transients (often with simultaneous intracellular depolarizations and multiple action potentials), together with high-frequency field oscillations ("VFO"), the latter sometimes at several hundred Hz [other terms include "ripples", "fast oscillations", and HFO or high-frequency oscillations]. This coexistence was observed in penicillin-induced epileptogenesis in cat hippocampus in vivo, in 1969 [7]; and not too long afterwards in the *in vitro* hippocampal slice by Philip Schwartzkroin and David Prince [32, 33]. Since then, coexisting large field transients, with superimposed VFO, have been observed in patient EEGs (for example, [42], and see also below), as well as in many experimental contexts ([24]; reviewed in [45]).

How can one account for the coexistence of these two field patterns, and their relation to normal brain activities, such as gamma (30–70 Hz) rhythms and physiological sharp waves? In this chapter, we shall note that putatively normal-appearing gamma can alternate with large synchronized bursts, providing a model for the transition between normal and abnormal neuronal population behaviors. Interestingly, in this model – and in many other situations, including in patients – there is a segment of VFO *prior* to the synchronized burst. We shall examine the somewhat surprising conditions in which, experimentally, VFO can occur alone; and we shall review the cellular mechanisms of one experimental type of gamma oscillations (which turns out to be related to VFO). Finally, we shall conclude with an hypothesis as to how the transition from relatively normal activities, to epileptic ones, might take place in situ. Imbalance between synaptic excitation and inhibition – the text-book explanation – provides partial, but not complete, understanding. Our hypothesis is testable, at least in part; and, if valid, the hypothesis may have clinical application.

An example of the alternation between "normal" gamma rhythm and epileptiform bursts. It was discovered in 1998 (Fisahn et al.) that stable (i.e. lasting hours) gamma oscillations could be induced in properly prepared hippocampal slices, simply by addition of a compound such as carbachol to the bath. Similar oscillations can be produced by other compounds, including kainate, in hippocampus, neocortex, entorhinal cortex, and cerebellum slices (reviewed in [45]). [We shall describe some of the cellular mechanisms below.] Interestingly, a high concentration of the metabotropic glutamate receptor agonist DHPG induces oscillations that alternate with epileptiform bursts, over periods of several seconds (Fig. 6.1). The amplitude of EPSPs in interneurons and in pyramidal cells evolves over the interburst periods, decreasing in interneurons, and increasing in pyramidal cells; and this explains, at least in part, the switch in behaviors [46]. Note, however, that field VFO actually precedes the epileptiform bursts (Fig. 6.1bii, and see also [24]).

Further examples of VFO associated with, and prior to, epileptic transients and "full-blown" electrographic seizures.

The slight advance of VFO, relative to epileptiform bursts, may be a quite general phenomenon (Fig. 6.2), occurring also in human tissue *in situ*, as well as in resected human tissue. Such observa-

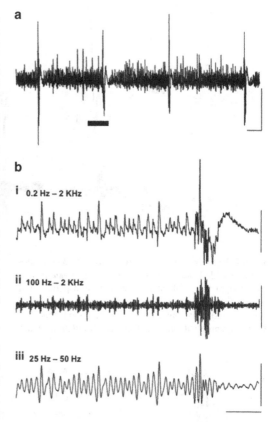

Fig. 6.1 Alternating gamma oscillation and synchronized epileptiform bursts. Rat hippocampal slice, CA3 region, bathed in 100 µM DHPG (a metabotropic glutamate receptor agonist), s. pyramidale field recordings. (**a**) long-duration trace showing 4 epileptiform bursts with interspersed gamma oscillations (~30 Hz). Scale bars 0.5 mV, 1 s. (**b**) the segment corresponding to the bar in (**a**) is expanded, and filtered to show broad-band (i), VFO (ii), and gamma (iii) signals. Note the brief VFO just prior to the epileptiform burst. Scale bars 0.5, 0.1, 0.2 mV; 200 ms (From Traub et al. [44], reproduced with permission)

tions suggest that perhaps VFO is really the "fundamental" event in epileptic bursts. Indeed, in resected human tissue, it has been shown that blockade of chemical synapses can eliminate the large field transients, while leaving VFO; whereas block of VFO with carbenoxolone also causes loss of the large transients [27]. At least in the experimental conditions there used, it was not possible to observe large transients without VFO, while the reverse could be observed. One wonders, therefore, if VFO at least contributes to the causation of the epileptiform bursts (Fig. 6.2).

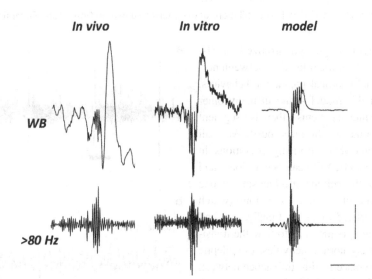

In vivo   **In vitro**   **model**

WB

>80 Hz

Fig. 6.2 VFO preceding epileptiform bursts: 3 examples. ("WB" = wide-band.) *Left, In vivo*, foramen ovale recording of right temporal interictal activity in a patient with mesial temporal sclerosis. *Middle, In vitro*, spontaneous field potential burst in resected temporal neocortex from the same patient. *Right*, model, simulation of network burst in multilayer neocortical circuit model, with multicompartment neurons interconnected by chemical synapses and by gap junctions. *Scale bars* 200 µV *in vivo*, 100 µV *in vitro*, arbitrary for model; 100 ms (From Roopun et al. [27], reproduced with permission)

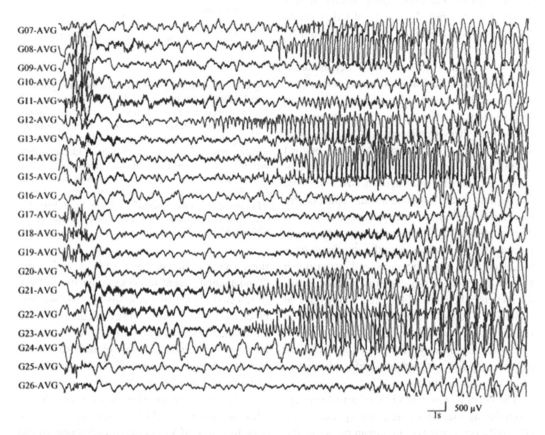

G07-AVG
G08-AVG
G09-AVG
G10-AVG
G11-AVG
G12-AVG
G13-AVG
G14-AVG
G15-AVG
G16-AVG
G17-AVG
G18-AVG
G19-AVG
G20-AVG
G21-AVG
G22-AVG
G23-AVG
G24-AVG
G25-AVG
G26-AVG

500 µV
1s

Fig. 6.3 Subdural grid ECoG recording of an electrographic seizure, preceded by a ~2 s generalized discharge, and then localized, low-amplitude VFO (e.g. G21-G23). Recordings from a child with a *right frontal* cortical dysplasia and intractable seizures. She responded well to surgery (From Traub et al. [42], reproduced with permission)

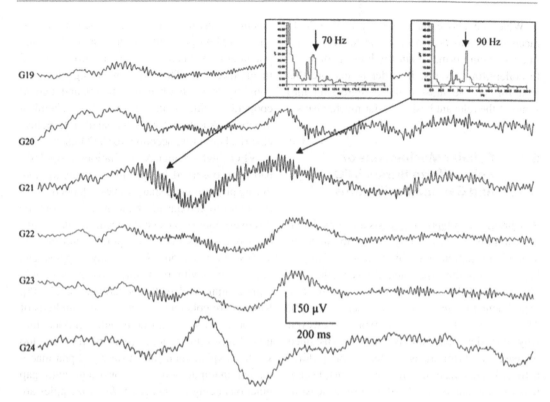

Fig. 6.4 Another run of preseizure VFO in ECoG. This example was recorded from the same patient whose ECoG was shown in Fig. 6.3, with the same subdural grid but different recording technique (From Traub et al. [42], reproduced with permission)

Fig. 6.5 *In vitro* ~200 Hz **ripples are strongly potentiated by tissue alkalinization.** Stratum pyramidale recordings of spontaneous VFO in the CA3 region of rat hippocampal slice. VFO occurs transiently in control conditions (*top*), but becomes nearly continuous after tissue alkalinization with 10 mM NH₄Cl. The effect is reversible (From Draguhn et al. [8], reproduced with permission)

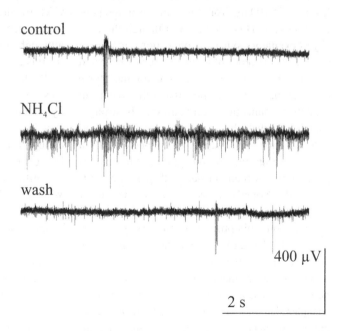

At times, VFO can be sustained for seconds prior to the onset of an electrographic seizure (but not, so far as we are aware, of an interictal burst). Figure 6.4 shows an example of this phenomenon, in an electrocorticographic (ECoG) recording.

What the above data suggest is that VFO mechanisms may provide a clue as to what is distinctive about normal brain rhythms, as opposed to epileptiform events. In order to explore this idea further, we must make a digression into some of the relevant basic cellular mechanisms.

6.3 Cellular Mechanisms of Epileptiform Bursts, VFO, and Gamma Oscillations

Synchronized epileptiform bursts are considered, traditionally, to arise from an imbalance in synaptic excitation and inhibition – an idea perhaps rooted in the experimental observation that blockade of $GABA_A$ receptors was an effective experimental means of inducing such bursts [7]. The imbalance idea does not explain, however, why one does not simply observe sustained increases in firing rates; instead, epileptiform activity is organized into transient events, lasting tens to hundreds of ms. Furthermore, at least *in vitro*, transient events can be elicited by stimulation of a small number of neurons, sometimes even one neuron [19], although there can be a latency of >100 ms from the stimulus to the population event. This occurs, even though the density of excitatory synaptic connections, *in vitro* in CA3, is of the order of a few per cent. Traub and Wong [39] were able to account for the above observations, if it were postulated that recurrent excitatory connections were sufficiently strong – specifically, that a burst of action potentials could induce a burst in a synaptically connected cell, in the relative absence of synaptic inhibition. This prediction was then verified with paired recordings [20]. Notably, however, the model under consideration did not account for the VFO superimposed on epileptiform bursts. Why is this important? Couldn't it be that the VFO is simply an irrelevant epiphenomenon?

We shall argue that the VFO is important, for a number of reasons, but in the present context, consider the following argument. "Strong" coupling between neurons appears to be important for a synchronized burst to develop. Suppose that gap junctions were to exist between principal neurons, with coupling powerful to allow a single action potential in one cell to evoke an action potential in another cell. This type of strong electrical coupling does actually exist [18, 48], and it could cooperate with recurrent excitatory chemical synapses. Additionally, as we shall note below, electrical coupling accounts for VFO itself.

VFO: high-frequency oscillations ("ripples") had been observed in the hippocampus *in vivo*, during physiological sharp waves [4], but distinctive clues to cellular mechanisms came from the discovery that ripples could occur *in vitro*, without sharp waves [8] – the ripples could then be studied in isolation. Remarkably, ripples can occur *in vitro* without chemical synapses, both in hippocampus and in the neocortex [8, 22, 46]. Ripples are coherent (*in vitro*) over hundreds of microns, and so are a true population phenomenon. Extracellular fields (tens of μV) are too small to explain them, and a variety of pharmacological manipulations are consistent with gap junctions being fundamental. *In vitro* ripples are also associated with spikelets [8, 46] which, in the hippocampus, are likely of axonal origin [28]. Dye-coupling exists between axons of nearby CA1 pyramidal cells [28], consistent with the occurrence of gap junctions between axons, although not providing definitive proof (by itself) for this concept.

We have shown that *in vitro* VFO, at frequencies up to about 250–300 Hz, can be explained by electrical coupling between axons under certain conditions: first, the coupling is strong enough for a spike in one axon to evoke a spike in a coupled axon (indirectly supported by data of Dhillon and Jones [6], Mercer et al. [18] and Wang et al. [48]); second, each axon couples, on average, to more than one other; finally, that spontaneous axonal action potentials occur at least sometimes. This model accounts for the admixture of spikes and spikelets during VFO, for continuous frequency transitions from gamma to almost 200 Hz, and for spatial patterns of VFO in the neocortex [5, 35, 40, 46]; and, most importantly, it accounts for the propensity of VFO to occur when chemical synapses are blocked. The model predicts that somatic action potentials during VFO are antidromic [2, 47].

Persistent gamma oscillations are traditionally viewed as arising simply from recurrent synaptic excitation to interneurons, and synaptic inhibition to pyramidal neurons. A number of pieces of experimental evidence indicate that the mechanisms are somewhat more complicated. First, while it is true that blockade of AMPA/kainate, or of $GABA_A$ receptors, will suppress persistent gamma, it is also true that persistent gamma is sensitive to gap junction blockade [11, 41, 42]. Second, the power spectrum of gamma oscillation fields reveals a peak at 70 or more Hz. This activity can be seen in Fig. 6.1bii. This faster peak is not simply a harmonic of the gamma activity, because the high frequency peak persists when gamma is abolished by synaptic receptor blockade [42], or when stratum oriens is separated from stratum pyramidale – in which case VFO persists in s. oriens [43]. Finally, pyramidal cell somata fire rarely during persistent gamma [11].

The above disparate and counter-intuitive observations are readily explained with a model that basically simulates persistent gamma as continuous VFO that is "chopped up" by recurrent synaptic inhibition – something that is possible if axonal gap junctions are not too far from perisomatic sources of inhibition [41]. The model thereby accounts for the pharmacology, the field potential profiles, and the rare somatic firing (the latter because the action potentials that drive the oscillation are generated in axons, and only some of these successfully propagate back to the soma as full spikes). The model predicts that, during persistent gamma, axons fire at higher rates than somata; and that somatic action potentials are antidromic: these predictions have been experimentally verified [10].

6.4   VFO and Origin of Seizures

Experimental VFO is potentiated by alkaline conditions. A relation between systemic (and presumably brain) pH has long been suspected,

with alkaline pH being epileptogenic: in absence and other seizure types associated with spike-wave [12, 23], and in febrile seizures and their experimental models [29, 30]. In addition, some drugs with anticonvulsant properties, are blockers of carbonic anhydrase (acetazolomide, topiramate, zonisamide) [21, 25]. Remarkably, alkaline pH strongly potentiates *in vitro* VFO (Fig. 6.5, [8, 46]). The effects on VFO are unlikely to result from actions of pH on synaptic transmission, as the effects can occur when synaptic transmission is effectively blocked [46]. A likely cause is the opening of gap junction channels by alkaline pH [36], although it has not been possible yet to prove this directly.

VFO and calcium. A class of experimental *in vitro* epilepsy models includes so-called field bursts and related phenomena, in which ionic manipulations are used to suppress synaptic transmission (lowering $[Ca^{2+}]$, use of Mn^{2+}), to increase neuronal excitability (for example, elevating extracellular $[K^+]$), and probably to open gap junctions with increased pH [13, 34, 38, 51]. Such field bursts likely (in our opinion) depend on gap junctions [26]. The occurrence of field bursts fits in with long-held hypotheses concerning a primary role for glia in epilepsy [9, 14, 37, 50]; and also with long-standing observations that afferent stimulation, as well as seizures themselves, can have significant effects on extracellular ion concentrations, including the lowering of $[Ca^{2+}]_o$ [15].

Experimental demonstration of gamma/VFO/ seizure evolution in alkaline conditions. Figure 6.6 illustrates a transition from VFO (~110 Hz) to electrographic seizure, suggestive of the human patient data of Figs. 6.3 and 6.4, although the data in Fig. 6.6 are from an *in vitro* hippocampal slice. The slice was bathed in an alkalinizing solution, and then a tetanic stimulus was delivered that evokes an epoch of so-called tetanic gamma, during which $[K^+]_o$ is expected to rise, and $[Ca^{2+}]_o$ to fall [50]. The gamma is followed by VFO (middle trace in Fig. 6.6), that turns into a brief electrographic seizure. We propose that synaptic excitation of interneurons is depressed during the VFO period, analogous to

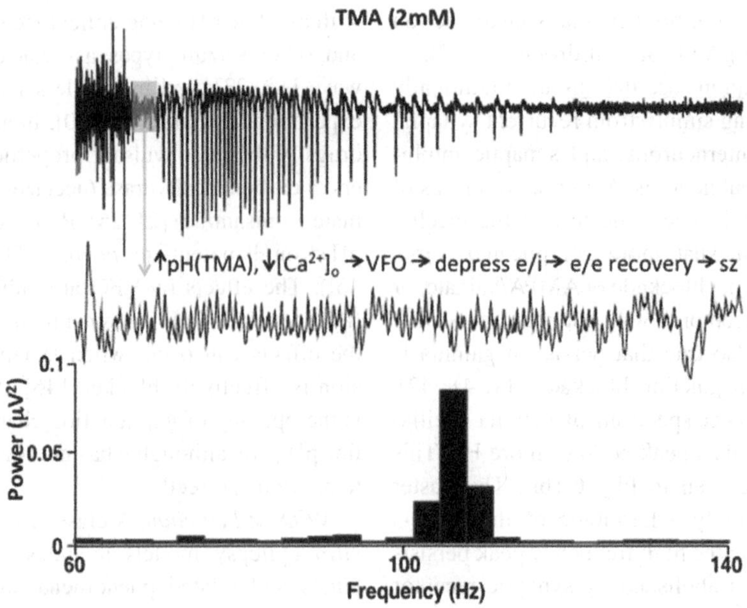

Fig. 6.6 VFO-electrographic seizure transition *in vitro*, in alkaline conditions. Rat hippocampal slices were alkalinized with 2 mM trimethylamine (TMA), and a tetanic stimulus given to s. radiatum of CA1. In s. pyramidale field potential recordings, this stimulation resulted (upper trace, 17 s of activity) in post-tetanic gamma at first, then VFO (1), then ~2 mV epileptiform field transients (2), then further VFO (3). The middle trace shows, on an expanded time scale (850 ms of activity), the ~0.1 mV VFO potential fluctuations, at ~110 Hz (power spectrum below). Our hypothesized sequence of events is shown above the VFO trace (see text for further details). "e/i", pyramidal cell-to-interneuron; "e/e", pyramidal cell-to-pyramidal cell. "sz", seizure (From Traub et al. [42], reproduced with permission)

what has been shown in the preparation of Fig. 6.1 [44]; such synaptic habituation during VFO remains, however, to be shown directly.

6.5 Conclusion and Hypothesis

To summarize some of these data then, our view is that high-frequency firing in the pyramidal cell axon plexus is what drives both VFO and persistent gamma oscillations. **VFO occur under specific extracellular conditions – which we hypothesize to be the initiating factor for seizure activity.** What is now required, we believe, is direct measurement of extracellular tissue parameters [16], in epileptic patients, perhaps now using MRI [1, 17], and preferably in conjunction with EEG or ECoG recordings. If such measurements do indeed indicate, for example, tissue alkalinization just prior to seizure onset, it will suggest alternative approaches to seizure prevention, and perhaps also better understanding of how present treatments – such as the ketogenic diet – are effective [3, 31].

Acknowledgement With deepest gratitude to Philip A. Schwartzkroin who (with Robert Wong and David Prince) introduced RDT to epilepsy research in 1977.

Other Acknowledgement Supported by IBM, NIH/NINDS, the Alexander von Humboldt Stiftung, Einstein Stiftung Berlin, the Hadwen Trust and the Wellcome Trust. We thank Andreas Draguhn, Dietmar Schmitz, Yoshio Okada, and Nikita Vladimirov for helpful discussions.

References

1. Angelovski G, Chauvin T, Pohmann R, Logothetis NK, Tóth E (2011) Calcium-responsive paramagnetic CEST agents. Bioorg Med Chem 19:1097–1105

2. Bähner F, Weiss EK, Birke G, Maier N, Schmitz D, Rudolph U, Frotscher M, Traub RD, Both M, Draguhn A (2011) Cellular correlate of assembly formation in oscillating hippocampal networks in vitro. Proc Natl Acad Sci U S A 108:E607–E616

3. Bough KJ, Rho JM (2007) Anticonvulsant mechanisms of the ketogenic diet. Epilepsia 48:43–58

4. Buzsáki G, Horváth Z, Urioste R, Hetke J, Wise K (1992) High-frequency network oscillation in the hippocampus. Science 256:1025–1027

5. Cunningham MO, Roopun AK, Schofield IS, Whittaker RG, Duncan R, Russell A, Jenkins A, Nicholson C, Whittington MA, Traub RD (2012) Glissandi: transient fast electrocorticographic oscillation of steadily increasing frequency, explained by temporally increasing gap junction conductance. Epilepsia 53:1205–1214

6. Dhillon A, Jones RSG (2000) Laminar differences in recurrent excitatory transmission in the rat entorhinal cortex in vitro. Neuroscience 99:413–422

7. Dichter M, Spencer WA (1969) Penicillin-induced interictal discharges from the cat hippocampus. I. Characteristics and topographical features. J Neurophysiol 32:649–662

8. Draguhn A, Traub RD, Schmitz D, Jefferys JGR (1998) Electrical coupling underlies high-frequency oscillations in the hippocampus in vitro. Nature 394:189–192

9. Duffy S, MacVicar BA (1999) Modulation of neuronal excitability by astrocytes. In: Delgado-Escueta AV, Wilson WA, Olsen RW, Porter RJ (eds) Jasper's basic mechanisms of the epilepsies, vol 79, 3rd edn, Advances in neurology. Lippincott, Philadelphia, pp 573–581

10. Dugladze T, Schmitz D, Whittington MA, Vida I, Gloveli T (2012) Segregation of axonal and somatic activity during fast network oscillations. Science 336:1458–1461

11. Fisahn A, Pike FG, Buhl EH, Paulsen O (1998) Cholinergic induction of network oscillations at 40 Hz in the hippocampus in vitro. Nature 394:186–189

12. Foerster O (1924) Hyperventilationsepilepsie. Dtsch Z Nervenheilkd 83:347–356

13. Haas HL, Jefferys JGR (1984) Low-calcium field burst discharges of CA1 pyramidal neurones in rat hippocampal slices. J Physiol 354:185–201

14. Heinemann U, Gabriel S, Schuchmann S, Eder C (1999) Contribution of astrocytes to seizure activity. In: Delgado-Escueta AV, Wilson WA, Olsen RW, Porter RJ (eds) Jasper's basic mechanisms of the epilepsies, vol 79, 3rd edn, Advances in neurology. Lippincott, Philadelphia, pp 583–590

15. Heinemann U, Louvel J (1983) Changes in $[Ca^{2+}]_o$ and $[K^+]_o$ during repetitive electrical stimulation and during pentetrazol induced seizure activity in the sensorimotor cortex of cats. Pflugers Arch 398:310–317

16. Javaheri S, Clendening A, Papadakis N, Brody JS (1981) Changes in brain surface pH during acute isocapnic metabolic acidosis and alkalosis. J Appl Physiol 51:276–281

17. Magnotta VA, Heo HY, Dlouhy BJ, Dahdaleh NS, Follmer RL, Thedens DR, Welsh MJ, Wemmie JA (2012) Detecting activity-evoked pH changes in human brain. Proc Natl Acad Sci U S A 109:8270–8273

18. Mercer A, Bannister AP, Thomson AM (2006) Electrical coupling between pyramidal cells in adult cortical regions. Brain Cell Biol 35:13–27

19. Miles R, Wong RKS (1983) Single neurones can initiate synchronized population discharge in the hippocampus. Nature 306:371–373

20. Miles R, Wong RKS (1986) Excitatory synaptic interactions between CA3 neurones in the guinea-pig hippocampus. J Physiol 373:397–418

21. Mirza NS, Alfirevic A, Jorgensen A, Marson AG, Pirmohamed M (2011) Metabolic acidosis with topiramate and zonisamide: an assessment of its severity and predictors. Pharmacogenet Genomics 21:297–302

22. Nimmrich V, Maier N, Schmitz D, Draguhn A (2005) Induced sharp wave-ripple complexes in the absence of synaptic inhibition in mouse hippocampal slices. J Physiol 563:663–670

23. Nims LF, Gibbs EL, Lennox WG, Williams D (1940) Adjustment of acid-base balance of patients with petit mal epilepsy to overventilation. Arch Neurol Psychiatr 43:262–269

24. Pais I, Hormuzdi SG, Monyer H, Traub RD, Wood IC, Buhl EH, Whittington MA, LeBeau FEN (2003) Sharp wave-like activity in the hippocampus in vitro in mice lacking the gap junction protein connexin 36. J Neurophysiol 89:2046–2054

25. Panayiotopoulos CP (2001) Treatment of typical absence seizures and related epileptic syndromes. Paediatr Drugs 3:379–403

26. Perez-Velazquez JL, Valiante TA, Carlen PL (1994) Modulation of gap junctional mechanisms during calcium-free induced field burst activity: a possible role for electrotonic coupling in epileptogenesis. J Neurosci 14:4308–4317

27. Roopun AK, Simonotto JD, Pierce ML, Jenkins A, Schofield I, Kaiser M, Whittington MA, Traub RD, Cunningham MO (2010) A non-synaptic mechanism underlying interictal discharges in human epileptic neocortex. Proc Natl Acad Sci U S A 107:338–343

28. Schmitz D, Schuchmann S, Fisahn A, Draguhn A, Buhl EH, Petrasch-Parwez RE, Dermietzel R, Heinemann U, Traub RD (2001) Axo-axonal coupling: a novel mechanism for ultrafast neuronal communication. Neuron 31:831–840

29. Schuchmann S, Schmitz D, Rivera C, Vanhatalo S, Salmen B, Mackie K, Sipilä ST, Voipio J, Kaila K (2006) Experimental febrile seizures are precipitated by a hyperthermia-induced respiratory alkalosis. Nat Med 12:817–823

30. Schuchmann S, Hauck S, Henning S, Grüters-Kieslich A, Vanhatalo S, Schmitz D, Kaila K (2011) Respiratory alkalosis in children with febrile seizures. Epilepsia 52:1949–1955

31. Schwartzkroin PA (1999) Mechanisms underlying the anti-epileptic efficacy of the ketogenic diet. Epilepsy Res 37:171–180

32. Schwartzkroin PA, Prince DA (1977) Penicillin-induced epileptiform activity in the hippocampal *in vitro* preparation. Ann Neurol 1:463–469

33. Schwartzkroin PA, Prince DA (1978) Cellular and field potential properties of epileptogenic hippocampal slices. Brain Res 147:117–130

34. Schweitzer JS, Wang H, Xiong ZQ, Stringer JL (2000) pH Sensitivity of non-synaptic field bursts in the dentate gyrus. J Neurophysiol 84:927–933

35. Simon A, Traub RD, Vladimirov N, Jenkins A, Nicholson C, Whittaker R, Schofield I, Clowry GJ, Cunningham MO, Whittington MA (2014) Gap junction networks can generate both ripple-like and fast-ripple-like oscillations. Eur J Neurosci 39:46–60

36. Spray DC, Harris AL, Bennett MVL (1981) Gap junctional conductance is a simple and sensitive function of intracellular pH. Science 211:712–715

37. Sypert GW, Ward AA Jr (1974) Changes in extracellular potassium activity during neocortical propagated seizures. Exp Neurol 45:19–41

38. Taylor CP, Dudek FE (1984) Synchronization without active chemical synapses during hippocampal after-discharges. J Neurophysiol 52:143–155

39. Traub RD, Wong RKS (1982) Cellular mechanism of neuronal synchronization in epilepsy. Science 216:745–747

40. Traub RD, Schmitz D, Jefferys JGR, Draguhn A (1999) High-frequency population oscillations are predicted to occur in hippocampal pyramidal neuronal networks interconnected by axoaxonal gap junctions. Neuroscience 92:407–426

41. Traub RD, Bibbig A, Fisahn A, LeBeau FEN, Whittington MA, Buhl EH (2000) A model of gamma-frequency network oscillations induced in the rat CA3 region by carbachol *in vitro*. Eur J Neurosci 12:4093–4106

42. Traub RD, Whittington MA, Buhl EH, LeBeau FEN, Bibbig A, Boyd S, Cross H, Baldeweg T (2001) A possible role for gap junctions in generation of very fast EEG oscillations preceding the onset of, and perhaps initiating, seizures. Epilepsia 42:153–170

43. Traub RD, Cunningham MO, Gloveli T, LeBeau FEN, Bibbig A, Buhl EH, Whittington MA (2003) GABA-enhanced collective behavior in neuronal axons underlies persistent gamma-frequency oscillations. Proc Natl Acad Sci USA 100:11047–11052

44. Traub RD, Pais I, Bibbig A, LeBeau FEN, Buhl EH, Monyer H, Whittington MA (2005) Transient depression of excitatory synapses on interneurons contributes to epileptiform bursts intermixed with gamma oscillations in the mouse hippocampal slice. J Neurophysiol 94:1225–1235

45. Traub RD, Whittington MA (2010) Cortical oscillations in health and disease. Oxford University Press, New York

46. Traub RD, Duncan R, Russell AJC, Baldeweg T, Tu Y, Cunningham MO, Whittington MA (2010) Spatiotemporal patterns of electrocorticographic very fast oscillations (>80 Hz) consistent with a network model based on electrical coupling between principal neurons. Epilepsia 51:1587–1597

47. Vladimirov N, Tu Y, Traub RD (2013) Synaptic gating at axonal branches, and sharp-wave ripples with replay: a simulation study. Eur J Neurosci 38:3435–3447

48. Wang Y, Barakat A, Zhou H (2010) Electrotonic coupling between pyramidal neurons in the neocortex. PLoS One 5:e10253

49. Ward AA Jr (1978) Glia and epilepsy. In: Schoffeniels E, Franck G, Tower GB, Hertz L (eds) Dynamic properties of glia cells. Oxford, Pergamon, pp 413–427

50. Whittington MA, Doheny HC, Traub RD, LeBeau FEN, Buhl EH (2001) Differential expression of synaptic and non-synaptic mechanisms during stimulus-induced gamma oscillations *in vitro*. J Neurosci 21:1727–1738

51. Yaari Y, Konnerth A, Heinemann U (1983) Spontaneous epileptiform activity of CA1 hippocampal neurons in low extracellular calcium solutions. Exp Brain Res 51:153–156

Is There Such a Thing as "Generalized" Epilepsy?

Gilles van Luijtelaar, Charles Behr, and Massimo Avoli

Abstract

The distinction between generalized and partial epilepsies is probably one, if not the most, pregnant assertions in modern epileptology. Both absence and generalized tonic-clonic seizures, the prototypic seizures found in generalized epilepsies, are classically seen as the result of a rapid, synchronous recruitment of neuronal networks resulting in impairment of consciousness and/or convulsive semiology. The term generalized also refers to electroencephalographic presentation, with bilateral, synchronous activity, such as the classical 3 Hz spike and wave discharges of typical absence epilepsy. However, findings obtained from electrophysiological and functional imaging studies over the last few years, contradict this view, showing a rather focal onset for most of the so-called generalized seizure types. Therefore, we ask here the question whether "generalized epilepsy" does indeed exist.

Keywords

Idiopathic generalized epilepsies • Absence seizures • Generalized tonic clonic seizures • Myoclonic juvenile seizures • Spike and wave discharges • Genetic absence models

G. van Luijtelaar, Ph.D.
Donders Centre for Cognition, Radboud University Nijmegen, Montessorilaan 3, 6525 HR, Nijmegen, The Netherlands

C. Behr, M.D.
Montreal Neurological Institute and Departments of Neurology and Neurosurgery and of Physiology, McGill University, 3801 University Street, Montréal, QC H3A 2B4, Canada

M. Avoli, M.D., Ph.D. (✉)
Montreal Neurological Institute and Departments of Neurology and Neurosurgery and of Physiology, McGill University, 3801 University Street, Montréal, QC H3A 2B4, Canada

Department of Experimental Medicine, Faculty of Medicine and Odontoiatry, Sapienza Università di Roma, Rome, Italy
e-mail: massimo.avoli@mcgill.ca

H.E. Scharfman and P.S. Buckmaster (eds.), *Issues in Clinical Epileptology: A View from the Bench*, Advances in Experimental Medicine and Biology 813, DOI 10.1007/978-94-017-8914-1_7, © Springer Science+Business Media Dordrecht 2014

7.1   Background

The concept of generalized and partial seizures dates back to conflict during the last century between "universalizers" and "localizers", the former defending a holistic integrated view of brain function against the "centrencephalic system" of Penfield and Jasper (reviewed in [4]). It was Hughlings Jackson [64] who propose a distinction between generalized and partial seizures. Only much later was the term "generalized" epilepsy itself first employed by Gastaut [20]. Indeed, we will often refer in this chapter to generalized *versus* partial "seizures" rather than "epilepsies" since some experimental results may *stricto sensu* not be applicable to human epilepsy classification and thus they remain seizure-related material. However, since generalized epilepsies are defined as such because of the "generalized" nature of their concomitant seizures, any suspicion with regard to the "generalized" nature of these seizures, will automatically challenge the "generalized" nature of the corresponding epilepsy and *vice versa*.

Generalized seizures are characterized by sudden, often unexpected, manifestations (presumably reflecting the involvement of the entire brain, or at least a large part of the brain) compared to the slower, clinically heterogeneous partial seizures where the patient often remains conscious, at least at the beginning of the seizure. With the development of EEG recordings, this assertion received a formidable confirmation [7]. The electroencephalographic manifestations accompanying generalized absence seizures consist of highly stereotyped pattern of bilateral synchronous, regular and rhythmic spike and wave (SW) discharges at 2.5–4 Hz in children, juveniles and adults, lasting from a few seconds up to 30 s. In contrast, scalp EEG recordings reveal sustained diffused, synchronous, discharges during the *tonic* phase, and interrupted bursts during the *clonic* stage in generalized tonic-clonic seizures (GCTS).

7.2   Evolution of the Classification

Classification in epileptology is a work-in-progress and a simple examination of the past 50 years reveals how the Jacksonian dogma has evolved. In the 1969 classification [19], emphasis was put on the distinction between "seizure that are generalized from the beginning and those that are focal or partial at onset and become generalized secondarily". In the 1981 classification, it was proposed that generalized seizures have electroclinical patterns that "presumably reflect neuronal discharge which is widespread in both hemispheres", underlying a conceptual shift from bilateral "onset" to bilateral "spread" [57]. In 1989, a new classification postulated that partial localization-related epilepsies are "epileptic disorders in which seizure semiology or findings at investigation disclose a localized origin of the seizures", whereas generalized epilepsies are defined by initially bilateral ictal encephalographic patterns [56]. The "generalized" designation was essentially an electroclinical feature, which was discarded in 2010 when terminology was revised.

Indeed, in 2001 and 2006, an ILAE Task Force debated the relevance of this conceptual dichotomy [15, 16]. Recognizing that it was out of date with regard to pathophysiological advances in epileptology, the commission decided to keep its core concept in the 2010 revised classification for convenience. Hence it was proposed that "Generalized and focal are redefined for seizures as occurring in and rapidly engaging bilaterally distributed networks (generalized) and within networks limited to one hemisphere and either discretely localized or more widely distributed (focal)" [6]. Today this convenient scheme, presumably aimed at distinguishing between epilepsies that are recommended for surgery ("surgical") and others ("non-surgical" epilepsies), represents the first issue to be addressed when diagnosing a person with epilepsy. However, it deserves re-evaluation in the light of recent advances obtained from clinical and basic research studies.

7.3 Absence Epilepsy: From the Centrencephalon to the Thalamo-cortical Loop

There is presently compelling evidence based on brain imaging, EEG recording and signal analysis techniques that a key element of generalized epilepsies, the sudden involvement of the whole brain, is highly disputable in typical absence epilepsy [4]. Paradoxically, one of the first electrophysiological studies of the pathophysiology of generalized SW discharges: the Jasper and Droogleever-Fortuyn paper, already suggested a mechanism for focal onset of absence seizures [34]. These authors succeeded in inducing typical 3 Hz SW discharges by local 3 Hz stimulation in the midline and intralaminar nuclei of the thalamus. However, in spite of this evidence, the Montreal school [53] preferred the integrative hypothesis of the centrencephalic system presumably influenced by the recent discovery of the reticular formation [47]. Later, Gloor's team further explored the corticoreticular nature of generalized SW discharges, using the feline generalized penicillin epilepsy model, introduced by Prince and Farrell [55], and established the link between sleep spindles and SW discharges (reviewed in [35]). Those data were still consistent with the thalamo-cortical origin for SW activity and did not challenge or question the "generalized" character of absence epilepsy. Indeed, in vitro studies in ferret brain slices and computational models confirmed later that thalamo-cortical oscillations could be driven by an intrathalamic circuit and revealed that some cellular properties of thalamic cells, which are involved in sleep spindles, most likely contribute to SW generation [29].

During the 1980s, in parallel with the growing interest for a genetic etiology for the so-called "idiopathic" epilepsies, animal models with genetic inheritance of absence epilepsy were described. Specifically, both "genetic absence epilepsy in rats from Strasbourg" (GAERS) [73] and "Wistar Albino Glaxo/Rijswijk" (WAG/Rij) rats [68] were identified. For the first time these models provided the opportunity to directly test hypotheses in animals presenting with spontaneous absence seizures [13, 43, 69]. Both the in vitro studies mentioned above [29] and the evidence obtained from genetic models led to the idea of a thalamo-cortico-thalamic network in which the typical SW discharges could elicit spontaneously. This network included the thalamic reticular nucleus in which inhibitory interneurons trigger GABAergic IPSPs on thalamic relay cells. T-type Ca^{2+} current are deinactivated, as a consequence of hyperpolarization causing a burst of action potentials that in turn excited both reticular thalamic and cortical cells [5, 14, 31, 32, 52]. Enhanced T-type Ca^{2+} currents were recorded in GAERS reticular thalamic cells and thalamic relay cells [67]. In addition, subtle abnormalities in GABAergic transmission were found in the reticular thalamic nucleus of GAERS compared to control rats [8]. However in vivo studies using a different model, namely a feline Lennox-Gastaut model, allowed Steriade and coworkers to pinpoint a cortical trigger for the SW discharge, reopening the controversy about whether the cortex or the thalamus were to be responsible for the generation of SW discharges [62].

7.4 A Neocortical "Focus" as Trigger of Generalized Absence Seizures

More than 50 years of research were necessary to decrypt the pathophysiology of the thalamo-cortical loop in absence seizures, meaning how this neuronal circuit could "jump" from the physiological sleep spindle to the pathological SW discharge. But the real trigger for an absence seizure was hiding somewhere else, and with its discovery the concept of "generalized" absence epilepsy ended. This discovery began in the 1960s, when it was reported that there was focal neocortical initiation of absence seizures. Marcus and Watson [41] discovered that bilateral application of proconvulsant drugs to the frontal cortices could produce a pattern of generalized SW discharges similar to what was observed during an absence seizure. In the 1970 and 1980s, topographical EEG studies performed in patients with absence epilepsy confirmed that SW discharges did not occupy the entire cortex. The *wave* component of the SW discharge was characterized by a maximum localized in frontal areas [58].

These data were soon confirmed in the GAERS model with intracellular recordings of thalamic relay cells where an excitatory drive (EPSCs) was shown, presumably originating from the cortex [54]. The role of the cortex in initiating SW discharges was further established *in vitro* by demonstrating that cortico-thalamic input strength is critical for thalamo-cortical rhythmic activity and for changing a spindle into a SW oscillation [9]. But the evidence that was most compelling ultimately came from experiments performed in the WAG/Rij model where cortico-sub-cortical multiple-site EEG signals were studied using non-linear association analysis [45]. The authors identified a consistent initiation zone in the peri-oral region of the somatosensory cortex (Fig. 7.1), along with a leading role of the cortical projections to the thalamus lasting for the initial 500 ms of the SW discharge. Furthermore, the high degree of bilateral synchronization, characteristic of generalized SW discharges in absence epilepsy, appeared to rely mainly on cortico-cortical connectivity, as indicated by non-linear analysis of inter, intra and thalamo-cortical relationships.

Both initiating and leading roles of the neocortex were further suggested by data showing that SW discharges could be recorded in this structure without concomitant SW activity in thalamus, while the opposite situation never occurred [45, 70]. However, once a seizure evolved, both structures oscillated in concert, suggesting a stereotypical scenario where the *primum movens* involves the peri-oral somatosensory cortex, which secondarily "switches on" the thalamo-cortical loop. The primary role of neocortex in the initiation of SW activity was later confirmed in GAERS rats; in these experiments, Manning and co-workers [40] found that local application of the anti-absence drug ethosuximide has maximal efficacy when this drug is applied into the primary somatosensory peri-oral region while its infusion into the thalamus produced only minor and delayed reduction in SW discharges. Thus, SW discharge initiation in the two major genetic models of absence epilepsy occurs in the same restricted area of the somatosensory cortex. Why and how, however, remain to be clarified.

7.5 Focal Cortical Origin of Absence Seizures in Humans

Human studies on ictal generalized discharges in absence epilepsy have shown that patterns of activation and deactivation identified by fMRI are restricted to some cortical (medial frontal cortex, precuneus, lateral parietal, and frontal cortex) and subcortical regions (thalamus, brainstem) [1, 11]. Due to the limited temporal resolution of fMRI, these results did not allow clear confirmation of a cortical initiation site. However, high-resolution EEG and MEG studies in combination with advanced signal analytical techniques have confirmed the existence of a preferential cortical origin of SW discharges. Localized sources were detected either in the frontal cortex, orbito-frontal, medial temporal or parietal lobe [27, 65, 74]. In addition, a rather localized preictal SW rhythm of low frequency (3 Hz) was detected in atypical absence patients [24], whereas a reproducible topography of locally synchronous cortical sources with increased local connectivity was described in a multifocal network, comprising the right prefrontal mesial, left orbito-frontal and left lateral post-central area [2].

Absence seizures may indeed appear "bilateral and synchronous" (and thus "generalized") in EEG recordings because of the highly connected inter- and intra-cortical networks that are sustained by a cortico-thalamic-cortical loop thus leading to oscillatory activity. However, the SW discharges appear to originate from specific cortical areas. The velocity of spread between hemispheres is presumably based on extensive monosynaptic inter-hemispheric connections via the corpus callosum (Fig. 7.2). Evidence for the role of callosal interhemispheric connectivity came from callosotomy experiments resulting in the disruption of the bilateral and synchronous SW discharges in several absence seizure and genetic animal models of absence epilepsy [41, 42, 48, 72].

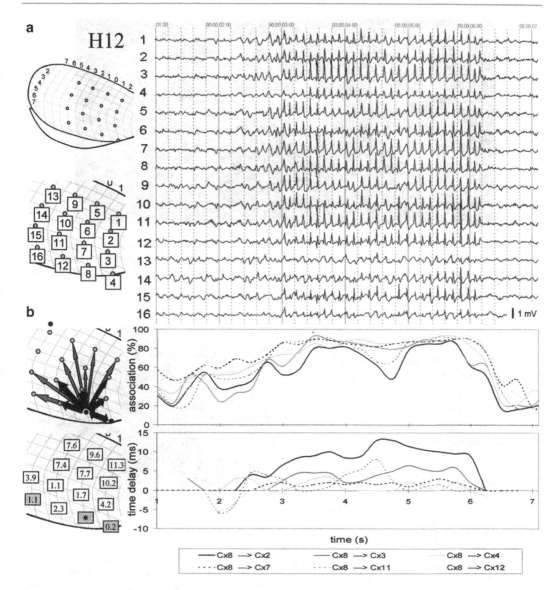

Fig. 7.1 Evolution of the intra-hemispheric cortico-cortical association strength h^2 (a non-linear correlation coefficient between two signals was calculated for all electrode combinations as a function of time of shift between the signals) and time delays of the local field potentials signals between electrode pairs. A cortical grid covering a major part of the somatosensory area was used for electrographical seizure recording in 16–22 month old WAG/Rij rats with spontaneous occurring spike-wave discharges. (**a**) *Left*: Electrode positions (*top*) and electrode labels (*bottom*) on the somatosensory cortex of rat H12. The numbers on the *top graph* refer to coordinates based on the rat's anatomical brain atlas of Paxinos and Watson. *Right side* refers to frontal. (**a**) *Right*: A typical 3 s lasting electrographical spike-wave discharge recorded (with negativity up) with the cortical grid that covers a great part of the lateral neocortex with position of the electrodes and their labels on the *left*. (**b**) Time courses of the cortico-cortical nonlinear associations (*top panel*) and time delays (*bottom panel*) for several sites (as indicated by the *black arrows* on the *left*) with respect to the focal site (electrode 8). The association and time delays were assessed for successive 50 % overlapping 500 msec epochs. For comparison the pictures on the *left* depict the average overall associations (*top*) and the average overall time delays (*bottom*; in milliseconds). There is a gradual increase in association strength before the start of the seizure and a steep drop in association strength at the end. Before the seizure, time delays are inconsistent, and there is often a zero time lag. During the seizure, time delays are always positive indicating a delay at the different electrode positions compared to position 8, although the magnitude of the delay can vary [45]

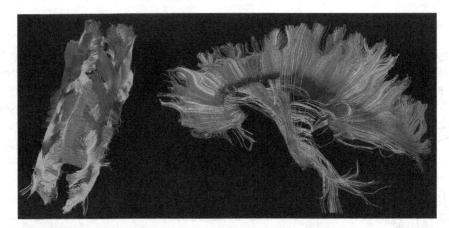

Fig. 7.2 Diffusion tensor imaging (DTI) of the human corpus callosum colored by end point location. *Left*: viewed from the *top*; Anterior side points to the *bottom*. *Right*: lateral view with anterior part pointing to the *left*. Notice the massive cross- hemispheric projections through which seizures might get quickly "generalized" (From Tromp [66]. Reprinted with permission from the author)

The cortex is endowed with a variety of excitatory neocortical projection neurons that play a role in the quick information transfer between homotopical regions of the two hemispheres via long myelinated axons through the corpus callosum. Homotopic regions include the somato-sensory cortex [75], providing an anatomical explanation for fast spread and bilateral involvement shown by the electrophysiological results [45]. In addition to their role in integrating homotopic neocortical regions, callosal projection neurons are also responsible for information transfer within each hemisphere [17]. Finally, there is an abundant and widespread thalamo- cortico-thalamic network; the descending projections to the thalamus "are estimated to outnumber thalamo-cortical ones by an order of magnitude" [61]. The extensiveness of the cortico-thalamo-cortical network, visualized with white matter tractography, can be appreciated in Fig. 7.3. The speed of involvement of the thalamus and cortical spread after local cortical initiation is undoubtedly mediated through these massive networks.

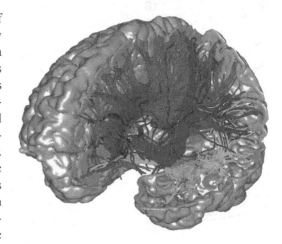

Fig. 7.3 Large-scale model of mammalian thalamo-cortical system based on DTI scans. The massive reciprocal connections between cortex and thalamus are responsible for the quick propagation of SW discharges and other electroencephalographic markers of "generalized" epilepsies from their cortical sites of origin to the thalamus and back. In the illustration, left frontal, parietal, and a part of temporal cortex have been cut to show only a small fraction of white-matter fibers, color-coded according to their destination. *Red*: projections to the frontal cortex, *blue* to parietal cortex, *green* to temporal cortex implying different sources of reentrant axonal fibers connecting one part of the cortex to another [30]

7.6 Generalized Convulsive Seizures

Aside from the archetypical absence epilepsy, a rather heterogeneous ensemble of syndromes constitute the group of primary (idiopathic) generalized epilepsies such as Juvenile Myoclonic Epilepsy, Juvenile Absence epilepsy, and Lennox-Gastaut syndrome, to mention only a few. If growing evidence points to the focal onset for typical absence epilepsy, what about other types of generalized epilepsies? Our pathophysiological understanding of idiopathic generalized epilepsies mainly relies on the ability of the aforementioned and incriminated networks to generate paroxysmal discharges. This susceptibility would thus result from the combination of a paroxysm-inducing mechanism such as arousal ("dyshormia") or photosensitivity and a genetically-prone network [50]. Myoclonus, absence and ultimately GTCS would thus represent a crescendo of clinical manifestations related to this genetic predisposition for generalized paroxysms.

7.6.1 Generalized Myoclonic Seizures

Myoclonus, on the one hand, can be either focal or generalized, and of either cortical, thalamo-cortical, reticular reflex or negative nature, i.e., characterized by the inhibition of muscular activity [49]. In idiopathic generalized epilepsy, such as juvenile myoclonic epilepsy and absence myoclonic epilepsy, myoclonus is supposedly of thalamo-cortical nature and is associated with generalized EEG discharges. Interestingly mild peri-oral myoclonus has also been described in typical absence epilepsy [26], thus concerning the same cortical regions supposedly driving SW discharges in the genetic absence models [45].

Experimentally, myoclonic seizures can be triggered in rodents, either by electrical or pharmacological stimulation by $GABA_A$ receptor antagonists such as bicuculline, picrotoxin and pentylenetetrazole or flurothyl; any of these procedures induce GTCSs with an initial, variable myoclonic phase, thus being slightly different from human generalized myoclonic epilepsy. Local application of most pro-convulsant drugs onto the cortex also elicits myoclonus of focal origin [71]. In a genetic model such as the photosensitive *Papio papio* baboons, generalized myoclonic discharges appear to start in the fronto-rolandic cortex [18]. In human idiopathic generalized epilepsies, evidence for asymmetry, asynchrony and ultimately focal onset of EEG generalized discharges has been gathered through the years; patients with heterogeneous primary generalized epilepsy have been studied using repetitive EEG showing the consistent presence of focal features [39, 50]. A restricted cortical network has been described during typical "generalized" 4–6 Hz seizure propagation in juvenile myoclonic patients; this includes regions of frontal and temporal cortex [28]. Another study using Jerk-locked averaging in JME patients pinpointed a frontal cortical generator [51]. The association of JME with some particular personality type, as described by Janz [33], has been related to fronto-cortical disturbances. Neuropsychological studies confirmed verbal and visual memory impairment along with disturbed visuospatial processing and working memory alteration [63].

Neuropathological studies have also revealed microdysgenesis in idiopathic generalized epilepsy [44]. Hence, these results highlight an early cortical involvement in juvenile myoclonic epilepsy. FMRI studies in patients with idiopathic generalized epilepsy, again due to the poor temporal resolution of fMRI, have failed to demonstrate early focal activation. However, a consistent pattern of thalamic activation and cortical default-mode network deactivation were described during idiopathic generalized epilepsy [22, 38, 46], suggesting a common pathophysiology for generalized SW discharges among idiopathic generalized epilepsies. Suspension of default-mode network represent the earliest BOLD signal change to be observed and may thus hide a more discrete cortical onset, whereas later thalamic activation account for the sustained SW discharge. And indeed, BOLD pattern recorded during photoparoxysmal generalized discharges, a rather cortical electroencephalographic trait of photosensible generalized epilepsy, does not concern thalamus [46].

7.6.2 Primary Generalized Tonic Clonic Seizures

Primary GTCS, the third major type of seizure present in generalized epilepsy is classically differentiated from secondary GTCS that occur in partial epilepsy. Use of the same term, for both primary and secondary GTCS seems contradictory despite clinical similarities, since pathophysiology is obviously rather different. MEG studies have shown some discrepancies among the two types of GCTS with regards to levels of close and distant ictal synchronization. Distant synchrony appears higher in primary GTCS whereas increased local synchrony is reported in secondary GCTS [23]. Additionally interhemispheric coherence during secondary GCTS is surprisingly low [21], and variable during the time-course of a seizure [36]. A single-photon positron emission tomography study comparing spontaneous GCTS with electroconvulsive-therapy-induced GCTS showed specific fronto-parieto-temporal along with thalamic activation in bilaterally electroconvulsive-therapy-induced GCTS [10]. Infantile "generalized" spasms can be focal in its etiology [12], and even involve the brainstem, as it is clear from generalized symmetric seizures in hemispherectomized children [37].

7.7   Conclusive Remarks

Focal versus generalized epilepsy is a classical dichotomy inherited from the Jacksonian era, and somewhat confirmed by standard EEG. However, increasing evidence from both structural and functional imaging studies has been gathered though the years to call into question the concept of generalized epilepsy. Recent studies on rodent genetic absence models [40, 45] prove that absence epilepsy, considered as the prototype for generalized epilepsy, may originate from a rather focal, cortico-frontal region. In human, recordings using high density EEG/MEG studies with proper signal analytical techniques [27, 28, 60, 74] also revealed focal features. The traditional view of a widespread recruitment in absence seizures has also been contradicted by fMRI studies demonstrating a rather restricted network of activation and deactivation, mainly corresponding to alterations in the default-mode network [1, 3, 25, 46, 59]. Refuting the concept of generalized epilepsy however, remains almost impossible *per se*, as it consists of hundreds of different epileptic syndromes. Looking for a focal onset in all those syndromes is not realistic and probably unnecessary.

Most studies on "generalized epilepsies" mainly include patients with idiopathic generalized epilepsy as this group represent the majority of so-called generalized epilepsy.

Overtaking the "centrencephalic" theory of the past century bares new ideas about the nature of generalized epilepsies. It is a safe bet that avoiding the use of the term "generalized epilepsy" will benefit for the next generation of epileptologists and patients. Future classifications based on networks properties, along with more specific information about etiology may decrease the emphasis on the classical electroclinical distinction of partial vs. generalized epilepsy. Nowadays the remaining distinction will thus reflect differences in terms of spreading velocity properties of the underlying network rather than of that network size itself. But as conceptual evolution has a tendency to spread rather slowly in the medical community, one can predict that epilepsy will remain to be characterized as either partial or generalized for some time to come.

Acknowledgments We dedicate this manuscript to Phil Schwartzkroin, a friend and colleague, who had a great impact on neurobiological research and on our own studies during the last four decades. He created opportunities for others to promote new concepts and theories. His challenging approach to epilepsy research continues to motivate many of us, and it is indeed mirrored by the question addressed here.

References

1. Aghakhani Y, Bagshaw AP, Benar CG, Hawco C, Andermann F, Dubeau F et al (2004) fMRI activation during spike and wave discharges in idiopathic generalized epilepsy. Brain J Neurol 127 (Pt 5):1127–1144
2. Amor F, Baillet S, Navarro V, Adam C, Martinerie J, Quyen MV (2009) Cortical local and long-range synchronization interplay in human absence seizure initiation. NeuroImage 45(3):950–962
3. Archer JS, Abbott DF, Waites AB, Jackson GD (2003) fMRI "deactivation" of the posterior cingulate during generalized spike and wave. NeuroImage 20(4):1915–1922
4. Avoli M (2012) A brief history on the oscillating roles of thalamus and cortex in absence seizures. Epilepsia 53(5):779–789

5. Bal T, von Krosigk M, McCormick DA (1995) Synaptic and membrane mechanisms underlying synchronized oscillations in the ferret lateral geniculate nucleus in vitro. J Physiol 483(Pt 3):641–663

6. Berg AT, Berkovic SF, Brodie MJ, Buchhalter J, Cross JH, van Emde BW et al (2010) Revised terminology and concepts for organization of seizures and epilepsies: report of the ILAE Commission on Classification and Terminology, 2005–2009. Epilepsia 51(4):676–685

7. Berger H (1933) Über das Elektrenkephalogramm des Menschen. Arch Für Psychiatr Nervenkrankh 100(1): 301–320

8. Bessaih T, Bourgeais L, Badiu CI, Carter DA, Toth TI, Ruano D et al (2006) Nucleus-specific abnormalities of GABAergic synaptic transmission in a genetic model of absence seizures. J Neurophysiol 96(6):3074–3081

9. Blumenfeld H, McCormick DA (2000) Corticothalamic inputs control the pattern of activity generated in thalamocortical networks. J Neurosci Off J Soc Neurosci 20(13):5153–5162

10. Blumenfeld H, Westerveld M, Ostroff RB, Vanderhill SD, Freeman J, Necochea A et al (2003) Selective frontal, parietal, and temporal networks in generalized seizures. NeuroImage 19(4):1556–1566

11. Carney PW, Masterton RA, Harvey AS, Scheffer IE, Berkovic SF, Jackson GD (2010) The core network in absence epilepsy. Differences in cortical and thalamic BOLD response. Neurology 75(10):904–911

12. Chugani HT, Shields WD, Shewmon DA, Olson DM, Phelps ME, Peacock WJ (1990) Infantile spasms: I. PET identifies focal cortical dysgenesis in cryptogenic cases for surgical treatment. Ann Neurol 27(4):406–413

13. Danober L, Deransart C, Depaulis A, Vergnes M, Marescaux C (1998) Pathophysiological mechanisms of genetic absence epilepsy in the rat. Prog Neurobiol 55(1):27–57

14. Deschenes M, Paradis M, Roy JP, Steriade M (1984) Electrophysiology of neurons of lateral thalamic nuclei in cat: resting properties and burst discharges. J Neurophysiol 51(6):1196–1219

15. Engel J Jr (2006) Report of the ILAE classification core group. Epilepsia 47(9):1558–1568

16. Engel J Jr, International League Against E (2001) A proposed diagnostic scheme for people with epileptic seizures and with epilepsy: report of the ILAE Task Force on Classification and Terminology. Epilepsia 42(6):796–803

17. Fame RM, MacDonald JL, Macklis JD (2011) Development, specification, and diversity of callosal projection neurons. Trends Neurosci 34(1):41–50

18. Fischer-Williams M, Poncet M, Riche D, Naquet R (1968) Light-induced epilepsy in the baboon, Papio papio: cortical and depth recordings. Electroencephalogr Clin Neurophysiol 25(6):557–569

19. Gastaut H (1969) Clinical and electroencephalographical classification of epileptic seizures. Epilepsia 10(Suppl):2–13

20. Gastaut H, Gastaut Y, Roger J, Roger A (1955) Statistical study of the different electroclinical types of epilepsy. Marseille Med 92(10):653–662

21. Gotman J (1987) Interhemispheric interactions in seizures of focal onset: data from human intracranial recordings. Electroencephalogr Clin Neurophysiol 67(2):120–133

22. Gotman J, Grova C, Bagshaw A, Kobayashi E, Aghakhani Y, Dubeau F (2005) Generalized epileptic discharges show thalamocortical activation and suspension of the default state of the brain. Proc Natl Acad Sci U S A 102(42):15236–15240

23. Guevara R, Velazquez JL, Nenadovic V, Wennberg R, Senjanovic G, Dominguez LG (2005) Phase synchronization measurements using electroencephalographic recordings: what can we really say about neuronal synchrony? Neuroinformatics 3(4): 301–314

24. Gupta D, Ossenblok P, van Luijtelaar G (2011) Space-time network connectivity and cortical activations preceding spike wave discharges in human absence epilepsy: a MEG study. Med Biol Eng Comput 49(5):555–565

25. Hamandi K, Salek-Haddadi A, Laufs H, Liston A, Friston K, Fish DR et al (2006) EEG-fMRI of idiopathic and secondarily generalized epilepsies. NeuroImage 31(4):1700–1710

26. Hirsch E, Panayiotopoulos T (2005) Childhood absence epilepsy and related syndromes. In: Roger J, Bureau M, Genton P, Tassinari C, Wolf P (eds) Epileptic syndromes in infancy, childhood, and adolescence. John Libbey Eurotext, Paris, pp 315–335

27. Holmes MD, Brown M, Tucker DM (2004) Are "generalized" seizures truly generalized? Evidence of localized mesial frontal and frontopolar discharges in absence. Epilepsia 45(12):1568–1579

28. Holmes MD, Quiring J, Tucker DM (2010) Evidence that juvenile myoclonic epilepsy is a disorder of frontotemporal corticothalamic networks. NeuroImage 49(1):80–93

29. Huguenard JR, McCormick DA (2007) Thalamic synchrony and dynamic regulation of global forebrain oscillations. Trends Neurosci 30(7):350–356

30. Izhikevich EM, Edelman GM (2008) Large-scale model of mammalian thalamocortical systems. Proc Natl Acad Sci U S A 105(9):3593–3598

31. Jahnsen H, Llinas R (1984) Electrophysiological properties of guinea-pig thalamic neurones: an in vitro study. J Physiol 349:205–226

32. Jahnsen H, Llinas R (1984) Ionic basis for the electro-responsiveness and oscillatory properties of guinea-pig thalamic neurones in vitro. J Physiol 349:227–247

33. Janz D (1985) Epilepsy with impulsive petit mal (juvenile myoclonic epilepsy). Acta Neurol Scand 72(5):449–459

34. Jasper H, Droogleever-Fortuyn J (1946) Experimental studies on the functional anatomy of petit mal epilepsy. Res Publ Assoc Res Nerv Ment Dis 26:272–298

35. Kostopoulos GK (2000) Spike-and-wave discharges of absence seizures as a transformation of sleep spindles: the continuing development of a hypothesis. Clin Neurophysiol 111(Suppl 2):S27–S38

36. Le Van Quyen M, Foucher J, Lachaux J, Rodriguez E, Lutz A, Martinerie J et al (2001) Comparison of Hilbert transform and wavelet methods for the analysis of neuronal synchrony. J Neurosci Methods 111(2):83–98

37. Lhatoo S, Lüders H (2006) The semiology and pathophysiology of the secondary generalized tonic-clonic seizures. In: Hirsch E, Andermann F, Chauvel P, Engel J, Lopes da Silva F, Lüders H (eds) Generalized seizures: from clinical phenomenology to underlying systems and networks. John Libbey Eurotext, Montrouge, pp 229–245

38. Liao W, Zhang Z, Mantini D, Xu Q, Ji GJ, Zhang H et al (2013) Dynamical intrinsic functional architecture of the brain during absence seizures. Brain Struct Funct. doi:10.1007/s00429-013-0619-2

39. Lombroso CT (1997) Consistent EEG, focalities detected in subjects with primary generalized epilepsies monitored for two decades. Epilepsia 38(7):797–812

40. Manning JP, Richards DA, Leresche N, Crunelli V, Bowery NG (2004) Cortical-area specific block of genetically determined absence seizures by ethosuximide. Neuroscience 123(1):5–9

41. Marcus EM, Watson CW (1966) Bilateral synchronous spike wave electrographic patterns in the cat. Interaction of bilateral cortical foci in the intact, the bilateral cortical-callosal, and adiencephalic preparation. Arch Neurol 14(6):601–610

42. Marcus EM, Watson CW (1968) Symmetrical epileptogenic foci in monkey cerebral cortex. Mechanisms of interaction and regional variations in capacity for synchronous discharges. Arch Neurol 19(1):99–116

43. Marescaux C, Vergnes M, Depaulis A (1992) Genetic absence epilepsy in rats from Strasbourg – a review. J Neural Transm Suppl 35:37–69

44. Meencke HJ, Janz D (1984) Neuropathological findings in primary generalized epilepsy: a study of eight cases. Epilepsia 25(1):8–21

45. Meeren HK, Pijn JP, Van Luijtelaar EL, Coenen AM, Lopes da Silva FH (2002) Cortical focus drives widespread corticothalamic networks during spontaneous absence seizures in rats. J Neurosci 22(4):1480–1495

46. Moeller F, Stephani U, Siniatchkin M (2013) Simultaneous EEG and fMRI recordings (EEG-fMRI) in children with epilepsy. Epilepsia 54(6):971–982

47. Moruzzi G, Magoun HW (1949) Brain stem reticular formation and activation of the EEG. Electroencephalogr Clin Neurophysiol 1(4):455–473

48. Musgrave J, Gloor P (1980) The role of the corpus callosum in bilateral interhemispheric synchrony of spike and wave discharge in feline generalized penicillin epilepsy. Epilepsia 21(4):369–378

49. Myoclonus and epilepsy in childhood. Commission on Pediatric Epilepsy of the International League Against Epilepsy (1997) Epilepsia 38(11):1251–1254

50. Niedermeyer E (1996) Primary (idiopathic) generalized epilepsy and underlying mechanisms. Clin EEG 27(1):1–21

51. Panzica F, Rubboli G, Franceschetti S, Avanzini G, Meletti S, Pozzi A et al (2001) Cortical myoclonus in Janz syndrome. Clin Neurophysiol 112(10):1803–1809

52. Parri HR, Crunelli V (1998) Sodium current in rat and cat thalamocortical neurons: role of a non-inactivating component in tonic and burst firing. J Neurosci 18(3):854–867

53. Penfield W (ed) (1957) Consciousness and centrencephalic organization. Premier Congres International des Sciences Neurologiques, Bruxelles

54. Pinault D, Leresche N, Charpier S, Deniau JM, Marescaux C, Vergnes M et al (1998) Intracellular recordings in thalamic neurones during spontaneous spike and wave discharges in rats with absence epilepsy. J Physiol 509(Pt 2):449–456

55. Prince DA, Farrell D (1969) "Centrencephalic" spike wave discharges following parenteral penicillin injection in the cat. Neurology 19:309–310

56. Proposal for revised classification of epilepsies and epileptic syndromes. Commission on Classification and Terminology of the International League Against Epilepsy (1989) Epilepsia 30(4):389–399

57. Proposal for revised clinical and electroencephalographic classification of epileptic seizures. From the Commission on Classification and Terminology of the International League Against Epilepsy (1981) Epilepsia 22(4):489–501

58. Rodin E, Ancheta O (1987) Cerebral electrical fields during petit mal absences. Electroencephalogr Clin Neurophysiol 66(6):457–466

59. Salek-Haddadi A, Lemieux L, Merschhemke M, Friston KJ, Duncan JS, Fish DR (2003) Functional magnetic resonance imaging of human absence seizures. Ann Neurol 53(5):663–667

60. Stefan H, Lopes da Silva FH (2013) Epileptic neuronal networks: methods of identification and clinical relevance. Front Neurol 4:8

61. Steriade M (2001) The GABAergic reticular nucleus: a preferential target of corticothalamic projections. Proc Natl Acad Sci U S A 98(7):3625–3627

62. Steriade M, Contreras D (1995) Relations between cortical and thalamic cellular events during transition from sleep patterns to paroxysmal activity. J Neurosci Off J Soc Neurosci 15(1 Pt 2):623–642

63. Swartz BE, Simpkins F, Halgren E, Mandelkern M, Brown C, Krisdakumtorn T et al (1996) Visual working memory in primary generalized epilepsy: an 18FDG-PET study. Neurology 47(5):1203–1212

64. Taylor J, Holmes GS, Walshe FS (1931) Selected writings of John Hughlings Jackson. Hodder and Stoughton, London

65. Tenney JR, Fujiwara H, Horn PS, Jacobson SE, Glauser TA, Rose DF (2013) Focal corticothalamic sources during generalized absence seizures: a MEG

study. Epilepsy Res 106(1–2):113–122. doi:10.1016/j. eplepsyres.2013.05.006

66. Tromp D. University of Wisconsin-Madison. http:// brainimaging.waisman.wisc.edu/~tromp/resources. html#images

67. Tsakiridou E, Bertollini L, de Curtis M, Avanzini G, Pape HC (1995) Selective increase in T-type calcium conductance of reticular thalamic neurons in a rat model of absence epilepsy. J Neurosci Off J Soc Neurosci 15(4):3110–3117

68. van Luijtelaar EL, Coenen AM (1986) Two types of electrocortical paroxysms in an inbred strain of rats. Neurosci Lett 70(3):393–397

69. van Luijtelaar G, Sitnikova E (2006) Global and focal aspects of absence epilepsy: the contribution of genetic models. Neurosci Biobehav Rev 30(7):983–1003. doi:10.1016/j.neubiorev.2006.03.002

70. van Luijtelaar G, Sitnikova E, Littjohann A (2011) On the origin and suddenness of absences in genetic absence models. Clin EEG Neurosci 42(2):83–97

71. Veliskova J, Velisek L (2006) Animal models of myoclonic seizures and epilepsies. In: Hirsch E, Andermann F, Chauvel P, Engel J, Lopez da Silva F, Luders H (eds) Generalized seizures: from clinical phenomenology to underlying systems and networks. John Libbey Eurotext, Montrouge, pp 147–161

72. Vergnes M, Marescaux C, Lannes B, Depaulis A, Micheletti G, Warter JM (1989) Interhemispheric desynchronization of spontaneous spike-wave discharges by corpus callosum transection in rats with petit mal-like epilepsy. Epilepsy Res 4(1):8–13

73. Vergnes M, Marescaux C, Micheletti G, Reis J, Depaulis A, Rumbach L et al (1982) Spontaneous paroxysmal electroclinical patterns in rat: a model of generalized non-convulsive epilepsy. Neurosci Lett 33(1):97–101

74. Westmijse I, Ossenblok P, Gunning B, van Luijtelaar G (2009) Onset and propagation of spike and slow wave discharges in human absence epilepsy: a MEG study. Epilepsia 50(12):2538–2548. doi:10.1111/j.1528-1167.2009.02162.x

75. Yorke CH Jr, Caviness VS Jr (1975) Interhemispheric neocortical connections of the corpus callosum in the normal mouse: a study based on anterograde and retrograde methods. J Comp Neurol 164(2):233–245. doi:10.1002/cne.901640206

Part II

Synaptic Plasticity

Are There Really "Epileptogenic" Mechanisms or Only Corruptions of "Normal" Plasticity?

Giuliano Avanzini, Patrick A. Forcelli, and Karen Gale

Abstract

Plasticity in the nervous system, whether for establishing connections and networks during development, repairing networks after injury, or modifying connections based on experience, relies primarily on highly coordinated patterns of neural activity. Rhythmic, synchronized bursting of neuronal ensembles is a fundamental component of the activity-dependent plasticity responsible for the wiring and rewiring of neural circuits in the CNS. It is therefore not surprising that the architecture of the CNS supports the generation of highly synchronized bursts of neuronal activity in non-pathological conditions, even though the activity resembles the ictal and interictal events that are the hallmark symptoms of epilepsy. To prevent such natural epileptiform events from becoming pathological, multiple layers of homeostatic control operate on cellular and network levels. Many data on plastic changes that occur in different brain structures during the processes by which the epileptogenic aggregate is constituted have been accumulated but their role in counteracting or promoting such processes is still controversial. In this chapter we will review experimental and clinical evidence on the role of neural plasticity in the development of epilepsy. We will address questions such as: is epilepsy a progressive disorder? What do we know about mechanism(s) accounting for progression? Have we reliable biomarkers of epilepsy-related plastic processes? Do seizure-associated plastic changes protect against injury and aid in recovery? As a necessary premise we will consider the value of seizure-like activity in the context of normal neural development.

Keywords

Seizure • Epilepsy • Epileptogenesis • Neural plasticity • Synaptic plasticity

G. Avanzini (✉)
Fondazione I.RC.C.S. Istituto Neurologico
Carlo Besta, Via Celoria 11, 20133 Milan, Italy
e-mail: avanzini@istituto-besta.it

P.A. Forcelli, Ph.D. • K. Gale, Ph.D.
Department of Pharmacology and Physiology,
Georgetown University, Washington, DC 20057, USA
e-mail: paf22@georgetown.edu; galek@georgetown.edu

H.E. Scharfman and P.S. Buckmaster (eds.), *Issues in Clinical Epileptology: A View from the Bench*,
Advances in Experimental Medicine and Biology 813, DOI 10.1007/978-94-017-8914-1_8,
© Springer Science+Business Media Dordrecht 2014

The first evidence for experimentally induced plastic changes in nervous system was provided by Minea in 1907. In his thesis, he described metamorphic phenomena in sensory neurons provoked by the compression and transplantation of ganglia into various organs [84, quoted by Marinesco 1909]. In the years that have passed, the study of neural plasticity has become a very important line of research in the neurosciences, with the aim of uncovering the neurobiological bases for the exquisite capability of the nervous system to adapt to environmental changes. Plasticity in the nervous system, whether for establishing connections and networks during development, repairing networks after injury, or modifying connections based on experience, relies primarily on highly coordinated patterns of neural activity. Rhythmic, synchronized bursting of neuronal ensembles is a fundamental component of the activity-dependent plasticity responsible for the wiring and rewiring of neural circuits in the CNS. It is therefore not surprising that the architecture of the CNS supports the generation of highly synchronized bursts of neuronal activity in non-pathological conditions, even though the activity resembles the ictal and interictal events that are the hallmark symptoms of epilepsy. To prevent such natural epileptiform events from becoming pathological, multiple layers of homeostatic control operate on cellular and network levels. While there are extensive data concerning the plastic changes that occur in brain structures during the process of epileptogenesis, the role of these changes in counteracting or promoting epileptogenic processes remains controversial.

In this chapter we will review experimental and clinical evidence on the role of neural plasticity in the development of epilepsy. We will address questions such as: is epilepsy a progressive disorder? What do we know about mechanism(s) accounting for progression? Have we reliable biomarkers of epilepsy-related plastic processes? Do seizure-associated plastic changes protect against injury and aid in recovery? Moreover, as a necessary premise we will consider the value of seizure-like activity in the context of normal neural development.

8.1 Modeling and Remodeling of Network Architecture During Development

The architecture of neuronal connectivity in the CNS is shaped through a process of functional validation. During postnatal development, this process depends primarily on environmental input and sensory stimulation, while during prenatal development, spontaneous patterned activity is largely generated intrinsically. Neuronal activity is essential for guiding synapse formation, remodeling, and elimination, so as to establish optimal connectivity. For example, well in advance of eye opening, embryonic retinal ganglion cells generate rhythmic bursts of action potentials in both rodent [83, 127] and primate [70]; this highly correlated bursting is required for establishing retinotopic maps in the connections across the neuraxis. This activity is highly synchronized within populations of neighboring neurons, and propagates throughout the visual pathway [2], so that waves of stimulation in defined regions of the retina can then coordinate the activity-dependent refinement of corresponding eye-specific layers in the lateral geniculate nucleus [83, 127]. Moreover, the spontaneous bursting is relayed via thalamocortical projections to visual cortex, where it can shape the emerging ocular dominance columns [52]. A similar pattern of correlated bursting activity occurs pre-hearing in the developing auditory system, from the level of the cochlear ganglion cells to the brainstem auditory pathways [57, 58]. These spontaneous, highly synchronized and propagated rhythmic bursting patterns, which share many characteristics of ictal and interictal events, are a classic example of the developmental principle that neurons that fire together, wire together [23, 47].

Another classic example of highly synchronized rhythmic bursting of neuronal populations in utero takes place in spinal cord motoneurons. In fact, the spontaneous waves of hypersynchronous activity in this system have been characterized as epileptiform activity [101]. These discharge patterns propagate through the spinal cord, triggering transregional synchronization and fast rhythmic repetitive limb movements described as clonus and convulsive-like in nature [14, 94]. While clearly a normal and adaptive feature of prenatal development, a similar pattern of activity in a postnatal organism would be considered pathological.

Before discussing the role of plastic mechanisms in epileptogenesis, we will review the evidence for a progressive course of epilepsies. For the purposes of this chapter we will rely to the following arbitrary definitions:

Epileptic mechanisms responsible for seizure generation consist of changes in cellular excitability leading to excessive, disordered discharges underlying ictal manifestation.

Epileptogenic mechanisms responsible for epilepsy, i.e. an enduring propensity to generate epileptic seizures [32] consist of some hypothetical neurobiological processes leading to a permanently dysfunction of the neuronal network/system.

8.2 Epilepsy as a Progressive Disorder

Clinical and experimental observations suggest that an acute "initial event" (e.g. traumatic, infectious) can set in motion a series of degenerative, regenerative and inflammatory changes resulting in a permanent epileptic neuronal aggregate. A crucial role is attributed to the epileptic activity both in the initial event (e.g. febrile seizure/status) and in the ensuing process leading to chronic epilepsy. Evidence in some patients for a progressive increase in the risk for seizures with increasing number of seizures is currently quoted in support of Gower's statement that seizures beget seizures [24]. Indeed, in support of the notion that seizures beget seizures, Hauser and Lee [44] found a significant increase in risk for subsequent seizures with increasing seizure number in a population of patients who are generally considered to have a good prognosis for going into remission: those with unprovoked seizures of unknown cause, normal neurological examination,

and normal EEG. It is worth saying, however, that there are many types of epilepsies that do not progress, in spite of seizure repetition. These non-progressive epileptic syndromes include benign childhood epilepsies with centrotemporal spikes [136], benign occipital epilepsies, childhood and juvenile absence epilepsies [137], juvenile myoclonic epilepsy, benign familial neonatal, infant and neonatal-infant epilepsies [124], and autosomal dominant nocturnal frontal lobe epilepsies. For many of these syndromes seizure activity decreases or disappears with age [40, 136].

A progressive course toward drug refractoriness can be observed only in some types of human epilepsies currently grouped under the definition of epileptic encephalopathies (EEs) [6]. As for mesial temporal lobe epilepsy, a role of repeated seizures for inducing progressive cumulative alterations in neural circuits, resulting in progression of epilepsy severity, has been assumed (based on experimental results with rodent kindling models) but never demonstrated. Patients with mesial temporal lobe epilepsy (MTLE) and EEs substantially contribute to the 30–40 % of patients with epilepsy who show drug resistance [27]; this population represents the main unsolved problem in clinical epileptology. This explains the great investment in research lines aimed at unraveling the neural mechanisms responsible for these types of epilepsies and the need to elaborate strategies capable of preventing their development.

The latent period. In several instances the natural history of MTLE indicates an initial precipitating event as an underlying cause of a chronic epilepsy [79, 80]. In some cases (trauma, infection, autoimmune process), the initial event is associated with repeated seizures often presenting as status epilepticus (SE). Moreover, a significant proportion of patients with mesial temporal sclerosis and MTLE had antecedents of complex febrile seizures in early childhood [19]. Between the initial event and the onset of the chronic epilepsy there is an interval of variable duration currently referred to as latent period. During this period, biological changes may occur that are considered to substantiate the epileptogenic process. For these reasons much interest is focused on animal models that reproduce the typical sequence of initial event-latent period-chronic epilepsies.

The pilocarpine and kainic acid models are both obtained by the acute administration of a chemoconvulsant agent (pilocarpine or kainic acid) to rodents; this treatment induces a state of prolonged SE, followed by spontaneous recurrent seizures beginning after a variable latency (15–20 days) [18, 92, 128]. It must be said that that the existence of the latent period in chemoconvulsant rodent models has been disputed [118] and that, in view of the inter-individual variability of its duration, the possibility that it simply reflects the outer fringe of a probabilistic spread seizure latencies must taken into account. Indeed an impressive bulk of published results suggest that during the latent period several changes occur in hippocampal structures that are associated with the alteration of excitability and synchronization and may hypothetically account for epileptogenesis. For example, axonal sprouting, synaptic reorganization, gene and protein expression, neurogenesis, gliosis and functional glial alterations, inflammation, and angiogenesis have all been suggested to contribute to epileptogenesis in these models (see [125] for a review). Interestingly these changes can also be found in temporal lobe tissue samples from patients who underwent epilepsy surgery for drug refractory MTLE (see [22] for a review). Obviously, the fact that these changes occur during an ongoing epileptogenic process does not prove that they are necessary contributors to disease pathogenesis until the prevention of any of them is unequivocally proved to prevent the development of later epilepsy. For example, the mossy fiber sprouting that characteristically occurs after SE and is thought to be a hallmark feature of the post-SE neuroplasticity, has been demonstrated not to be necessary for the development of spontaneous recurrent seizures [25, 48, 86, 95, 140]. Moreover the histopathological alterations observed in pilocarpine and kainic models are not limited to the mesial temporal lobe structures, raising a question about their validity as MTLE models (e.g., [17, 21, 66, 129]).

From the clinical standpoint, it is not clear that MTLE results from a process that is sustained or facilitated by epileptic activity. The analysis of the natural history of MTLE cannot answer the question of whether unfavorable outcomes are due to the persistence of epileptic activity (which is usually undetectable in the latent period between the initial event and the chronic phase), or if it is instead a product of the underlying neuroplasticity set in motion by the initial event. While previous prospective longitudinal analyses have been inconclusive [113], the results of the ongoing FEBSTAT (Consequences of Prolonged Febrile Seizures) prospective study [72, 90] may clarify this issue. Several studies have shown that only symptomatic SEs correlate with brain damage and late epilepsy [45, 64, 103, 114]. This makes it impossible to point to a necessary role of epileptic activity above and beyond (or apart from) the role of underlying lesion causing SE for the initiation and maintenance of the epileptogenic process [46].

The statement "seizures beget seizures" implies a role of epileptic activity not only in initiating the epileptogenic process but also in maintaining it, suggesting a further progression of epilepsy toward a more severe state. Whereas there is some evidence consistent with possible acute seizure-associated epileptogenic changes (see review in [15]), a subsequent correlation of recurrent seizures and progression to the clinico-pathological picture of drug refractory MTLE with hippocampal sclerosis has not been unequivocally demonstrated. A prospective analysis of 103 patients with newly diagnosed focal epilepsy [107] showed a decrease in hippocampal volume after 1–3 years in 13 % of patients. Here again, the coexistence of uncontrolled seizures and hippocampal atrophy in a limited subset of patients is insufficient to prove a clear causal relationship between recurrent seizures and atrophy. In fact, sporadic reports on acute, possibly inflammatory, damage to the hippocampus (localized limbic encephalitis; [11] and febrile SE [113]) suggest that acute seizures and hippocampal damage can develop within weeks/months, resulting in a MTLE pattern [59, 121] with little, if any, signs of further progression of structural damage as detected by imaging, in spite of the persistence of seizure activity [71].

Animal models of MTLE based on acute induction of SE by pilocarpine and kainic acid do not clarify this issue, because once the spontaneous recurrent seizures have fully developed in the late chronic period, they do not tend to worsen. In fact, in a similar model of focal tetanus toxin-induced spontaneous seizures in hippocampus, the spontaneous seizures last for about 6 weeks and then tend to subside [55]. More recent data obtained in an animal model of post-traumatic epilepsy, indicates that following brain injury induced in mice or rats by a controlled cortical impact (CCI), once spontaneous seizures appear, they maintain a fairly constant frequency and severity and do not appear to worsen over an extended time period [13].

It is likely that numerous types of plasticity accompany the process by which epilepsy develops. For example, in addition to changes that occur in association with the primary site of epileptogenesis, there is experimental evidence for secondary epileptogenesis occurring in distant sites as a result of the abnormal activation of synaptic projections coming from the primary site (primary focus). Thus, the primary focus induces similar paroxysmal behavior (secondary focus) in the cellular elements of the otherwise normal network [87]. This may explain some of the cases in which epileptic seizures appear to become more severe, as a result of recruiting additional pathways into the network of seizure propagation.

At the same time, some of the plasticity associated with repeated seizure activity may give rise to compensatory processes that serve to limit or prevent the development of chronic state of seizures or epilepsy. Similarly, other aspects of the plasticity may compensate for the original damage that triggered the seizure activity. These types of plasticity and their mechanisms have been investigated in experimental animals in which repeated brief seizures (induced by electrical kindling stimulation, focal tetanus toxin, or electroshock treatments) cause little or no neuronal damage [10, 38, 56, 68, 78, 82, 130]. These repeated seizures have been shown to activate a host of genomic responses in the adult brain,

ranging from immediate early genes, genes for neurotrophic factors and neuropeptides, as well as multiple alterations in the regulation of neurotransmitters and their receptors in various brain regions [5, 12, 33, 41, 53, 65, 67, 73, 85, 88, 89, 97, 102, 104, 106, 110, 117, 126, 139]. In most cases, these responses include the induction or modulation of numerous trophic and neuroprotective factors such as bFGF[(FGF-2) [41]], NGF and BDNF [5], GDNF [3], and heparin-binding EGF-like growth factor [115] in various brain regions. These factors are responsible for triggering neuroplasticity, synaptogenesis and even neurogenesis, at the same time that they confer resistance to injury.

8.3   Neuroprotection, Repair and Recovery After Brain Injury

A consequence of the induction of trophic and neuroprotective factors following repeated non-injurious seizures is a dramatic neuroprotective state in which the seizure-exposed animals become resistant to neuronal damage as evidenced by histopathology and sensitive molecular markers of cell death [63, 66, 78, 98]. Because neuroprotection requires multiple seizure treatments over several days (a single treatment is not protective), there is likely to be a cumulative buildup of resistance to injury. This suggests that seizures in the adult brain may be an endogenous therapeutic mechanism to recruit trophic cascades that promote neuronal survival and recovery in the face of degenerative insults or traumatic injury.

It is noteworthy that the type of seizures that are effective in conferring a neuroprotective effect on forebrain regions are seizures that last only several seconds and engage limbic forebrain networks either through kindling [63, 98] or by minimal electroshock administered via corneal electrodes [66, 78]. These stimuli produce characteristic signs of limbic-motor seizures (facial and forelimb clonus with rearing in the rodent), typically without the tonic-clonic motor manifestations characteristic of seizures that have spread to brainstem seizure-generating sites. In sharp contrast, repeated exposure to seizures that only engage brainstem seizure circuitry (and not forebrain limbic networks) and evoke only tonic-clonic motor responses (without limbic-motor seizures) [16] fails to confer neuroprotection and may even worsen neural injury [4]. These observations emphasize the network-specificity of the protective responses.

Consistent with this concept of network specificity, the regional induction of mRNA for bFGF is very different following maximal electroshock seizures as compared to low-intensity (minimal) electroshock seizures, even when the two types of seizures are induced by corneal electrodes [33]. The minimal seizures increased bFGFmRNA levels by 350 % in entorhinal cortex by 5 h, whereas at the same timepoint after maximal seizures, the increase was only 200 %. Similarly, the increase in bFGF mRNA in hippocampus was greater after minimal seizures than after maximal seizures. In contrast, maximal seizures, but not minimal seizures, induced increases in bFGF mRNA in striatum and cerebellum [33]. This suggests that the minimal seizures, which are more selective for activating limbic forebrain networks may be more efficacious in triggering neuroplastic changes in those networks as compared with more generalized seizures. The circuit-specificity of seizure-induced neuroprotection also indicates that nonspecific responses such as stress associated with repeated seizures, various endocrine changes and other nonspecific physiological responses to seizures cannot account for the neuroprotection that appears to be selective for seizures involving limbic forebrain activation. Instead, adaptive changes restricted to the network through which the seizures propagate appear to be required for the neuroprotective state.

The fact that exogenous infusion of bFGF directly into hippocampus can protect against excitotoxic neuronal injury [75, 76] indicates that an increase in bFGF protein, as observed after several days of electroshock seizures [41] may account for a component of seizure-induced neuroprotection. It is also likely that multiple neuroprotective adaptive responses are engaged by repeated seizures and that the relative importance of any given factor may vary with cell type and brain region.

Because very brief seizures are remarkably protective even in the complete absence any evidence of injury or cellular stress, it may be that seizures serve to trigger activity-dependent mechanisms of neuroplasticity that recapitulate the injury-resistant and resilient conditions characteristic of development. In the face of insults to the nervous system, transient recurrent seizure activity could serve to attenuate neurodegeneration and promote regrowth and remodeling in the network affected by an insult. An especially robust demonstration of this type of protective action comes from a study in which daily electroshock seizures were administered to adrenalectomized rats [78]. Removal of the adrenal glands in adult rats leads to a highly selective apoptotic degeneration of the dentate granule cells in the hippocampus [74, 119, 120], reflecting the fact that these neurons are directly or indirectly dependent upon adrenal corticosteroids [39]. These seizures completely prevented the dentate granule cell degeneration, while sham treatments or daily exposure to restraint stress did not alter the profile of degeneration seen in the adrenalectomized animals.

Moreover, daily exposure to brief seizures has been shown to accelerate recovery of function following cortical damage, probably by enhancing post-injury plasticity in local and distant networks connected to the site of injury [43, 50]. Thus, clinical, and possibly subclinical, seizures that occur transiently during the post-traumatic period may serve an adaptive function: to reduce injury, promote repair and trigger compensatory plasticity. If this is the case, then the frequent procedure of placing patients on prophylactic anticonvulsant therapy immediately following either head injury or neurosurgical intervention may potentially retard or diminish functional recovery [49, 111]. In this context, seizures may be analogous to fever—a symptom that may have adaptive and protective value in specific pathological settings; and like fever, seizures can become maladaptive and injurious in their own right in the rare cases where they go beyond a self-limiting state and evolve into a chronic epileptic condition.

During nervous system development, a dynamic continuum between neuronal plasticity and neuronal death exists and coordinated synaptic stimulation can 'protect' or 'select' the population of neurons or synapses that will be maintained. In an analogous fashion, seizure evoked stimulation may allow the sparing of otherwise vulnerable populations of neurons in the injured adult brain, a phenomenon that we have referred to as 'excitotrophic' [35]. This could account for the therapeutic benefits of controlled administration of electroshock seizures in various neurodegenerative disorders including Parkinson's and Huntington's Disease [1, 8, 9, 30, 34, 60, 62, 91, 99, 108, 122], but it remains to be determined if the seizures slow the disease progression. Currently electroshock seizures are used in the treatment of bipolar affective disorders and their therapeutic impact may derive from neuro-protective actions [77]. Further characterization of the mechanisms contributing to the neuroprotective impact of brief seizure episodes may generate novel strategies of neuroprotection and recovery of function following excitotoxic insults and other forms of injury to the central nervous system.

8.4   Controlled Patterns of Hyper-synchronous Discharge in Certain Subcortical Networks

As Stevens had observed, "Rapid neuronal discharge (bursting), although typical of epileptic discharge, is part of the normal brain repertoire and does not necessarily signal pathology" [123]. These events can be distinguished from pathological seizure activity in that they occur in highly confined areas and/or during highly restricted time periods (such as during certain phases of sleep). The networks of the limbic system, brainstem, and diencephalon are organized in such a way as to give rise to highly synchronized bursting in discrete nuclei in association with certain physiological states or functions such as parturition, growth hormone release, milk ejection, ovulation, and orgasm (see discussion in [123]). In association with the estrus cycle, neuronal spiking and coordinated burst discharges in nuclei of the basal hypothalamus and forebrain limbic preoptic area appear to coordinate the cyclical and pulsatile release of reproductive hormones [61]. Stevens [123] described this type of highly regulated hypersynchronous discharge as "microseizures" that serve to augment signal transmission for critical species-specific survival functions, but do not propagate beyond highly restricted circuitry due to surrounding inhibitory control. In conditions in which the inhibitory control mechanisms are compromised, such microseizures could potentially propagate beyond their physiologically appropriate boundaries. The fact that certain phases of sleep [54, 96, 116] or hormonal cycles [7, 26, 31, 112] are associated with increased vulnerability to seizures may be a reflection of this natural fluctuation in physiological microseizure discharge.

The limbic system network is also organized to generate synchronized, reverberatory discharge characteristic of Hebbian cell assemblies [47] that instantiate memories via temporal lobe circuitry. This core feature of associative learning is reminiscent of the activity-dependent plasticity that drives the shaping of neuronal connectivity during development. But in the case of learning in the mature brain, there is a selective strengthening of specific synaptic connections within a network in a highly defined spatiotemporal pattern. The limbic network comprised of the hippocampus, amygdala, mediodorsal thalamus, and piriform and rhinal cortices is especially suited to the amplification of discharge patterns via reverberatory loops between the nuclei. The ability to amplify repetitive discharge originating at one site, such as occurs during the process of kindling from the amygdala or other sites within the limbic network, is a reflection of the propensity for activity-dependent plasticity in this system. In fact, kindling has been used as a model of learning and memory [36], especially because once an animal is fully kindled, the remodeling of the network that supports the kindled state is relatively permanent [37]. At the same time, it is curious that the network would be poised to amplify the stimulation to the point that propagated seizures emerge, considering the fact that these seizures have no clear adaptive function. Perhaps by providing near-physiological stimulation in a repetitive manner, the kindling process lures the network into an amplification process until it becomes hijacked by the long-lasting modifications associated with the repeated seizures.

The transfer effect in kindling, in which pre-kindling from one area reduces the number of stimulations necessary to kindle from another area, may likewise reflect plasticity within the limbic seizure network (or the recruitment of new components to the network, as discussed above in the context of secondary seizure foci). However, transfer of kindling within the limbic system appears to require other networks (e.g., brainstem). For example, transfer of amygdala kindling is impaired by transection of or damage to midbrain and brainstem [20, 42, 51, 132]. This suggests that repeated limbic stimulation can alter the functions of other networks, either by actively recruiting them into a transfer process, or perhaps by disrupting endogenous seizure-suppressive functions of these extra-limbic circuits.

The fact that the process of kindling can be retarded or suppressed by the occurrence of generalized convulsive seizures [69, 93, 100] emphasizes the importance of homeostatic mechanisms for the process. The same is true for kindling using chemoconvulsants or corneal electroshock [28, 29, 105, 109, 138]: repeated minimal (threshold) limbic seizures become amplified over time, while repeated maximal seizures induce a seizure-resistance over time.

It is, however, essential to recognize that the vulnerability to kindling is highly species-specific, with the rate of kindling taking days to weeks in rodents [37, 81], weeks to months in cats [37, 42, 135], and months to years (with a relatively low success rate) in primates [37, 131, 133, 134]. These species differences may reflect the extent to which the limbic network is under inhibitory control from an increasingly elaborated frontal cortex. Thus, the kindling phenomenon, which

has been most thoroughly characterized in the rodent, may not readily generalize to humans or to human clinical conditions.

8.5 Conclusions

Abundant evidence supports a neuroplasticity-inducing action of seizures and seizure-like events in the CNS. However, the extent to which the neuroplasticity serves an adaptive function vs. a maladaptive function depends on the context in which the seizure activity occurs. During fetal CNS maturation, neuroplasticity induced by naturally-occurring ictal activity and seizure-like phenomena promotes the formation of neural connections. Similarly, in the aftermath of injury in the mature CNS, limited seizure activity may promote neural repair and compensation and serve a neuroprotective role. However, the circumstances in which seizure activity serves a "normal" function typically involve seizure activity that is highly limited temporally and/or spatially (e.g., in specific circuitry, during specific developmental stages, or within a short period post injury). In the small percentage of cases in which the seizure activity does not remain highly controlled and limited, it becomes pathological, repeatedly interrupting normal function with maladaptive, and even potentially injurious consequences. The epilepsies represent this type of pathological seizure occurrence, and it is likely that some of the associated neuroplasticity impairs normal CNS function. Whether the neuroplasticity is also an essential component of the process of epileptogenesis remains to be determined, but since chronic epilepsy occurs only in a small percentage of individuals, we first need to understand the unique features that render those individuals susceptible, and whether the unique features change the nature of the seizure-induced plasticity in those individuals.

We can therefore conclude that epileptogenic mechanisms may indeed be a corruption of normal, adaptive neuroplasticity. If the normal neuroplasticity associated with limited, controlled seizure activity is largely helpful, turning pathophysiological in rare circumstances, this raises several challenging questions for future epilepsy research to address:

1. Do adaptive and maladaptive neuroplasticity differ, and if so, how?
2. If there are distinctions between adaptive and maladaptive neuroplasticity, can we selectively prevent the maladaptive with compromising the adaptive?
3. Which control mechanisms that normally prevent seizures from becoming repetitive and self-sustaining become compromised in individuals susceptible to epilepsy? Is it possible that it is the failure of these control mechanisms, rather than neuroplasticity, that is essential for epileptogenesis?

Acknowledgments The Authors would like to express their pleasure in contributing to this book published in honor of Phil Schwartzkroin, a good friend and great scientist. Since his earliest publications in the 1970s Phil's work has been a point of reference for all people working in epileptology. Moreover, his expert and wise support of other people's research as editor of Epilepsia will not be forgotten by the international epilepsy community. We look forward to benefiting for many years from his commitment to the development of our discipline.

References

1. Abrams R (1989) ECT for Parkinson's disease. Am J Psychiatry 146:1391–1393
2. Ackman JB, Burbridge TJ, Crair MC (2012) Retinal waves coordinate patterned activity throughout the developing visual system. Nature 490:219–225
3. Anastasía A, Wojnacki J, de Erausquin GA, Mascó DH (2011) Glial cell-line derived neurotrophic factor is essential for electroconvulsive shock-induced neuroprotection in an animal model of Parkinson's disease. Neuroscience 195:100–111
4. André V, Ferrandon A, Marescaux C, Nehlig A (2000) Electroshocks delay seizures and subsequent epileptogenesis but do not prevent neuronal damage in the lithium-pilocarpine model of epilepsy. Epilepsy Res 42:7–22
5. Angelucci F, Aloe L, Jiménez-Vasquez P, Mathé AA (2002) Electroconvulsive stimuli alter the regional concentrations of nerve growth factor, brain-derived neurotrophic factor, and glial cell line-derived neurotrophic factor in adult rat brain. J ECT 18:138–143
6. Avanzini G, Capovilla G (eds) (2013) Epileptic encephalopathies: proceedings of the international Sicilian workshop. Epilepsia 54:1–50
7. Badawy RAB, Vogrin SJ, Lai A, Cook MJ (2013) Are patterns of cortical hyperexcitability altered in catamenial epilepsy? Ann Neurol 74:743–757
8. Balldin J, Edén S, Granérus AK, Modigh K, Svanborg A, Wålinder J, Wallin L (1980) Electroconvulsive therapy in Parkinson's syndrome with "on-off" phenomenon. J Neural Transm 47:11–21
9. Beale MD, Kellner CH, Gurecki P, Pritchett JT (1997) ECT for the treatment of Huntington's disease: a case study. Convuls Ther 13:108–112
10. Bertram EH 3rd, Lothman EW (1993) Morphometric effects of intermittent kindled seizures and limbic status epilepticus in the dentate gyrus of the rat. Brain Res 603:25–31
11. Bien CG, Urbach H, Schramm J, Soeder BM, Becker AJ, Voltz R, Vincent A, Elger CE (2007) Limbic encephalitis as a precipitating event in adult-onset temporal lobe epilepsy. Neurology 69:1236–1244
12. Blümcke I, Becker AJ, Klein C, Scheiwe C, Lie AA, Beck H, Waha A, Friedl MG, Kuhn R, Emson P, Elger C, Wiestler OD (2000) Temporal lobe epilepsy associated up-regulation of metabotropic glutamate receptors: correlated changes in mGluR1 mRNA and protein expression in experimental animals and human patients. J Neuropathol Exp Neurol 59:1–10
13. Bolkvadze T, Pitkänen A (2012) Development of post-traumatic epilepsy after controlled cortical impact and lateral fluid-percussion-induced brain injury in the mouse. J Neurotrauma 29:789–812
14. Bradley NS, Ryu YU, Lin J (2008) Fast locomotor burst generation in late stage embryonic motility. J Neurophysiol 99:1733–1742
15. Briellmann RS, Wellard RM, Jackson GD (2005) Seizure-associated abnormalities in epilepsy: evidence from MR imaging. Epilepsia 46:760–766
16. Browning RA, Nelson DK (1985) Variation in threshold and pattern of electroshock-induced seizures in rats depending on site of stimulation. Life Sci 37:2205–2211
17. Castro OW, Furtado MA, Tilelli CQ, Fernandes A, Pajolla GP, Garcia-Cairasco N (2011) Comparative neuroanatomical and temporal characterization of FluoroJade-positive neurodegeneration after status epilepticus induced by systemic and intra-hippocampal pilocarpine in Wistar rats. Brain Res 1374:43–55
18. Cavalheiro EA, Riche DA, Le Gal La Salle G (1982) Long-term effects of intrahippocampal kainic acid injection in rats: a method for inducing spontaneous recurrent seizures. Electroencephalogr Clin Neurophysiol 53:581–589
19. Cendes F (2004) Febrile seizures and mesial temporal sclerosis. Curr Opin Neurol 17:161–164

20. Chiba S, Wada JA (1997) The effect of electrolytic lesioning of the midbrain prior to amygdala kindling in rats. Neurosci Lett 227:83–86
21. Clifford DB, Olney JW, Maniotis A, Collins RC, Zorumski CF (1987) The functional anatomy and pathology of lithium-pilocarpine and high-dose pilocarpine seizures. Neuroscience 23:953–968
22. De Lanerolle NC, Lee TS, Spencer DD (2012) Histopathology of human epilepsy. In: Noebels JL, Avoli M, Rogawski MA, Olsen RW, Delgado-Escueta AV (eds) Jasper's basic mechanisms of the epilepsies. Oxford, New York, pp 387–404
23. Doidge N (2007) The brain that changes itself: stories of personal triumph from the frontiers of brain science. Viking, New York
24. Eadie MJ (2011) William Gowers' interpretation of epileptogenic mechanisms: 1880–1906. Epilepsia 52:1045–1051
25. Ebert U, Löscher W (1995) Differences in mossy fibre sprouting during conventional and rapid amygdala kindling of the rat. Neurosci Lett 190:199–202
26. Edwards HE, Burnham WM, Mendonca A, Bowlby DA, MacLusky NJ (1999) Steroid hormones affect limbic afterdischarge thresholds and kindling rates in adult female rats. Brain Res 838:136–150
27. Engel J Jr (1998) Etiology as a risk factor for medically refractory epilepsy: a case for early surgical intervention. Neurology 51:1243–1244
28. Essig CF (1965) Repeated electroconvulsions resulting in elevation of pentylenetetrazole seizure threshold. Int J Neuropharmacol 4:201–204
29. Essig CF, Flanary HG (1966) The importance of the convulsion in occurrence and rate of development of electroconvulsive threshold elevation. Exp Neurol 14:448–452
30. Faber R, Trimble MR (1991) Electroconvulsive therapy in Parkinson's disease and other movement disorders. Mov Disord 6:293–303
31. Finocchi C, Ferrari M (2011) Female reproductive steroids and neuronal excitability. Neurol Sci 32(Suppl 1):S31–S35
32. Fisher RS, van Emde Boas W, Blume W, Elger C, Genton P, Lee P, Engel J Jr (2005) Epileptic seizures and epilepsy: definitions proposed by the International League Against Epilepsy (ILAE) and the International Bureau for Epilepsy (IBE). Epilepsia 46:470–472
33. Follesa P, Gale K, Mocchetti I (1994) Regional and temporal pattern of expression of nerve growth factor and basic fibroblast growth factor mRNA in rat brain following electroconvulsive shock. Exp Neurol 127:37–44
34. Fregni F, Simon DK, Wu A, Pascual-Leone A (2005) Non-invasive brain stimulation for Parkinson's disease: a systematic review and meta-analysis of the literature. J Neurol Neurosurg Psychiatry 76:1614–1623
35. Gale K (2004) Epilepsy and seizures: excitotoxicity or excitotrophicity? In: Ferrarese C, Beal MF (eds) Excitotoxicity in neurological diseases. Springer, Boston, pp 137–170
36. Goddard GV, Douglas RM (1975) Does the engram of kindling model the engram of normal long term memory? Can J Neurol Sci 2:385–394
37. Goddard GV, McIntyre DC, Leech CK (1969) A permanent change in brain function resulting from daily electrical stimulation. Exp Neurol 25:295–330
38. Gombos Z, Spiller A, Cottrell GA, Racine RJ, McIntyre Burnham W (1999) Mossy fiber sprouting induced by repeated electroconvulsive shock seizures. Brain Res 844:28–33
39. Gould E, Woolley CS, McEwen BS (1990) Short-term glucocorticoid manipulations affect neuronal morphology and survival in the adult dentate gyrus. Neuroscience 37:367–375
40. Guerrini R, Pellacani S (2012) Benign childhood focal epilepsies. Epilepsia 53(Suppl 4):9–18
41. Gwinn RP, Kondratyev A, Gale K (2002) Time-dependent increase in basic fibroblast growth factor protein in limbic regions following electroshock seizures. Neuroscience 114:403–409
42. Hamada K, Wada JA (1998) Amygdaloid kindling in brainstem bisected cats. Epilepsy Res 29:87–95
43. Hamm RJ, Pike BR, Temple MD, O'Dell DM, Lyeth BG (1995) The effect of postinjury kindled seizures on cognitive performance of traumatically brain-injured rats. Exp Neurol 136:143–148
44. Hauser WA, Lee JR (2002) Do seizures beget seizures? Prog Brain Res 135:215–219
45. Hauser WA, Rich SS, Annegers JF, Anderson VE (1990) Seizure recurrence after a 1st unprovoked seizure: an extended follow-up. Neurology 40:1163–1170
46. Haut SR, Velišková J, Moshé SL (2004) Susceptibility of immature and adult brains to seizure effects. Lancet Neurol 3(10):608–617
47. Hebb D (1949) The organization of behavior: a neuropsychological theory. Wiley, New York
48. Heng K, Haney MM, Buckmaster PS (2013) High-dose rapamycin blocks mossy fiber sprouting but not seizures in a mouse model of temporal lobe epilepsy. Epilepsia 54:1535–1541
49. Hernandez TD, Holling LC (1994) Disruption of behavioral recovery by the anti-convulsant phenobarbital. Brain Res 635:300–306
50. Hernandez TD, Schallert T (1988) Seizures and recovery from experimental brain damage. Exp Neurol 102:318–324
51. Hirayasu Y, Wada JA (1993) The effect of brainstem bisection prior to the amygdala kindling in rats. Brain Res 610:354–357
52. Huberman AD, Speer CM, Chapman B (2006) Spontaneous retinal activity mediates development of ocular dominance columns and binocular receptive fields in v1. Neuron 52:247–254
53. Hughes PE, Alexi T, Walton M, Williams CE, Dragunow M, Clark RG, Gluckman PD (1999) Activity and injury-dependent expression of induc-

ible transcription factors, growth factors and apoptosis-related genes within the central nervous system. Prog Neurobiol 57:421–450

54. Jaseja H (2004) Purpose of REM sleep: endogenous anti-epileptogenesis in man – a hypothesis. Med Hypotheses 62:546–548

55. Jefferys JG (1992) Mechanism of tetanus toxin in neuronal cell death. Trends Pharmacol Sci 13:13–14

56. Jefferys JG, Evans BJ, Hughes SA, Williams SF (1992) Neuropathology of the chronic epileptic syndrome induced by intrahippocampal tetanus toxin in rat: preservation of pyramidal cells and incidence of dark cells. Neuropathol Appl Neurobiol 18:53–70

57. Jones TA, Jones SM, Paggett KC (2001) Primordial rhythmic bursting in embryonic cochlear ganglion cells. J Neurosci Off J Soc Neurosci 21:8129–8135

58. Jones TA, Leake PA, Snyder RL, Stakhovskaya O, Bonham B (2007) Spontaneous discharge patterns in cochlear spiral ganglion cells before the onset of hearing in cats. J Neurophysiol 98:1898–1908

59. Kaiboriboon K, Hogan RE (2002) Hippocampal shape analysis in status epilepticus associated with acute encephalitis. AJNR Am J Neuroradiol 23:1003–1006

60. Kant R, Bogyi AM, Carosella NW, Fishman E, Kane V, Coffey CE (1995) ECT as a therapeutic option in severe brain injury. Convuls Ther 11:45–50

61. Kawakami M, Terasawa E, Ibuki T (1970) Changes in multiple unit activity of the brain during the estrous cycle. Neuroendocrinology 6:30–48

62. Kellner CH, Beale MD, Pritchett JT, Bernstein HJ, Burns CM (1994) Electroconvulsive therapy and Parkinson's disease: the case for further study. Psychopharmacol Bull 30:495–500

63. Kelly ME, McIntyre DC (1994) Hippocampal kindling protects several structures from the neuronal damage resulting from kainic acid-induced status epilepticus. Brain Res 634:245–256

64. Kho LK, Lawn ND, Dunne JW, Linto J (2006) First seizure presentation: do multiple seizures within 24 hours predict recurrence? Neurology 67:1047–1049

65. Klapstein GJ, Meldrum BS, Mody I (1999) Decreased sensitivity to Group III mGluR agonists in the lateral perforant path following kindling. Neuropharmacology 38:927–933

66. Kondratyev A, Sahibzada N, Gale K (2001) Electroconvulsive shock exposure prevents neuronal apoptosis after kainic acid-evoked status epilepticus. Brain Res Mol Brain Res 91:1–13

67. Kondratyev A, Ved R, Gale K (2002) The effects of repeated minimal electroconvulsive shock exposure on levels of mRNA encoding fibroblast growth factor-2 and nerve growth factor in limbic regions. Neuroscience 114:411–416

68. Kotloski R, Lynch M, Lauersdorf S, Sutula T (2002) Repeated brief seizures induce progressive hippocampal neuron loss and memory deficits. Prog Brain Res 135:95–110

69. Kragh J, Bruhn T, Woldbye DD, Bolwig TG (1993) Electroconvulsive shock (ECS) does not facilitate the development of kindling. Prog Neuropsychopharmacol Biol Psychiatry 17:985–989

70. Kuljis RO, Rakic P (1990) Hypercolumns in primate visual cortex can develop in the absence of cues from photoreceptors. Proc Natl Acad Sci U S A 87:5303–5306

71. Lewis DV, Barboriak DP, MacFall JR, Provenzale JM, Mitchell TV, VanLandingham KE (2002) Do prolonged febrile seizures produce medial temporal sclerosis? Hypotheses, MRI evidence and unanswered questions. Prog Brain Res 135:263–278

72. Lewis DV, Shinnar S, Hesdorffer DC, Bagiella E, Bello JA, Chan S, Xu Y, Macfall J, Gomes WA, Moshé SL, Mathern GW, Pellock JM, Nordli DR Jr, Frank LM, Provenzale J, Shinnar RC, Epstein LG, Masur D, Litherland C, Sun S, FEBSTAT StudyTeam (2014) Hippocampal sclerosis after febrile status epilepticus: the FEBSTAT study. Ann Neurol 75(2):178–185

73. Li S, Reinprecht I, Fahnestock M, Racine RJ (2002) Activity-dependent changes in synaptophysin immunoreactivity in hippocampus, piriform cortex, and entorhinal cortex of the rat. Neuroscience 115:1221–1229

74. Liposits Z, Kalló I, Hrabovszky E, Gallyas F (1997) Ultrastructural pathology of degenerating "dark" granule cells in the hippocampal dentate gyrus of adrenalectomized rats. Acta Biol Hung 48:173–187

75. Liu Z, D'Amore PA, Mikati M, Gatt A, Holmes GL (1993) Neuroprotective effect of chronic infusion of basic fibroblast growth factor on seizure-associated hippocampal damage. Brain Res 626:335–338

76. Liu Z, Holmes GL (1997) Basic fibroblast growth factor is highly neuroprotective against seizure-induced long-term behavioural deficits. Neuroscience 76:1129–1138

77. Manji H (2003) Depression, III: treatments. Am J Psychiatry 160:24

78. Masco D, Sahibzada N, Switzer R, Gale K (1999) Electroshock seizures protect against apoptotic hippocampal cell death induced by adrenalectomy. Neuroscience 91:1315–1319

79. Mathern GW, Adelson PD, Cahan LD, Leite JP (2002) Hippocampal neuron damage in human epilepsy: Meyer's hypothesis revisited. Prog Brain Res 135:237–251

80. Mathern GW, Babb TL, Leite JP, Pretorius K, Yeoman KM, Kuhlman PA (1996) The pathogenic and progressive features of chronic human hippocampal epilepsy. Epilepsy Res 26:151–161

81. McIntyre DC, Kelly ME, Dufresne C (1999) FAST and SLOW amygdala kindling rat strains: comparison of amygdala, hippocampal, piriform and perirhinal cortex kindling. Epilepsy Res 35:197–209

82. McIntyre DC, Poulter MO, Gilby K (2002) Kindling: some old and some new. Epilepsy Res 50:79–92

83. McLaughlin T, Torborg CL, Feller MB, O'Leary DDM (2003) Retinotopic map refinement requires spontaneous retinal waves during a brief critical period of development. Neuron 40:1147–1160
84. Minea J (1909) Cercetari Experimentale asupra. Variatiunilor Morfogice ale Neuronului sensitiv. Thesis, Bucharest, Brozer (quoted by Marinesco G, La cellule nerveuse, 2 vol., Octave Doin et fils, Paris, 1909)
85. Mody I (1999) Synaptic plasticity in kindling. Adv Neurol 79:631–643
86. Mohapel P, Armitage LL, Gilbert TH, Hannesson DK, Teskey GC, Corcoran ME (2000) Mossy fiber sprouting is dissociated from kindling of generalized seizures in the guinea-pig. Neuroreport 11:2897–2901
87. Morrell F, deToledo-Morrell L (1999) From mirror focus to secondary epileptogenesis in man: an historical review. Adv Neurol 81:11–23
88. Naylor P, Stewart CA, Wright SR, Pearson RC, Reid IC (1996) Repeated ECS induces GluR1 mRNA but not NMDAR1A-G mRNA in the rat hippocampus. Brain Res Mol Brain Res 35:349–353
89. Nibuya M, Morinobu S, Duman RS (1995) Regulation of BDNF and trkB mRNA in rat brain by chronic electroconvulsive seizure and antidepressant drug treatments. J Neurosci Off J Soc Neurosci 15:7539–7547
90. Nordli DR Jr (2012) Epileptic encephalopathies in infants and children. J Clin Neurophysiol 29:420–424
91. Obiora O, McCormick LM, Karim Y, Gonzales P, Beeghly J (2012) Maintenance electroconvulsive therapy in a patient with multiple system atrophy and bipolar disorder. J ECT 28:e1–e2
92. Olney JW, Rhee V, Ho OL (1974) Kainic acid: a powerful neurotoxic analogue of glutamate. Brain Res 77:507–512
93. Onat FY, Aker RG, Gurbanova AA, Ateş N, van Luijtelaar G (2007) The effect of generalized absence seizures on the progression of kindling in the rat. Epilepsia 48(Suppl 5):150–156
94. Oppenheim RW (1975) The role of supraspinal input in embryonic motility: a re-examination in the chick. J Comp Neurol 160:37–50
95. Osawa M, Uemura S, Kimura H, Sato M (2001) Amygdala kindling develops without mossy fiber sprouting and hippocampal neuronal degeneration in rats. Psychiatry Clin Neurosci 55:549–557
96. Parrino L, Smerieri A, Spaggiari MC, Terzano MG (2000) Cyclic alternating pattern (CAP) and epilepsy during sleep: how a physiological rhythm modulates a pathological event. Clin Neurophysiol 111(Suppl 2):S39–S46
97. Pekary AE, Meyerhoff JL, Sattin A (2000) Electroconvulsive seizures modulate levels of thyrotropin releasing hormone and related peptides in rat hypothalamus, cingulate and lateral cerebellum. Brain Res 884:174–183
98. Penner MR, Pinaud R, Robertson HA (2001) Rapid kindling of the hippocampus protects against neural damage resulting from status epilepticus. Neuroreport 12:453–457
99. Popeo D, Kellner CH (2009) ECT for Parkinson's disease. Med Hypotheses 73:468–469
100. Post RM, Putnam F, Contel NR, Goldman B (1984) Electroconvulsive seizures inhibit amygdala kindling: implications for mechanisms of action in affective illness. Epilepsia 25:234–239
101. Provine RR (1993) Natural priorities for developmental study: neuroembryological perspectives of motor development. In: Kalverboer AF, Hopkins B, Geuze R (eds) Motor development in early and later childhood. Cambridge University Press, Cambridge, pp 51–73
102. Racine RJ, Adams B, Osehobo P, Fahnestock M (2002) Neural growth, neural damage and neurotrophins in the kindling model of epilepsy. Adv Exp Med Biol 497:149–170
103. Ramos Lizana J, Cassinello Garciá E, Carrasco Marina LL, Vázquez López M, Martín González M, Muñoz Hoyos A (2000) Seizure recurrence after a first unprovoked seizure in childhood: a prospective study. Epilepsia 41:1005–1013
104. Reti IM, Baraban JM (2000) Sustained increase in Narp protein expression following repeated electroconvulsive seizure. Neuropsychopharmacology 23:439–443
105. Sacks J, Glaser N (1941) Changes in susceptibility to the convulsant action of metrazol. J Pharm 73:239–295
106. Sakhi S, Sun N, Wing LL, Mehta P, Schreiber SS (1996) Nuclear accumulation of p53 protein following kainic acid-induced seizures. Neuroreport 7:493–496
107. Salmenperä T, Könönen M, Roberts N, Vanninen R, Pitkänen A, Kälviäinen R (2005) Hippocampal damage in newly diagnosed focal epilepsy: a prospective MRI study. Neurology 64:62–68
108. Sandyk R (1986) ECT, opioid system, and motor response in Parkinson's disease. Biol Psychiatry 21:235–236
109. Sangdee P, Turkanis SA, Karler R (1982) Kindling-like effect induced by repeated corneal electroshock in mice. Epilepsia 23:471–479
110. Sato K, Kashihara K, Morimoto K, Hayabara T (1996) Regional increases in brain-derived neurotrophic factor and nerve growth factor mRNAs during amygdaloid kindling, but not in acidic and basist growth factor mRNAs. Epilepsia 37:6–14
111. Schallert T, Hernandez TD, Barth TM (1986) Recovery of function after brain damage: severe and chronic disruption by diazepam. Brain Res 379:104–111
112. Scharfman HE, MacLusky NJ (2006) The influence of gonadal hormones on neuronal excitability, seizures, and epilepsy in the female. Epilepsia 47:1423–1440
113. Scott RC, King MD, Gadian DG, Neville BGR, Connelly A (2003) Hippocampal abnormalities after

prolonged febrile convulsion: a longitudinal MRI study. Brain J Neurol 126:2551–2557

114. Shinnar S, Berg AT, Moshe SL, O'Dell C, Alemany M, Newstein D, Kang H, Goldensohn ES, Hauser WA (1996) The risk of seizure recurrence after a first unprovoked afebrile seizure in childhood: an extended follow-up. Pediatrics 98:216–225

115. Shishido Y, Tanaka T, Piao Y, Araki K, Takei N, Higashiyama S, Nawa H (2006) Activity-dependent shedding of heparin-binding EGF-like growth factor in brain neurons. Biochem Biophys Res Commun 348:963–970

116. Shouse MN (1988) Sleep deprivation increases thalamocortical excitability in the somatomotor pathway, especially during seizure-prone sleep or awakening states in feline seizure models. Exp Neurol 99:664–677

117. Simonato M, Molteni R, Bregola G, Muzzolini A, Piffanelli M, Beani L, Racagni G, Riva M (1998) Different patterns of induction of FGF-2, FGF-1 and BDNF mRNAs during kindling epileptogenesis in the rat. Eur J Neurosci 10:955–963

118. Sloviter RS (2008) Hippocampal epileptogenesis in animal models of mesial temporal lobe epilepsy with hippocampal sclerosis: the importance of the "latent period" and other concepts. Epilepsia 49(Suppl 9):85–92

119. Sloviter RS, Dean E, Neubort S (1993) Electron microscopic analysis of adrenalectomy-induced hippocampal granule cell degeneration in the rat: apoptosis in the adult central nervous system. J Comp Neurol 330:337–351

120. Sloviter RS, Valiquette G, Abrams GM, Ronk EC, Sollas AL, Paul LA, Neubort S (1989) Selective loss of hippocampal granule cells in the mature rat brain after adrenalectomy. Science 243:535–538

121. Sokol DK, Demyer WE, Edwards-Brown M, Sanders S, Garg B (2003) From swelling to sclerosis: acute change in mesial hippocampus after prolonged febrile seizure. Seizure J Br Epilepsy Assoc 12:237–240

122. Stern MB (1991) Electroconvulsive therapy in untreated Parkinson's disease. Mov Disord 6:265

123. Stevens JR (1995) Clozapine: the Yin and Yang of seizures and psychosis. Biol Psychiatry 37:425–426

124. Striano P, de Falco FA, Minetti C, Zara F (2009) Familial benign nonprogressive myoclonic epilepsies. Epilepsia 50(Suppl 5):37–40

125. Sutula TP (2004) Mechanisms of epilepsy progression: current theories and perspectives from neuroplasticity in adulthood and development. Epilepsy Res 60:161–171

126. Tongiorgi E, Armellin M, Giulianini PG, Bregola G, Zucchini S, Paradiso B, Steward O, Cattaneo A, Simonato M (2004) Brain-derived neurotrophic factor mRNA and protein are targeted to discrete dendritic laminas by events that trigger epileptogenesis. J Neurosci Off J Soc Neurosci 24:6842–6852

127. Triplett JW, Owens MT, Yamada J, Lemke G, Cang J, Stryker MP, Feldheim DA (2009) Retinal input instructs alignment of visual topographic maps. Cell 139:175–185

128. Turski L, Ikonomidou C, Turski WA, Bortolotto ZA, Cavalheiro EA (1989) Review: cholinergic mechanisms and epileptogenesis. The seizures induced by pilocarpine: a novel experimental model of intractable epilepsy. Synapse 3:154–171

129. Turski WA, Cavalheiro EA, Schwarz M, Czuczwar SJ, Kleinrok Z, Turski L (1983) Limbic seizures produced by pilocarpine in rats: behavioural, electroencephalographic and neuropathological study. Behav Brain Res 9:315–335

130. Tuunanen J, Pitkänen A (2000) Do seizures cause neuronal damage in rat amygdala kindling? Epilepsy Res 39:171–176

131. Uemura S, Kimura H, Kashiba A, Kumashiro H, Wada JA (1988) Bifunctional roles of catecholamines in the development of amygdala kindling demonstrated by continuous intra-amygdala infusion of 6-hydroxydopamine. Brain Res 448:162–166

132. Wada JA, Hamada K (1999) Role of the midline brainstem in feline amygdaloid kindling. Epilepsia 40:669–676

133. Wada JA, Mizoguchi T, Osawa T (1978) Secondarily generalized convulsive seizures induced by daily amygdaloid stimulation in rhesus monkeys. Neurology 28:1026–1036

134. Wada JA, Osawa T (1976) Spontaneous recurrent seizure state induced by daily electric amygdaloid stimulation in Senegalese baboons (Papio papio). Neurology 26:273–286

135. Wada JA, Sata M (1974) Generalized convulsive seizures induced by daily electrical stimulation of the amygdala in cats. Correlative electrographic and behavioral features. Neurology 24:565–574

136. Wirrell EC (1998) Benign epilepsy of childhood with centrotemporal spikes. Epilepsia 39(Suppl 4):S32–S41

137. Wirrell EC (2003) Natural history of absence epilepsy in children. Can J Neurol Sci 30:184–188

138. Wláz P, Potschka H, Löscher W (1998) Frontal versus transcorneal stimulation to induce maximal electroshock seizures or kindling in mice and rats. Epilepsy Res 30:219–229

139. Zhang LX, Smith MA, Kim SY, Rosen JB, Weiss SR, Post RM (1996) Changes in cholecystokinin mRNA expression after amygdala kindled seizures: an in situ hybridization study. Brain Res Mol Brain Res 35:278–284

140. Zhang X, Cui S-S, Wallace AE, Hannesson DK, Schmued LC, Saucier DM, Honer WG, Corcoran ME (2002) Relations between brain pathology and temporal lobe epilepsy. J Neurosci Off J Soc Neurosci 22:6052–6061

When and How Do Seizures Kill Neurons, and Is Cell Death Relevant to Epileptogenesis?

9

Ray Dingledine, Nicholas H. Varvel, and F. Edward Dudek

Abstract

The effect of seizures on neuronal death and the role of seizure-induced neuronal death in acquired epileptogenesis have been debated for decades. Isolated brief seizures probably do not kill neurons; however, severe and repetitive seizures (i.e., status epilepticus) certainly do. Because status epilepticus both kills neurons and also leads to chronic epilepsy, neuronal death has been proposed to be an integral part of acquired epileptogenesis. Several studies, particularly in the immature brain, have suggested that neuronal death is not necessary for acquired epileptogenesis; however, the lack of neuronal death is difficult if not impossible to prove, and more recent studies have challenged this concept. Novel mechanisms of cell death, beyond the traditional concepts of necrosis and apoptosis, include autophagy, phagoptosis, necroptosis, and pyroptosis. The traditional proposal for why neuronal death may be necessary for epileptogenesis is based on the *recapitulation of development hypothesis*, where a loss of synaptic input from the dying neurons is considered a critical signal to induce axonal sprouting and synaptic-circuit reorganization. We propose a second hypothesis – the *neuronal death pathway hypothesis*, which states that the biochemical pathways causing programmed neurodegeneration, rather than neuronal death *per se*, are responsible for or contribute to epileptogenesis. The reprogramming of neuronal death pathways – if true – is proposed to derive from necroptosis or pyroptosis. The proposed new hypothesis may inform on why neuronal death seems closely linked to epileptogenesis, but may not always be.

Keywords

Neurodegeneration • Epilepsy • Necrosis • Apoptosis • Autophagy • Phagoptosis • Necroptosis • Pyroptosis

R. Dingledine (✉) • N.H. Varvel
Department of Pharmacology, Emory University
School of Medicine, Atlanta, GA 30322, USA
e-mail: rdingledine@pharm.emory.edu;
nicholas.h.varvel@emory.edu

F.E. Dudek
Department of Neurosurgery, University of Utah
School of Medicine, Salt Lake City, UT 84108, USA
e-mail: ed.dudek@hsc.utah.edu

H.E. Scharfman and P.S. Buckmaster (eds.), *Issues in Clinical Epileptology: A View from the Bench*,
Advances in Experimental Medicine and Biology 813, DOI 10.1007/978-94-017-8914-1_9,
© Springer Science+Business Media Dordrecht 2014

9.1 Seizures and Neuronal Death: When, Where, and What?

Debates and controversies concerning the interplay among seizures, neuronal death and epilepsy continue to occur. Over several decades, many epilepsy researchers have focused on various aspects of the issue of whether seizures cause neuronal death, and conversely, whether neuronal death is necessary and/or sufficient to cause epilepsy. For example, a classic – yet still ongoing – debate is the degree to which GABAergic interneurons are lost in tissue from patients and animal models of temporal lobe epilepsy, and the consequence of such loss. In spite of the longevity and intensity of the previous debates, the relationship between seizures, neuronal death and epilepsy remains one of the most disputed in translational neuroscience, particularly as it relates to possible mechanisms of acquired epileptogenesis and the clinical interactions and consequences of seizures and neuronal death. We will discuss, as the title implies, two important and longstanding hypotheses of contemporary epilepsy research – important because the degree to which seizures cause brain damage and the hypothetical role of neuronal death in the development of epilepsy are inter-related and could underlie the often quoted statement "Seizures beget seizures" [33]. These two issues are not "black and white"; rather, they probably form an interactive continuum and are quite complicated; and furthermore, technical limitations and interpretational difficulties plague any analysis of them. The key questions include *when* do seizures kill neurons, *where* in the brain are neurons most susceptible to seizure activity, and *what* is the identity of the neurons that are preferentially killed? Answers to a fourth issue – *"How* do seizures kill neurons?" – may hold a key to understanding at least one component of epileptogenesis, as described below. We will begin with a brief summary of some of the key questions and controversial topics; then, we will review more recent views of the many possible mechanisms whereby seizures may kill neurons; and finally, we will conclude with a brief discussion of some of the ongoing issues and controversies in this area.

9.1.1 When

A large and long-standing body of experimental and clinical data indicates that some types of seizures lead to neuronal death, while other types do not. In either experimental animals or humans, whenever seizures are long enough in duration and occur repetitively for prolonged periods, some neurons – particularly in adults – are killed. In terms of the temporal features of the seizures that are thought to cause neuronal death, relatively brief seizures – such as typical *absence* seizures in children (usually lasting 5–10 s) – do not appear to cause overt brain damage. However, the more prolonged seizures characteristic of temporal lobe epilepsy, such as the traditional complex partial seizures (i.e., dyscognitive focal seizures) that may progress to tonic-clonic convulsive seizures, are much more likely to lead to neuronal loss [84]. Finally, the prolonged and repetitive seizures that define status epilepticus typically cause brain damage, often with extensive neuronal death [10, 15, 29, 40, 57, 62, 67, 85]. Interestingly, however, status epilepticus in the immature brain causes far less neuronal death [16, 38, 59, 68, 74, 75, 80, 81], and appears less likely to cause epileptogenesis [51, 74]. The long-standing observation that experimental status epilepticus in laboratory animals, mostly rodents, leads to a chronic epileptic state raises the following question: Is the occurrence of neuronal death during status epilepticus a critical part of the epileptogenesis? In terms of epilepsy, one could view seizure clusters, where some of the interseizure intervals are much shorter than the typical interseizure intervals [36, 37], as essentially a reduced form of status epilepticus. The difference between status epilepticus and a seizure cluster in a patient with epilepsy is not always so clear. Thus, a fundamental question in clinical epilepsy is: Do the spontaneous recurrent seizures kill neurons – particularly when the seizures occur in clusters? If so, under what conditions does this contribute to a worsening of epilepsy? Are seizure clusters a particular concern in terms of neuronal death and brain injury? These are some of the unanswered questions that are both clinically important and can theoretically be addressed with animal models.

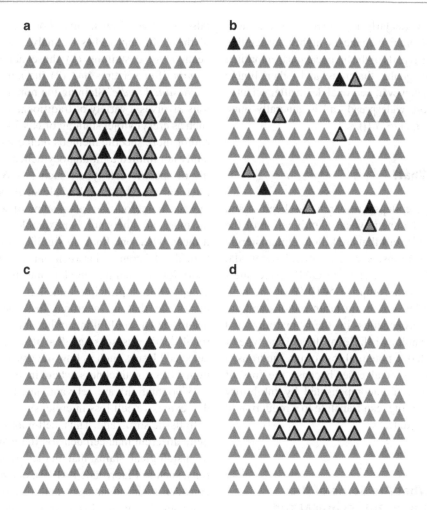

Fig. 9.1 Schematic diagrams showing hypothetical relationships of neuronal populations after a brain insult that activates cellular mechanisms of neuronal death. In the four panels of the figure, two or three populations of neurons are depicted in a schematic manner. Dead neurons (*filled black triangles*) are shown within a network of live and completely-normal neurons (*filled red triangles*). Among these two populations of cells is another group of neurons, which form the core of this hypothesis; these neurons have undergone only the initial steps of a neuronal-death and/or are under the molecular influence of the neuronal death process (*black triangular outline with red stiples* inside). (**a**) Focal neuronal loss. A small cluster of dead neurons is shown to be clumped together within a network of normal neurons, as would be expected to occur during an infarct. Between these two completely different neuronal populations is the group of neurons

that are hypothetically epileptogenic, because they have undergone the first part of a neuronal-death process and/ or are under the molecular influence of the neuronal death process. (**b**) Diffuse neuronal loss. Using the same code to define the members of the neuronal population, this diagram illustrates scattered neuronal loss, as would be expected to occur after status epilepticus (vs an infarct in (**a**)). (**c**) Occurrence of neuronal death without generation of neurons altered or influenced by death-process mechanisms, which theoretically represents the occurrence of frank brain damage without subsequent epilepsy. (**d**) Absence of neuronal death after a brain insult, but with the presence of death-pathway neurons. In this case, the death-pathway neurons are hypothesized to become epileptogenic, and they generate spontaneous recurrent seizures without the prior occurrence of overt neuronal death

9.1.2   Where?

If we focus on the seizures that characterize temporal lobe epilepsy and other forms of severe acquired epilepsy (e.g., after hypoxic-ischemic

encephalopathy), many specific areas appear to be particularly prone to seizure-induced neuronal death. Depending on the etiology, neuronal death can be relatively circumscribed, as with an infarct (Fig. 9.1a), or it can be diffuse (Fig. 9.1b).

Seizures, particularly repetitive seizures, cause substantial brain damage in highly susceptible areas, such as parts of the hippocampus, entorhinal cortex, amygdala, thalamus and other limbic structures; however, neuronal death after seizures can be more widespread and is generally quite variable (e.g., [24, 77]).

9.1.3   What?

A focus in epilepsy research has been – and remains – the unequivocal identification of the type(s) of neurons that are killed: glutamatergic principal neurons, such as cortical pyramidal cells, and subpopulations of GABAergic interneurons, which comprise 5–10 % of the neurons in epilepsy-relevant brain regions and are highly heterogeneous in their anatomy and electrophysiology [4]. Regardless of the type of brain insult, the potential loss of interneurons is obviously a special case, because the loss of interneurons, if uncompensated by inhibitory axonal sprouting, can translate to a reduction in GABAergic tone.

9.2   What Are Some of the Important Technical and Experimental-Design Issues?

The challenges and controversies concerning how to evaluate whether neuronal death has occurred and how to quantify it are substantial. Even when one only considers a fraction of the methodological and protocol-related issues, the additional factors involving "what, where, and when" of neuronal death ("how" is discussed below) add further complexity to the potential analyses and interpretations. Additional disagreement surrounds the question "What is a seizure?" and the problem of what comprises an adequate animal model of acquired epilepsy.

An important issue in regard to considerations of neuronal death in epilepsy – as with most other research – involves the complimentary concerns of false positives (specificity) and

false negatives (sensitivity). For example, two of the main approaches to analyzing neuronal death involve staining (1) those neurons that *remain* after seizures and (2) the neurons that are *destined to die*. Staining the remaining neurons involves a variety of traditional techniques such as cresyl violet staining of Nissl substance, and/or more specific methods including but certainly not limited to immunocytochemical staining of specific cell types, such as GABAergic interneurons. This most basic level of methodology has numerous caveats – some of which are obvious, and others not. For example, what does it mean when one finds no significant (i.e., statistical) difference between an experimental condition or animal model and the control group? On first principles, one has to ask: Does this mean that no neuronal loss (death) has occurred? Or, could it mean the amount of neuronal death was so small that it could not be detected? Issues such as how the tissue was sectioned (section thickness, where in the brain, but also orientation) are relevant, not to mention that extensive cell loss in epilepsy is associated with tissue shrinkage. Thus, cell number can be quite different than cell density. In regard to use of histological stains that mark "dying" neurons, such as FluoroJade B (FJB), one must also consider their advantages and disadvantages. For example, one has to question our confidence that they will actually die – can FJB-labeled neurons remain viable for a prolonged period before death? If we assume that all of the FJB neurons are going to die, or even most of them, then this approach has the important advantage that it can reveal situations in which only a small fraction of the neurons will die, which is simply not feasible with stains that mark the "remaining" neurons. Another issue, however, is that the FJB technique will only stain neurons that are dying at that particular time; so therefore, euthanasia, fixation, and staining must be performed at the appropriate time; neurons could have died at other times, and their death would not be detected with FJB [94]. Thus, although it is quite difficult to quantify neuronal loss, it is even more difficult – if not impossible – to show that neuronal loss has not occurred.

9.3 Mechanisms of Seizure-Induced Neuron Injury

In order to explore how seizures could kill neurons, it is first necessary to review cell-death pathways; our understanding of them has expanded well beyond the traditional mechanisms of *apoptosis* and *necrosis*. This seemingly simple endeavor is complicated by the observation that some of the newly identified cell-death pathways share criteria used for identification. For clarity, we classify cell-death processes as non-inflammatory (apoptosis, autophagy, phagoptosis) and inflammatory (necrosis, necroptosis, pyroptosis) (Table 9.1).

9.3.1 Apoptosis

A controlled, programmed process of packaging internal components of the cell for clearing by phagocytes characterizes the apoptotic process. As such, intracellular molecules with the potential to activate immune responses are disposed of rapidly, without initiating an immune response [2]. Apoptosis is also characterized by chromatin and cytoplasmic condensation, plasma membrane blebbing, formation of apoptotic bodies as well as fragmentation of cellular compartments and DNA. Apoptosis occurs naturally during development and serves as a means to facilitate cellular turnover in healthy tissue, and also in response to hormone withdrawal [47]. This programed series of events is reliant upon the effector functions of activated caspases −3, −6, and −7, which enzymatically cleave intracellular organelles, proteins and DNA. The degraded cellular corpse is then packaged in preparation for phagocytosis by macrophages or

microglia [26]. Processing of intracellular compartments and subsequent removal of cellular debris during apoptosis does not result in a secondary inflammatory response in surrounding tissue as inflammatory mediators are largely sequestered and degraded [2]. Changes in mitochondrial membrane permeability [50] and release of mitochondrial proteins are also observed [87]. Another characteristic of apoptotic cells is the exposure of phosphatidylserine on the extracellular leaflet of their plasma membrane. While phosphatidylserine is normally found exclusively on the cytoplasmic side of the plasma membrane, apoptotic cells present phosphatidylserine on the extracellular surface to serve as an "eat-me" signal for neighboring phagocytes [32, 70]. Cellular shrinkage, likely due to caspase-mediated proteolysis of cytoskeletal proteins, also typifies apoptotic cells [49].

9.3.2 Autophagy

Although autophagy usually serves a protective role, in extreme stress conditions it can contribute to cell death. In similar fashion as apoptosis, autophagic pathways also progress in a series of cellular steps that involve programmed degradation of cellular components. However, intracellular autophagic, largely non-caspase, enzymes are responsible for degradation of organelles or other cytoplasmic proteins within double-membrane vesicles known as autophagosomes [54]. The autophagosome then fuses with intracellular lysosomes to facilitate degradation of the contents within the autophagosome by acid hydrolases. In contrast to apoptosis, caspase activation is not required and chromatin condensation is minor [11].

Table 9.1 Six mechanisms of cell death

Death process	Programmed?	Inflammatory lysis?	Effector	Shape Δ	TUNEL?
Necrosis	No	Yes	Non-caspase	Swell	No
Necroptosis	Yes	Yes	TNF-α RIPK1	Swell	No
Pyroptosis	Yes	Yes	Caspase-1	Swell	Yes
Autophagy	Yes	No	Lysosomes	?	No
Phagoptosis	Yes	No	Microglia	No	No
Apoptosis	Yes	No	Caspase-3/6/7	Shrink	Yes

In addition to contributing to the death of a cell, autophagic mechanisms also contribute to cellular function and homeostatic maintenance. For example, in the immune system, antigen-presenting cells utilize autophagy to digest intact proteins, creating smaller antigens for subsequent presentation to T lymphocytes [21, 54]. Moreover, mice deficient in proteins involved in autophagy develop spontaneous neurodegeneration [35, 48]. Taken together, these findings indicate that, in addition to cell death, autophagy mediates an important role in the organism's response to pathogens as well as maintenance of cellular homeostasis.

9.3.3 Phagoptosis

Many of the identified physiological cell death pathways involve phagocytosis of either whole cells doomed to die or of fractured cellular components. As such, the process of phagocytosis has been viewed as a secondary event, occurring after the death of the cell [70]. However, the process of phagocytosis can also kill living cells. Recent studies have identified a pathway, termed "phagoptosis", wherein phagocytes such as activated microglia actively contribute to the death of viable neurons and other cells [8]. Similar to apoptosis, the otherwise viable cell presents "eat-me" signals, such as phosphatidylserine, on the outer leaflet of its cellular membrane. The "eat-me" signals are then recognized by nearby phagocytes, and cellular uptake ensues followed by digestion of the viable cells. Importantly, cell death can be prevented during phagoptosis by inhibiting phagocytosis [28, 60]. This is because "eat-me" signal exposure is transient and reverses when phagocytosis is prevented. Therefore, neuronal insults not severe enough to initiate apoptotic pathways might be a trigger for phagoptosis due to the temporary exposure of eat-me signals on stressed but viable neurons [8, 28].

9.3.4 Necrosis

In contrast to these non-inflammatory modes of neuron death, during necrosis cells lyse, effectively spilling their internal contents into the interstitial fluid and releasing molecules that can initiate inflammatory cascades. This uncontrolled release of intracellular molecules can potentially damage surrounding tissue and cells [76, 79]. Necrotic cell death is typically initiated by extreme physiological stress or trauma that kills cells quickly. Biochemically, caspase is not involved. Morphologically, condensation or digestion of internal cellular compartments is not observed. Instead, organelles and the entire cell undergo extensive swelling. The cell eventually bursts, spilling its internal contents into the surrounding environment, triggering robust inflammation in the neighboring tissue [45].

9.3.5 Necroptosis

While necrosis leads to an uncontrolled cellular death, a variant of necrosis, which has some controllable features, has recently been described. This programmed pathway, termed necroptosis, exhibits characteristics of both programmed cell death and necrosis. The main characteristic distinguishing necrosis from necroptosis is that the latter is initiated by TNF-α and other death receptor activators, which promote the assembly of receptor-interacting protein kinase 1 (RIP1) with RIP3 [86]. Thus, kinase activity controls necroptosis [45]. Interestingly, RIP1 and RIP3 assemble into a functional kinase-containing cell-death complex only in the absence of functional caspase 8 [25, 46]. While the physiological impact of necroptosis is currently under investigation, it is conceivable this pathway may be relevant in the event caspase activity is impeded and thus canonical apoptosis is not possible.

9.3.6 Pyroptosis

Perhaps the most extreme example of inflammation-related cell death is pyroptosis (i.e., caspase 1-dependent programmed cell death). While this form of cell death was first described after infectious stimuli, such as *Salmonella* and *Shigella* infection [6, 18], caspase-1-dependent cell death also occurs in myocytes after myocardial infarction [27] and in the central nervous system [53, 103]. The primary distinguishing feature of pyroptosis is the formation of the inflammasome, an intracellular multimolecular complex that is required for the activation of inflammatory caspases, particularly caspase 1. The activated inflammasome culminates with production of enzymatically active caspase 1, which in turn mediates the maturation and secretion of active IL-1β and IL-18 [2]. Secreted pro-inflammatory cytokines can subsequently influence nearby cells with potentially adverse consequences, such as blood-brain barrier breakdown and possible leukocyte entry into the brain. Although TUNEL-positive breaks in cellular DNA typify both apoptosis and pyroptosis, the latter is entirely reliant upon caspase-1 [7, 17]. This is important for classification purposes because caspase-1 is not involved in apoptosis. Mitochondrial release of cytochrome c, a hallmark of apoptosis, also does not occur during pyroptosis. In contrast to the coordinated packaging of intracellular components observed in apoptosis, cellular lysis and release of inflammatory effector molecules occur during pyroptosis [26].

9.4   How Might Seizure-Induced Neuronal Injury Promote Epileptogenesis?

9.4.1   Overview of Two Competing Hypotheses

We envision two conceptually distinct answers to this question. *First*, maladaptive new circuits among neurons could form to replace synapses *lost* during neuronal death. This mechanism, potentially involving axonal sprouting within excitatory pathways and amplified by loss of inhibitory interneurons, has been described in numerous previous studies and can be termed the *"recapitulation of development"* hypothesis. If replacement of lost synapses is the critical factor underlying this mechanism, then neuronal death

would seem to be an essential component of the process. *Second*, rather than neuron death *per se* being responsible, molecular signals from upstream pathways that mediate some of the more newly recognized forms of cell death might underlie or contribute to epileptogenesis. We call this the *"neuronal death pathway"* hypothesis. We will focus on potential roles for IL-1β and TNF-α. We will also consider whether the inflammasome pathways (caspase-1 activation leading to synthesis of IL-1β and IL18), normally considered a feature of myeloid cells and innate immunity, might be involved in epilepsy-related neurodegeneration.

In some cases focal inflammation produced by lytic cell death, perhaps involving only a small number of neurons undetectable by normal Nissl stains (e.g., Fig. 9.1b), could promote increased neuronal excitability and perhaps synchronous activity. However, in the absence of any neuronal death (Fig. 9.1d), how might inflammatory cascades be initiated? Understanding how microglia, the innate immune cells of the CNS, respond to injurious or danger signals may provide insights into this undoubtedly complex process. Microglia in the intact, healthy brain continuously palpate the surrounding tissue for subtle disturbances [61], and can rapidly respond to tissue injury or danger signals by altering morphology, proliferating and expressing a wide variety of inflammatory cytokines and chemokines [19, 69]. Microglial activation can be initiated by injured neurons through the release of molecules collectively known as alarmins [4].

One well-characterized alarmin, prostaglandin E2, is released by highly active neurons in a COX-2-dependent process. Cyclooxygenase 2 (COX-2) is rapidly upregulated in hippocampal pyramidal cells and dentate granule cells after seizures [55, 73, 98], but the impact of *neuronal* COX-2 has remained elusive because astrocytes, endothelial cells and probably other cell types in the CNS also express COX-2. To determine the role of neuronal COX-2 after status epilepticus, a neuron specific conditional knockout mouse was utilized wherein principal neurons of the hippocampus, dentate granule cells, amygdala, thalamus and layer-specific neurons in the piriform and neocortex (layer 5) are devoid of COX-2, while the remaining cell types of the CNS still express functional protein [43, 72]. Interestingly, conditional ablation of COX-2 from neurons resulted in less severe damage to hippocampal neurons after status epilepticus produced by pilocarpine. The intensity of status epilepticus was not diminished in the COX-2 conditional knockouts, as judged by the temporal evolution of behavioral seizures and by cortical EEG [78], making it unlikely that neuroprotection was caused by a less severe seizure episode. Neuroprotection was accompanied by reduction in multiple markers of neuroinflammation as well as preserved integrity of the blood-brain barrier, suggesting that neuronal COX-2 mediates a broad deleterious role after status epilepticus. These findings provide strong evidence that the neuron itself can contribute to the neuroinflammatory milieu [78]. The beneficial effects of the conditional ablation of COX-2 from principal forebrain neurons were completely recapitulated by systemic administration of a novel antagonist of EP2, a receptor for PGE2 [44].

Injured neurons might indirectly contribute to inflammation after status epilepticus through cell-to-cell signaling with microglia. Multiple lines of evidence indicate that the local microenvironment plays an important role in regulating the microglial phenotype wherein microglia activation is constitutively inhibited by repressive forces [34, 65, 69]. For example, surface proteins on microglia, such as CD200R and CX3CR1 (the fracktalkine receptor), normally interact with the neuronal surface protein ligands, CD200 and CX3CL1 (fracktalkine), respectively [14, 42]. If interactions between CD200R and CD200 [42, 102] or CX3CR1 and CX3CL1 [3, 13] are disrupted by signals released during neuronal damage or distress, then microglia are unleashed from this constitutive state of inhibition and a more florid microglial response ensues. Enhanced microglial activation is likely attributed to the presence of ITIM motifs (immunoreceptor tyrosine-based inhibitory motif) on both CD200R and CX3CR1 as these motifs function as activators for SHP-1 and SHP-2 phosphatases that can repress further inflammatory signaling [5]. Indeed, CX3CR1-deficient mice exhibit microglia-mediated neurotoxicity,

through enhanced IL-1β secretion, after immune challenge [14]. Interestingly, altered expression of CX3CL1 has been reported in both epileptic patients and animals models after status epilepticus [97].

In addition to the above-mentioned studies, viable neurons might also induce inflammatory cascades. Studies in *Drosophila melanogaster* originated this concept, wherein damaged cells, prevented from dying, release mitotic signals that prompt neighboring cells to divide. Cells in the wing of flies were triggered to die by X-rays, but they were blocked from completing the death process by expression of anti-apoptotic proteins. The authors describe the resulting cells as "undead". The neighboring cells divide in an apparent attempt to fill the void in the tissue expected to be left by the dying cells [64]. Do similar situations occur in human disease? Interestingly, neuronal populations expected to degenerate in the brains of Alzheimer's Disease (AD) patients re-express proteins typically encountered in a mitotic cell cycle [12, 58, 92, 93]. Importantly, DNA replication accompanies cell cycle entry [99]. Transgenic mouse models of AD also recapitulate neuronal cell cycle entry [88, 101], suggesting that the same "stressors" that provoke neuronal cell cycle entry in the human AD brain are phenocopied in the mice. However, cycling neurons exhibit little atrophy [100] and robust neuronal loss is absent in AD mice [41, 56], indicating that re-expression of mitotic proteins and DNA synthesis in a postmitotic neuron is not sufficient to induce death, at least in the lifetime of the mouse. It has been proposed that cycling neurons also might send out mitotic signals, pressuring otherwise healthy neurons to enter the "undead" state [39].

9.4.2   Inflammatory Pathways and Epileptogenesis

How might inflammatory signaling upstream of neurodegeneration increase excitability and subsequent synchronicity? Immune responses in the brain are initiated, maintained and terminated by soluble effector proteins known as cytokines. Although a strong correlation between seizures and elevated inflammatory cytokines or their mRNA transcripts has been reported [90], emerging experimental evidence indicates that inflammatory cytokines can in turn alter neuronal excitability and synchronicity by modulating receptor function and expression [31, 89]. For example, the pro-inflammatory cytokine TNF-α has also been shown to promote the recruitment of AMPA receptors to postsynaptic membranes. Interestingly, the recruited receptors preferentially lack the GluR2 subunit [52, 63, 82] and consequently the calcium conductance underlying EPSPs is increased. Additionally, TNF-α causes endocytosis of GABA$_A$ receptors from the cellular surface, decreasing inhibitory synaptic strength [82]. Taken together these findings demonstrate that TNFα can have a profound impact on circuit homeostasis in a manner that can provoke the pathogenesis of seizures.

In addition to TNF-α, multiple lines of evidence directly implicate IL-1β in lowering the seizure threshold, and perhaps in epileptogenesis. First, hippocampal application of IL-1β can increase seizure intensity threefold. This proconvulsant effect is attributed to IL-1β-mediated engagement of Src-family kinases in hippocampal neurons. The activated kinases subsequently phosphorylate the NR2B subunit of the NMDA receptor, leading to seizure exacerbation [1]. Second, IL-1β can inhibit calcium currents through protein kinase C, at least at low concentrations [66]. Finally, IL-1β can also inhibit GABA$_A$ receptor current, which could underlie neuronal hyperexcitability [95]. These studies, coupled with the findings that pharmacological treatments targeting IL-1β or its activation result in robust anticonvulsant effects [20, 71, 90, 91], indicate that inflammation might play an important role in epileptogenesis and is a viable therapeutic target class.

9.5   Implications of the New Concepts on Neuronal Death for Epileptogenesis

The long-standing *recapitulation-of-development* hypothesis essentially states that neuronal death in acquired epilepsy is linked to a re-activation of developmental processes, which replace the synapses lost through neuronal death [30]. Initially, most experimental and clinical epileptologists viewed this hypothesis as "mossy fiber sprouting", which causes the formation of new recurrent excitatory circuits among dentate granule cells. This hypothesis is discussed by Buckmaster [9] and a more general view would be that neuronal death in many areas of the brain, particularly in seizure-sensitive regions, causes multiple networks to form new local excitatory circuits [22, 23, 83]. The data reviewed above suggest a new hypothesis, the *neuronal-death-pathway* hypothesis, whereby the biochemical pathways causing programmed neurodegeneration, rather than neuronal death *per se*, are responsible for or contribute to epileptogenesis. This hypothesis is consistent with the view that frank brain damage (i.e., cases where obvious neuronal death has occurred) leads to epilepsy, and further, that the likelihood of developing intractable epilepsy is linked somehow to the severity of the brain injury. In addition, however, this hypothesis may begin to explain why brain injuries that clearly induce neuronal death do not always appear to lead to epilepsy, since the critical hypothetical mechanism for acquired epileptogenesis would be the linkage between the to-be-defined mechanisms *within the pathways responsible for neuronal death*, as opposed to neuronal death itself (Fig. 9.1c). The identification of these hypothetical processes is an area ripe for future investigation. Finally, this hypothesis could also explain how epilepsy may occur when neuronal death is absent or appears minimal (Fig. 9.1d). It is conceivable that these molecular mechanisms may be aborted or reversed before neuronal death actually occurs, for example, so that specific signaling molecules direct some of the surrounding neurons toward an epileptogenic phenotype, even though the processes of neuronal death may not reach completion. The key point here is the proposal or hypothesis that molecular/genetic signals from neurons that are on a "death pathway" could initiate epileptogenesis independent of the final outcome (i.e., neuronal death).

9.6 Concepts and Conclusions

Although much has been learned about when seizures do kill neurons and the conditions when they appear to cause less damage, it is extraordinarily difficult to *rule out* that neuronal death has occurred after seizures. One problem is both a conceptual and technical one, namely, showing that something has not occurred is particularly challenging, if not impossible. We simply do not know if a threshold exists whereby a few, brief seizures – possibly in the seizure-resistant immature brain – cause absolutely no neuronal death. In terms of the question, "When?", there is no way to show that neuronal death has not occurred during and/or after seizures, except to count the remaining neurons in control and experimental groups; however, the potential error – even in well-powered studies, can be 10 % or more ([10] [see Table 1 and Fig. 2A-B]; [96] [see Fig. 6]) – and yet a loss of just a few percent of the neurons within a brain structure could have a substantive epileptogenic effect. If one considers the problem of "Where?", it becomes obvious that the answer is "Almost anywhere!". For the animal models of repetitive seizures and status epilepticus – whether induced by hypoxia, pilocarpine, or some other precipitating insult – numerous seizure-sensitive areas of the brain show neuronal loss, and the structure could be different for individuals within a similarly-treated cohort of animals, further supporting the idea that it is extremely difficult to exclude a role of neuronal death. In terms of, "What types of neurons may be lost?", excluding loss of part of the critical interneuron pool generally requires specific staining techniques, such as immunohistochemistry with stereology (e.g., [10]). As important, however, is the discovery of new neuronal death pathways that could lead to neuron loss in ways that have previously not been appreciated. This latter set of observations opens up the possibility that a gateway to seizure-induced neuronal loss involves signaling pathways that represent or are influenced by early neuron-death pathways. Thus, we propose that – in addition to the previously proposed *recapitulation-of-development*

mechanisms – another hypothesis could be the *neuronal-death-pathway* hypothesis, whereby the early steps of neuronal death generate signals that promote epileptogenesis even if the neurons ultimately do not die. An attractive feature of this hypothesis is that it could lend itself to classification by molecular markers that reflect these neuronal pathway molecules. This hypothesis might also explain why neuronal death seems so important to acquired epileptogenesis, yet might in some cases be unnecessary.

Acknowledgments Ray Dingledine first met Phil Schwartzkroin in 1975 just after Phil had returned to Stanford from his postdoctoral stint with Per Andersen in Oslo, little realizing that he would follow in Phil's footsteps by joining Per's lab just 2 years later. Ray was in the act of returning some equipment that he had "borrowed" from David Prince's lab to complete his graduate work, recognizing that it would never be missed in the vast warehouse of the Prince lab. Phil was then, as he is now, a very gracious yet intense scientist, and Ray values his friendship.

Ed Dudek also met Phil in about 1975; we have never worked together, but we have had numerous critical if not intense discussions on a variety of topics. Ed deeply appreciates and respects his thoughtful and constructive criticisms and his kind encouragement; Phil's traits and his actions have made all of us better scientists and professional colleagues.

Other Acknowledgements This work is supported by National Institutes of Health awards NS076775, NS074169, and P20 NS080185 (RD) and NS058158 (RD), and NS079135, NS079274, and NS086364 (FED). We thank Jonas Neher (Tuebingen, Germany) and Jeff Ekstrand (Salt Lake City, Utah) for helpful comments as well as Vicki Skelton (Salt Lake City, Utah) for figure preparation and assistance with the manuscript.

References

1. Balosso S, Maroso M, Sanchez-Alavez M, Ravizza T, Frasca A, Bartfai T, Vezzani A (2008) A novel non-transcriptional pathway mediates the proconvulsive effects of interleukin-1beta. Brain 131: 3256–3265
2. Bergsbaken T, Fink SL, Cookson BT (2009) Pyroptosis: host cell death and inflammation. Nat Rev Microbiol 7:99–109
3. Bhaskar K, Konerth M, Kokiko-Cochran ON, Cardona A, Ransohoff RM, Lamb BT (2010) Regulation of tau pathology by the microglial fractalkine receptor. Neuron 68:19–31

4. Bianchi ME (2007) DAMPs, PAMPs and alarmins: all we need to know about danger. J Leukoc Biol 81:1–5
5. Billadeau DD, Leibson PJ (2002) ITAMs versus ITIMs: striking a balance during cell regulation. J Clin Invest 109:161–168
6. Boise LH, Collins CM (2001) Salmonella-induced cell death: apoptosis, necrosis or programmed cell death? Trends Microbiol 9:64–67
7. Brennan MA, Cookson BT (2000) Salmonella induces macrophage death by caspase-1-dependent necrosis. Mol Microbiol 38:31–40
8. Brown GC, Neher JJ (2012) Eaten alive! Cell death by primary phagocytosis: 'phagoptosis'. Trends Biochem Sci 37:325–332
9. Buckmaster PS (2014) Does mossy fiber sprouting give rise to the epileptic state? In: Issues in clinical epileptology: a view from the bench. Springer, Dordrecht, pp 161–168
10. Buckmaster PS, Dudek FE (1997) Neuron loss, granule cell axon reorganization, and functional changes in dentate gyrus of epileptic kainate-treated rats. J Comp Neurol 385:385–404
11. Bursch W (2001) The autophagosomal-lysosomal compartment in programmed cell death. Cell Death Differ 8:569–581
12. Busser J, Geldmacher DS, Herrup K (1998) Ectopic cell cycle proteins predict the sites of neuronal cell death in Alzheimer's disease brain. J Neurosci 18:2801–2807
13. Cardona AE, Huang D, Sasse ME, Ransohoff RM (2006) Isolation of murine microglial cells for RNA analysis or flow cytometry. Nat Protoc 1:1947–1951
14. Cardona AE, Pioro EP, Sasse ME, Kostenko V, Cardona SM, Dijkstra IM, Huang D, Kidd G, Dombrowski S, Dutta R, Lee JC, Cook DN, Jung S, Lira SA, Littman DR, Ransohoff RM (2006) Control of microglial neurotoxicity by the fractalkine receptor. Nat Neurosci 9:917–924
15. Cavalheiro EA, Leite JP, Borolotto ZA, Turski WA, Ikonomidou C, Turski L (1991) Long-term effects of pilocarpine in rats: structural damage of the brain triggers kindling and spontaneous recurrent seizures. Epilepsia 32:778–782
16. Cavalheiro EA, Silva DF, Turski WA, Calderazzo-Filho LS, Bortolotto ZA, Turski L (1987) The susceptibility of rats to pilocarpine-induced seizures is age-dependent. Brain Res 465:43–58
17. Chen Y, Smith MR, Thirumalai K, Zychlinsky A (1996) A bacterial invasin induces macrophage apoptosis by binding directly to ICE. EMBO J 15:3853–3860
18. Cookson BT, Brennan MA (2001) Pro-inflammatory programmed cell death. Trends Microbiol 9:113–114
19. Davalos D, Grutzendler J, Yang G, Kim JV, Zuo Y, Jung S, Littman DR, Dustin ML, Gan WB (2005) ATP mediates rapid microglial response to local brain injury in vivo. Nat Neurosci 8:752–758
20. De Simoni MG, Perego C, Ravizza T, Moneta D, Conti M, Marchesi F, De Luigi A, Garattini S, Vezzani A (2000) Inflammatory cytokines and related genes are induced in the rat hippocampus by limbic status epilepticus. Eur J Neurosci 12:2623–2633
21. Dengjel J, Schoor O, Fischer R, Reich M, Kraus M, Muller M, Kreymborg K, Altenberend F, Brandenburg J, Kalbacher H, Brock R, Driessen C, Rammensee HG, Stevanovic S (2005) Autophagy promotes MHC class II presentation of peptides from intracellular source proteins. Proc Natl Acad Sci U S A 102:7922–7927
22. Dudek FE, Staley KJ (2011) The time course of acquired epilepsy. Implications for therapeutic intervention to suppress epileptogenesis. Neurosci Lett 497:240–246
23. Dudek FE, Sutula TP (2007) Chapter 41: Epileptogenesis in the dentate gyrus: a critical perspective. In: Scharfman HE (ed) Progress in brain research, vol 163. Elsevier B.V., Amsterdam, pp 755–773
24. Ekstrand JJ, Pouliot W, Scheerlinck P, Dudek FE (2011) Lithium pilocarpine-induced status epilepticus in postnatal day 20 rats results in greater neuronal injury in ventral versus dorsal hippocampus. Neuroscience 192:699–707
25. Feoktistova M, Geserick P, Kellert B, Dimitrova DP, Langlais C, Hupe M, Cain K, MacFarlane M, Hacker G, Leverkus M (2011) cIAPs block Ripoptosome formation, a RIP1/caspase-8 containing intracellular cell death complex differentially regulated by cFLIP isoforms. Mol Cell 43:449–463
26. Fink SL, Cookson BT (2005) Apoptosis, pyroptosis, and necrosis: mechanistic description of dead and dying eukaryotic cells. Infect Immun 73:1907–1916
27. Frantz S, Ducharme A, Sawyer D, Rohde LE, Kobzik L, Fukazawa R, Tracey D, Allen H, Lee RT, Kelly RA (2003) Targeted deletion of caspase-1 reduces early mortality and left ventricular dilatation following myocardial infarction. J Mol Cell Cardiol 35:685–694
28. Fricker M, Neher JJ, Zhao JW, Thery C, Tolkovsky AM, Brown GC (2012) MFG-E8 mediates primary phagocytosis of viable neurons during neuroinflammation. J Neurosci 32:2657–2666
29. Fujikawa DG (1996) The temporal evolution of neuronal damage from pilocarpine-induced status epilepticus. Brain Res 725:11–22
30. Galanopoulou AS, Moshe SL (2014) Does epilepsy cause a reversion to immature function? In: Issues in clinical epileptology: a view from the bench. Springer, Dordrecht
31. Galic MA, Riazi K, Pittman QJ (2012) Cytokines and brain excitability. Front Neuroendocrinol 33:116–125
32. Gardai SJ, McPhillips KA, Frasch SC, Janssen WJ, Starefeldt A, Murphy-Ullrich JE, Bratton DL, Oldenborg PA, Michalak M, Henson PM (2005)

Cell-surface calreticulin initiates clearance of viable or apoptotic cells through trans-activation of LRP on the phagocyte. Cell 123:321–334

33. Gowers WR (1881) Epilepsy and other chronic convulsive diseases: their causes, symptoms and treatment. J & A Churchill, London

34. Hanisch UK, Kettenmann H (2007) Microglia: active sensor and versatile effector cells in the normal and pathologic brain. Nat Neurosci 10:387–1394

35. Hara T, Nakamura K, Matsui M, Yamamoto A, Nakahara Y, Suzuki-Migishima R, Yokoyama M, Mishima K, Saito I, Okano H, Mizushima N (2006) Suppression of basal autophagy in neural cells causes neurodegenerative disease in mice. Nature 441:885–889

36. Haut SR (2006) Seizure clustering. Epilepsy Behav 8:50–55

37. Haut SR, Shinnar S, Moshé SL, O'Dell C, Legatt AD (1999) The association between seizure clustering and convulsive status epilepticus in patients with intractable complex partial seizures. Epilepsia 40:1832–1834

38. Haut SR, Velíšková J, Moshé SL (2004) Susceptibility of immature and adult brains to seizure effects. Lancet Neurol 3:608–617

39. Herrup K, Yang Y (2007) Cell cycle regulation in the postmitotic neuron: oxymoron or new biology? Nat Rev Neurosci 8:368–378

40. Hirsch E, Baram TZ, Sneed OC III (1992) Otogenic study of lithiumpilocarpine-induced status epilepticus in rats. Brain Res 583:120–126

41. Hock BJ Jr, Lamb BT (2001) Transgenic mouse models of Alzheimer's disease. Trends Genet 17:S7–S12

42. Hoek RM, Ruuls SR, Murphy CA, Wright GJ, Goddard R, Zurawski SM, Blom B, Homola ME, Streit WJ, Brown MH, Barclay AN, Sedgwick JD (2000) Down-regulation of the macrophage lineage through interaction with OX2 (CD200). Science 290:1768–1771

43. Hoesche C, Sauerwald A, Veh RW, Krippl B, Kilimann MW (1993) The 5′-flanking region of the rat synapsin I gene directs neuron-specific and developmentally regulated reporter gene expression in transgenic mice. J Biol Chem 268: 26494–26502

44. Jiang J et al (2013) Inhibition of the prostaglandin receptor EP2 following status epilepticus reduces delayed mortality and brain inflammation. Proc Natl Acad Sci U S A 110:3591–3596

45. Kaczmarek A, Vandenabeele P, Krysko DV (2013) Necroptosis: the release of damage-associated molecular patterns and its physiological relevance. Immunity 38:209–223

46. Kaiser WJ, Upton JW, Long AB, Livingston-Rosanoff D, Daley-Bauer LP, Hakem R, Caspary T, Mocarski ES (2011) RIP3 mediates the embryonic lethality of caspase-8-deficient mice. Nature 471:368–372

47. Kerr JF, Wyllie AH, Currie AR (1972) Apoptosis: a basic biological phenomenon with wide-ranging implications in tissue kinetics. Br J Cancer 26:239–257

48. Komatsu M, Waguri S, Chiba T, Murata S, Iwata J, Tanida I, Ueno T, Koike M, Uchiyama Y, Kominami E, Tanaka K (2006) Loss of autophagy in the central nervous system causes neurodegeneration in mice. Nature 441:880–884

49. Kothakota S, Azuma T, Reinhard C, Klippel A, Tang J, Chu K, McGarry TJ, Kirschner MW, Koths K, Kwiatkowski DJ, Williams LT (1997) Caspase-3-generated fragment of gelsolin: effector of morphological change in apoptosis. Science 278:294–298

50. Kroemer G, Reed JC (2000) Mitochondrial control of cell death. Nat Med 6:513–519

51. Kubova H, Mares P, Suchomelova L, Brozek G, Druga R, Pitkanen A (2004) Status epilepticus in immature rats leads to behavioural and cognitive impairment and epileptogenesis. Eur J Neurosci 19:3255–3265

52. Leonoudakis D, Zhao P, Beattie EC (2008) Rapid tumor necrosis factor alpha-induced exocytosis of glutamate receptor 2-lacking AMPA receptors to extrasynaptic plasma membrane potentiates excitotoxicity. J Neurosci 28:2119–2130

53. Liu XH, Kwon D, Schielke GP, Yang GY, Silverstein FS, Barks JD (1999) Mice deficient in interleukin-1 converting enzyme are resistant to neonatal hypoxic-ischemic brain damage. J Cereb Blood Flow Metab 19:1099–1108

54. Ma Y, Galluzzi L, Zitvogel L, Kroemer G (2013) Autophagy and cellular immune responses. Immunity 39:211–227

55. Marcheselli VL, Bazan NG (1996) Sustained induction of prostaglandin endoperoxide synthase-2 by seizures in hippocampus. Inhibition by a platelet-activating factor antagonist. J Biol Chem 271:24794–24799

56. McGowan E, Eriksen J, Hutton M (2006) A decade of modeling Alzheimer's disease in transgenic mice. Trends Genet 22:281–289

57. Mello LE, Cavalheiro EA, Tan AM, Kupfer WR, Pretorius JK, Babb TL, Finch DM (1993) Circuit mechanisms of seizures in the pilocarpine model of chronic epilepsy: cell loss and mossy fiber sprouting. Epilepsia 34:985–995

58. Nagy Z, Esiri M, Cato A, Smith A (1997) Cell cycle markers in the hippocampus in Alzheimer's disease. Acta Neuropathol (Berl) 94:6–15

59. Nairismagi J, Pitkanen A, Kettunen MI, Kauppinen RA, Kubova H (2006) Status epilepticus in 12-day-old rats leads to temporal lobe neurodegeneration and volume reduction: a histologic and MRI study. Epilepsia 47:479–488

60. Neher JJ et al (2013) Phagocytosis executes delayed neuronal death after focal brain ischemia. Proc Natl Acad Sci U S A 110:E4098–4107

61. Nimmerjahn A, Kirchhoff F, Helmchen F (2005) Resting microglial cells are highly dynamic surveillants of brain parenchyma in vivo. Science 308:1314–1318

62. Obenaus A, Esclapez M, Houser CR (1993) Loss of glutamate decarboxylase mRNA-containing neurons in the rat dentate gyrus following pilocarpine-induced seizures. J Neurosci 13:4470–4485

63. Ogoshi F, Yin HZ, Kuppumbatti Y, Song B, Amindari S, Weiss JH (2005) Tumor necrosis-factor-alpha (TNF-α) induces rapid insertion of Ca2+–permeable alpha-amino-3-hydroxyl-5-methyl-4-isoxazole-propionate (AMPA)/kainate (Ca-A/K) channels in a subset of hippocampal pyramidal neurons. Exp Neurol 193:384–393

64. Perez-Garijo A, Martin FA, Morata G (2004) Caspase inhibition during apoptosis causes abnormal signalling and developmental aberrations in drosophila. Development 131:5591–5598

65. Perry VH, Teeling J (2013) Microglia and macrophages of the central nervous system: the contribution of microglia priming and systemic inflammation to chronic neurodegeneration. Semin Immunopathol 35:601–612

66. Plata-Salaman CR, ffrench-Mullen JM (1994) Interleukin-1 beta inhibits Ca2+ channel currents in hippocampal neurons through protein kinase C. Eur J Pharmacol 266:1–10

67. Pollard H, Charriaut-Marlangue C, Cantagrel S, Represa A, Robain O, Moreau J, Ben-Ari Y (1994) Kainate-induced apoptotic cell death in hippocampal neurons. Neuroscience 63:7–18

68. Priel MR, dos Santos NF, Cavalheiro EA (1996) Developmental aspects of the pilocarpine model of epilepsy. Epilepsy Res 26:115–121

69. Ransohoff RM, Cardona AE (2010) The myeloid cells of the central nervous system parenchyma. Nature 468:253–262

70. Ravichandran KS (2011) Beginnings of a good apoptotic meal: the find-me and eat-me signaling pathways. Immunity 35:445–455

71. Ravizza T, Lucas SM, Balosso S, Bernardino L, Ku G, Noe F, Malva J, Randle JC, Allan S, Vezzani A (2006) Inactivation of caspase-1 in rodent brain: a novel anticonvulsive strategy. Epilepsia 47:1160–1168

72. Rempe D, Vangeison G, Hamilton J, Li Y, Jepson M, Federoff HJ (2006) Synapsin I Cre transgene expression in male mice produces germline recombination in progeny. Genesis 44:44–49

73. Sandhya TL, Ong WY, Horrocks LA, Farooqui AA (1998) A light and electron microscopic study of cytoplasmic phospholipase A2 and cyclooxygenase-2 in the hippocampus after kainate lesions. Brain Res 788:223–231

74. Sankar R, Shin DH, Liu H, Mazarati A, Vasconcelos A, Wasterlain CG (1998) Patterns of status epilepticus-induced neuronal injury during development and long-term consequences. J Neurosci 18:8382–8393

75. Sankar R, Shin DH, Wasterlain CG (1997) Serum neuron-specific enolase is a marker for neuronal damage following status epilepticus in the rat. Epilepsy Res 28:129–136

76. Scaffidi P, Misteli T, Bianchi ME (2002) Release of chromatin protein HMGB1 by necrotic cells triggers inflammation. Nature 418:191–195

77. Scholl EA, Dudek FE, Ekstrand JJ (2013) Neuronal degeneration is observed in multiple regions outside the hippocampus after lithium pilocarpine-induced status epilepticus in the immature rat. Neuroscience 12:45–59

78. Serrano GE, Lelutiu N, Rojas A, Cochi S, Shaw R, Makinson CD, Wang D, FitzGerald GA, Dingledine R (2011) Ablation of cyclooxygenase-2 in forebrain neurons is neuroprotective and dampens brain inflammation after status epilepticus. J Neurosci 31:14850–14860

79. Shi Y, Evans JE, Rock KL (2003) Molecular identification of a danger signal that alerts the immune system to dying cells. Nature 425:516–521

80. Sperber EF, Haas KZ, Stanton PK, Moshe SL (1991) Resistance of the immature hippocampus to seizure-induced synaptic reorganization. Brain Res Dev Brain Res 60:88–93

81. Stafstrom CE, Thompson JL, Holmes GL (1992) Kainic acid seizures in the developing brain: status epilepticus and spontaneous recurrent seizures. Brain Res Dev Brain Res 65:227–236

82. Stellwagen D, Beattie EC, Seo JY, Malenka RC (2005) Differential regulation of AMPA receptor and GABA receptor trafficking by tumor necrosis factor-alpha. J Neurosci 25:3219–3228

83. Sutula TP, Dudek FE (2007) Chapter 29: Unmasking recurrent excitation generated by mossy fiber sprouting in the epileptic dentate gyrus: an emergent property of a complex system. In: Scharfman HE (ed) Progress in brain research, vol 163. Elsevier B.V., Amsterdam, pp 541–563

84. Sutula T, Pitkanen A (2002) Do seizures damage the brain? Elsevier Science, Amsterdam

85. Turski WA, Cavalheiro EA, Schwarz M, Czuczwar SJ, Kleinrok Z, Turski L (1983) Limbic seizures produced by pilocarpine in rats: behavioural, electroencephalographic and neuropathological study. Behav Brain Res 9:315–335

86. Vandenabeele P, Galluzzi L, Vanden Berghe T, Kroemer G (2010) Molecular mechanisms of necroptosis: an ordered cellular explosion. Nat Rev Mol Cell Biol 11:700–714

87. Van Loo G, Demol H, van Gurp M, Hoorelbeke B, Schotte P, Beyaert R, Zhivotovsky B, Gevaert K, Declercq W, Vandekerckhove J, Vandenabeele P (2002) A matrix-assisted laser desorption ionization post-source decay (MALDI-PSD) analysis of proteins released from isolated liver mitochondria treated with recombinant truncated Bid. Cell Death Differ 9:301–308

88. Varvel NH, Bhaskar K, Patil AR, Pimplikar SW, Herrup K, Lamb BT (2008) Abeta oligomers induce neuronal cell cycle events in Alzheimer's disease. J Neurosci 28:10786–10793

89. Vezzani A, French J, Bartfai T, Baram TZ (2011) The role of inflammation in epilepsy. Nat Rev Neurol 7:31–40

90. Vezzani A, Granata T (2005) Brain inflammation in epilepsy: experimental and clinical evidence. Epilepsia 46:1724–1743

91. Vezzani A, Moneta D, Conti M, Richichi C, Ravizza T, De Luigi A, De Simoni MG, Sperk G, Andell-Jonsson S, Lundkvist J, Iverfeldt K, Bartfai T (2000) Powerful anticonvulsant action of IL-1 receptor antagonist on intracerebral injection and astrocytic overexpression in mice. Proc Natl Acad Sci U S A 97:11534–11539

92. Vincent I, Jicha G, Rosado M, Dickson D (1997) Aberrant expression of mitotic cdc2/cyclin B1 kinase in degenerating neurons of Alzheimer's disease brain. J Neurosci 17:3588–3598

93. Vincent I, Rosado M, Davies P (1996) Mitotic mechanisms in Alzheimer's disease? J Cell Biol 132:413–425

94. Wang L, Liu Y–H, Huang Y–G, Chen L–W (2008) Time-course of neuronal death in the mouse pilocar-pine model of chronic epilepsy using Fluoro-Jade C staining. Brain Res 1241:157–167

95. Wang S, Cheng Q, Malik S, Yang J (2000) Interleukin-1beta inhibits gamma-aminobutyric acid type A (GABA(A)) receptor current in cultured hippocampal neurons. J Pharmacol Exp Ther 292:497–504

96. Williams PA, Wuarin JP, Dou P, Ferraro DJ, Dudek FE (2002) Reassessment of the effects of cyclohexi-mide on mossy fiber sprouting and epileptogenesis

in the pilocarpine model of temporal lobe epilepsy. J Neurophysiol 88:2075–2087

97. Xu Y, Zeng K, Han Y, Wang L, Chen D, Xi Z, Wang H, Wang X, Chen G (2012) Altered expression of CX3CL1 in patients with epilepsy and in a rat model. Am J Pathol 180:1950–1962

98. Yamagata K, Andreasson KI, Kaufmann WE, Barnes CA, Worley PF (1993) Expression of a mitogen-inducible cyclooxygenase in brain neurons: regula-tion by synaptic activity and glucocorticoids. Neuron 11:371–386

99. Yang Y, Geldmacher DS, Herrup K (2001) DNA rep-lication precedes neuronal cell death in Alzheimer's disease. J Neurosci 21:2661–2668

100. Yang Y, Herrup K (2005) Loss of neuronal cell cycle control in ataxia-telangiectasia: a unified disease mechanism. J Neurosci 25:2522–2529

101. Yang Y, Varvel NH, Lamb BT, Herrup K (2006) Ectopic cell cycle events link human Alzheimer's disease and amyloid precursor protein transgenic mouse models. J Neurosci 26:775–784

102. Zhang S, Wang XJ, Tian LP, Pan J, Lu GQ, Zhang YJ, Ding JQ, Chen SD (2011) CD200-CD200R dysfunction exacerbates microglial activation and dopaminergic neurodegeneration in a rat model of Parkinson's disease. J Neuroinflammation 8:154

103. Zhang WH, Wang X, Narayanan M, Zhang Y, Huo C, Reed JC, Friedlander RM (2003) Fundamental role of the Rip2/caspase-1 pathway in hypoxia and ischemia-induced neuronal cell death. Proc Natl Acad Sci U S A 100:16012–16017

How Is Homeostatic Plasticity Important in Epilepsy?

10

John W. Swann and Jong M. Rho

Abstract

Maintaining physiological variables within narrow operating limits by homeostatic mechanisms is a fundamental property of most if not all living cells and organisms. In recent years, research from many laboratories has shown that the activity of neurons and neural circuits are also homeostatically regulated. Here, we attempt to apply concepts of homeostasis in general, and more specifically synaptic homeostatic plasticity, to the study of epilepsy. We hypothesize that homeostatic mechanisms are actively engaged in the epileptic brain. These processes attempt to re-establish normal neuronal and network activity, but are opposed by the concurrent mechanisms underlying epileptogenesis. In forms of intractable epilepsy, seizures are so frequent and intense that homeostatic mechanisms are unable to restore normal levels of neuronal activity. In such cases, we contend that homeostatic plasticity mechanisms nevertheless remain active. However, their continuing attempts to reset neuronal activity become maladaptive and results in dyshomeostasis with neurobehavioral consequences. Using the developing hippocampus as a model system, we briefly review experimental results and present a series of arguments to propose that the cognitive neurobehavioral comorbidities of childhood epilepsy result, at least in part, from unchecked homeostatic mechanisms.

Keywords

Homeostasis • Synaptic plasticity • Seizures • Synapses • Epileptogenesis • Learning and memory

J.W. Swann (✉)
The Cain Foundation Laboratories,
The Jan and Dan Duncan Neurological
Research Institute, Texas Children's Hospital,
Baylor College of Medicine, 1250 Moursund St,
Houston, TX 77030, USA
e-mail: jswann@bcm.edu

J.M. Rho
Departments of Pediatrics and Clinical
Neurosciences, Alberta Children's Hospital,
Calgary, AB, Canada

Faculty of Medicine, University of Calgary,
Calgary, AB, Canada

H.E. Scharfman and P.S. Buckmaster (eds.), *Issues in Clinical Epileptology: A View from the Bench*,
Advances in Experimental Medicine and Biology 813, DOI 10.1007/978-94-017-8914-1_10,
© Springer Science+Business Media Dordrecht 2014

Homeostasis or the maintenance of a physiological state – despite external or internal disturbances that would be expected to alter that state – is a biological concept central to both animal and human physiology [27]. This concept was first introduced nearly 150 years ago by Claude Bernard who demonstrated the ability of organisms to maintain a relatively constant internal environment and who stated that the maintenance of "*le milieu interieur*" was essential for life [3]. In 1929, Walter Cannon extended these concepts and coined the term homeostasis or "similar state" [7]. In adopting a systems level approach to physiology, Cannon suggested that coordinated adjustments of interacting systems through feedback systems result in the maintenance of physiological parameters such as body temperature and circulating oxygen levels within a set range.

Today, the concept of homeostasis is fundamental to all studies of physiology from the level of individual cells to that of entire organisms. Indeed, it is such a well-accepted concept that it is taken for granted that the underlying basic mechanisms are essential for survival. Homeostasis includes, but is not limited to, the regulation of blood pH, circulating levels of glucose, body temperature, interstitial level of O_2 and CO_2 as well as critical electrolytes such as Na^+, Cl^-, K^+ and Ca^{+2}. In keeping with the work of Cannon, feedback systems are thought to be the primary regulatory mechanisms underlying homeostasis. In general, these feedback systems must first detect a change in a parameter that needs to be held within narrow limits – also referred to as a set point. If it deviates from these limits, the system activates mechanisms to return the parameter to the set point. Such systems consist of a sensor that is able to measure changes in the parameter, an integrator that compares the detected information to the desired set point and an effector that generates the compensatory response to return the parameter to homeostasis. Circulating levels of Ca^{+2} are a good example of such a feedback system. When blood Ca^{+2} falls below its set point, Ca^{+2}-sensing receptors in cells of the parathyroid gland are activated. This results in the release of parathyroid hormone which acts to increase circulating Ca^{+2} levels. Multiple effector mechanisms

are induced, including increased absorption from the gastrointestinal tract and reabsorption from urine. Another important source of Ca^{+2} is bone. Parathyroid hormone increases the activity of bone-degrading osteoblasts which release Ca^{+2} from bone and thereby return circulating Ca^{+2} to its set point.

10.1 Neuronal Homeostatic Plasticity

Over the past 15 years, a great deal of evidence has accumulated to suggest that the physiological activity of neurons and neuronal networks are homeostatically regulated [33, 34]. Neuronal networks of the central nervous system (CNS) are highly dynamic. This is easily observed in variations in human and animal EEG recordings over a 24-h period. For example, the dramatic alterations in recordings at transitions from NREM sleep to the awake state reflect marked changes in the activity of individual neurons and the operations of their networks. At these times, an organism in interacting with its environment will store information for future use. Hebbian synaptic plasticity mechanisms such as long-term potentiation (LTP) and long-term depression (LTD) are widely thought to underlie the processes for learning and the storage of memories. However, for some time, theoretical neurophysiologists have recognized that Hebbian plasticity should destabilize and consequently interfere with network operations [1]. The idea behind this claim is that once a group of excitatory synapses undergo a use-dependent form of plasticity – like LTP – they will in the future produce larger excitatory post-synaptic potentials (EPSPs) which will more likely induce action potentials in the postsynaptic neuron. This in turn results in even larger EPSPs and more neuronal firing, and a self-perpetuating cascade of ever-increasing synaptic strengthening and ultimately increasing network excitability.

Homeostatic synaptic plasticity has been proposed as a stabilizing mechanism to counter the potential run-away excitation of Hebbian plasticity [33]. As would be expected, this relatively new field of homeostatic plasticity has borrowed concepts from other forms of physiological homeostasis. For example, synaptic homeostasis has been defined as "a form of plasticity that acts to stabilize the activity of a neuron or neuronal circuit around some *set point* value" [33]. Possibly the best demonstration of homeostatic plasticity comes from studies of dissociated cultures of CNS neurons. In these models, when the networks of cortical neurons are pharmacologically challenged by application of a γ-aminobutyric acid type A (GABA$_A$) receptor antagonist such as bicuculline, the firing rates of individual neurons initially increase. However, over a period of many hours to days, firing rates return to their original rate, which is interpreted to be the homeostatic set point of affected neurons and their networks. Similarly, when activity is suppressed, firing rates are initially very low but are restored over time [35].

There have been many studies of the cellular and molecular events underlying this form of neuronal plasticity. It has become clear that there are numerous mechanisms that can act independently to regulate post-synaptic and pre-synaptic strength as well as mechanisms that operate at the level of individual synapses, and others that act in parallel but on a more global scale. The most studied of this type on neuronal plasticity is synaptic homeostasis – particularly at excitatory, glutamatergic synapses. In these studies, following periods of pharmacologically-induced heighted neuronal activity, the amplitude or strength of miniature excitatory post-synaptic currents (mEPSCs) has been shown to decrease at the times when neuronal firing rates had returned to their set point [35]. These alterations in mEPSCs were also found to parallel decreases in the amplitude of evoked synaptic events. Much evidence has emerged suggesting that synaptic homeostasis results from post-synaptic alterations in glutamatergic subunit expression and localization. Under some experimental conditions, a decrease in the number of glutamatergic synapses has also been demonstrated [13]. In other cases and circumstances, pre-synaptic changes in transmitter release have been reported [4, 6].

In addition to synaptic homeostasis of glutamatergic synapses, the function of inhibitory synapses and inhibitory interneurons appear to be homeostatically regulated, as are the intrinsic excitability properties of individual neurons. As might be expected, synaptic inhibition is regulated in the opposite direction of excitation. For instance, when neuronal activity is experimentally depressed as the amplitude of mEPSCs in pyramidal cells is increased, miniature inhibitory post-synaptic currents (mIPSCs) decrease in strength [15, 20]. Both pre-synaptic and post-synaptic changes appear to contribute to homeostatic regulation of synaptic inhibition. The variety of mechanisms underlying this form of regulation may be a reflection of the diversity of inhibitory synapses and inhibitory interneurons in the CNS. For instance, when ascending activity to the visual cortex is experimentally lowered (in an attempt to mimic activity suppression *in vitro*), the amplitude of inhibitory synapses onto layer 4 pyramidal cells from fast spiking interneurons is reduced [22]. However, inhibitory synapses from other interneuronal subtypes appear to be stronger although fewer in number.

In terms of the intrinsic excitability of neurons, it has been shown that when the activity of cultured neurons is experimentally suppressed, the intrinsic excitability of excitatory neurons is enhanced [9]. So these neurons are able to generate more action potentials in response to a given synaptic input than untreated control neurons. Thus, at the same time that excitatory synaptic transmission is increased and synaptic inhibition in decreased, alterations in the expression and function of ion channels (likely both inward and outward voltage-dependent currents) further enhance neuronal excitability in attempting to re-establish normal neuronal and ultimately network activity.

In summary, numerous studies over the past 15 years have not only repeatedly demonstrated the ability of neurons to homeostatically adapt to experimentally-induced alterations in their activity but have also shown that the cells have a wide array of mechanisms at their disposal to stabilize their activity in the face of forces like Hebbian synaptic plasticity that can potentially lead to neuronal and network instability.

10.2   Homeostatic Plasticity Versus Epileptogenesis

The neuroscience community has learned a great deal from studies of the basic mechanisms of homeostatic synaptic plasticity. This has been propelled by the use of relatively simple culture systems that are amenable to rigorous experimental manipulations and testing of hypotheses. A number of other studies have been performed *in vivo* in attempts to extend information from *in vitro* studies to the intact CNS. However, how these results impact our understanding of epilepsy is now only beginning to be explored. For at least the past 15 years, a large proportion of experimental epilepsy research efforts have been focused on understanding the mechanisms underlying epileptogenesis, a process that at least superficially appears to be the antithesis of homeostasis. The term epileptogenesis has been defined as a chronic process by which normal brain

is transformed into tissue capable of generating spontaneous recurrent seizures [17]. In the acquired epilepsies (e.g. following traumatic brain injury), the seizure-prone state is thought to arise from a progressive series of molecular, cellular and circuit changes that evolve over time.

Results from long-term continuous video-EEG recordings in several animal models of acquired epilepsy have emphasized the progressive nature of epileptogenesis [11]. Within a week after injury, nonconvulsive seizures are first observed. A week thereafter they become convulsive. Seizure frequency can gradually increase nearly tenfold over the ensuing 3–4 months. Such results are consistent with the idea first proposed by Gowers in 1888 that "seizures beget seizures" [14]. Potential steps in the progression of epileptogenesis are illustrated by the positive feedback system in Fig. 10.1. Here, seizures are envisioned to induce a cascade of molecular and cellular events that lead to sprouting of glutamatergic synapses and other forms of network reorganization that further enhance network excitability and the genesis of more seizures with increasing frequency. Juxtaposed to this is a diagram of the negative feedback loop that is thought to characterize

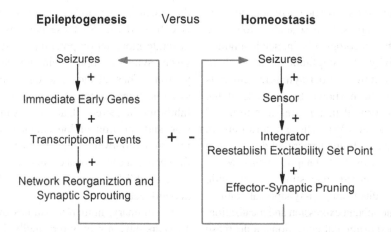

Fig. 10.1   Diagram outlining the hypothesized opposing forces of epileptogenesis and homeostasis. A positive feedback loop is envisioned to mediate epileptogenesis. Examples of some of the potential molecular events are named that lead to the network reorganization and synaptic sprouting that is thought to contribute to recurring seizures.

Homeostasis is suggested to oppose epileptogenesis through negative feedback loops that are designed to re-establish normal neuronal and neural circuit excitability. Pruning of glutamatergic synapses is but one example of an effector mechanism that would reduce seizure generation homeostasis in general and homeostatic plasticity more specifically. In this context, seizures are envisioned to activate sensory processes – just as increased neuronal activity is thought to during synaptic homeostasis. A molecular integrator compares neuronal activity to a set point value and induces changes in an effector, which in this example results in the pruning of excitatory synapses [13] and which in turn would be predicted to reduce neuronal and network excitability and reduce seizure frequency.

If homeostasis is such a fundamental property of animal and human physiology, why does it apparently fail in epilepsy? One possibility is that it does not always fail since seizures do spontaneously remit in some forms of epilepsy without any apparent reason. Many of the so-called benign epilepsies of infancy and childhood carry a favorable prognosis. In these instances, children are simply said to "outgrow their seizures". The mechanisms accounting for these observations are unknown. One possibility is that as the brain matures the developmental factors that enhance seizure susceptibility are no longer operant. Alternatively, ongoing homeostatic mechanisms may play a significant role in these remissions. In contrast, in more severe and intractable epilepsy, it seems possible that homeostatic mechanisms are actively engaged in epilepsy but in many cases the precipitating injury (or in the case of genetic forms of epilepsy, i.e., the consequences of gene mutations) are so severe that homeostatic mechanisms are simply unable to re-establish neuronal activity to the desired set point and seizure progression continues unabated by the processes underlying epileptogenesis. However, if this were the case, then as seizures recur, homeostatic mechanisms would also be repeatedly induced in an attempt to re-establish normal neuronal excitability and network stability.

There are a number of observations made in animal models of epilepsy that appear somewhat paradoxical in that molecular and anatomical changes observed would be expected to prevent seizures, not promote them. Increases in $GABA_A$ receptor subunits [16] and potassium channel expression [26, 28] as well as dendritic spine loss in hippocampal and neocortical pyramidal cells [5, 30] are just a few examples of alterations that researchers have sometimes referred to as "paradoxical" and possibly "compensatory" responses to on-going seizure activity. It is not hard to imagine that there are many other such paradoxical findings that remain unpublished since they could not be explained in the context of epileptogenesis or the seizures that were being studied. However, such observations could be indicators of homeostatic processes taking place.

10.3 Homeostasis and Seizures During Brain Development

The developing brain is well known to be highly susceptible to seizures. However, during the first 2–3 weeks of life in rats and mice, in general neither prolonged seizures nor recurrent seizures usually lead to the genesis of epilepsy later in life (but this remains unclear and controversial in the clinical setting). Nonetheless, these seizures are not without significant consequences since numerous studies have shown that they routinely produce deficits in learning and memory – particularly in spatial learning [21]. Several labs have begun to explore the underlying mechanisms. For example, recent studies of hippocampal CA1 pyramidal cells suggest that place cell function is impacted by recurrent early-life seizures. Place cells are thought to provide an animal with a spatial map of its environment and serve as surrogate markers for spatial memory. Among other observations, investigators have shown that place cells are unable to form stable maps in animals that experienced early-life seizures [19]. Further, rats exposed to early prolonged hyperthermia-induced seizures exhibit a significant increase in hippocampal T2 MRI signal intensity which is associated with spatial memory deficits [10].

In exploring the underlying mechanisms of learning and memory deficits, investigators have understandably focused on alterations in central excitatory and inhibitory synapses. Early-life seizures have been shown to profoundly affect synaptic signaling through both glutamate and GABA$_A$ receptors. For example, hypoxia-induced seizures in postnatal day 10 rats results in a decrease in silent N-methyl-D-aspartate (NMDA) synapses and an attenuation of hippocampal LTP that persists into adulthood [36].

Changes in inhibitory neurotransmission can also play an important role in processes critical for learning and memory. Enhancement of GABA$_A$ receptor signaling is known to impair LTP, and studies have shown increased inhibition after early-life seizures. For example, after both hyperthermia- and kainate-induced seizures, there is enhanced paired-pulse inhibition in the hippocampus [29] and selective increases in specific GABA$_A$ receptor subunits, notably the α1 subunit, after status epilepticus induced by either lithium-pilocarpine or kainate at postnatal day 20 [25]. However, it should be noted that GABA$_A$ receptor subunit changes following seizure activity are age-dependent as are the responses to agonists. Importantly, GABA$_A$ receptor activation in neonatal neurons results in membrane depolarization, in contrast to the normal hyperpolarizing response seen in mature neurons – a result of differential expression of the cation-co-transporters KCC2 and NKCC1 in early post-natal brain development which establishes the transmembrane chloride electrochemical gradient [2]. Early-life seizures have been reported to promote the developmental switch from depolarizing to hyperpolarizing, one consequence of which may be impaired spatial learning and memory [12].

In addition to molecular receptor changes affecting both excitatory and inhibitory neurotransmission, a number of studies have reported decreases not only in dendritic spine density but also dendrite length and branching complexity in hippocampal pyramidal cells [18, 23]. Similar abnormalities in dendrite morphology have been reported in human epilepsy [30]. However, in experimental studies of seizure induction in early-life, dendritic changes have been observed after a series of seizures that do not lead to epilepsy later in life and presumably do not induce a significant epileptogenic process. For instance,

when 15 brief (~3 min in duration) seizures are induced over a 5 -day period (3 seizures per day) in 1 week old mice, dendrite length and branching complexity are reduced by 25 % compared to control mice, and as adults these same mice are learning impaired [23]. Changes in CA1 dendrite arborization are observed within 1 week after the last seizure and have been shown to be the result of dendrite growth suppression. Very similar observations of growth suppression have been made in hippocampal slice cultures [24]. In these instances, slice cultures from 5 day-old mice are grown under conditions that produce recurring seizure-like activity. Within 24 h of initiating epileptiform activity, CA1 pyramidal cell dendrites are shorter in length and have fewer branches than pyramidal cells from sister control cultures. Moreover, over time, while dendrites in control slices continue to grow, dendrites in slices that are undergoing seizure-like activity do not. Similar to the studies of synaptic homeostatic plasticity in dissociated cultures discussed earlier, mEPSC amplitude and frequency (recorded in pyramidal cells) are reduced in slice cultures following a few days of treatment [31]. Remarkably, a very recent report has shown that similar changes in excitatory synaptic transmission and dendrite arborization can be observed after only a few hours of synchronized epileptiform activity [8]. Collectively, these results suggest that seizures may not only suppress on-going dendrite growth but acutely may even induce a retraction of growing dendritic branches.

One interpretation of these results is that the seizures *in vivo* and seizure-like activity *in vitro* are activating homeostatic mechanisms in attempts to limit neuronal excitability, re-establish network excitability *in vitro* and prevent the occurrence of future seizures. However, by limiting the branching complexity of hippocampal pyramidal cell dendrites, the number of excitatory glutamatergic synapse present on dendrites should also be reduced. Indeed, biochemical results have consistently shown reduced expression of markers for glutamatergic synapses, such as PSD95, in the hippocampus taken from mice that have experienced recurring early-life seizures and in slice culture that have undergone chronic epileptiform activity [31, 32]. With a reduction in the number of glutamatergic synapses, one might predict deficits in hippocampal-based learning and memory. This is because these synapses are well known to undergo Hebbian forms of synaptic plasticity, such as LTP and LTD, which are though to contribute in important ways to the formation of memories (see earlier discussion). Thus, in attempting to re-establish normal neuronal and network excitability, homeostatic mechanisms may also limit an animal's capacity to learn since some of the anatomical substrates for learning have been eliminated.

In such situations, homeostatic mechanisms could become maladaptive or dyshomeostatic, where these mechanisms are driven to such extremes that they have undesirable consequences. Regulating the circulating levels of Ca^{+2} that was discussed earlier provides an example of such a phenomenon. In some clinical situations, blood Ca^{+2} levels can fall below its set point for prolonged periods of time. Dietary deficiency is one cause of low blood Ca^{+2}. Under these circumstances, calcium sensing cells in the parathyroid gland release parathyroid hormone which activates osteoblasts in bone resulting in the release of Ca^{+2} into the blood in attempt to restore circulating Ca^{+2}. However, intense activation by parathyroid hormone will eventually lead to bone dissolution, cavitations of the skeleton and increased susceptibility to bone fractures. Similarly, it seems possible that uncontrolled seizures may induce neuronal homeostatic responses that in attempting to limit neuronal hyperexcitability results in impaired synaptic plasticity and learning deficits.

It is thought that synaptic homeostatic plasticity and Hebbian synaptic plasticity are normally complementary processes. While Hebbian plasticity occurs from moment-to-moment, homeostatic mechanisms occur more slowly, over hours and days and function to prevent runaway excitation or inhibition but do not to interfere with rapid information transfer and storage. However, in epilepsy where abnormal – and often extreme neuronal hyperexcitability – exists, homeostatic mechanisms appear unable to reset neuronal excitability levels to something approaching normal. But by continually attempting to reset normal levels, homeostasis may be driven to such extremes that it limits Hebbian plasticity and interferes with information processing.

10.4   Concluding Remarks

At this time, some may not be convinced that homeostatic mechanisms are active in the epileptic brain and more direct evidence is needed to support the notion that the cognitive neurobehavioral comorbidities of epilepsy are at least in part a consequence of homeostasis and homeostatic imbalance. Currently, the challenge is in developing ways to study such hypothetical seizure-induced homeostatic mechanisms in relative isolation and in greater detail without the confound of the myriad molecular, cellular and genetic processes that are active in epileptogenesis. The developing hippocampus may serve as a useful model system in this regard since at least under some experimental conditions homeostatic mechanisms appear to predominate over mechanisms of epileptogenesis. The future may provide better experimental opportunities and researchers should be prepared to exploit them. Ultimately, a full understanding of the molecular mechanisms underlying seizure-induced homeostasis will be required. It seems that employment of relatively simple *in vitro* systems (e.g. dissociated or slice cultures) would accelerate discovery. However, key findings *in vitro* will always need to be validated *in vivo*. Under some experimental situations (e.g. the prolonged seizures of status epilepticus) neuronal injury and death may occur and should be avoided if possible. Being able to discriminate between injury-induced changes and homeostatic-induced mechanisms will be critical. However, currently neuroscience researchers have a wealth of new and powerful cellular and molecular tools at their disposal that should make such studies possible. Live time-lapse imaging of neurons in which molecular biomarkers of suspected key contributors of seizure-induced homeostasis can be visualized is but one example of the types of experiments that should be possible. Once homeostatic mechanisms have been well characterized and ways to selective eliminate them have been discovered, returning to more complex situations where epileptogenesis and homeostatic plasticity co-exist will be important not only to definitively prove that homeostasis is active in epilepsy but also to understand the costs and benefits of suppressing or enhancing these homeostatic processes.

Acknowledgement From John Swann: I remember first meeting you at the 1981 SfN Meeting. At that point, I had inadvertently stumbled into epilepsy research but became intrigued by how little was known about the basic mechanisms of the childhood epilepsies. You had recently published your first papers on epileptiform activity in immature hippocampal slices and I was in the midst of somewhat similar experiments as a newly minted independent investigator. During our conversation, you encouraged me to continue my line of investigation even though we were potential competitors. Your generous gesture contributed importantly to my commitment to epilepsy research and I have valued you as a friend and colleague throughout the intervening years. I think this book in many ways reflects the positive influence you have had on the careers of so many scientists – in epilepsy and the neurosciences more generally. This is a legacy to be proud of and emulated by your students and the future generations of their students.

From Jong Rho: We first met while I was interviewing for my first faculty position at the University of Washington and the Seattle Children's Hospital in 1994. Your remarkable presence and the opportunity to launch an independent research career under your guidance were critical factors in my decision to relocate there after my post-doctoral training at the NIH. Without your constant support and mentorship, I would not have been able to secure my first research grant through the Epilepsy Foundation of America and a career development award from the NIH. The high point of my tenure in Seattle was working with you in creating the Pediatric Epilepsy Research Center, and to jointly delve into the mechanisms of ketogenic diet action in epileptic brain – an interest that has remained to this day. You have always been a role model for me, and have continued to exemplify the highest standards of scientific integrity, intellectual rigor, selflessness and humility, and most important of all, the humanity and love of those with whom you worked. This volume is a testament to all of that and more.

Other acknowledgement Work in the Swann Lab has been supported by grants from NIH-NINDS and CURE, and in the Rho Lab by grants from NIH-NINDS and Canadian Institutes for Health Research.

References

1. Abbott LF, Nelson SB (2000) Synaptic plasticity: taming the beast. Nat Neurosci 3(Suppl):1178–1183
2. Ben-Ari Y (2013) The developing cortex. Handb Clin Neurol 111:417–426
3. Bernard C (1865) Introduction à l'étude de la médecine expérimentale. J.B. Baillière et fils, Paris
4. Branco T, Staras K, Darcy KJ, Goda Y (2008) Local dendritic activity sets release probability at hippocampal synapses. Neuron 59:475–485
5. Brewster AL, Lugo JN, Patil VV, Lee WL, Qian Y, Vanegas F, Anderson AE (2013) Rapamycin reverses status epilepticus-induced memory deficits and dendritic damage. PLoS ONE 8:e57808
6. Burrone J, O'Byrne M, Murthy VN (2002) Multiple forms of synaptic plasticity triggered by selective suppression of activity in individual neurons. Nature 420:414–418
7. Cannon WB (1929) Organization for physiological homeostasis. Physiol Rev 9:399–431
8. Casanova JR, Nishimura M, Le JT, Lam TT, Swann JW (2013) Rapid hippocampal network adaptation to recurring synchronous activity: a role for calcineurin. Eur J Neurosci 38:3115–3127
9. Desai NS, Rutherford LC, Turrigiano GG (1999) Plasticity in the intrinsic excitability of cortical pyramidal neurons. Nat Neurosci 2:515–520
10. Dube CM, Zhou JL, Hamamura M, Zhao Q, Ring A, Abrahams J, McIntyre K, Nalcioglu O, Shatskih T, Baram TZ, Holmes GL (2009) Cognitive dysfunction after experimental febrile seizures. Exp Neurol 215:167–177
11. Dudek FE, Staley KJ (2012) The time course and circuit mechanisms of acquired epileptogenesis. In: Jasper's basic mechanisms of the epilepsies, 4th edn. Oxford University Press, Oxford, pp 405–415
12. Galanopoulou AS (2008) Dissociated gender-specific effects of recurrent seizures on GABA signaling in CA1 pyramidal neurons: role of GABA(A) receptors. J Neurosci 28:1557–1567
13. Goold CP, Nicoll RA (2010) Single-cell optogenetic excitation drives homeostatic synaptic depression. Neuron 68:512–528
14. Gowers WR (1888) Epilepsy and other chronic convulsive disorders: their causes, symptoms and treatment. J & A Churchill, London
15. Hartman KN, Pal SK, Burrone J, Murthy VN (2006) Activity-dependent regulation of inhibitory synaptic transmission in hippocampal neurons. Nat Neurosci 9:642–649
16. Houser CR, Zhang N, Peng Z (2012) Alterations in the distribution of GABAA receptors in epilepsy. In: Jasper's basic mechanisms of the epilepsies, 4th edn. Oxford University Press, Oxford, pp 532–544
17. Jacobs MP, Leblanc GG, Brooks-Kayal A, Jensen FE, Lowenstein DH, Noebels JL, Spencer DD,

Swann JW (2009) Curing epilepsy: progress and future directions. Epilepsy Behav 14:438–445

18. Jiang M, Lee CL, Smith KL, Swann JW (1998) Spine loss and other persistent alterations of hippocampal pyramidal cell dendrites in a model of early-onset epilepsy. J Neurosci 18:8356–8368

19. Karnam HB, Zhou JL, Huang LT, Zhao Q, Shatskikh T, Holmes GL (2009) Early life seizures cause long-standing impairment of the hippocampal map. Exp Neurol 217:378–387

20. Kilman V, Rossum MC, Turrigiano GG (2002) Activity deprivation reduces miniature IPSC amplitude by decreasing the number of postsynaptic GABA(A) receptors clustered at neocortical synapses. J Neurosci 22:1328–1337

21. Kleen JK, Scott RC, Lenck-Santini PP, Holmes GL (2012) Cognitive and behavioral co-morbidities of epilepsy. In: Jasper's basic mechanisms of the epilepsies, 4th edn. Oxford University Press, Oxford, pp 915–929

22. Maffei A, Nelson SB, Turrigiano GG (2004) Selective reconfiguration of layer 4 visual cortical circuitry by visual deprivation. Nat Neurosci 7:1353–1359

23. Nishimura M, Gu X, Swann JW (2011) Seizures in early life suppress hippocampal dendrite growth while impairing spatial learning. Neurobiol Dis 44:205–214

24. Nishimura M, Owens J, Swann JW (2008) Effects of chronic network hyperexcitability on the growth of hippocampal dendrites. Neurobiol Dis 29:267–277

25. Raol YH, Zhang G, Lund IV, Porter BE, Maronski MA, Brooks-Kayal AR (2006) Increased GABA(A)-receptor alpha1-subunit expression in hippocampal dentate gyrus after early-life status epilepticus. Epilepsia 47:1665–1673

26. Sheehan JJ, Benedetti BL, Barth AL (2009) Anticonvulsant effects of the BK-channel antagonist paxilline. Epilepsia 50:711–720

27. Sherwood L, Klandorf H, Yancey PH (2013) Animal physiology: from gene to organisms. Brooks/Cole, Cengage Learning, Belmont

28. Shruti S, Clem RL, Barth AL (2008) A seizure-induced gain-of-function in BK channels is associated with elevated firing activity in neocortical pyramidal neurons. Neurobiol Dis 30:323–330

29. Sutula T, Cavazos J, Golarai G (1992) Alteration of long-lasting structural and functional effects of kainic acid in the hippocampus by brief treatment with phenobarbital. J Neurosci 12:4173–4187

30. Swann J, Al-Noori S, Jiang M, Lee CL (2000) Spine loss and other dendritic abnormalities in epilepsy. Hippocampus 10:617–625

31. Swann JW, Le JT, Lam TT, Owens J, Mayer AT (2007) The impact of chronic network hyperexcitability on developing glutamatergic synapses. Eur J Neurosci 26:975–991

32. Swann JW, Le JT, Lee CL (2007) Recurrent seizures and the molecular maturation of hippocampal and neocortical glutamatergic synapses. Dev Neurosci 29:168–178

33. Turrigiano G (2012) Homeostatic synaptic plasticity: local and global mechanisms for stabilizing neuronal function. Cold Spring Harb Perspect Biol 4:a005736

34. Turrigiano G (2011) Too many cooks? Intrinsic and synaptic homeostatic mechanisms in cortical circuit refinement. Annu Rev Neurosci 34:89–103

35. Turrigiano GG, Leslie KR, Dasai NS, Rutherford LC, Nelson SB (1998) Activity-dependent scaling of quantal amplitude in neocortical neurons. Nature 391:845–846

36. Zhou C, Lippman JJ, Sun H, Jensen FE (2011) Hypoxia-induced neonatal seizures diminish silent synapses and long-term potentiation in hippocampal CA1 neurons. J Neurosci 31:18211–18222

Is Plasticity of GABAergic Mechanisms Relevant to Epileptogenesis?

11

Helen E. Scharfman and Amy R. Brooks-Kayal

Abstract

Numerous changes in GABAergic neurons, receptors, and inhibitory mechanisms have been described in temporal lobe epilepsy (TLE), either in humans or in animal models. Nevertheless, there remains a common assumption that epilepsy can be explained by simply an insufficiency of GABAergic inhibition. Alternatively, investigators have suggested that there is hyperinhibition that masks an underlying hyperexcitability. Here we examine the status epilepticus (SE) models of TLE and focus on the dentate gyrus of the hippocampus, where a great deal of data have been collected. The types of GABAergic neurons and $GABA_A$ receptors are summarized under normal conditions and after SE. The role of GABA in development and in adult neurogenesis is discussed. We suggest that instead of "too little or too much" GABA there is a complexity of changes after SE that makes the emergence of chronic seizures (epileptogenesis) difficult to understand mechanistically, and difficult to treat. We also suggest that this complexity arises, at least in part, because of the remarkable plasticity of GABAergic neurons and $GABA_A$ receptors in response to insult or injury.

Keywords

GABA • $GABA_A$ receptor • Interneuron • $\alpha1$ subunit • Chloride channel • Granule cell • Adult neurogenesis • Status epilepticus • Febrile seizures • Aberrant neurogenesis • Ectopic granule cell

H.E. Scharfman (✉)
The Nathan S. Kline Institute for Psychiatric Research, Orangeburg, NY, USA

Departments of Child & Adolescent Psychiatry, Physiology & Neuroscience, and Psychiatry, New York University Langone Medical Center, New York, NY, USA
e-mail: hscharfman@nki.rfmh.org

A.R. Brooks-Kayal
Departments of Pediatrics, Neurology and Pharmaceutical Sciences, University of Colorado School of Medicine, Aurora, CO 80045, USA

Skaggs School of Pharmacy and Pharmaceutical Sciences, Aurora, CO 80045, USA

Children's Hospital Colorado, Aurora, CO 80045, USA

H.E. Scharfman and P.S. Buckmaster (eds.), *Issues in Clinical Epileptology: A View from the Bench*,
Advances in Experimental Medicine and Biology 813, DOI 10.1007/978-94-017-8914-1_11,
© Springer Science+Business Media Dordrecht 2014

11.1   Introduction

In the nineteenth century, the idea that epilepsy was a brain disorder arose as a consequence of the relatively new discipline of neurology. In the latter half of the twentieth century, many studies showed that chemicals such as penicillin, a GABA$_A$ receptor (GABA$_A$R) antagonist, caused experimental seizures or epileptiform activity when applied to the neocortex of animals. Philip Schwartzkroin played a major role in the development and refinement of these ideas by the use of the hippocampal slice preparation [131, 132, 152]. One view that emerged was that epilepsy might be caused by defects in inhibition, which was supported by pharmacological experiments showing that several anticonvulsants, such as the barbiturates and benzodiazepines, exerted their actions by facilitating the actions of GABA at GABA$_A$Rs [88, 109].

The idea that epilepsy is caused by insufficient GABAergic inhibition has developed more support as it has become clear that some types of GABAergic neurons are vulnerable in animal models of epilepsy, or lost in tissue resected surgically from patients with intractable epilepsy [78, 126, 127]. In addition, mutations in the subunits of the GABA$_A$ receptor have been identified as a basis of some genetic epilepsy syndromes, such as Genetic Epilepsy with Febrile Seizures+ (GEFS+) which can be caused by a point mutation in the *GABRG$_2$* gene which normally encodes the γ subunit of the GABA$_A$R [4, 159]. However, many arguments have also been made that epilepsy cannot be explained solely by a defect in GABAR-mediated inhibition. Some of the opposing views have come from studies of GABAergic agonists, which exacerbate some types of seizures instead of inhibiting them. For example, drugs that enhance GABAergic inhibition increase absence seizures instead of suppressing them. The explanation is related to the actions of GABA at GABA$_B$ receptors on thalamocortical relay cells. By enhancing the actions of GABA to hyperpolarize relay cells, T-type Ca^{2+} current in relay cells are strongly deinactivated, leading to more robust bursts of action potentials in relay cells when the hyperpolarizations end; these rebound bursts drive the thalamocortical oscillation [58, 141].

In the last 20 years, a wealth of new information about GABA and GABARs has been published using animal models of epilepsy and clinical research. One of the complexities that has emerged is the plasticity of GABAergic mechanisms. This plasticity is remarkable because it involves many aspects of GABAergic transmission: the numbers of GABAergic neurons and the locations of their axons; the synthesis, release and uptake of GABA; and alterations in GABA receptors. Although the contribution of GABAergic mechanisms, and their plasticity, to epilepsy is still an area of active research, it seems unlikely that there is simply too little GABA in epilepsy – or too much. Instead, GABAergic transmission is very different in epilepsy compared to the normal brain. This concept, that GABAergic inhibition is not simply deficient in epilepsy, is consistent with the relatively normal function of individuals with epilepsy during the interictal state.

We discuss below the basic characteristics of GABAergic transmission in the normal and epileptic condition to clarify this idea. For the epileptic condition, we focus on temporal lobe epilepsy (TLE) where this concept appears to be particularly relevant. We also focus on the dentate gyrus (DG) in animal models where status epilepticus (SE) is used to produce spontaneous recurrent seizures and simulate acquired TLE. The reason for this focus is that the data that are available for this context are extensive. However, these models have been criticized because they do not simulate all aspects of TLE.

Most of the discussion below addresses the ways that GABAergic circuitry are changed by SE and alterations in GABA$_A$Rs in DG granule cells (GCs). Presynaptic GABA$_A$Rs and effects of GABA$_A$Rs on other cell types are also important to consider in the context of the DG and epilepsy, and are reviewed elsewhere [70]. Regulation of GABA$_A$Rs by phosphorylation also has implications for the dynamics of GABAergic transmission in epilepsy; effects relevant to the DG are discussed below and additional issues are described elsewhere [83, 155]. Finally, GABA$_B$Rs clearly have a role in epilepsy, but are outside the scope of this discussion and readers are referred to excellent reviews published previously [14, 84].

11.2   GABAergic Transmission in the Normal Adult Dentate Gyrus (DG)

11.2.1   GABAergic Neurons in the DG of the Adult Rodent

Figure 11.1 illustrates the fundamental circuitry of the DG in the normal adult rodent [2]. The principal cell of the DG is the granule cell (GC), which uses glutamate as its primary neurotransmitter, but also has the capacity to synthesize GABA, especially after seizures (discussed further below). GCs also synthesize numerous peptides that are packaged in dense core vesicles and behave as co-transmitters [55]. The peptides are numerous: dynorphin [25], leu-enkephalin [153], brain-derived neurotrophic factor [125], and others. The major afferent input to the GCs is the perforant path projection from entorhinal cortical neurons in layer II [161]. The GCs form the major output from the DG, the "mossy fiber" pathway, which innervates neurons in the hilus and area CA3 [2]. There is another glutamatergic neuron in the DG, located in the hilus, which is called a mossy cell (for reviews see [53, 126]). The major afferent input to mossy cells comes from the GCs, and mossy cells project to GCs and GABAergic neurons within the DG [126].

There are many other types of neurons in the DG, and they use GABA as a neurotransmitter. Most of the GABAergic neurons have an axon that projects primarily in the area surrounding the cell body, similar to other cortical circuits where most of the GABAergic neurons are local interneurons. However, there are several subtypes of DG interneurons that also have axons that project to distant areas of the DG, such as the contralateral DG [34, 49]. Like GCs, GABAergic neurons of the DG also use peptides as co-transmitters [55, 138], and after seizures, some of the peptides in

GCs are the same peptides as those in GABAergic neurons (e.g., neuropeptide Y; NPY; [120]).

The primary type of GABAergic neuron in the DG is the basket cell, which makes basket-like endings around GC somata. It initially was described as a pyramidal-shaped neuron with somata at the base of the GC layer (on the border of the GC layer and the hilus) but the location, somatic morphology and other characteristics are actually diverse [115]. Furthermore, some of the basket cells with pyramidal shaped somata have axons that project to the contralateral DG [49]. There also is variation in neuropeptide content in pyramidal-shaped GABAergic neurons, ranging from paravalbumin, cholecystokinin, to substance P [55, 81, 139]. Electrophysiologically, these cells also vary, although they fit the general characteristics of interneurons because they have a very large afterhyperpolarization following single action potentials [115]. They inhibit their postsynaptic targets by opening chloride channels of GABA$_A$Rs at the soma. Because the resting potential of GCs is close to the reversal potential for chloride or hyperpolarized to it, chloride entry depolarizes the GC rather than hyperpolarizing it, shunting currents that would otherwise reach threshold for action potential (AP) generation; for this reason, "shunting inhibition" is probably the main inhibitory effect of basket cells, rather than hyperpolarization.

Another very important inhibitory cell type also inhibits AP generation of GCs, but is slightly different because it primarily innervates the axon hillock, rather than the somata of GCs. This cell type, the axo-axonic cell, is similar to chandelier cells in neocortex [142] in that chandelier-type endings envelope the axon hillock of GCs. The cell bodies of axo-axonic cells are variable and many types of neuropeptides are co-localized with GABA. The intrinsic electrophysiology of axo-axonic cells is consistent with fast-spiking interneurons [22].

Another type of DG interneuron is the so-called HIPP cell, named because it has a *Hi*lar cell body and projects to the outer 2/3 of the molecular layer, where the *p*erforant *p*ath projection terminates. This neuronal subtype usually expresses somatostatin and NPY [145] and has

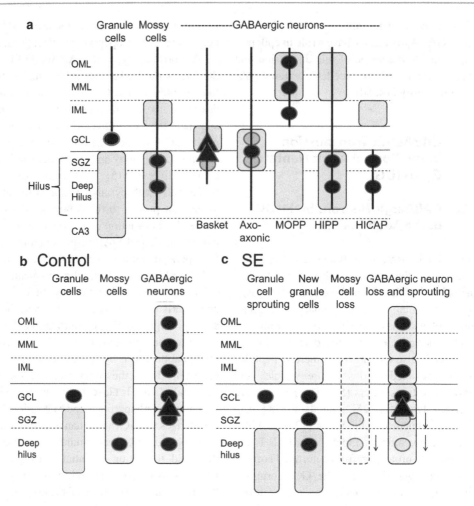

Fig. 11.1 DG circuitry in the normal adult rodent and following status epilepticus (SE). (a) Circuitry of the normal rodent DG is shown schematically. Cell bodies outlined in *green* are glutamatergic; those cells outlined in *red* are GABAergic. *Black circles* indicate the primary location of the somata; *grey circles* are secondary locations. *Gray rectangles* indicate the location of the axon terminals. Abbreviations of the lamina of the DG are as follows: *OML* outer molecular layer, *MML* middle molecular layer, *IML* inner molecular layer, *GCL* granule cell layer, *SGZ* subgranular zone. MOPP, molecular layer cell body, axon in the terminal field of the perforant path; HIPP, hilar cell body, axon in the terminal field of the perforant path. HICAP, hilar cell body, axon in the terminal field of the commissural/associational projection (Adapted from Freund and Buzsaki [42]). (**b**) A summary of a. (**c**) Changes in the DG circuitry following SE are diagrammed. After SE, changes are as follows: GC axons sprout into the IML; newborn GCs are born and some migrate into the hilus and GCL; many mossy cells are lost (indicated by the *arrow, light cell body color* and *dotted line* around the axon plexus); some GABAergic neurons are lost and others sprout into several layers (For references, see text)

axon collaterals primarily in the molecular layer [52], with a less dense projection in the hilus [35]. It has been suggested that it inhibits the EPSPs produced by the perforant path input, presumably by innervating GC dendrites and shunting EPSPs traveling to the GC soma. HIPP cells may also inhibit glutamate release from perforant path terminals because they make synapses on the terminals [80]. The electrophysiology of HIPP cells is characteristic of interneurons generally [44], but it has been noted that they are relatively slow spiking [2, 115] and have a pronounced 'sag' in response to hyperpolarizing current commands [89]. This cell type has attracted a lot of attention in epilepsy research because these cells are relatively vulnerable to insults or injury [116, 126]. Several mechanisms have been proposed for their vulnerability, such as STAT3 expression [29]. It has also been shown that p75[NTR] receptors are present on the septocholinergic terminals that innervate the HIPP cells, and can cause their death when the septocholinergic pathway is lesioned [37, 38].

Analysis of the numbers of GABAergic neurons using immunocytochemical markers and stereological techniques has led to estimations that the majority of DG interneurons are basket cells or axo-axonic cells, which express parvalbumin or CCK. The other major subtype of DG interneuron is hilar HIPP cells, which co-express GABA and NPY or somatostatin (for reviews see [55, 81]). However, many other types of DG interneurons exist: MOPP cells [28], ivy cells and neurogliaform cells [3] and hilar neurons that innervate the inner molecular layer (HICAP cells; [51, 52]).

The major afferents to DG interneurons are the perforant path, GCs, and mossy cells. In addition, there is extrinsic input from the ascending serotoninergic, cholinergic, and noradrenergic nuclei. The primary effects appear to be inhibitory [41]. In addition, there are additional inputs to the DG from areas outside the hippocampus that are not well understood functionally, such as the supramammillary input [74]. Many neuromodulators, such as endocannabinoids, have been shown to exert striking effects in the DG [40], but how all the neuromodulators act in concert in the awake behaving animal is still unclear.

11.2.2 GABA Receptors in the Normal Adult GC

Post-synaptic $GABA_A Rs$ mediate most fast synaptic inhibition in the forebrain (Fig. 11.2). $GABA_A Rs$ are heteromeric protein complexes composed of multiple subunits that form ligand-gated, anion-selective channels whose properties are modulated by barbiturates, benzodiazepines, zinc, ethanol, anesthetics and neurosteroids. There are several different $GABA_A R$ subunit families and multiple subtypes exist within each of these subtypes (α1-6, β1-4, γ1-3, δ, ϵ, π, Φ). The most common $GABA_A R$ is the $\alpha1\beta2\gamma2$ subtype, but multiple subtype combinations exist and they vary in different brain regions and cell types, and during different times in development [73, 111, 134]. Subunit composition of $GABA_A Rs$ plays a major role in determining the intrinsic properties of each channel, including affinity for GABA, kinetics, conductance, allosteric modulation, probability of channel opening, interaction with modulatory proteins, and subcellular distribution [77, 97, 134]. For example, alterations in the α-subtype results in differences in receptor kinetics, membrane localization and $GABA_A R$ modulation by benzodiazepines and zinc [87, 97, 140, 154]. In the GC, $GABA_A Rs$ that contain $\alpha1$ subunits paired with $\gamma2$ subunits are sensitive to benzodiazepines and generally located at the synapse, contributing to phasic inhibition, a term that refers to the effects of GABA released at GABAergic synapses that binds to postsynaptic receptors located at the synaptic cleft. These effects are primarily related to increased conductance when chloride channels open, and hyperpolarization of postsynaptic membrane potential when chloride influx occurs. However, as mentioned above, when the postsynaptic membrane potential is hyperpolarized relative to E_{Cl^-}, which may occur in GCs, there is a depolarization. $GABA_A Rs$ that contain $\alpha4$ subunits have unique pharmacological properties, such as insensitivity to benzodiazepines and increased sensitivity to zinc blockade. Receptors containing $\alpha4$ subunits are most often found with the δ rather than the γ subunit in combination with $\alpha\beta$. These $\alpha4\beta\delta$ $GABA_A Rs$ are localized to extrasynaptic sites and contribute to tonic inhibition, which refers to the basal inhibitory current produced by low concentrations of extracellular GABA that are present outside of the synapse (resulting from diffusion from synaptic to extrasynaptic space). Under physiological conditions, only a minor population of $\alpha4\beta\gamma2$ $GABA_A Rs$ are found at synapses of GABAergic neurons on

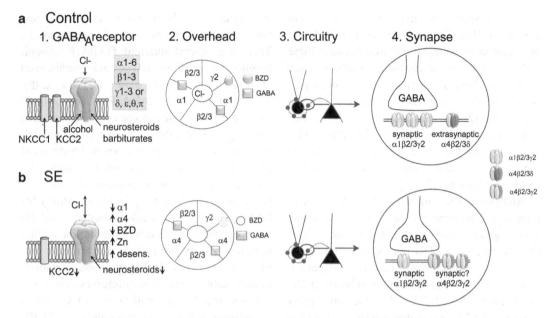

Fig. 11.2 GABA$_A$ receptor subunits in dentate gyrus (DG) granule cells (GCs) in the normal adult rodent and following SE. (a) Control conditions. (*1*) The subunits of the GABA$_A$ receptor (GABA$_A$R) are diagrammed, with sites of modulation noted. The location of the K$^+$Cl$^-$ cotransporters NKCC1 and KCC2 are depicted schematically. (*2*) An overhead view of a typical GABA$_A$R in a normal adult GC. It has α1, β2/3 and γ2 subunits with two sites for GABA and a benzodiazepine (BZD) site for modulation. (*3*) The prototypical GABAergic neuron in the DG is the basket cell (triangle) which has an axon that encircles GC somata, making periodic GABAergic synapses. (*4*) A schematic of the GABAergic synapse in control conditions has synaptic α1β2/3γ2 receptors and extrasynaptic receptors that contain different subunits (α4β2/3δ). **(b)** After SE, KCC2 expression decreases and the direction of chloride flux may change as a result. The expression of α1 subunits decrease and α4 subunits increase. Other changes are altered sensitivity to modulators. (*2*) One of the changes in the GABA$_A$Rs in the DG after SE is loss of benzodiazepine sensitivity. (*3*) The pyramidal basket cell and its basket plexus appears to be similar after SE, although other GABAergic neurons are altered, and there may be changes in expression of various peptides. (*4*) The GABAergic synapse after SE has fewer α1 subunits and increased α4 subunits, which may become perisynaptic (indicated by a ?) (References are listed in the text. Parts 1–2 of this figures were adapted from Jacob et al. [59])

GCs, where they are proposed to affect both the rise time and decay of synaptic currents [71].

11.2.3 Regulation of [Cl]$_i$ in Early Development and Its Relevance to TLE

One of the characteristics of GABAergic inhibition at GABA$_A$Rs that has implications for epilepsy – and has been studied extensively in the hippocampus in TLE – is the regulation of chloride flux through the GABA$_A$R. The direction of chloride flux is regulated by many factors, and one source of regulation that has attracted a great deal of attention is the K$^+$-Cl$^-$ cotransporters

KCC2 and NKCC1. KCC2 extrudes chloride normally, and NKCC1 transports chloride into the cell [7]. In early life, KCC2 expression is low and there is a relatively high concentration of intracellular chloride; chloride efflux occurs when GABA binds to the GABA$_A$R, leading to a depolarization [8, 27]. After maturation, KCC2 expression increases and this leads to a lower [Cl$^-$]$_i$ and chloride influx when GABA binds to GABA$_A$Rs, leading to a hyperpolarization [106]. As mentioned above, an exception is the GC, which has a resting potential (−70 to −80 mV) that is usually negative to E$_{Cl^-}$. Therefore, in early life, a strong depolarization of GCs by GABA is predicted, and a smaller depolarization in adulthood compared to adulthood.

The idea that GABA is depolarizing in early postnatal life has recently been contested because most data that led to the idea were collected in slices where truncation of neuronal processes leads to elevated $[Cl^-]_i$ [15]. However, *in vivo* studies have been conducted that are consistent with a depolarizing action of GABA in pyramidal neurons in neonatal life [9]. It remains to be determined exactly at what age these depolarizing effects end; in rodents it seems likely to be the first or second postnatal week [9, 15].

In the DG, one might expect that the switch from depolarizing to hyperpolarizing effects of GABA would not be as important because GABA typically has a depolarizing effect on GCs regardless. However, the size of the depolarization will be substantially greater if KCC2 expression is low, and moreover, there are many cells besides GCs in the DG that will be affected; only the GC has a very high resting potential. There are also many types of GABAergic inhibition, not only postsynaptic. If the GABA$_A$R is presynaptic, for example, the net effect could very different if the terminal is depolarized or hyperpolarized by GABA.

There is also another process in the DG that is likely to be affected if the effects of GABA "switch" from depolarizing to hyperpolarizing – the maturation of GCs that are born postnatally, i.e., postnatal or "adult" neurogenesis [67]. GABA is a critical regulator of the maturation and migration of immature neurons in early life [24, 160]. GABA also influences maturation and migration of adult-born GCs [36]. In acquired TLE this is potentially important because animal models of TLE have shown that there is a large increase in proliferation of adult-born GCs after seizures [90], and the young GCs often mismigrate (discussed further below). It has been suggested that these mismigrated GCs contribute to chronic seizures (discussed further below).

11.3   Alterations in GABAergic Transmission in Animal Models of TLE

There are many types of TLE, and one of the ways to classify the types is based on whether the epilepsy appears to have been "acquired."

The term 'acquired' indicates that an insult or injury occurred prior to seizures and is likely to have caused the epilepsy. Acquired TLE has been simulated in laboratory animals by various insults or injuries that lead to a pattern of brain damage that is typical of TLE, called mesial temporal sclerosis (MTS; [127]). In general, MTS involves loss of a large number of CA1 and CA3 pyramidal cells, with sparing of CA2 and GCs. Many hilar neurons are lost, and these include both mossy cells and HIPP cells [116]. Notably, there are individuals with acquired TLE that do not have this classic description of MTS, and animal models vary in the extent they simulate MTS [127]. However, the pattern has been the focus of the most research in TLE, based on the assumption that this general pattern of neuropathology causes TLE or is very important to TLE.

One method that leads to a MTS-like pattern of neuropathology in adult rodents is induction of SE, either by injection of a chemoconvulsant such as kainic acid or pilocarpine, or electrical stimulation of hippocampus [31, 85, 95]. Here we will focus primarily on the SE models to study TLE in adult rodents, and use the data from SE models to address changes in GABAergic inhibition. We suggest that these changes involve plasticity of GABAergic mechanisms rather than simply an erosion or increase in the effects of GABA.

11.3.1   Alterations in GABAergic Neurons After SE

Early observations that GABAergic neurons were decreased in neocortical epileptic foci produced by alumina gel in monkeys supported ideas that disinhibition may be the cause of epilepsy [100–102], particularly because the reduction in GABAergic neurons preceded epilepsy [56, 103]. Chandelier cells appeared to be one of the subtypes that was affected, and it was suggested that loss of the chandelier subtype of GABAergic neuron would be most likely to cause disinhibition of cortical pyramidal cells because loss of only a few axo-axonic cells would substantially change the number of GABAergic terminals at the axon hillock [33].

However, as more animal models were examined, there was less enthusiasm for the idea that disinhibition was the fundamental cause of seizures. In seizure-sensitive gerbils [93], the audiogenic seizure model [110], and kainic acid model [32], GABAergic neurons were not always decreased [54]. In fact, some GABAergic neurons increased their axon arbors, exhibiting axon sprouting (discussed further below). When GABA$_A$R-mediated inhibition was examined, it was often strong rather than weak [11]. Therefore, even if some changes in these animal models involve disinhibition acutely, GABAergic neurons and GABA$_A$R-dependent inhibition often show recovery and plasticity.

In the DG, an alternative hypothesis to disinhibition was suggested to address an animal model of TLE in which the perforant path of adult rats was stimulated electrically to simulate the precipitating insult in TLE. In this animal model, a 24 h period of intermittent perforant path stimulation in urethane-anesthetized rats led to a loss of 'paired-pulse' inhibition. Based on the results from these experiments, investigators suggested that the basket cells, (defined by parvalbumin expression) were spared but there was loss of HIPP cells (defined by somatostatin expression) and mossy cells [135]. Because mossy cells appeared to be decreased in numbers, and there were suggestions in the literature that they innervated basket cells, it was hypothesized that the parvalbumin-expressing basket cells lost afferent input from mossy cells and became 'dormant' and this led to disinhibition of GCs [136]. The hypothesis became known as 'the dormant basket cell hypothesis.' It was suggested that the hypothesis explained epileptogenesis in acquired TLE: if an early insult or injury led to loss of vulnerable mossy cells and HIPP cells, but GCs and basket cells were spared, the result would be disinhibition of GCs [6, 75].

However, later studies led to some doubt that this hypothesis could explain acquired TLE [12]. An alternative hypothesis – the 'irritable mossy cell hypothesis' – suggested that mossy cells could cause GC hyperexcitability because the mossy cells, which project directly to GCs, developed increased excitability. This hypothesis was

developed on the basis of recordings from mossy cells in slices after post-traumatic injury [113, 114], another type of precipitating insult that leads to TLE. In addition, mossy cell hyperexcitability was shown subsequently in slices from epileptic rats after SE [128].

A result that argued against these two hypotheses came from studies of animals with chronic epilepsy after kainic acid-induced SE. These experiments showed that there was an increase in paired-pulse inhibition of GCs, not a decrease [139]. In addition, slices from animals after SE did not exhibit spontaneous seizure-like activity, suggesting they had intact inhibition rather than weak inhibition. This was unlikely to be due to the differences in the SE model since 'irritable mossy cells' were observed, at least in one study of SE [128]. In slices, exposure of slices to GABA$_A$R antagonists led to seizure-like activity that was more prolonged in slices from animals that had SE than slices from control rats. From these experiments, it was suggested that slices from animals with SE were hyperexcitable but it was normally masked by GABA$_A$R-mediated inhibition [129, 147]. In slices from humans with intractable TLE, there was enhanced sensitivity to bicuculline [39]. These observations and others led to the idea that increased inhibition was present to compensate for underlying hyperexcitability [147, 162]. Although in some cases the studies of animals with SE and intractable TLE reflect differences in the models or the subtypes of TLE, here the data from different models and humans was consistent, making the observations compelling.

Although an attractive idea, GABAergic inhibition in the animal models of SE does not necessarily seem to be too strong, masking underlying hyperexcitability. For example, interneurons exhibit axonal sprouting in the DG in animal models of TLE [5, 32, 151]. It is not clear that they simply extend their output, inhibiting more glutamatergic neurons than normal, because they innervate inhibitory neurons as well [137]. Interneurons develop abnormal glutamatergic input from sprouting of the GCs into the inner molecular layer (mossy fiber sprouting; for review see [19]). The evidence for this is based on staining of the mossy fibers with Timm stain [137]. Electron microscopy of the mossy fiber boutons in the inner molecular layer supported the idea that the sprouted mossy fibers activate GABAergic basket cells [43]. In further support of this idea, it was suggested that normal mossy fibers in the hilus and area CA3 primarily innervate GABAergic neurons and primarily have an inhibitory effect on CA3 [1]. Moreover, GCs express GABA as well as glutamate after SE [50] and GABA release from GCs can be inhibitory [158] although the latest studies suggest this may be limited to GCs at an early stage of development [23]. The vast majority of studies show that GCs in normal hippocampus excite their target cells [60, 122, 156]. In addition, when mossy fiber synapses in the epileptic rat were quantified in the inner molecular layer, the majority were located on GCs, not interneurons [19, 20].

One way to reconcile the different data is to suggest that mossy fibers have a large dynamic range, with filopodia that excite interneurons and massive boutons that excite principal cells. The outcome may depend on recent activity, which can potentially upregulate GABA expression, or alter the peptide content of the massive boutons so that they are more excitatory [123]. Other hypotheses suggest that mossy fibers can be inhibitory to area CA3 pyramidal cells depending on the firing mode of GCs – after bursts of GC action potentials, excitation of pyramidal cells is transiently suppressed [82].

As our experimental techniques improved, our understanding of the underlying changes became clearer. For example, initial assays to assess inhibition measured paired-pulse inhibition which uses extracellular recordings and is not an extremely reliable measurement, because small changes in the stimulating or recording sites can alter the extent of inhibition even in the same preparation [157]. As patch clamp recordings developed, more indices of pre- and postsynaptic GABAergic inhibition became possible, and the results have shown that the GABAergic system in the DG is changed in diverse ways after SE, not always consistent with disinhibition of GCs, and not always consistent with hyperinhibition (Fig. 11.1b, c).

11.3.2 Alterations in GABA Receptors in GCs After SE

During SE, inhibitory GABAergic synaptic transmission in the DG becomes compromised, presumably due to the dramatic increase in activation of GABAergic neurons. Miniature inhibitory post-synaptic currents (mIPSCs) are reduced in GCs and the number of active $GABA_A Rs$ per GC decreases [26, 47, 86] via enhanced clathrin-dependent $GABA_A R$ internalization [48, 59]. In vitro studies using hippocampal neurons, stimulated with a buffer containing low magnesium to induce spontaneous recurrent epileptiform discharges, showed a large decrease in GABA-gated chloride currents that correlated with reduced cell surface expression and intracellular accumulation of $GABA_A Rs$ [13, 48]. In vivo studies using chemoconvulsants have shown that SE promotes a rapid reduction in the number of physiologically active $GABA_A Rs$ in GCs that correlated with a reduction in the level of $\beta2/\beta3$ and $\gamma2$ immunoreactivity present in the vicinity of a presynaptic marker [86]. In fact, SE appears to trigger subunit specific events to regulate the trafficking of $GABA_A Rs$ by promoting the dephosphorylation of $\beta3$ subunits [47, 150]. Decreased phosphorylation of $\beta3$ increases the interaction of $GABA_A Rs$ with the clathrin-adaptor protein 2 (AP2), facilitating the recruitment of $GABA_A Rs$ into clathrin-coated pits and promoting their removal from the plasma membrane [47, 150]. In hippocampal slices obtained from mice after SE, increased $GABA_A R$ phosphorylation or blockade of normal AP2 function resulted in $GABA_A R$ accumulation at the plasma membrane and increased synaptic inhibition [150].

Alterations in $GABA_A R$ subunit composition occur subsequent to SE in a number of animal models, and there is evidence that these changes my contribute to epileptogenesis [18, 72, 76, 92, 144, 166]. SE results in changes in the expression and membrane localization (i.e., extrasynaptic vs. synaptic) of several $GABA_A R$ subunits (e.g., $\alpha1$, $\alpha4$, $\gamma2$, and δ) in GCs. Beginning soon after SE and continuing until and after the animals become epileptic, these alterations are associated with changes in phasic and tonic $GABA_A R$-

mediated inhibition, and in GABA$_A$R pharmacology [21, 30, 45]. After pilocarpine-induced SE, GABA$_A$R α1 subunit mRNA expression decreases, and GABA$_A$R α4 subunit mRNA expression increases [18]. Changes in GABA$_A$R function and subunit expression have also been observed in neurons from surgically resected hippocampus of patients with intractable TLE; [17, 143]. These alterations are associated with an increase in α4γ2 containing receptors, a reduction in α1γ2 containing receptors in the DG [76], and shift of α4-containing receptors from extrasynaptic to synaptic and perisynaptic locations, which is likely to be related to the appearance of α4$\beta$$\gamma$2 receptors [146, 166]. Changes in expression and localization of α-subunits associated with changes in synaptic GABA$_A$R composition result in a number of changes in synaptic inhibition in GCs, including diminished benzodiazepine sensitivity, enhanced zinc sensitivity, reduced neurosteroid modulation, and diminished phasic inhibition in dendrites [21, 30, 45, 146]. Preventing the reduction in GABA$_A$R subunit α1 expression after SE via viral-mediated transfer of an α1 subunit transgene in adult rodents reduced subsequent epilepsy development, resulting in a three-fold increase in the mean time to the first spontaneous seizure, and a decrease to 39 % of AAV-α1-injected rats developing spontaneous seizures in the first 4 weeks after SE compared to 100 % of rats receiving sham injections [99]. Together, these data support a role for GABA$_A$R α-subunit changes in the process of epileptogenesis.

Receptors containing α4 subunits are most often found with the δ rather than the γ subunit in combination with $\alpha$$\beta$. These α4$\beta$$\delta$ GABA$_A$Rs are localized to extrasynaptic sites and contribute to tonic inhibition. Under physiological conditions, only a minor population of α4$\beta$$\gamma$2 GABARs are found within GABAergic synapses on GCs, where they are proposed to affect both the rise time and decay of synaptic currents [71]. In parallel with the decrease in α1 subunit expression in GCs after SE, there is a marked increase in α4 subunit expression that results in an increase in the abundance of α4γ2-containing receptors in synaptic and perisynaptic locations [146, 166] (see Fig. 11.2), along with the reduction in

α1γ2-containing receptors [76]. The α4$\beta$$\gamma$2 receptors may contribute to epileptogenesis, as α4-containing GABA$_A$Rs have been shown to desensitize rapidly, especially when assembled with β3 subunits [71]. In addition, GABA$_A$Rs containing the α4 subunit are very sensitive to zinc blockade, as are GABA$_A$Rs on GCs in the epileptic brain [21, 30]. Zinc containing mossy fiber terminals sprout from the granule cell layer of the hippocampus onto other GCs and into CA3, likely depositing zinc onto the newly formed α4$\beta$$\gamma$2 receptors causing a decreased response to GABA. Collectively these alterations may contribute to epilepsy development, pharmacoresistance and further epilepsy progression.

GABA$_A$R subunit alterations after SE are regulated by increased synthesis of brain-derived neurotrophic factor (BDNF) and activation of its receptors (TrkB and p75) that control a number of down-stream pathways, including Janus kinase (JAK)/Signal Transducer and Activators of Transcription (STAT), protein kinase C, and mitogen activated protein kinase (MAPK; [76, 107, 108]). BDNF is known to enhance cAMP response element binding protein (CREB) phosphorylation through binding to TrkB receptors [105, 163], and is also a potent regulator of inducible cAMP response element repressor (ICER) synthesis [57]. Using chromatin immunoprecipitation (ChIP) and DNA pulldown studies, it has been determined that there is increased binding of pCREB and ICER to the GABARα1 gene promoter (*GABRA1-p*) in DG after SE [76]. BDNF regulation of ICER expression is mediated by JAK/STAT pathway activation, specifically activation of pJAK2 and pSTAT3 [76]. pSTAT3 association with the STAT-recognition site on the ICER promoter is enhanced after SE in DG and inhibition of JAK/STAT signaling pathway with pyridone 6 (P6) in primary hippocampal cultures and *in vivo* in DG prior to SE blocks both ICER induction and decreased transcription of *GABRA1* [76]. These findings suggest a specific signaling cascade involving BDNF, JAK/STAT, and CREB that is critical to the reported decreases in α1 subunit levels following SE and may contribute to epileptogenesis. Increases in GABARα4 subunit are transcriptionally regulated by BDNF activation of the TrkB

Fig. 11.3 Regulation
of GABA$_A$ receptor
expression after SE.
BDNF regulates the final
composition of GABA$_A$Rs
by differentially altering
the expression of α1 and
α4 subunits. Both in vivo
and in vitro evidence
suggest that increased
levels of BDNF following
SE activate at least two
different signaling
pathways: JAK/STAT and
PKC/MAPK, resulting in
the down-regulation of α1
subunits and the up-regulation of α4 subunits,
respectively (Reproduced
from Gonzalez and
Brooks-Kayal [46])

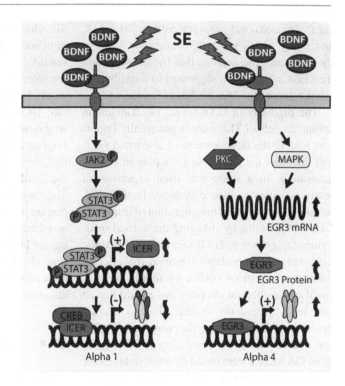

receiver which leads to upregulation of the early
growth response factor (Egr3) pathway via a
PKC/MAPK-dependent pathway [107]. Egr3
association with the early-growth response-
recognition (ERE) site on the *GABRA4* promoter
is enhanced after SE in DG [107] (See Fig. 11.3).

11.3.3 Regulation of GABA in Early Development and Its Relevance to TLE

One of the themes in studies of animal models of
TLE is the idea that the myriad of changes in hippocampal structure and function that have been
described are associated with a recapitulation of
development that is caused by the epileptogenic
insult. A robust example is the dramatic increase
in the rate of adult neurogenesis in the DG after
epileptogenic insults like SE. First noted by
Bengzon et al [10] using stimulus-evoked
afterdischarges, and Parent et al. [90] after
pilocarpine-induced SE, the increase in the rate of
adult neurogenesis after seizures, and particularly
SE (in adult rodents), has been reproduced by
many laboratories in response to virtually all

epileptogenic insults: kindling, kainic acid or
electrically-induced SE, or traumatic brain injury
[121, 124].

Initially it was suggested that many of the
neurons that are born after SE do not survive
long-term [90] which has also been shown by
others [96] but a substantial fraction of newborn
neurons can survive in some animal models, and
these mismigrate into the hilar region, where they
are called hilar ectopic GCs (hEGCs; [119]). Other
adult-born GCs migrate correctly but develop
abnormal dendrites in the hilus, called hilar basal
dendrites [104, 133]. These neurons also appear
to survive long-term and can be generated for a
long-time after SE [62]. Another subset of GC
that develops after SE and is abnormal develops
an enlarged cell body (hypertrophy; [98]). The
abnormal GCs are potentially important because
they contribute to mossy fiber sprouting, particularly hEGCs [69, 94, 119]. HEGCs participate
in seizures in vivo [130] and their numbers
are correlated with chronic seizure frequency
[79]. Manipulations that reduce hEGC number
reduce chronic seizure frequency after SE
[63], although selective deletion of hEGCs is not
yet possible. The hEGCs display a variety of

electrophysiological characteristics [61, 118, 164, 165] which are unlike normal GCs. For these reasons, the neurons that hypertrophy, and the hEGCs, have been suggested to contribute to seizure generation [63, 68, 98, 117, 119].

The plasticity of GABAergic mechanisms in animal models of TLE plays a potentially important role in the development of abnormal GCs, and therefore the role these GCs play in seizure generation. In a study that used experimental febrile seizures to induce epilepsy later in life, febrile seizures caused mismigration of immature GCs into the hilus by changing the normal regulation of migration by GABA acting at GABA$_A$Rs. This study was important in showing that altering the normal effect of GABA by febrile seizures could cause aberrant circuitry that would persist long-term, potentially contributing to seizure generation. Interestingly, the way that GABA was altered was in the expression of GABA$_A$Rs; more GABA$_A$Rs were found by western blot after febrile seizures. In response to increased depolarization by GABA, immature GCs migrated opposite to their normal direction, into the hilus instead of the GC layer. Knockdown of NKCC1 could block the formation of hEGCs and reduce the long-term effects [68]. The studies of Koyama and colleagues and Swijsen et al. [149], who also studied febrile seizures, both found increased β2/3subunits occurred in newborn GCs after febrile seizures [149]. Changes in α3 subunits were also noted by Swijsen et al. [148]. The results suggest that febrile seizures lead to long-lasting changes in the expression of GABA$_A$Rs in the DG, and in the GCs that were born after febrile seizures. These effects could lead to life-long reduction in limbic seizure threshold. They also may contribute to the comorbidities in TLE, such as depression [16, 66], a psychiatric condition where adult neurogenesis in the DG has been shown to play a critical role [112].

Another study of adult rodents is also relevant to the formation of aberrant GCs in TLE. This study used pilocarpine-induced SE in adult rodents to ask how KCC2 is altered immediately after SE. The investigators showed that there was a downregulation of KCC2 in the DG after

SE which would make GCs (both mature and immature GCs) depolarize more in response to GABA [91]. If the results of Koyama et al. [68] are correct, greater depolarization by GABA would be likely to foster mismigration of immature GCs. A similar phenomenon may explain why newborn neurons after SE, in the adult, mismigrate for long distances –it has been described that they migrate from the subgranular zone to the border of the hilus and area CA3 [118]. Together the new information about [Cl$^-$]$_i$ regulation are providing potential mechanisms underlying acquired epileptogenesis in the immature and mature brain. Although a great deal more information will be necessary before new treatments can be developed based on the new hypotheses, NKCC1 antagonists are already in clinical trial [64, 65].

11.4   Summary

In the DG, the robust plasticity of GCs has been of avid interest because they upregulate numerous proteins and exhibit robust sprouting of their axons after seizures. Although extensive studies of GABA in the DG have been made in TLE, the remarkable plasticity of GABAergic mechanisms is often not considered as much as development of disinhibition or hyperinhibition. Here we suggest that there are numerous pre- and postsynaptic changes in GABAergic transmission, even if one only addresses GABAergic synapses on GCs and GABA$_A$ receptors. Taken together, this plasticity leads to more complexity of GABAergic transmission in the epileptic brain, not simply an increase or decrease. The idea that GABAergic inhibition is dramatically altered, rather than increased or decreased, is consistent with the diversity of results of past studies. Therefore, this perspective helps address some of the conflicts in the past. It also provides a different and potentially more accurate perspective that will facilitate antiseizure drug development.

Acknowledgements This article is dedicated to Philip A. Schwartzkroin, a pioneer and leader in epilepsy research, esteemed mentor, and outstanding colleague.

Other Acknowledgements Supported by R01 NS-081203, R21 MH-090606 and the New York State Office of Mental Health (HES) and R01 NS-038595, R01 NS-051710, R01-NS-050393 and grants from Citizens United for Research in Epilepsy, Department of Defense and the American Epilepsy Society (ABK).

References

1. Acsady L, Kamondi A, Sik A, Freund T, Buzsaki G (1998) GABAergic cells are the major postsynaptic targets of mossy fibers in the rat hippocampus. J Neurosci 18:3386–3403
2. Amaral DG, Scharfman HE, Lavenex P (2007) The dentate gyrus: fundamental neuroanatomical organization (dentate gyrus for dummies). Prog Brain Res 163:3–22
3. Armstrong C, Krook-Magnuson E, Soltesz I (2012) Neurogliaform and ivy cells: a major family of nNOS expressing GABAergic neurons. Front Neural Circuit 6:23
4. Baulac S, Huberfeld G, Gourfinkel-An I, Mitropoulou G, Beranger A, Prud'homme JF, Baulac M, Brice A, Bruzzone R, LeGuern E (2001) First genetic evidence of GABA(A) receptor dysfunction in epilepsy: a mutation in the γ2-subunit gene. Nat Genet 28:46–48
5. Bausch SB (2005) Axonal sprouting of GABAergic interneurons in temporal lobe epilepsy. Epilepsy Behav 7:390–400
6. Bekenstein JW, Lothman EW (1993) Dormancy of inhibitory interneurons in a model of temporal lobe epilepsy. Science 259:97–100
7. Ben-Ari Y (2002) Excitatory actions of GABA during development: the nature of the nurture. Nat Rev Neurosci 3:728–739
8. Ben-Ari Y, Khalilov I, Kahle KT, Cherubini E (2012) The GABA excitatory/inhibitory shift in brain maturation and neurological disorders. Neuroscientist 18:467–486
9. Ben-Ari Y, Woodin MA, Sernagor E, Cancedda L, Vinay L, Rivera C, Legendre P, Luhmann HJ, Bordey A, Wenner P, Fukuda A, van den Pol AN, Gaiarsa JL, Cherubini E (2012) Refuting the challenges of the developmental shift of polarity of GABA actions: GABA more exciting than ever! Front Cell Neurosci 6:35
10. Bengzon J, Kokaia Z, Elmer E, Nanobashvili A, Kokaia M, Lindvall O (1997) Apoptosis and proliferation of dentate gyrus neurons after single and intermittent limbic seizures. Proc Natl Acad Sci U S A 94:10432–10437
11. Bernard C, Cossart R, Hirsch JC, Esclapez M, Ben-Ari Y (2000) What is GABAergic inhibition? How is it modified in epilepsy? Epilepsia 41(Suppl 6):S90–S95
12. Bernard C, Esclapez M, Hirsch JC, Ben-Ari Y (1998) Interneurones are not so dormant in temporal lobe epilepsy: a critical reappraisal of the dormant basket cell hypothesis. Epilepsy Res 32:93–103
13. Blair RE, Sombati S, Lawrence DC, McCay BD, DeLorenzo RJ (2004) Epileptogenesis causes acute and chronic increases in GABA(A) receptor endocytosis that contributes to the induction and maintenance of seizures in the hippocampal culture model of acquired epilepsy. J Pharmacol Exp Ther 310:871–880
14. Bowery NG (2010) Historical perspective and emergence of the GABA$_B$ receptor. Adv Pharmacol 58:1–18
15. Bregestovski P, Bernard C (2012) Excitatory GABA: how a correct observation may turn out to be an experimental artifact. Front Pharmacol 3:65
16. Brooks-Kayal AR, Bath KG, Berg AT, Galanopoulou AS, Holmes GL, Jensen FE, Kanner AM, O'Brien TJ, Whittemore VH, Winawer MR, Patel M, Scharfman HE (2013) Issues related to symptomatic and disease-modifying treatments affecting cognitive and neuropsychiatric comorbidities of epilepsy. Epilepsia 54(Suppl 4):44–60
17. Brooks-Kayal AR, Shumate MD, Jin H, Lin DD, Rikhter TY, Holloway KL, Coulter DA (1999) Human neuronal γ-aminobutyric acid$_a$ receptors: coordinated subunit mRNA expression and functional correlates in individual dentate granule cells. J Neurosci 19:8312–8318
18. Brooks-Kayal AR, Shumate MD, Jin H, Rikhter TY, Coulter DA (1998) Selective changes in single cell GABA(A) receptor subunit expression and function in temporal lobe epilepsy. Nat Med 4:1166–1172
19. Buckmaster PS (2012) Mossy fiber sprouting in the dentate gyrus. In: Noebels JL, Avoli M, Rogawski MA, Olsen RW, Delgado-Escueta AV (eds) Jasper's basic mechanisms of the epilepsies, 4th edn. Oxford, New York, pp 416–431
20. Buckmaster PS, Zhang GF, Yamawaki R (2002) Axon sprouting in a model of temporal lobe epilepsy creates a predominantly excitatory feedback circuit. J Neurosci 22:6650–6658
21. Buhl E, Otis T, Mody I (1996) Zinc-induced collapse of augmented inhibition by GABA in a temporal lobe epilepsy model. Science 271:369–373
22. Buhl EH, Han ZS, Lorinczi Z, Stezhka VV, Karnup SV, Somogyi P (1994) Physiological properties of anatomically identified axo-axonic cells in the rat hippocampus. J Neurophysiol 71:1289–1307
23. Caiati MD (2013) Is GABA co-released with glutamate from hippocampal mossy fiber terminals? J Neurosci 33:1755–1756
24. Cancedda L, Fiumelli H, Chen K, Poo MM (2007) Excitatory GABA action is essential for morphological maturation of cortical neurons in vivo. J Neurosci 27:5224–5235
25. Chavkin C (2000) Dynorphins are endogenous opioid peptides released from granule cells to act neurohumorly and inhibit excitatory neurotransmission in the hippocampus. Prog Brain Res 125:363–367

26. Chen JW, Naylor DE, Wasterlain CG (2007) Advances
    in the pathophysiology of status epilepticus. Acta
    Neurol Scand Suppl. 186:7–15
27. Cherubini E, Griguoli M, Safiulina V, Lagostena L
    (2011) The depolarizing action of GABA controls
    early network activity in the developing hippocam-
    pus. Mol Neurobiol 43:97–106
28. Chittajallu R, Kunze A, Mangin JM, Gallo V (2007)
    Differential synaptic integration of interneurons in
    the outer and inner molecular layers of the develop-
    ing dentate gyrus. J Neurosci 27:8219–8225
29. Choi YS, Lin SL, Lee B, Kurup P, Cho HY, Naegele
    JR, Lombroso PJ, Obrietan K (2007) Status
    epilepticus-induced somatostatinergic hilar interneu-
    ron degeneration is regulated by striatal enriched pro-
    tein tyrosine phosphatase. J Neurosci 27:2999–3009
30. Cohen AS, Lin DD, Quirk GL, Coulter DA (2003)
    Dentate granule cell GABA(A) receptors in epileptic
    hippocampus: enhanced synaptic efficacy and altered
    pharmacology. Eur J Neurosci 17:1607–1616
31. Curia G, Longo D, Biagini G, Jones RS, Avoli M
    (2008) The pilocarpine model of temporal lobe epi-
    lepsy. J Neurosci Methods 172:143–157
32. Davenport CJ, Brown WJ, Babb TL (1990)
    GABAergic neurons are spared after intrahippocam-
    pal kainate in the rat. Epilepsy Res 5:28–42
33. DeFelipe J (1999) Chandelier cells and epilepsy.
    Brain 122(Pt 10):1807–1822
34. Deller T (1998) The anatomical organization of the
    rat fascia dentata: new aspects of laminar organiza-
    tion as revealed by anterograde tracing with phaseo-
    lus vulgaris-luecoagglutinin (PHAL). Anat Embryol
    (Berl) 197:89–103
35. Deller T, Leranth C (1990) Synaptic connections of
    neuropeptide Y (NPY) immunoreactive neurons in
    the hilar area of the rat hippocampus. J Comp Neurol
    300:433–447
36. Dieni CV, Chancey JH, Overstreet-Wadiche LS
    (2012) Dynamic functions of GABA signaling during
    granule cell maturation. Front Neural Circuits 6:113
37. Dougherty KD, Milner TA (1999) Cholinergic septal
    afferent terminals preferentially contact neuropep-
    tide Y-containing interneurons compared to
    parvalbumin-containing interneurons in the rat den-
    tate gyrus. J Neurosci 19:10140–10152
38. Dougherty KD, Milner TA (1999) P75ntr immuno-
    reactivity in the rat dentate gyrus is mostly within
    presynaptic profiles but is also found in some
    astrocytic and postsynaptic profiles. J Comp
    Neurol 407:77–91
39. Franck JE, Pokorny J, Kunkel DD, Schwartzkroin
    PA (1995) Physiologic and morphologic characteris-
    tics of granule cell circuitry in human epileptic hip-
    pocampus. Epilepsia 36:543–558
40. Frazier CJ (2007) Endocannabinoids in the dentate
    gyrus. Prog Brain Res 163:319–337
41. Freund TF (1992) Gabaergic septal and serotonergic
    median raphe afferents preferentially innervate
    inhibitory interneurons in the hippocampus and den-
    tate gyrus. Epilepsy Res Suppl 7:79–91
42. Freund TF, Buzsaki G (1996) Interneurons of the
    hippocampus. Hippocampus 6:347–470
43. Frotscher M, Jonas P, Sloviter RS (2006) Synapses
    formed by normal and abnormal hippocampal mossy
    fibers. Cell Tissue Res 326:361–367
44. Fu LY, van den Pol AN (2007) Gaba excitation in
    mouse hilar neuropeptide Y neurons. J Physiol
    579:445–464
45. Gibbs J, Shumate M, Coulter D (1997) Differential
    epilepsy-associated alterations in postsynaptic
    GABA$_A$ receptor function in dentate granule and ca1
    neurons. J Neurophysiol 77:1924–1938
46. Gonzalez MI, Brooks-Kayal A (2011) Altered
    GABA(A) receptor expression during epileptogene-
    sis. Neurosci Lett 497:218–222
47. Goodkin HP, Joshi S, Mtchedlishvili Z, Brar J,
    Kapur J (2008) Subunit-specific trafficking of
    GABA(A) receptors during status epilepticus. J
    Neurosci 28:2527–2538
48. Goodkin HP, Yeh JL, Kapur J (2005) Status epilepti-
    cus increases the intracellular accumulation of
    GABA(A) receptors. J Neurosci 25:5511–5520
49. Goodman JH, Sloviter RS (1992) Evidence for com-
    missurally projecting parvalbumin-immunoreactive
    basket cells in the dentate gyrus of the rat.
    Hippocampus 2:13–21
50. Gutierrez R, Heinemann U (2006) Co-existence of
    GABA and Glu in the hippocampal granule cells:
    implications for epilepsy. Curr Top Med Chem
    6:975–978
51. Halasy K, Somogyi P (1993) Subdivisions in the
    multiple GABAergic innervation of granule cells in
    the dentate gyrus of the rat hippocampus. Eur J
    Neurosci 5:411–429
52. Han ZS, Buhl EH, Lorinczi Z, Somogyi P (1993) A
    high degree of spatial selectivity in the axonal and
    dendritic domains of physiologically identified
    local-circuit neurons in the dentate gyrus of the rat
    hippocampus. Eur J Neurosci 5:395–410
53. Henze DA, Buzsaki G (2007) Hilar mossy cells:
    functional identification and activity in vivo. Prog
    Brain Res 163:199–216
54. Houser CR (1991) GABA neurons in seizure disor-
    ders: a review of immunocytochemical studies.
    Neurochem Res 16:295–308
55. Houser CR (2007) Interneurons of the dentate gyrus:
    an overview of cell types, terminal fields and neuro-
    chemical identity. Prog Brain Res 163:217–232
56. Houser CR, Harris AB, Vaughn JE (1986) Time
    course of the reduction of GABA terminals in a
    model of focal epilepsy: a glutamic acid decarboxyl-
    ase immunocytochemical study. Brain Res 383:
    129–145
57. Hu Y, Russek SJ (2008) BDNF and the diseased ner-
    vous system: a delicate balance between adaptive
    and pathological processes of gene regulation. J
    Neurochem 105:1–17
58. Huguenard JR (1999) Neuronal circuitry of thalamo-
    cortical epilepsy and mechanisms of antiabsence
    drug action. Adv Neurol 79:991–999

59. Jacob TC, Moss SJ, Jurd R (2008) GABA(A) receptor trafficking and its role in the dynamic modulation of neuronal inhibition. Nat Rev Neurosci 9:331–343

60. Jaffe DB, Gutierrez R (2007) Mossy fiber synaptic transmission: communication from the dentate gyrus to area ca3. Prog Brain Res 163:109–132

61. Jakubs K, Nanobashvili A, Bonde S, Ekdahl CT, Kokaia Z, Kokaia M, Lindvall O (2006) Environment matters: synaptic properties of neurons born in the epileptic adult brain develop to reduce excitability. Neuron 52:1047–1059

62. Jessberger S, Zhao C, Toni N, Clemenson GD Jr, Li Y, Gage FH (2007) Seizure-associated, aberrant neurogenesis in adult rats characterized with retrovirus-mediated cell labeling. J Neurosci 27:9400–9407

63. Jung KH, Chu K, Kim M, Jeong SW, Song YM, Lee ST, Kim JY, Lee SK, Roh JK (2004) Continuous cytosine-b-D-arabinofuranoside infusion reduces ectopic granule cells in adult rat hippocampus with attenuation of spontaneous recurrent seizures following pilocarpine-induced status epilepticus. Eur J Neurosci 19:3219–3226

64. Kahle KT, Staley KJ (2008) The bumetanide-sensitive Na-K-2Cl cotransporter NKCC1 as a potential target of a novel mechanism-based treatment strategy for neonatal seizures. Neurosurg Focus 25:E22

65. Kahle KT, Staley KJ (2012) Neonatal seizures and neuronal transmembrane ion transport. In: Noebels JL, Avoli M, Rogawski MA, Olsen RW, Delgado-Escueta AV (eds) Jasper's basic mechanisms of the epilepsies, 4th edn. Oxford, New York, pp 1066–1076

66. Kanner AM (2013) Epilepsy, depression and anxiety disorders: a complex relation with significant therapeutic implications for the three conditions. J Neurol Neurosurg Psychiatry 84:e1

67. Kempermann G (2006) Adult neurogenesis. Oxford University Press, Oxford

68. Koyama R, Tao K, Sasaki T, Ichikawa J, Miyamoto D, Muramatsu R, Matsuki N, Ikegaya Y (2012) GABAergic excitation after febrile seizures induces ectopic granule cells and adult epilepsy. Nat Med 18:1271–1278

69. Kron MM, Zhang H, Parent JM (2010) The developmental stage of dentate granule cells dictates their contribution to seizure-induced plasticity. J Neurosci 30:2051–2059

70. Kullmann DM, Ruiz A, Rusakov DM, Scott R, Semyanov A, Walker MC (2005) Presynaptic, extrasynaptic and axonal GABA$_A$ receptors in the CNS: where and why? Prog Biophys Mol Biol 87:33–46

71. Lagrange AH, Botzolakis EJ, Macdonald RL (2007) Enhanced macroscopic desensitization shapes the response of α4 subtype-containing GABA$_A$ receptors to synaptic and extrasynaptic GABA. J Physiol 578:655–676

72. Lauren HB, Pitkanen A, Nissinen J, Soini SL, Korpi ER, Holopainen IE (2003) Selective changes in γ-aminobutyric acid type a receptor subunits in the hippocampus in spontaneously seizing rats

with chronic temporal lobe epilepsy. Neurosci Lett 349:58–62

73. Laurie DJ, Wisden W, Seeburg PH (1992) The distribution of thirteen GABA$_A$ receptor subunit mRNAs in the rat brain. Iii. Embryonic and postnatal development. J Neurosci 12:4151–4172

74. Leranth C, Hajszan T (2007) Extrinsic afferent systems to the dentate gyrus. Prog Brain Res 163:63–84

75. Lothman EW, Bertram EH 3rd, Kapur J, Bekenstein JW (1996) Temporal lobe epilepsy: studies in a rat model showing dormancy of GABAergic inhibitory interneurons. Epilepsy Res Suppl 12:145–156

76. Lund IV, Hu Y, Raol YH, Benham RS, Faris R, Russek SJ, Brooks-Kayal AR (2008) BDNF selectively regulates GABA$_A$ receptor transcription by activation of the JAK/STAT pathway. Sci Signal 1:ra9

77. Macdonald RL, Olsen RW (1994) GABA$_A$ receptor channels. Annu Rev Neurosci 17:569–602

78. Magloczky Z, Freund TF (2005) Impaired and repaired inhibitory circuits in the epileptic human hippocampus. Trends Neurosci 28:334–340

79. McCloskey DP, Hintz TM, Pierce JP, Scharfman HE (2006) Stereological methods reveal the robust size and stability of ectopic hilar granule cells after pilocarpine-induced status epilepticus in the adult rat. Eur J Neurosci 24:2203–2210

80. Milner TA, Veznedaroglu E (1992) Ultrastructural localization of neuropeptide Y-like immunoreactivity in the rat hippocampal formation. Hippocampus 2:107–125

81. Morgan RJ, Santhakumar V, Soltesz I (2007) Modeling the dentate gyrus. Prog Brain Res 163:639–658

82. Mori M, Gahwiler BH, Gerber U (2007) Recruitment of an inhibitory hippocampal network after bursting in a single granule cell. Proc Natl Acad Sci U S A 104:7640–7645

83. Moss SJ, Smart TG (1996) Modulation of amino acid-gated ion channels by protein phosphorylation. Int Rev Neurobiol 39:1–52

84. Mott DD, Lewis DV (1994) The pharmacology and function of central GABA$_B$ receptors. Int Rev Neurobiol 36:97–223

85. Nadler JV, Perry BW, Cotman CW (1978) Intraventricular kainic acid preferentially destroys hippocampal pyramidal cells. Nature 271:676–677

86. Naylor DE, Liu H, Wasterlain CG (2005) Trafficking of GABA(A) receptors, loss of inhibition, and a mechanism for pharmacoresistance in status epilepticus. J Neurosci 25:7724–7733

87. Nusser Z, Mody I (2002) Selective modulation of tonic and phasic inhibitions in dentate gyrus granule cells. J Neurophysiol 87:2624–2628

88. Olsen RW (1981) The GABA postsynaptic membrane receptor-ionophore complex. Site of action of convulsant and anticonvulsant drugs. Mol Cell Biochem 39:261–279

89. Paredes MF, Greenwood J, Baraban SC (2003) Neuropeptide Y modulates a g protein-coupled inwardly rectifying potassium current in the mouse hippocampus. Neurosci Lett 340:9–12

90. Parent JM, Yu TW, Leibowitz RT, Geschwind DH, Sloviter RS, Lowenstein DH (1997) Dentate granule cell neurogenesis is increased by seizures and contributes to aberrant network reorganization in the adult rat hippocampus. J Neurosci 17:3727–3738

91. Pathak HR, Weissinger F, Terunuma M, Carlson GC, Hsu FC, Moss SJ, Coulter DA (2007) Disrupted dentate granule cell chloride regulation enhances synaptic excitability during development of temporal lobe epilepsy. J Neurosci 27:14012–14022

92. Peng Z, Huang CS, Stell BM, Mody I, Houser CR (2004) Altered expression of the delta subunit of the $GABA_A$ receptor in a mouse model of temporal lobe epilepsy. J Neurosci 24:8629–8639

93. Peterson GM, Ribak CE (1987) Hippocampus of the seizure-sensitive gerbil is a specific site for anatomical changes in the GABAergic system. J Comp Neurol 261:405–422

94. Pierce JP, McCloskey DP, Scharfman HE (2011) Morphometry of hilar ectopic granule cells in the rat. J Comp Neurol 519:1196–1218

95. Pitkanen A, Moshe s, Schwartzkroin PA (2006) Models of seizures and epilepsy. Elsevier, New York

96. Porter BE, Maronski M, Brooks-Kayal AR (2004) Fate of newborn dentate granule cells after early life status epilepticus. Epilepsia 45:13–19

97. Pritchett D, Sontheimer H, Shivers B, Ymer S, Kellenmann H, Schofield P, Seeburg P (1989) Importance of a novel $GABA_A$ receptor subunit for benzodiazepine pharmacology. Nature 338:582–585

98. Pun RY, Rolle IJ, Lasarge CL, Hosford BE, Rosen JM, Uhl JD, Schmeltzer SN, Faulkner C, Bronson SL, Murphy BL, Richards DA, Holland KD, Danzer SC (2012) Excessive activation of mTOR in postnatally generated granule cells is sufficient to cause epilepsy. Neuron 75:1022–1034

99. Raol YH, Lund IV, Bandyopadhyay S, Zhang G, Roberts DS, Wolfe JH, Russek SJ, Brooks-Kayal AR (2006) Enhancing GABA(A) receptor α 1 subunit levels in hippocampal dentate gyrus inhibits epilepsy development in an animal model of temporal lobe epilepsy. J Neurosci 26:11342–11346

100. Ribak CE (1985) Axon terminals of GABAergic chandelier cells are lost at epileptic foci. Brain Res 326:251–260

101. Ribak CE, Harris AB, Vaughn JE, Roberts E (1979) Inhibitory, GABAergic nerve terminals decrease at sites of focal epilepsy. Science 205:211–214

102. Ribak CE, Hunt CA, Bakay RA, Oertel WH (1986) A decrease in the number of GABAergic somata is associated with the preferential loss of GABAergic terminals at epileptic foci. Brain Res 363:78–90

103. Ribak CE, Joubran C, Kesslak JP, Bakay RA (1989) A selective decrease in the number of GABAergic somata occurs in pre-seizing monkeys with alumina gel granuloma. Epilepsy Res 4:126–138

104. Ribak CE, Shapiro LA, Yan XX, Dashtipour K, Nadler JV, Obenaus A, Spigelman I, Buckmaster PS (2012) Seizure-induced formation of basal dendrites on granule cells of the rodent dentate gyrus. In: Noebels JL, Avoli M, Rogawski MA, Olsen RW, Delgado-Escueta AV (eds) Jasper's basic mechanisms of the epilepsies, 4th edn. National Center for Biotechnology Information, Bethesda

105. Riccio A, Alvania RS, Lonze BE, Ramanan N, Kim T, Huang Y, Dawson TM, Snyder SH, Ginty DD (2006) A nitric oxide signaling pathway controls CREB-mediated gene expression in neurons. Mol Cell 21:283–294

106. Rivera C, Voipio J, Payne JA, Ruusuvuori E, Lahtinen H, Lamsa K, Pirvola U, Saarma M, Kaila K (1999) The K^+/Cl^- co-transporter KCC2 renders GABA hyperpolarizing during neuronal maturation. Nature 397:251–255

107. Roberts DS, Hu Y, Lund IV, Brooks-Kayal AR, Russek SJ (2006) Brain-derived neurotrophic factor (BDNF)-induced synthesis of early growth response factor 3 (Egr3) controls the levels of type A GABA receptor α4 subunits in hippocampal neurons. J Biol Chem 281:29431–29435

108. Roberts DS, Raol YH, Bandyopadhyay S, Lund IV, Budreck EC, Passini MA, Wolfe JH, Brooks-Kayal AR, Russek SJ (2005) Egr3 stimulation of gabra4 promoter activity as a mechanism for seizure-induced up-regulation of GABA(A) receptor α4 subunit expression. Proc Natl Acad Sci U S A 102:11894–11899

109. Roberts E (1986) Failure of GABAergic inhibition: a key to local and global seizures. Adv Neurol 44:319–341

110. Roberts RC, Ribak CE, Oertel WH (1985) Increased numbers of GABAergic neurons occur in the inferior colliculus of an audiogenic model of genetic epilepsy. Brain Res 361:324–338

111. Rudolph U, Mohler H (2006) GABA-based therapeutic approaches: $GABA_A$ receptor subtype functions. Curr Opin Pharmacol 6:18–23

112. Sahay A, Hen R (2008) Hippocampal neurogenesis and depression. Novartis Found Symp 289:152–160; discussion 60–64, 93–95

113. Santhakumar V, Bender R, Frotscher M, Ross ST, Hollrigel GS, Toth Z, Soltesz I (2000) Granule cell hyperexcitability in the early post-traumatic rat dentate gyrus: the 'irritable mossy cell' hypothesis. J Physiol 524(Pt 1):117–134

114. Santhakumar V, Ratzliff AD, Jeng J, Toth Z, Soltesz I (2001) Long-term hyperexcitability in the hippocampus after experimental head trauma. Ann Neurol 50:708–717

115. Scharfman HE (1995) Electrophysiological diversity of pyramidal-shaped neurons at the granule cell layer/hilus border of the rat dentate gyrus recorded in vitro. Hippocampus 5:287–305

116. Scharfman HE (1999) The role of nonprincipal cells in dentate gyrus excitability and its relevance to animal models of epilepsy and temporal lobe epilepsy. Adv Neurol 79:805–820

117. Scharfman HE (2004) Functional implications of seizure-induced neurogenesis. Adv Exp Med Biol 548:192–212

118. Scharfman HE, Sollas AE, Berger RE, Goodman JH, Pierce JP (2003) Perforant path activation of ectopic granule cells that are born after pilocarpine-induced seizures. Neuroscience 121:1017–1029

119. Scharfman HE, Goodman JH, Sollas AL (2000) Granule-like neurons at the hilar/CA3 border after status epilepticus and their synchrony with area CA3 pyramidal cells: functional implications of seizure-induced neurogenesis. J Neurosci 20:6144–6158

120. Scharfman HE, Gray WP (2006) Plasticity of neuropeptide y in the dentate gyrus after seizures, and its relevance to seizure-induced neurogenesis. EXS: 95:193–211

121. Scharfman HE, Gray WP (2007) Relevance of seizure-induced neurogenesis in animal models of epilepsy to the etiology of temporal lobe epilepsy. Epilepsia 48(Suppl 2):33–41

122. Scharfman HE, Kunkel DD, Schwartzkroin PA (1990) Synaptic connections of dentate granule cells and hilar neurons: results of paired intracellular recordings and intracellular horseradish peroxidase injections. Neuroscience 37:693–707

123. Scharfman HE, Maclusky NJ (2014) Differential regulation of BDNF, synaptic plasticity and sprouting in the hippocampal mossy fiber pathway of male and female rats. Neuropharmacology 76:696–708

124. Scharfman HE, McCloskey DP (2009) Postnatal neurogenesis as a therapeutic target in temporal lobe epilepsy. Epilepsy Res 85:150–161

125. Scharfman HE, Mercurio TC, Goodman JH, Wilson MA, MacLusky NJ (2003) Hippocampal excitability increases during the estrous cycle in the rat: a potential role for brain-derived neurotrophic factor. J Neurosci 23:11641–11652

126. Scharfman HE, Myers CE (2012) Hilar mossy cells of the dentate gyrus: a historical perspective. Front Neural Circuits 6:106

127. Scharfman HE, Pedley TA (2006) Temporal lobe epilepsy. In: Gilman S (ed) The neurobiology of disease. Academic, New York

128. Scharfman HE, Smith KL, Goodman JH, Sollas AL (2001) Survival of dentate hilar mossy cells after pilocarpine-induced seizures and their synchronized burst discharges with area CA3 pyramidal cells. Neuroscience 104:741–759

129. Scharfman HE, Sollas AL, Berger RE, Goodman JH (2003) Electrophysiological evidence of monosynaptic excitatory transmission between granule cells after seizure-induced mossy fiber sprouting. J Neurophysiol 90:2536–2547

130. Scharfman HE, Sollas AL, Goodman JH (2002) Spontaneous recurrent seizures after pilocarpine-induced status epilepticus activate calbindin-immunoreactive hilar cells of the rat dentate gyrus. Neuroscience 111:71–81

131. Schneiderman JH, Schwartzkroin PA (1982) Effects of phenytoin on normal activity and on penicillin-induced bursting in the guinea pig hippocampal slice. Neurology 32:730–738

132. Schwartzkroin PA, Prince DA (1977) Penicillin-induced epileptiform activity in the hippocampal in vitro preparation. Ann Neurol 1:463–469

133. Shapiro LA, Ribak CE, Jessberger S (2008) Structural changes for adult-born dentate granule cells after status epilepticus. Epilepsia 49(Suppl 5):13–18

134. Sieghart W (2006) Structure, pharmacology, and function of $GABA_A$ receptor subtypes. Adv Pharmacol 54:231–263

135. Sloviter RS (1987) Decreased hippocampal inhibition and a selective loss of interneurons in experimental epilepsy. Science 235:73–76

136. Sloviter RS (1991) Permanently altered hippocampal structure, excitability, and inhibition after experimental status epilepticus in the rat: the "dormant basket cell" hypothesis and its possible relevance to temporal lobe epilepsy. Hippocampus 1:41–66

137. Sloviter RS (1992) Possible functional consequences of synaptic reorganization in the dentate gyrus of kainate-treated rats. Neurosci Lett 137:91–96

138. Sloviter RS, Nilaver G (1987) Immunocytochemical localization of GABA-, cholecystokinin-, vasoactive intestinal polypeptide-, and somatostatin-like immunoreactivity in the area dentata and hippocampus of the rat. J Comp Neurol 256:42–60

139. Sloviter RS, Zappone CA, Harvey BD, Frotscher M (2006) Kainic acid-induced recurrent mossy fiber innervation of dentate gyrus inhibitory interneurons: possible anatomical substrate of granule cell hyper-inhibition in chronically epileptic rats. J Comp Neurol 494:944–960

140. Smith SS, Shen H, Gong QH, Zhou X (2007) Neurosteroid regulation of GABA(A) receptors: focus on the $\alpha4$ and delta subunits. Pharmacol Ther 116:58–76

141. Snead OC 3rd (1995) Basic mechanisms of generalized absence seizures. Ann Neurol 37:146–157

142. Somogyi P, Freund TF, Hodgson AJ, Somogyi J, Beroukas D, Chubb IW (1985) Identified axo-axonic cells are immunoreactive for GABA in the hippocampus and visual cortex of the cat. Brain Res 332:143–149

143. Sperk G, Drexel M, Pirker S (2009) Neuronal plasticity in animal models and the epileptic human hippocampus. Epilepsia 50:29–31

144. Sperk G, Furtinger S, Schwarzer C, Pirker S (2004) GABA and its receptors in epilepsy. Adv Exp Med Biol 548:92–103

145. Sperk G, Hamilton T, Colmers WF (2007) Neuropeptide Y in the dentate gyrus. Prog Brain Res 163:285–297

146. Sun C, Mtchedlishvili Z, Erisir A, Kapur J (2007) Diminished neurosteroid sensitivity of synaptic inhibition and altered location of the $\alpha4$ subunit of GABA(A) receptors in an animal model of epilepsy. J Neurosci 27:12641–12650

147. Sutula TP, Dudek FE (2007) Unmasking recurrent excitation generated by mossy fiber sprouting in the

epileptic dentate gyrus: an emergent property of a complex system. Prog Brain Res 163:541–563

148. Swijsen A, Avila A, Brone B, Janssen D, Hoogland G, Rigo JM (2012) Experimental early-life febrile seizures induce changes in GABA(A)R-mediated neurotransmission in the dentate gyrus. Epilepsia 53:1968–1977

149. Swijsen A, Brone B, Rigo JM, Hoogland G (2012) Long-lasting enhancement of GABA(A) receptor expression in newborn dentate granule cells after early-life febrile seizures. Dev Neurobiol 72:1516–1527

150. Terunuma M, Xu J, Vithlani M, Sieghart W, Kittler J, Pangalos M, Haydon PG, Coulter DA, Moss SJ (2008) Deficits in phosphorylation of GABA(A) receptors by intimately associated protein kinase C activity underlie compromised synaptic inhibition during status epilepticus. J Neurosci 28:376–384

151. Thind KK, Yamawaki R, Phanwar I, Zhang G, Wen X, Buckmaster PS (2010) Initial loss but later excess of GABAergic synapses with dentate granule cells in a rat model of temporal lobe epilepsy. J Comp Neurol 518:647–667

152. Van Duijn H, Schwartzkroin PA, Prince DA (1973) Action of penicillin on inhibitory processes in the cat's cortex. Brain Res 53:470–476

153. Van Kempen TA, Kahlid S, Gonzalez AD, Spencer-Segal JL, Tsuda MC, Ogawa S, McEwen BS, Waters EM, Milner TA (2013) Sex and estrogen receptor expression influence opioid peptide levels in the mouse hippocampal mossy fiber pathway. Neurosci Lett 552:66–70

154. Vicini S (1991) Pharmacologic significance of the structural heterogenetity of the GABA$_A$ receptor-chloride ion channel complex. Neuropsychopharmacology 4:9–15

155. Vithlani M, Moss SJ (2009) The role of GABA$_A$R phosphorylation in the construction of inhibitory synapses and the efficacy of neuronal inhibition. Biochem Soc Trans 37:1355–1358

156. von Kitzing E, Jonas P, Sakmann B (1994) Quantal analysis of excitatory postsynaptic currents at the hippocampal mossy fiber-CA3 pyramidal cell synapse. Adv Second Messenger Phosphoprotein Res 29:235–260

157. Waldbaum S, Dudek FE (2009) Single and repetitive paired-pulse suppression: a parametric analysis and assessment of usefulness in epilepsy research. Epilepsia 50:904–916

158. Walker MC, Ruiz A, Kullmann DM (2001) Monosynaptic GABAergic signaling from dentate to CA3 with a pharmacological and physiological profile typical of mossy fiber synapses. Neuron 29:703–715

159. Wallace RH, Scheffer IE, Barnett S, Richards M, Dibbens L, Desai RR, Lerman-Sagie T, Lev D, Mazarib A, Brand N, Ben-Zeev B, Goikhman I, Singh R, Kremmidiotis G, Gardner A, Sutherland GR, George AL Jr, Mulley JC, Berkovic SF (2001) Neuronal sodium-channel α1-subunit mutations in generalized epilepsy with febrile seizures plus. Am J Hum Genet 68:859–865

160. Wang DD, Kriegstein AR (2009) Defining the role of GABA in cortical development. J Physiol 587:1873–1879

161. Witter MP (2007) The perforant path: projections from the entorhinal cortex to the dentate gyrus. Prog Brain Res 163:43–61

162. Wu K, Leung LS (2001) Enhanced but fragile inhibition in the dentate gyrus in vivo in the kainic acid model of temporal lobe epilepsy: a study using current source density analysis. Neuroscience 104:379–396

163. Ying SW, Futter M, Rosenblum K, Webber MJ, Hunt SP, Bliss TV, Bramham CR (2002) Brain-derived neurotrophic factor induces long-term potentiation in intact adult hippocampus: requirement for ERK activation coupled to CREB and upregulation of Arc synthesis. J Neurosci 22:1532–1540

164. Zhan RZ, Nadler JV (2009) Enhanced tonic GABA current in normotopic and hilar ectopic dentate granule cells after pilocarpine-induced status epilepticus. J Neurophysiol 102:670–681

165. Zhan RZ, Timofeeva O, Nadler JV (2010) High ratio of synaptic excitation to synaptic inhibition in hilar ectopic granule cells of pilocarpine-treated rats. J Neurophysiol 104:3293–3304

166. Zhang N, Wei W, Mody I, Houser CR (2007) Altered localization of GABA(A) receptor subunits on dentate granule cell dendrites influences tonic and phasic inhibition in a mouse model of epilepsy. J Neurosci 27:7520–7531

Do Structural Changes in GABA Neurons Give Rise to the Epileptic State?

12

Carolyn R. Houser

Abstract

Identifying the role of GABA neurons in the development of an epileptic state has been particularly difficult in acquired epilepsy, in part because of the multiple changes that occur in such conditions. Although once questioned, there is now considerable evidence for loss of GABA neurons in multiple brain regions in models of acquired epilepsy. This loss can affect several cell types, including both somatostatin- and parvalbumin-expressing interneurons, and the cell type that is most severely affected can vary among brain regions and models. Because of the diversity of GABA neurons in the hippocampus and cerebral cortex, resulting functional deficits are unlikely to be compensated fully by remaining GABA neurons of other subtypes. The fundamental importance of GABA neuron loss in epilepsy is supported by findings in genetic mouse models in which GABA neurons appear to be decreased relatively selectively, and increased seizure susceptibility and spontaneous seizures develop. Alterations in remaining GABA neurons also occur in acquired epilepsy. These include alterations in inputs or receptors that could impair function, as well as morphological reorganization of GABAergic axons and their synaptic connections. Such axonal sprouting could be compensatory if normal circuits are reestablished, but the creation of aberrant circuitry could contribute to an epileptic condition. The functional effects of GABA neuron alterations

C.R. Houser (✉)
Department of Neurobiology and Brain Research
Institute, David Geffen School of Medicine
at the University of California, Los Angeles,
10833 Le Conte Ave., Los Angeles, CA 90095, USA

Research Service, Veterans Administration Greater
Los Angeles Healthcare System, West Los Angeles,
Los Angeles, CA 90073, USA
e-mail: houser@mednet.ucla.edu

H.E. Scharfman and P.S. Buckmaster (eds.), *Issues in Clinical Epileptology: A View from the Bench*,
Advances in Experimental Medicine and Biology 813, DOI 10.1007/978-94-017-8914-1_12,
© Springer Science+Business Media Dordrecht 2014

thus may include not only reductions in GABAergic inhibition but also excessive neuronal synchrony and, potentially, depolarizing GABAergic influences. The combination of GABA neuron loss and alterations in remaining GABA neurons provides likely, though still unproven, substrates for the epileptic state.

Keywords

Inhibition • Plasticity • Seizures • Sprouting • Somatostatin • Parvalbumin

Abbreviations

CCK	Cholecystokinin
eGFP	Enhanced green fluorescent protein
eYFP	Enhanced yellow fluorescent protein
GABA	Gamma aminobutyric acid
GAD	Glutamic acid decarboxylase
NPY	Neuropeptide Y
PV	Parvalbumin
SOM	Somatostatin
s. oriens	Stratum oriens
TLE	Temporal lobe epilepsy
uPAR	Urokinase plasminogen activator receptor

12.1 Introduction

The question of whether GABA neuron loss gives rise to the epileptic state in acquired epilepsy has persisted for many years. Indeed, even the occurrence of GABA neuron loss was questioned at one time. While progress has been made, and a loss of GABA neurons has been convincingly identified in humans with temporal lobe epilepsy (TLE) and in many related animal models, determining the functional consequences of GABA neuron loss in epilepsy remains a major challenge. This review will focus on some of the complexities associated with interneuron loss and their role in epilepsy and suggest that GABA neuron loss could indeed give rise to the epileptic state through both direct and indirect routes.

12.2 Loss of GABA Neurons Is a Consistent Finding in Models of Acquired Epilepsy

Interneuron loss is one of the most frequently observed alterations in models of TLE, and the consistency of GABA neuron loss provides a solid base for suggesting the potential importance of this alteration in creating an epileptic state. Although loss of GABA neurons has been identified in multiple brain regions, loss of somatostatin (SOM)-expressing GABA neurons in the hilus of the dentate gyrus remains one of the clearest and most consistent findings (Fig. 12.1a, b). Loss of these GABA neurons has been found in virtually all models of acquired epilepsy, including kindling, status epilepticus and traumatic brain injury models [7, 25, 32, 44, 45]. Importantly, loss of SOM/GABA neurons in the dentate hilus is also found in human TLE, as part of the broader loss of neurons in typical hippocampal sclerosis, as well as in pathological conditions with more limited cell loss such as end-folium sclerosis [14, 35, 42, 48, 49].

SOM neurons in stratum oriens (s. oriens) of CA1 are also among the vulnerable interneurons in several models of acquired epilepsy [1, 11, 27, 36]. As SOM neurons in both the hilus and s. oriens provide GABAergic innervation of dendrites of granule cells and pyramidal cells, respectively, they are ideally positioned to control excitability of the principal cells directly at the sites of their major excitatory inputs. This pattern of GABA neuron loss has led to the suggestion that

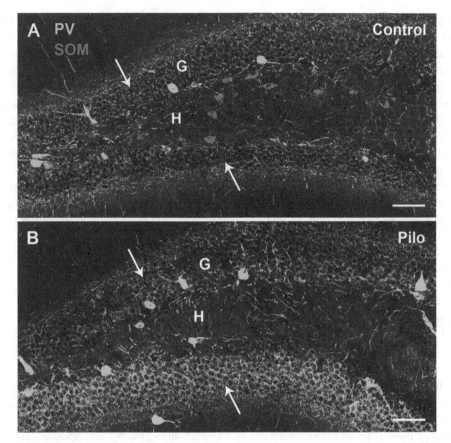

Fig. 12.1 Comparisons of somatostatin (SOM)- and parvalbumin (PV)-labeled neurons in the rostral dentate gyrus of control (a) and pilocarpine (Pilo)-treated (b) mice. (**a**) In the control dentate gyrus, cell bodies of SOM neurons are located primarily within the hilus (H) whereas those of PV neurons are positioned predominantly along the base of the granule cell layer (G). PV-labeled axon terminals are concentrated in perisomatic locations within the granule cell layer (*arrows*) while SOM terminal fields are located in dendritic regions in the outer molecular layer (not shown). (**b**) In the pilocarpine-treated mouse at 2 months after status epilepticus, a severe loss of SOM neurons is evident in the hilus whereas many PV neurons and their axon terminals (*arrows*) in the granule cell layer are preserved. Scale bars, 100 μm

interneurons that provide dendritic innervation are more vulnerable to damage in epilepsy than those which provide primarily perisomatic innervation, such as basket cells and axo-axonic cells, many of which express the calcium binding protein parvalbumin (PV). Electrophysiological findings of decreased dendritic inhibition, with preservation of somatic inhibition, in pyramidal cells of CA1 in models of recurrent seizures support this idea [11].

While the distinction between dendritic and perisomatic innervation provides a useful framework for considering GABA neuron loss, the differences in vulnerability among the broad types of interneurons are not clear-cut, and additional complexities exist. While PV-expressing neurons are a major source of perisomatic innervation of pyramidal cells in the hippocampus, cholecystokinin (CCK)-expressing interneurons also provide perisomatic innervation of these neurons in CA1 [3, 23]. In a mouse pilocarpine model of recurrent seizures, this CCK innervation appeared to be decreased while the PV innervation was preserved [51]. This could create an imbalance in perisomatic control that could favor synchronizing actions of PV basket cells, with loss of major modulatory inputs from CCK neurons.

Perisomatic innervation includes both basket cells and axo-axonic cells [22], and these cell types could be affected differentially in epilepsy. Decreased innervation of the axon initial segment by PV-expressing axo-axonic neurons has been found in the hippocampus and cerebral cortex in several epilepsy conditions, suggesting that axo-axonic cells could be more vulnerable than basket cells [13, 15, 41].

Thus loss of PV neurons can occur and has been described in the dentate gyrus in several animal models, without distinctions between basket cells and axo-axonic cells [1, 25, 29]. However, when both PV and SOM neurons have been studied in the same animals, the loss of PV-expressing neurons in the dentate gyrus is generally less severe than that of SOM interneurons [7] (Fig. 12.1a, b).

A decrease in numbers of PV-expressing interneurons has now been identified in several other regions of the hippocampal formation where their loss could be particularly important in regulating activity within the broader hippocampal circuit. A significant decrease in PV neurons has been identified in layer II of the entorhinal cortex where loss of these neurons could contribute to increased excitability of the perforant path input to the dentate gyrus [30]. Interestingly, lower densities of PV-containing neurons have also been identified in the subiculum, where their loss could lead to increased excitability of this major output region of the hippocampal formation, a region that is otherwise generally well preserved [2, 16].

Thus decreases in both SOM and PV neurons can occur in epilepsy, and the particular pattern of loss may vary among epilepsy models, species, brain regions and even rostral-caudal levels of the hippocampal formation. The types of GABA neurons that are affected remain important as they will determine the specific functional effects. However, loss of GABA neurons remains a unifying theme.

Despite strong evidence for GABA neuron loss in many brain regions, direct relationships to the epileptic state have been difficult to demonstrate. This could be in part because GABA neuron loss does not occur in isolation in most forms of acquired or lesion-induced epilepsy. In human TLE and related epilepsy models, extensive cell loss can occur in many regions, including CA1 and CA3 as well as the dentate hilus and extra-hippocampal regions. This neuronal loss generally involves both principal cells and interneurons, making it more difficult to link GABA neuron loss directly to development of an epileptic state. However, findings in several genetic mouse models provide support for the importance of GABA neuron loss in the development of epilepsy, and these findings also have relevance for acquired epilepsy.

12.3 Selective Loss of GABA Neurons Can Lead to an Epileptic State in Genetic Models

Some of the strongest evidence for loss of GABA neurons giving rise to the epileptic state has come from genetically-modified mice in which GABA neurons are selectively affected, and increased seizure susceptibility and spontaneous seizures occur. In mice with loss of the *Dlx1* gene, a transcription factor that regulates development of GABAergic interneurons originating in the medial ganglionic eminence, there is a time-dependent reduction in the number of interneurons in the cerebral cortex and hippocampus and development of an epilepsy phenotype [10]. SOM and calretinin-expressing neurons were reduced in number whereas PV-expressing neurons appeared to be unaffected. Because the loss of GABA neurons is apparently selective in these mice, the findings provide strong support for loss of GABA neurons giving rise to an epileptic state and also suggest that the loss of GABA neurons does not need to be extensive. In these mice, behavioral seizures were selectively induced by mild stressors by 2 months of age when there was an approximately 22 % reduction in GAD67-labeled neurons in the cerebral cortex and 24 and 29 % reduction in the dentate gyrus and CA1 respectively. Comparable or even greater GABA neuron loss has been observed in the hippocampal formation in models of acquired epilepsy [37, 50].

Similarly, in mice with mutation of the gene encoding urokinase plasminogen activator receptor (*uPAR*), a key component in hepatocyte growth factor activation and function, interneuron migration is altered, and the mice have a nearly complete loss of PV neurons in the anterior cingulate and parietal cortex [40]. These mice also developed spontaneous myoclonic seizures and increased susceptibility to pharmacologically-induced convulsions. Thus the apparently selective loss of either of two major groups of interneurons supports the importance of GABA neuron loss in the development of epilepsy.

12.4 Remaining GABA Neurons Could Play a Critical Role in Development of Epilepsy

Despite clear evidence for loss of GABA neurons in virtually all models of acquired epilepsy and human TLE, some GABA neurons invariably remain, and alterations in these neurons could contribute to the creation of an epileptic condition. Indeed, it may be difficult to separate the effects of loss of GABA neurons from altered function of remaining neurons as the initial loss of GABA neurons may be a stimulus for the subsequent changes in remaining GABA neurons. Critical changes may include impaired function of remaining GABA neurons and morphological reorganization of remaining interneurons that could lead to altered or aberrant circuitry.

12.4.1 Impaired Function of Remaining Interneurons

Remaining GABA neurons often appear particularly prominent in tissue from animal models of epilepsy, and the preservation of some GABA neurons has suggested that GABA neuron loss may be of limited importance in establishing the epileptic state. However, the function of remaining GABA neurons could be altered, leading to

inadequate control of principal cell activity. In the dentate gyrus and hippocampus, the functional state of remaining basket cells has been debated for many years. Specific details of the "dormant basket cell" hypothesis [46, 47] have been questioned, including the role of hilar mossy cell loss in reducing basket cell activity [5, 18]. However, the broad suggestion that basket cells and other GABAergic neurons might be functioning suboptimally due to decreased or impaired excitatory input remains plausible. Recent studies have identified deficits in basket cell function in the dentate gyrus that could indicate a decrease in excitatory afferent input or reduction of the readily releasable pool of synaptic vesicles, in association with an increased failure rate at basket cell to granule cell synapses [53].

Alterations in the receptors and channels of remaining GABA neurons also could reduce the activity of these neurons. In both the rat and mouse pilocarpine model, expression of the δ subunit of the $GABA_A$ receptor is increased in subgroups of GABA neurons in the dentate gyrus [38, 52]. As $GABA_A$ receptors expressing the δ subunit are responsible for the majority of tonic inhibition in these neurons [24], an increase in δ subunit expression in interneurons could reduce their excitability and impair inhibitory control of the network [38]. Recent studies have demonstrated that tonic inhibition is indeed enhanced in fast-spiking basket cells of the dentate gyrus at 1 week after pilocarpine-induced status epilepticus [52]. However, additional changes, including decreased KCC2 expression in the basket cells, appeared to compensate partially for the increased tonic inhibition of the basket cells, and dentate excitability was not increased. Nevertheless, simulation studies suggested that the changes in tonic inhibition, in combination with other recognized alterations in dentate gyrus circuitry in epilepsy models, could lead to increased granule cell firing and self-sustained seizure-like activity in a subset of simulated networks [52]. Thus occasional alterations in interneuron activity, when combined with other changes in the network, may be sufficient to overrule compensatory changes and lead to sporadic seizure activity.

While regulation of δ subunit-containing $GABA_A$ receptors by neurosteroids and other endogenous modulators could play important roles [21], changes in numerous other channels and receptors in remaining GABAergic interneurons could reduce their effectiveness and contribute to the epileptic state.

Functional alterations in interneurons and their relationship to seizure activity are demonstrated convincingly in genetic mouse models in which specific channels have been deleted relatively selectively in interneurons. As a key example, loss of the alpha subunit of the $Na_v1.1$ sodium channel, that is encoded by the *SCN1A* gene, impairs sodium currents more severely in GABAergic neurons than in pyramidal cells [8, 17, 34]. Such changes limit the ability of the inhibitory interneurons, including PV neurons, to fire action potentials at high frequency, and the animals develop spontaneous generalized seizures.

Similarly, loss of function of the $Ca_V2.1$ voltage-gated Ca^{2+} channel reduces GABA release from cortical PV neurons, and generalized seizures occur in mice with such loss [43]. While decreased expression of this calcium channel was found in both PV and SOM neurons, only the loss in fast spiking, presumably PV, interneurons led to spontaneous seizures. Compensation by N-type Ca^{2+} channels appeared to maintain function of the SOM interneurons but was insufficient for adequate function of the PV neurons.

Finally, elimination of the voltage-gated potassium channels of the Kv3 subfamily, that are particularly prominent in fast-spiking interneurons in the deep layers of the neocortex, led to an inability of these interneurons to fire at their normal high frequency and an increased susceptibility to seizures [31].

Thus in several genetic models, impairment of fast-spiking PV neurons, particularly a reduction in their ability to fire action potentials at high frequency, can lead to increased seizure susceptibility. Although these functional deficits are induced by specific genetic modifications, similar alterations in remaining GABA neurons may occur in acquired epilepsy, and even small functional impairment in remaining neurons could tip the balance toward seizure activity.

12.4.2 Morphological Reorganization of Remaining Interneurons

Clear demonstrations of loss of GABA neurons in acquired epilepsy have often been obscured by the plasticity of remaining interneurons. Remaining GABA neurons frequently express increased levels of GABA neuron markers, including the mRNA and protein of two isoforms of the GABA synthesizing enzyme, glutamic acid decarboxylase 65 and 67 (GAD65 and GAD67), as well as GABA [9, 19, 20]. Similarly the expression of peptides such as SOM and neuropeptide Y (NPY) within specific subclasses of GABA neurons are frequently upregulated [6, 33, 44]. These changes can be substantial and, during the chronic period, labeling of remaining GABA neurons can be quite strong and can suggest that either little loss of GABA neurons has occurred or that axons of remaining GABA neurons have sprouted [12]. It has remained particularly difficult to distinguish morphological growth and reorganization of GABAergic axons from an increase in GABAergic markers within remaining neurons [4, 27].

Additional support for sprouting of existing SOM neurons in the dentate gyrus has been obtained from mice that express enhanced green fluorescent protein (eGFP) in a subgroup of SOM neurons [54]. By studying the labeled interneurons in pilocarpine-treated mice, Buckmaster and colleagues demonstrated that SOM neurons that survive in the ventral (caudal) dentate gyrus can re-innervate the outer half of the dentate gyrus that was partially deafferented by loss of hilar SOM neurons. Such reorganization has generally been presumed to be compensatory as remaining GABA neurons are replacing the innervation of neurons of a similar type and function [26, 54].

Axonal reorganization of remaining GABA neurons can also create aberrant GABAergic circuitry as has been observed in the rostral dentate gyrus in the pilocarpine mouse model [39]. Apparent reinnervation of the dentate molecular layer was observed during the chronic period, but, in contrast to the previous study, few remaining SOM neurons were found in the rostral hilus. To determine if the innervation could be derived

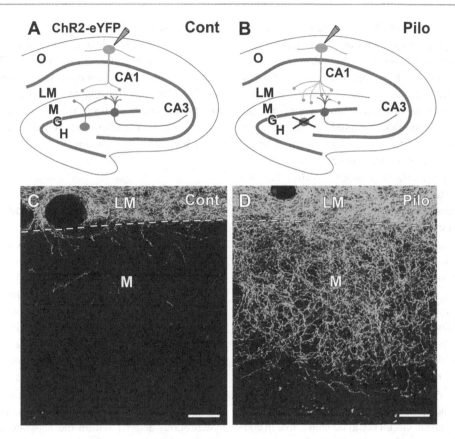

Fig. 12.2 Axonal reorganization of remaining somatostatin (SOM) neurons in pilocarpine (Pilo)-treated mice at 2 months after status epilepticus, illustrated schematically in (a, b) and in confocal images in (c, d). (**a**) This schematic illustrates the normal circuitry of SOM neurons in the hilus (*red*) and s. oriens (*green*) and the labeling protocol. In a control SOM-Cre mouse, selective labeling of SOM neurons in s. oriens (O) of CA1, by Cre-dependent AAV transfection of ChR2-eYFP, leads to labeling of their axon terminals that are confined to s. lacunosum-moleculare (LM). SOM neurons (*red*) in the hilus (H) innervate the outer molecular layer (M) of the dentate gyrus where they form synapses with dentate granule cells (*G, blue*). These hilar SOM neurons are not labeled by the injection in s. oriens. (**b**) In pilocarpine-treated mice, similar labeling of SOM neurons in s. oriens leads to axonal labeling not only in s. lacunosum-moleculare of CA1 but also in the molecular layer of the dentate gyrus, a region that was previously innervated by vulnerable SOM neurons (*red*) in the hilus. (**c**) In a control SOM-Cre mouse, eYFP-labeled axons are concentrated in s. lacunosum-moleculare (LM), and only a limited number of labeled fibers cross the hippocampal fissure (*dashed line*) to enter the molecular layer (M) of the dentate gyrus. (**d**) In a similarly transfected pilocarpine-treated mouse, numerous labeled fibers cross the hippocampal fissure and form an extensive plexus in the outer two-thirds of the dentate molecular layer, where they innervate dentate granule cells. Scale bars, 20 μm (Adapted from data in Peng et al. [39])

from other sources, SOM neurons in s. oriens of control and pilocarpine-treated SOM-Cre recombinase mice were selectively labeled with a viral vector containing Cre-dependent channel-rhodopsin2 (ChR2) fused to enhanced yellow fluorescent protein (eYFP). In control mice, the axons of many labeled SOM neurons in s. oriens formed a dense plexus of fibers in s. lacunosum-moleculare of CA1 (Fig. 12.2a, c). This plexus was sharply delineated by the hippocampal fissure, and relatively few fibers crossed the fissure to enter the adjacent molecular layer of the dentate gyrus (Fig. 12.2c). In contrast, in pilocarpine-treated mice, an extensive axonal plexus of eYFP-labeled fibers was evident in the outer two-thirds of the dentate gyrus during the chronic period (Fig. 12.2b, d). Thus SOM neurons in s. oriens exhibited an unexpected capacity for morphological growth and reorganization, and created an aberrant circuit between hippocampal interneurons in s. oriens and granule cells of the dentate gyrus. The reorganized axons formed symmetric synaptic contacts with presumptive granule cell dendrites and spines, and optogenetic stimulation demonstrated that activation of the reorganized neurons produced GABAergic inhibition in dentate granule cells [39].

The *in vivo* effects of the altered circuit are not known, but it is unlikely to provide normal control of granule cell activity. Because the reorganized fibers originated from GABA neurons in s. oriens of CA1, they would not receive the normal input from dentate granule cells that would be required for efficient feedback inhibition. However, strong activity of CA1 pyramidal cells could potentially activate these SOM neurons and produce inhibitory responses in the granule cells, although through an indirect circuit with presumably altered timing.

These results emphasize that the reemergence of a GABAergic axonal plexus does not necessarily indicate establishment of normal circuitry, and the reorganized circuit could be ineffective in controlling activity of principal cells. Such findings demonstrate yet another way in which GABA neuron loss could lead to altered inhibitory control and thus contribute to an epileptic state.

12.5 Replacement of GABA Neurons Supports Their Functional Importance in Epilepsy

Recent studies of transplantation of GABA neurons in the hippocampal formation of pilocarpine-treated mice support contributions of GABA neuron loss to the epileptic state [28]. After GABA neuron transplantation in the hippocampal formation, the number of spontaneous seizures in these mice was reduced, despite the maintained presence of mossy fiber sprouting in the inner molecular layer and, presumably, loss of mossy cells in the dentate hilus. While these findings are consistent with a loss of GABA neurons leading to the epileptic state, it remains possible that the transplanted GABA neurons could be counteracting other fundamental epilepsy-producing alterations through compensatory increases in inhibition.

12.6 GABA Neuron Loss Has Multiple Effects in Epilepsy

Loss of even a small fraction of GABA neurons can have profound functional effects due to the innervation of numerous principal cells by the expansive axonal plexus of many interneurons. However the effects of an initial loss of GABA neurons could be enhanced further by alterations of remaining GABA neurons. Despite having some basic compensatory effects, the remaining GABA neurons could contribute periodically to the epileptic state through multiple mechanisms. These could include creation of excessive synchronous activity within the network and an inability of aberrant GABAergic circuitry to respond appropriately when increased inhibitory control is required. While still speculative, there is increasing evidence that GABA neuron loss, through both direct and indirect mechanisms, could give rise to the epileptic state.

Acknowledgments I would like to express my deep gratitude to Phil Schwartzkroin for his generous spirit and support over many years; his insightful and thought-provoking questions that have stimulated and enhanced basic science research in the epilepsy field; and his deep commitment and service to the entire epilepsy community which will continue in many forms.

Other Acknowledgements This work was supported by National Institutes of Health Grant NS075245 and Veterans Affairs Medical Research Funds. I gratefully acknowledge the members of my laboratory, past and present, for their superb work and strong dedication to our studies of GABA neurons and epilepsy.

References

1. Andre V, Marescaux C, Nehlig A, Fritschy JM (2001) Alterations of hippocampal GABAergic system contribute to development of spontaneous recurrent seizures in the rat lithium-pilocarpine model of temporal lobe epilepsy. Hippocampus 11:452–468
2. Andrioli A, Alonso-Nanclares L, Arellano JI, DeFelipe J (2007) Quantitative analysis of parvalbumin-immunoreactive cells in the human epileptic hippocampus. Neuroscience 149:131–143
3. Armstrong C, Soltesz I (2012) Basket cell dichotomy in microcircuit function. J Physiol 590:683–694
4. Bausch SB (2005) Axonal sprouting of GABAergic interneurons in temporal lobe epilepsy. Epilepsy Behav 7:390–400
5. Bernard C, Esclapez M, Hirsch JC, Ben-Ari Y (1998) Interneurons are not so dormant in temporal lobe epilepsy: a critical reappraisal of the dormant basket cell hypothesis. Epilepsy Res 32:93–103
6. Boulland JL, Ferhat L, Tallak Solbu T, Ferrand N, Chaudhry FA, Storm-Mathisen J, Esclapez M (2007) Changes in vesicular transporters for gamma-aminobutyric acid and glutamate reveal vulnerability and reorganization of hippocampal neurons following pilocarpine-induced seizures. J Comp Neurol 503:466–485
7. Buckmaster PS, Dudek FE (1997) Neuron loss, granule cell axon reorganization, and functional changes in the dentate gyrus of epileptic kainate-treated rats. J Comp Neurol 385:385–404
8. Catterall WA, Kalume F, Oakley JC (2010) Na$_V$1.1 channels and epilepsy. J Physiol 588:1849–1859
9. Cavalheiro EA, Fernandes MJ, Turski L, Naffah-Mazzacoratti MG (1994) Spontaneous recurrent seizures in rats: amino acid and monoamine determination in the hippocampus. Epilepsia 35:1–11
10. Cobos I, Calcagnotto ME, Vilaythong AJ, Thwin MT, Noebels JL, Baraban SC, Rubenstein JL (2005) Mice lacking Dlx1 show subtype-specific loss of interneurons, reduced inhibition and epilepsy. Nat Neurosci 8:1059–1068
11. Cossart R, Dinocourt C, Hirsch JC, Merchan-Perez A, DeFelipe J, Ben-Ari Y, Esclapez M, Bernard C (2001) Dendritic but not somatic GABAergic inhibition is decreased in experimental epilepsy. Nat Neurosci 4:52–62
12. Davenport CJ, Brown WJ, Babb TL (1990) Sprouting of GABAergic and mossy fiber axons in dentate gyrus following intrahippocampal kainate in the rat. Exp Neurol 109:180–190
13. DeFelipe J (1999) Chandelier cells and epilepsy. Brain 122:1807–1822
14. de Lanerolle NC, Kim JH, Robbins RJ, Spencer DD (1989) Hippocampal interneuron loss and plasticity in human temporal lobe epilepsy. Brain Res 495:387–395
15. Dinocourt C, Petanjek Z, Freund TF, Ben-Ari Y, Esclapez M (2003) Loss of interneurons innervating pyramidal cell dendrites and axon initial segments in the CA1 region of the hippocampus following pilocarpine-induced seizures. J Comp Neurol 459:407–425
16. Drexel M, Preidt AP, Kirchmair E, Sperk G (2011) Parvalbumin interneurons and calretinin fibers arising from the thalamic nucleus reuniens degenerate in the subiculum after kainic acid-induced seizures. Neuroscience 189:316–329
17. Dutton SB, Makinson CD, Papale LA, Shankar A, Balakrishnan B, Nakazawa K, Escayg A (2012) Preferential inactivation of Scn1a in parvalbumin interneurons increases seizure susceptibility. Neurobiol Dis 49:211–220
18. Esclapez M, Hirsch JC, Khazipov R, Ben-Ari Y, Bernard C (1997) Operative GABAergic inhibition in the hippocampal CA1 pyramidal neurons in experimental epilepsy. Proc Natl Acad Sci U S A 94:12151–12156
19. Esclapez M, Houser CR (1999) Up-regulation of GAD65 and GAD67 in remaining hippocampal GABA neurons in a model of temporal lobe epilepsy. J Comp Neurol 412:488–505
20. Feldblum S, Ackermann RF, Tobin AJ (1990) Long-term increase of glutamate decarboxylase mRNA in a rat model of temporal lobe epilepsy. Neuron 5:361–371
21. Ferando I, Mody I (2012) GABA$_A$ receptor modulation by neurosteroids in models of temporal lobe epilepsies. Epilepsia 53(Suppl 9):89–101
22. Freund TF, Buzsaki G (1996) Interneurons of the hippocampus. Hippocampus 6:347–470
23. Freund TF, Katona I (2007) Perisomatic inhibition. Neuron 56:33–42
24. Glykys J, Peng Z, Chandra D, Homanics GE, Houser CR, Mody I (2007) A new naturally occurring GABA$_A$ receptor subunit partnership with high sensitivity to ethanol. Nat Neurosci 10:40–48
25. Gorter JA, van Vliet EA, Aronica E, Lopes da Silva FH (2001) Progression of spontaneous seizures after status epilepticus is associated with mossy fibre sprouting and extensive bilateral loss of hilar parvalbumin and somatostatin-immunoreactive neurons. Eur J Neurosci 13:657–669
26. Halabisky B, Parada I, Buckmaster PS, Prince DA (2010) Excitatory input onto hilar somatostatin interneurons is increased in a chronic model of epilepsy. J Neurophysiol 104:2214–2223
27. Houser CR, Esclapez M (1996) Vulnerability and plasticity of the GABA system in the pilocarpine model of spontaneous recurrent seizures. Epilepsy Res 26:207–218
28. Hunt RF, Girskis KM, Rubenstein JL, Alvarez-Buylla A, Baraban SC (2013) GABA progenitors grafted into the adult epileptic brain control seizures and abnormal behavior. Nat Neurosci 16:692–697
29. Kobayashi M, Buckmaster PS (2003) Reduced inhibition of dentate granule cells in a model of temporal lobe epilepsy. J Neurosci 23:2440–2452
30. Kumar SS, Buckmaster PS (2006) Hyperexcitability, interneurons, and loss of GABAergic synapses in

entorhinal cortex in a model of temporal lobe epilepsy. J Neurosci 26:4613–4623

31. Lau D, Vega-Saenz de Miera EC, Contreras D, Ozaita A, Harvey M, Chow A, Noebels JL, Paylor R, Morgan JI, Leonard CS, Rudy B (2000) Impaired fast-spiking, suppressed cortical inhibition, and increased susceptibility to seizures in mice lacking Kv3.2 K+ channel proteins. J Neurosci 20:9071–9085

32. Lowenstein DH, Thomas MJ, Smith DH, McIntosh TK (1992) Selective vulnerability of dentate hilar neurons following traumatic brain injury: a potential mechanistic link between head trauma and disorders of the hippocampus. J Neurosci 12:4846–4853

33. Marksteiner J, Sperk G (1988) Concomitant increase of somatostatin, neuropeptide Y and glutamate decarboxylase in the frontal cortex of rats with decreased seizure threshold. Neuroscience 26:379–385

34. Martin MS, Dutt K, Papale LA, Dube CM, Dutton SB, de Haan G, Shankar A, Tufik S, Meisler MH, Baram TZ, Goldin AL, Escayg A (2010) Altered function of the SCN1A voltage-gated sodium channel leads to gamma-aminobutyric acid-ergic (GABAergic) interneuron abnormalities. J Biol Chem 285:9823–9834

35. Mathern GW, Babb TL, Pretorius JK, Leite JP (1995) Reactive synaptogenesis and neuron densities for neuropeptide Y, somatostatin, and glutamate decarboxylase immunoreactivity in the epileptogenic human fascia dentata. J Neurosci 15:3990–4004

36. Morin F, Beaulieu C, Lacaille J–C (1998) Selective loss of GABA neurons in area CA1 of the rat hippocampus after intraventricular kainate. Epilepsy Res 32:363–369

37. Obenaus A, Esclapez M, Houser CR (1993) Loss of glutamate decarboxylase mRNA-containing neurons in the rat dentate gyrus following pilocarpine-induced seizures. J Neurosci 13:4470–4485

38. Peng Z, Huang CS, Stell BM, Mody I, Houser CR (2004) Altered expression of the δ subunit of the GABAₐ receptor in a mouse model of temporal lobe epilepsy. J Neurosci 24:8629–8639

39. Peng Z, Zhang N, Wei W, Huang CS, Cetina Y, Otis TS, Houser CR (2013) A reorganized GABAergic circuit in a model of epilepsy: evidence from optogenetic labeling and stimulation of somatostatin interneurons. J Neurosci 33:14392–14405

40. Powell EM, Campbell DB, Stanwood GD, Davis C, Noebels JL, Levitt P (2003) Genetic disruption of cortical interneuron development causes region- and GABA cell type-specific deficits, epilepsy, and behavioral dysfunction. J Neurosci 23:622–631

41. Ribak CE (1985) Axon terminals of GABAergic chandelier cells are lost at epileptic foci. Brain Res 326:251–260

42. Robbins RJ, Brines ML, Kim JH, Adrian T, de Lanerolle NC, Welsh S, Spencer DD (1991) A selective loss of somatostatin in the hippocampus of patients with temporal lobe epilepsy. Ann Neurol 29:325–332

43. Rossignol E, Kruglikov I, van den Maagdenberg AM, Rudy B, Fishell G (2013) Ca,2.1 ablation in cortical interneurons selectively impairs fast-spiking basket cells and causes generalized seizures. Ann Neurol 74:209–222

44. Schwarzer C, Williamson JM, Lothman EW, Vezzani A, Sperk G (1995) Somatostatin, neuropeptide Y, neurokinin B and cholecystokinin immunoreactivity in two chronic models of temporal lobe epilepsy. Neuroscience 69:831–845

45. Sloviter RS (1987) Decreased hippocampal inhibition and a selective loss of interneurons in experimental epilepsy. Science 235:73–76

46. Sloviter RS (1991) Permanently altered hippocampal structure, excitability, and inhibition after experimental status epilepticus in the rat: the "dormant basket cell" hypothesis and its possible relevance to temporal lobe epilepsy. Hippocampus 1:41–66

47. Sloviter RS, Zappone CA, Harvey BD, Bumanglag AV, Bender RA, Frotscher M (2003) "Dormant basket cell" hypothesis revisited: relative vulnerabilities of dentate gyrus mossy cells and inhibitory interneurons after hippocampal status epilepticus in the rat. J Comp Neurol 459:44–76

48. Sundstrom LE, Brana C, Gatherer M, Mepham J, Rougier A (2001) Somatostatin- and neuropeptide Y-synthesizing neurones in the fascia dentata of humans with temporal lobe epilepsy. Brain 124:688–697

49. Swartz BE, Houser CR, Tomiyasu U, Walsh GO, DeSalles A, Rich JR, Delgado-Escueta A (2006) Hippocampal cell loss in posttraumatic human epilepsy. Epilepsia 47:1373–1382

50. Thind KK, Yamawaki R, Phanwar I, Zhang G, Wen X, Buckmaster PS (2010) Initial loss but later excess of GABAergic synapses with dentate granule cells in a rat model of temporal lobe epilepsy. J Comp Neurol 518:647–667

51. Wyeth MS, Zhang N, Mody I, Houser CR (2010) Selective reduction of cholecystokinin-positive basket cell innervation in a model of temporal lobe epilepsy. J Neurosci 30:8993–9006

52. Yu J, Proddutur A, Elgammal FS, Ito T, Santhakumar V (2013) Status epilepticus enhances tonic GABA currents and depolarizes GABA reversal potential in dentate fast-spiking basket cells. J Neurophysiol 109:1746–1763

53. Zhang W, Buckmaster PS (2009) Dysfunction of the dentate basket cell circuit in a rat model of temporal lobe epilepsy. J Neurosci 29:7846–7856

54. Zhang W, Yamawaki R, Wen X, Uhl J, Diaz J, Prince DA, Buckmaster PS (2009) Surviving hilar somatostatin interneurons enlarge, sprout axons, and form new synapses with granule cells in a mouse model of temporal lobe epilepsy. J Neurosci 29:14247–14256

Does Mossy Fiber Sprouting Give Rise to the Epileptic State?

Paul S. Buckmaster

Abstract

Many patients with temporal lobe epilepsy display structural changes in the seizure initiating zone, which includes the hippocampus. Structural changes in the hippocampus include granule cell axon (mossy fiber) sprouting. The role of mossy fiber sprouting in epileptogenesis is controversial. A popular view of temporal lobe epileptogenesis contends that precipitating brain insults trigger transient cascades of molecular and cellular events that permanently enhance excitability of neuronal networks through mechanisms including mossy fiber sprouting. However, recent evidence suggests there is no critical period for mossy fiber sprouting after an epileptogenic brain injury. Instead, findings from stereological electron microscopy and rapamycin-delayed mossy fiber sprouting in rodent models of temporal lobe epilepsy suggest a persistent, homeostatic mechanism exists to maintain a set level of excitatory synaptic input to granule cells. If so, a target level of mossy fiber sprouting might be determined shortly after a brain injury and then remain constant. Despite the static appearance of synaptic reorganization after its development, work by other investigators suggests there might be continual turnover of sprouted mossy fibers in epileptic patients and animal models. If so, there may be opportunities to reverse established mossy fiber sprouting. However, reversal of mossy fiber sprouting is unlikely to be antiepileptogenic, because blocking its development does not reduce seizure frequency in pilocarpine-treated mice. The challenge remains to identify which, if any, of the many other structural changes in the hippocampus are epileptogenic.

Keywords

Dentate gyrus • Granule cell • Epilepsy • Epileptogenesis • Hilus • Seizure • Pilocarpine

P.S. Buckmaster (✉)
Departments of Comparative Medicine
and Neurology & Neurological Sciences,
Stanford University, Stanford, CA, USA
e-mail: psb@stanford.edu

H.E. Scharfman and P.S. Buckmaster (eds.), *Issues in Clinical Epileptology: A View from the Bench*,
Advances in Experimental Medicine and Biology 813, DOI 10.1007/978-94-017-8914-1_13,
© Springer Science+Business Media Dordrecht 2014

13.1   Introduction

Temporal lobe epilepsy is common and its under-lying mechanisms remain unclear [12]. Many patients have a history of an initial brain insult followed by a latent period [30]. A common view of temporal lobe epileptogenesis is that during the latent period, cascades of molecular and cellular events together alter the excitability of neuronal networks, ultimately causing spontaneous seizures [38]. According to this view, following an injury there is a critical period when tempo-rary treatment might permanently prevent network reorganization. Substantial network reorganiza-tion occurs in the hippocampus of many patients with temporal lobe epilepsy [29]. The hippocampus is prone to epileptic activity [40] and is a site of seizure initiation in patients [37, 45]. Therefore, it is logical to ask whether structural changes in the hippocampus give rise to the epileptic state and whether blocking the development of structural changes during a critical period would prevent epileptogenesis. Structural changes in the hippocampus of patients with temporal lobe epilepsy include specific patterns of neuron loss [29], including inhibitory interneurons [11], hypertrophy of some surviving interneurons [28], GABAergic axon sprouting [1], dispersion of granule cells to ectopic locations [19], exces-sive development of hilar basal dendrites on granule cells [53], and mossy fiber sprouting (reviewed in [2]).

Philip Schwartzkroin's research included work on mossy fiber sprouting. His laboratory's slice experiments on tissue resected to treat patients revealed a general correlation between mossy fiber sprouting and hyperexcitability [13]. In those experiments, intracellular labeling and elec-tron microscopy showed sprouted mossy fibers synapsing with dendrites of granule cells and interneurons. Those findings were supported and extended by experiments with kainate-treated rats that included evidence of autaptic synapses by sprouted mossy fibers [58]. However, Phil and colleagues cautioned that the results provided no direct evidence that mossy fiber sprouting was either necessary or sufficient for hyperexcitability [13].

Phil and colleagues characterized mossy fiber sprouting in other animal models, including p35-deficient mice with cortical dysplasia [57], different mouse strains treated with kainic acid [31], and infant monkeys after limbic status epilepticus [16, 56]. Phil's laboratory also helped localize the zinc transporter-3 to mossy fiber syn-aptic vesicle membranes [55], which has become a useful marker for visualizing mossy fiber sprouting.

Despite much investigation, the role of mossy fiber sprouting in epileptogenesis remains unclear and controversial. It has been proposed to be proepileptogenic [48], antiepileptogenic [44], and an epiphenomenon [15]. Recent evi-dence reviewed here raises questions about whether there is a critical period for mossy fiber sprouting after epileptogenic injuries and whether mossy fiber sprouting contributes to the generation of spontaneous seizures.

13.2   Is Mossy Fiber Sprouting a Homeostatic Mechanism?

Neuron loss in the hilus of the dentate gyrus is a common structural change in patients with tem-poral lobe epilepsy [29] that is replicated in animal models. Nadler et al. (1980) first showed that the excitotoxin kainic acid kills hilar neurons, whose axons degenerate in the inner third of the molecular layer into which mossy fibers later sprout. To quantify the initial loss of synapses onto granule cell proximal dendrites in the inner molecular layer and the later restoration of excit-atory synaptic input from sprouted mossy fibers, we used stereological electron microscopy to evaluate a rat model of temporal lobe epilepsy [49]. Tissue was obtained: (1) from rats 5 days after pilocarpine-induced status epilepticus to measure loss of synaptic input to granule cells before axon sprouting had occurred and (2) after mossy fiber sprouting was well established 3–6 months after status epilepticus. Numbers of granule cells were estimated from Nissl stained sections. Numbers of excitatory synapses in the molecular layer, where granule cell dendrites extend, were estimated in serial electron micrographs that had been processed by post-embedding immunocytochemistry for GABA to avoid counting inhibitory synapses. Subsequently, numbers of excitatory synapses per granule cell were calculated for each rat (Fig. 13.1a). Analysis of the inner third of the molecular layer revealed that the number of excitatory synapses per granule cell decreased to only 38 % of controls by 5 days after pilocarpine-induced status epilepticus (Fig. 13.1b). This substantial loss of synapses probably is attributable primarily to loss of hilar mossy cells. Mossy cells, which were first characterized electrophysiologically by intracellular recording and anatomical labeling techniques in Phil's laboratory [39], project most of their axon collaterals to the inner molecular layer of the dentate gyrus where they form glutamatergic synapses with proximal dendrites of granule cells [6, 8, 54]. Epileptogenic injuries, like status epilepticus, kill mossy cells [4] and thereby denervate proximal dendrites of granule cells. The extent of mossy cell loss correlates with the extent of mossy fiber sprouting [23]. However, mossy cell loss alone is insufficient to cause mossy fiber sprouting [24], and the molecular signals necessary for triggering mossy fiber sprouting are not yet known.

After initial loss, numbers of excitatory synapses per granule cell in the inner molecular layer partially rebound to 84 % of controls in rats 3–6 months after status epilepticus (Fig. 13.1b). This recovery probably is attributable primarily to mossy fiber sprouting [9], but other sources of excitatory synaptic input to the proximal dendrites of granule cells include surviving mossy cells and proximal CA3 pyramidal cells [60]. Synapses with proximal dendrites of granule cells at 3–6 months after status epilepticus are 1.3-times larger than in controls and twice as likely to be perforated [49]. Large, perforated synapses are likely to be functionally stronger than small, nonperforated synapses [14, 33, 35]. To maintain functional stability in the face of change, brains use an array of homeostatic mechanisms, including synaptic scaling [50]. Larger, stronger mossy fiber synapses in the inner molecular layer of epileptic rats might be a homeostatic mechanism to compensate for fewer synapses (84 % of controls, in this case).

Similarly, on granule cell distal dendrites in the outer two-thirds of the molecular layer, numbers of excitatory synapses decrease to 69 % of controls by 5 days after status epilepticus, but rebound to 101 % of controls by 3–6 months (Fig. 13.1b). With more complete recovery of synapse numbers, synapse size in the outer molecular does not change significantly [49]. Initial loss of synapses with distal dendrites of granule cells probably is attributable to partial loss of layer II entorhinal cortical neurons caused by status epilepticus [26]. And recovery of synapses probably is attributable to sprouting of axons of surviving layer II neurons [46]. Together, findings from the inner and outer molecular layer suggest a homeostatic mechanism maintains excitatory synaptic input to granule cells in response to synapse loss after an epileptogenic injury.

13.3 No Critical Period for Mossy Fiber Sprouting

If a homeostatic mechanism controls the number of excitatory synapses with granule cells, signals underlying that control might persist as long as a synaptic deficit continues. Persistent signals contrast with the view of a transient cascade of molecular and cellular events that peak and then diminish after a critical period following a brain injury. To address these issues, we determined whether mossy fiber sprouting would occur after a 2 month delay [27]. Rapamycin, which inhibits mossy fiber spouting [3], was administered to mice daily beginning 24 h after pilocarpine-induced status epilepticus. After 2 months, mossy fiber sprouting was suppressed almost by half in the rapamycin group compared to vehicle-treated controls (Fig. 13.1c). Another cohort was evaluated 6 months after the end of treatment, which was 8 months after status epilepticus. Mossy fiber sprouting was well developed in both vehicle- and rapamycin-treated mice, indicating that signals stimulating mossy fiber sprouting must have persisted for more than 2 months. These findings suggest there is no transient critical period for mossy fiber sprouting after an epileptogenic

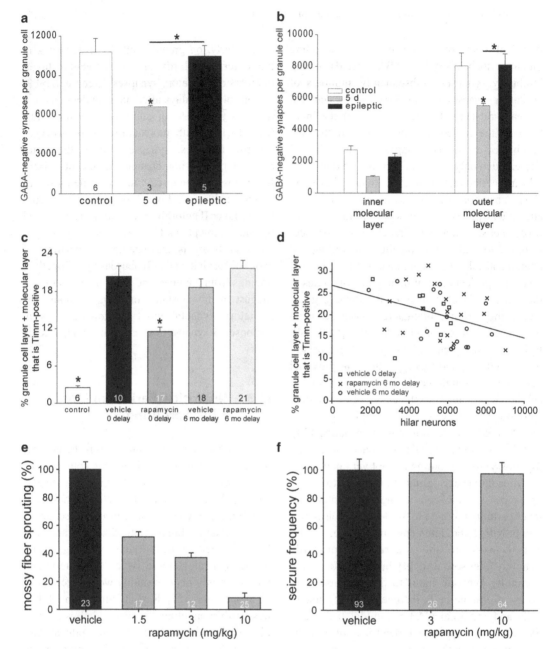

Fig. 13.1 Excitatory synapse loss and mossy fiber sprouting of granule cells in pilocarpine-treated rodent models of temporal lobe epilepsy. (**a**) Number of putative excitatory synapses per granule cell in control rats and in rats 5 days and 3–6 months (epileptic) after pilocarpine-induced status epilepticus. (**b**) Number of synapses with granule cell dendrites in the inner one-third and outer two-thirds of the molecular layer. Values represent mean ± sem. Sample size indicated at base of bars. Asterisks indicate differences from the control value unless specified by a horizontal line (p < 0.05, ANOVA, Student-Newman Keuls method). (**a**) and (**b**) from Thind et al. [49]. (**c**) Extent of mossy fiber sprouting in control mice and mice that experienced status epilepticus and were treated with vehicle or 1.5 mg/kg rapamycin every day for 2 months and then were perfused with no delay (0 delay) or after a 6 month delay (6 month delay) (From Lew and Buckmaster [27]). (**d**) Number of

large hilar neurons (>12 μm soma diameter) per hippocampus versus extent of mossy fiber sprouting in mice that experienced pilocarpine-induced status epilepticus and were treated with vehicle or rapamycin for 2 months and then evaluated immediately (vehicle 0 months) or after another 6 months (vehicle or rapamycin 6 months). A linear regression line is plotted (R = 0.34, p = 0.021, ANOVA). (**e**) Percent mossy fiber sprouting was calculated by subtracting the average percentage of the molecular layer plus granule cell layer that was Timm-positive in control mice and normalizing by the average value of mice that had experienced status epilepticus and were treated with vehicle for 2 months. Averages of all groups are significantly different from others (p < 0.05, ANOVA, Student-Newman-Keuls method). (**f**) Percent seizure frequency was calculated by normalizing by the average of the vehicle-treated group. (**e**) and (**f**) from Heng et al. [18]

brain injury. Instead, preventing mossy fiber sprouting might require long-term or continuous treatment. This scenario challenges the view of transient signaling cascades whose consequences could be permanently blocked by temporary treatment during a critical period.

One might question whether the precipitating injury in the mouse model was so severe that it maximally stimulated mossy fiber sprouting toward saturation levels despite the delay caused by rapamycin. However, an all-or-none "toggle-like" signal and saturation effect is inconsistent with the graded degree of mossy fiber sprouting among individual mice, which ranged over a factor of three and was correlated with the extent of hilar neuron loss (Fig. 13.1d). Wide ranges in mossy fiber sprouting between individuals were evident in vehicle-treated mice 2 months after status epilepticus and vehicle- and rapamycin-treated mice 8 months after status epilepticus, indicating that sprouting did not progressively develop toward saturated levels. These findings suggest that a target level of mossy fiber sprouting in an individual might be determined shortly after a brain injury and then remain constant.

Together, findings from stereological electron microscopy and rapamycin-delayed mossy fiber sprouting suggest a persistent, homeostatic mechanism exists to maintain a set level of excitatory synaptic input to granule cells. If mossy fiber sprouting were epileptogenic, this might be an example of a normally adaptive homeostatic mechanism that became pathogenic in response to an injury, which has been proposed previously as a theoretical possibility [10]. More generally, epileptogenesis might be an unintended side-effect of the brain's homeostatic mechanisms, which evolved to maintain function in the face of plasticity. Epileptogenic injuries might trigger changes so much more extensive than normal plasticity that they push homeostatic responses into a range that creates a network that generates spontaneous seizures. Phil proposed a similar idea [41]: "I believe that the brain has been designed to operate at a knife's edge. The evolutionary demand for plasticity – a key attribute of higher order learning, memory, and all those complex functions that are characteristic of the mammalian CNS – has necessitated a sacrifice in stability of neuronal function."

On the other hand, if epileptogenesis is maintained by homeostatic mechanisms gone awry, there may be opportunities to reverse established epilepsy-related structural abnormalities. Although mossy fiber sprouting appears to develop gradually, plateau, and then cease, there might instead be continual turnover. Mossy fibers in tissue from patients with temporal lobe epilepsy display evidence of continuing synaptic reorganization years after precipitating injuries [21, 32, 36]. At least some sprouted mossy fibers arise from adult generated granule cells [22, 25], which might continue to be generated long after precipitating injuries (but see [17]). To test whether mossy fiber sprouting could be reversed after it had established, we infused rapamycin focally into the dentate gyrus for 1 month beginning 2 months after pilocarpine-induced status epilepticus in rats, but there was no effect [3]. However, Huang et al. [20] reported that in chronically epileptic pilocarpine-treated rats, systemically administered rapamycin partially reversed already established mossy fiber sprouting. Moreover, grafts of CA3 pyramidal cells reduce mossy fiber sprouting even when implanted 45 days after kainate-treatment, during which time considerable mossy fiber sprouting is likely to have developed [42]. In addition, mild mossy sprouting generated by electroconvulsive shock was reported to decline over time [51]. Thus, more work is needed to test the reversibility of mossy fiber sprouting.

13.4   Mossy Fiber Sprouting Is Not Epileptogenic

Rapamycin also was used to test whether mossy fiber sprouting was epileptogenic. Systemic treatment with rapamycin at increasing doses to inhibit mossy fiber sprouting to increasing degrees had no effect on the frequency of spontaneous seizures in mice that had experienced pilocarpine-induced status epilepticus (Fig. 13.1e, f) [5, 18]. These findings suggest that mossy fiber sprouting is neither pro- nor antiepileptogenic, but instead is an epiphenomenon unrelated to seizure genesis. There are caveats with this conclusion, because rapamycin has side-effects [47], including suppression of axon sprouting by inhibitory GABAergic interneurons [7]. And rapamycin reduces seizure frequency in some rat models of temporal lobe epilepsy [20, 52, 59] but not all [43]. It remains unclear whether rapamycin's action in rats is antiseizure or antiepileptogenic. Nevertheless, the findings from the mouse studies suggest mossy fiber sprouting is not epileptogenic.

13.5   Conclusions

Patients with temporal lobe epilepsy display many structural changes, especially in the hippocampus. One possibility is that together numerous structural changes and other abnormalities all contribute partially to seizure generation. In that scenario, blocking the development of any one change, like mossy fiber sprouting, might have negligible effects on epileptogenesis. Another possibility is that some or perhaps even many epilepsy-related structural changes are not epileptogenic, including mossy fiber sprouting, and that seizure generation is attributable to one or two critical abnormalities whose importance has not yet been recognized. These alternate possibilities – many abnormalities each contributing partially versus one or two abnormalities primarily responsible for seizure generation–might require different therapeutic approaches, so it is important to distinguish between them. To do so, it will be useful to tap the ever-increasing knowledge base of molecular and cellular mechanisms underlying brain developmental processes and responses to injury. Creative application of ideas and reagents (for example, rapamycin), even from fields outside of epilepsy research, might yield useful approaches for specifically inhibiting or exacerbating individual epilepsy-related structural changes. Experimental manipulation of specific structural changes, one at a time, and rigorous measurement of effects on spontaneous seizures, might eventually reveal which, if any, are epileptogenic. If no single change alone appears to be responsible, then blockade of many or all could be

used to test whether together they make the brain epileptic or if the cause of seizures is unrelated to structural changes in the hippocampus.

Acknowledgements Dedicated to Philip A. Schwartzkroin, a trusted graduate advisor and outstanding research mentor who taught me so much over the years.

Other Acknowledgements Supported by NINDS/NIH.

References

1. Babb TL, Pretorius JK, Kupfer WR, Crandall PH (1989) Glutamate decarboxylase-immunoreactive neurons are preserved in human epileptic hippocampus. J Neurosci 9:2562–2574
2. Buckmaster PS (2012) Mossy fiber sprouting in the dentate gyrus. In: Noebels JL, Avoli M, Rogawski MA, Olsen RW, Delgado-Escueta AV (eds) Japser's basic mechanisms of the epilepsies, 4th edn. Oxford University Press, New York, pp 416–431
3. Buckmaster PS, Ingram EA, Wen X (2009) Inhibition of the mammalian target of rapamycin signaling pathway suppresses dentate granule cell axon sprouting in a rodent model of temporal lobe epilepsy. J Neurosci 29:8259–8269
4. Buckmaster PS, Jongen-Rêlo AL (1999) Highly specific neuron loss preserves lateral inhibitory circuits in the dentate gyrus of kainate-induced epileptic rats. J Neurosci 19:9519–9529
5. Buckmaster PS, Lew FH (2011) Rapamycin suppresses mossy fiber sprouting but not seizure frequency in a mouse model of temporal lobe epilepsy. J Neurosci 31:2337–2347
6. Buckmaster PS, Strowbridge BW, Kunkel DD, Schmiege DL, Schwartzkroin PA (1992) Mossy cell axonal projections to the dentate gyrus molecular layer in the rat hippocampal slice. Hippocampus 2:349–362
7. Buckmaster PS, Wen X (2011) Rapamycin suppresses axon sprouting by somatostatin interneurons in a mouse model of temporal lobe epilepsy. Epilepsia 52:2057–2064
8. Buckmaster PS, Wenzel HJ, Kunkel DD, Schwartzkroin PA (1996) Axon arbors and synaptic connections of hippocampal mossy cells in the rat in vivo. J Comp Neurol 366:270–292
9. Buckmaster PS, Zhang G, Yamawaki R (2002) Axon sprouting in a model of temporal lobe epilepsy creates a predominantly excitatory feedback circuit. J Neurosci 22:6650–6658
10. Davis GW, Goodman CS (1998) Genetic analysis of synaptic development and plasticity: homeostatic regulation of synaptic efficacy. Curr Opin Neurobiol 8:149–156
11. de Lanerolle NC, Kim JH, Robbins RJ, Spencer DD (1989) Hippocampal interneuron loss and plasticity in human temporal lobe epilepsy. Brain Res 495:387–395

12. Engel J Jr, Williamson PD, Wieser H-G (1997) Mesial temporal lobe epilepsy. In: Engel J Jr, Pedley TA (eds) Epilepsy: a comprehensive textbook. Lippincott-Raven, Philadelphia, pp 2417–2426

13. Franck JE, Pokorny J, Kunkel DD, Schwartzkroin PA (1995) Physiologic and morphologic characteristics of granule cell circuitry in human epileptic hippocampus. Epilepsia 36:543–558

14. Ganeshina O, Berry RW, Petralia RS, Nicholson DA, Geinesman Y (2004) Differences in the expression of AMPA and NMDA receptors between axospinous perforated and nonperforated synapses are related to the configuration and size of postsynaptic densities. J Comp Neurol 468:86–95

15. Gloor P (1997) The temporal lobe and limbic system. Oxford University Press, New York, pp 677–691

16. Gunderson VM, Dubach M, Szot P, Born DE, Wenzel JH, Maravilla KR, Zierath DK, Robbins CA, Schwartzkroin PA (1999) Development of a model of status epilepticus in pigtailed macaque infant monkeys. Dev Neurosci 21:352–364

17. Hattiangady B, Rao MS, Shetty AK (2004) Chronic temporal lobe epilepsy is associated with severely declined dentate neurogenesis in the adult hippocampus. Neurobiol Dis 17:473–490

18. Heng K, Haney MM, Buckmaster PS (2013) High-dose rapamycin blocks mossy fiber sprouting but not seizures in a mouse model of temporal lobe epilepsy. Epilepsia 54:1535–1541

19. Houser CR (1990) Granule cell dispersion in the dentate gyrus of humans with temporal lobe epilepsy. Brain Res 535:195–204

20. Huang X, Zhang H, Yang J, Wu J, McMahon J, Lin Y, Cao Z, Gruenthal M, Huang Y (2010) Pharmacological inhibition of the mammalian target of rapamycin pathway suppresses acquired epilepsy. Neurobiol Dis 40:193–199

21. Isokawa M, Levesque MF, Babb TL, Engel JE Jr (1993) Single mossy fiber axonal systems of human dentate granule cells studied in hippocampal slices from patients with temporal lobe epilepsy. J Neurosci 13:1511–1522

22. Jessberger S, Zhao C, Toni N, Clemenson GD Jr, Li Y, Gage FH (2007) Seizure-associated, aberrant neurogenesis in adult rats characterized with retrovirus-mediated cell labeling. J Neurosci 27:9400–9407

23. Jiao Y, Nadler JV (2007) Stereological analysis of GluR2-immunoreactive hilar neurons in the pilocarpine model of temporal lobe epilepsy: correlation of cell loss with mossy fiber sprouting. Exp Neurol 205:569–582

24. Jinde S, Zsiros V, Jiang Z, Nakao K, Pickel J, Kohno K, Belforte JE, Nakazawa K (2012) Hilar mossy cell degeneration causes transient dentate granule cell hyperexcitability and impaired pattern separation. Neuron 76:1189–1200

25. Kron MM, Zhang H, Parent JM (2010) The developmental stage of dentate granule cells dictates their contribution to seizure-induced plasticity. J Neurosci 30:2051–2059

26. Kumar SS, Buckmaster PS (2006) Hyperexcitability, interneurons, and loss of GABAergic synapses in entorhinal cortex in a model of temporal lobe epilepsy. J Neurosci 26:4613–4623

27. Lew F, Buckmaster PS (2011) Is there a critical period for mossy fiber sprouting in a mouse model of temporal lobe epilepsy? Epilepsia 52:2326–2332

28. Maglóczky Z, Wittner L, Borhegyi Z, Halász P, Vajda J, Czirják S, Freund TF (2000) Changes in the distribution and connectivity of interneurons in the epileptic human dentate gyrus. Neuroscience 96:7–25

29. Margerison JH, Corsellis JAN (1966) Epilepsy and the temporal lobes. Brain 89:499–530

30. Mathern GW, Pretorius JK, Babb TL (1995) Influence of the type of initial precipitating injury and at what age it occurs on course and outcome in patients with temporal lobe seizures. J Neurosurg 82:220–227

31. McKhann GM 2nd, Wenzel HJ, Robbins CA, Sosunov AA, Schwartzkroin PA (2003) Mouse strain differences in kainic acid sensitivity, seizure behavior, mortality, and hippocampal pathology. Neuroscience 122:551–561

32. Mikkonen M, Soininen H, Kälviäinen R, Tapiola T, Ylinen A, Vapalahti M, Paljärvi L, Pitkänen A (1998) Remodeling of neuronal circuitries in human temporal lobe epilepsy: increased expression of highly polysialylated neural cell adhesion molecular in the hippocampus and entorhinal cortex. Ann Neurol 44:923–934

33. Murthy VN, Schikorski T, Stevens CF, Zhu Y (2001) Inactivity produces increases in neurotransmitter release and synapse size. Neuron 32:673–682

34. Nadler JV, Perry BW, Cotman CW (1980) Selective reinnervation of hippocampal area CA1 and the fascia dentata after destruction of CA3-CA4 afferents with kainic acid. Brain Res 182:1–9

35. Nusser Z, Lujan R, Laube G, Roberts JDB, Molnar E, Somogyi P (1998) Cell type and pathway dependence of synaptic AMPA receptor number and variability in the hippocampus. Neuron 21:545–559

36. Proper EA, Oestreicher AB, Jansen GH, Veelen CWM, van Rijen PC, Gispen WH, de Graan PNE (2000) Immunohistochemical characterization of mossy fibre sprouting in the hippocampus of patients with pharmaco-resistant temporal lobe epilepsy. Brain 123:19–30

37. Quesney LF (1986) Clinical and EEG features of complex partial seizures of temporal lobe origin. Epilepsia 27(Suppl 2):S27–S45

38. Rakhade SN, Jensen FE (2009) Epileptogenesis in the immature brain: emerging mechanisms. Nat Rev Neurol 5:380–391

39. Scharfman HE, Schwartzkroin PA (1988) Electrophysiology of morphologically identified mossy cells of the dentate hilus recorded in guinea pig hippocampal slices. J Neurosci 8:3812–3821

40. Schwartzkroin PA (1994) Role of the hippocampus in epilepsy. Hippocampus 4:239–242

41. Schwartzkroin PA (1997) Origins of the epileptic state. Epilepsia 38:853–858

42. Shetty AK, Zaman V, Hattiangady B (2005) Repair of the injured adult hippocampus through graft-mediated modulation of the plasticity of the dentate gyrus in a rat model of temporal lobe epilepsy. J Neurosci 25:8391–8401

43. Sliwa A, Plucinska G, Bednarczyk J, Lukasiuk K (2012) Post-treatment with rapamycin does not prevent epileptogenesis in the amygdala stimulation model of temporal lobe epilepsy. Neurosci Lett 509:105–109

44. Sloviter RS (1992) Possible functional consequences of synaptic reorganization in the dentate gyrus of kainate-treated rats. Neurosci Lett 137:91–96

45. Spencer SS, Williamson PD, Spencer DD, Mattson RH (1987) Human hippocampal seizure spread studied by depth and subdural recording: the hippocampal commissure. Epilepsia 28:479–489

46. Steward O (1976) Reinnervation of dentate gyrus by homologous afferents following entorhinal cortical lesions in adult rats. Science 194:426–428

47. Swiech L, Perycz M, Malik A, Jaworski J (2008) Role of mTOR in physiology and pathology of the nervous system. Biochim Biophys Acta 1784:116–132

48. Tauck DL, Nadler JV (1985) Evidence of functional mossy fiber sprouting in hippocampal formation of kainic acid-treated rats. J Neurosci 5:1016–1022

49. Thind KK, Yamawaki R, Phanwar I, Zhang G, Wen X, Buckmaster PS (2010) Initial loss but later excess of GABAergic synapses with dentate granule cells in a rat model of temporal lobe epilepsy. J Comp Neurol 518:647–667

50. Turrigiano GG (2008) The self-tuning neuron: synaptic scaling of excitatory synapses. Cell 135:422–435

51. Vaidya VA, Siuciak JA, Du F, Duman RS (1999) Hippocampal mossy fiber sprouting induced by chronic electroconvulsive seizures. Neuroscience 89:157–166

52. van Vliet EA, Forte G, Holtman L, den Buerger JCG, Sinjewel A, de Vries HE, Aronica E, Gorter JA (2012) Inhibition of mammalian target of rapamycin reduces epileptogenesis and blood-brain barrier leakage but not microglia activation. Epilepsia 43:1254–1263

53. von Campe G, Spencer DD, de Lanerolle NC (1997) Morphology of dentate granule cells in the human hippocampus. Hippocampus 7:472–488

54. Wenzel HJ, Buckmaster PS, Anderson NL, Wenzel ME, Schwartzkroin PA (1997) Ultrastructural localization of neurotransmitter immunoreactivity in mossy cell axons and their synaptic targets in the rat dentate gyrus. Hippocampus 7:559–570

55. Wenzel HJ, Cole TB, Born DE, Schwartzkroin PA, Palmiter RD (1997) Ultrastructural localization of zinc transporter-3 (ZnT-3) to synaptic vesicle membranes within mossy fiber boutons in the hippocampus of mouse and monkey. Proc Natl Acad Sci U S A 94:12676–12681

56. Wenzel HJ, Born DE, Dubach MF, Gunderson VM, Maravilla KR, Robbins CA, Szot P, Zierath D, Schwartzkroin PA (2000) Morphological plasticity in an infant monkey model of temporal lobe epilepsy. Epilepsia 41(Suppl 6):S70–S75

57. Wenzel HJ, Robbins CA, Tsai LH, Schwartzkroin PA (2001) Abnormal morphological and functional organization of the hippocampus in a p35 mutant model of cortical dysplasia with spontaneous seizures. J Neurosci 21:983–998

58. Wenzel HJ, Woolley CS, Robbins CA, Schwartzkroin PA (2000) Kainic acid-induced mossy fiber sprouting and synapse formation in the dentate gyrus of rats. Hippocampus 10:244–260

59. Zeng L-H, Rensing NR, Wong M (2009) The mammalian target of rapamycin signaling pathway mediates epileptogenesis in a model of temporal lobe epilepsy. J Neurosci 29:6964–6972

60. Zhang W, Huguenard JR, Buckmaster PS (2012) Increased positive-feedback from hilar and CA3 neurons to granule cells in a rat model of temporal lobe epilepsy. J Neurosci 32:1183–1196

Does Brain Inflammation Mediate Pathological Outcomes in Epilepsy?

14

Karen S. Wilcox and Annamaria Vezzani

Abstract

Inflammation in the central nervous system (CNS) is associated with epilepsy and is characterized by the increased levels of a complex set of soluble molecules and their receptors in epileptogenic foci with profound neuromodulatory effects. These molecules activate receptor-mediated pathways in glia and neurons that contribute to hyperexcitability in neural networks that underlie seizure generation. As a consequence, exciting new opportunities now exist for novel therapies targeting the various components of the immune system and the associated inflammatory mediators, especially the IL-1β system. This review summarizes recent findings that increased our understanding of the role of inflammation in reducing seizure threshold, contributing to seizure generation, and participating in epileptogenesis. We will discuss preclinical studies supporting the hypothesis that pharmacological inhibition of specific proinflammatory signalings may be useful to treat drug-resistant seizures in human epilepsy, and possibly delay or arrest epileptogenesis.

Keywords

Inflammation • IL-1β • TNF-α • IL-6 • Reactive astrogliosis

K.S. Wilcox (✉)
Department of Pharmacology and Toxicology,
University of Utah, Salt Lake City, UT 84108, USA
e-mail: Karen.wilcox@hsc.utah.edu

A. Vezzani
Department of Neuroscience, IRCSS-Istituto
di Ricerche Farmacologiche "Mario Negri",
via G. La Masa 19, Milan 20156, Italy
e-mail: annamaria.vezzani@marionegri.it

14.1 Introduction

The state-of-the-art knowledge acquired in the last decade of experimental and clinical work indicates that cytokines and related molecules are increased in brain tissue after epileptogenic injuries or during seizures. In the experimental setting, these molecules, endowed with proinflammatory properties, contribute significantly to the generation

H.E. Scharfman and P.S. Buckmaster (eds.), *Issues in Clinical Epileptology: A View from the Bench*,
Advances in Experimental Medicine and Biology 813, DOI 10.1007/978-94-017-8914-1_14,
© Springer Science+Business Media Dordrecht 2014

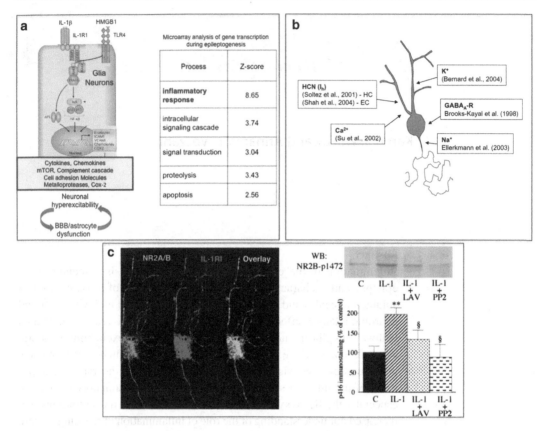

Fig. 14.1 *Schematic representation of the pathophysiologic outcomes of innate immunity activation in epilepsy.* Activation of innate immune signaling occurs in epilepsy also in the absence of infection, thus triggering the so-called "sterile" inflammatory cascade (**a**). Endogenous molecules (damage associated molecular patterns, DAMPs) such as IL-1β and the High Mobility Group Box 1 (*HMGB1*) protein are released by neurons and glia following epileptogenic inciting events, or during recurrent seizures. The activation of their cognate receptors (IL-1R type 1 and TLR4, respectively) upregulated in astrocytes triggers the NFkB-dependent inflammatory genes cascade, thus inducing various molecules with *proinflammatory* and *neuromodulatory* properties. The signaling activation in neurons increases excitability by provoking acquired channelopathies involving voltage-gated channels (*HCN1*) or AMPA and GABA-A receptor complexes (**b**), as well as by rapid activation of Src kinase inducing the phosphorylation of the NR2B subunit of the NMDA receptor thereby promoting neuronal Ca^{2+} influx (**c**). This chain of event contributes to the generation and establishment of an hyperexcitable neuronal network by direct receptor-mediated neuronal effects or indirectly by inducing astrocytes and BBB dysfunctions

and maintenance of a hyperexcitable neuronal network, thus decreasing seizure threshold (Fig. 14.1) and making the occurrence of a seizure more likely.

A key question that basic science has been addressing is how these proinflammatory molecules affect neuronal and glial functions. Answers to this question will increase our knowledge of the complex mechanistic aspects of hyperexcitability following inflammation and will be instrumental in highlighting novel targets for developing drugs and therapies that raise seizure threshold, prevent seizure generation after an inciting event, and inhibit their recurrence in chronic epilepsy.

14.1.1 Inflammatory Molecules as Neuromodulators

The presence of molecules with proinflammatory properties in brain specimens obtained from

Table 14.1 Inflammatory mediators in human epilepsies and experimental models

Clinical evidence

Inflammatory mediators are overexpressed in epileptogenic foci in human pharmacoresistant epilepsy of differing etiologies (e.g. RE, LE, MCD, mTLE)

Microglia and astrocytes are main sources of inflammatory mediators in brain tissue; neurons and endothelial cells of the blood brain barrier (BBB) also contribute to the generation of brain inflammation

Leukocyte extravasation in brain depends on the etiology of epilepsy

BBB damage is often detected together with brain inflammation

Experimental evidence

Recurrent seizures and epileptogenic brain injuries induce inflammatory mediators in astrocytes, microglia, neurons, and microvessels in brain areas involved in seizure onset and generalization

This phenomenon is long lasting and may exceed the initial precipitating event by days or weeks depending on the epilepsy model. It is inadequately controlled by anti-inflammatory mechanisms

In models of epileptogenesis, inflammation initiates before the development of epilepsy

Specific anti-inflammatory treatments reduce acute and chronic seizures and delay their time of onset

Transgenic mice with perturbed cytokine signaling show altered seizure susceptibility

Proinflammatory insults decrease seizure threshold (*acutely* and *long-term*)

patients with epilepsy has been described as "brain inflammation" (Table 14.1). However, there is emerging evidence that these molecules have neuromodulatory functions that activate signaling in neurons and glia that are different from those induced by the same molecules in leukocytes in the frame of a classical inflammatory response to infection. During infection, proinflammatory cytokines and related molecules are released during innate immunity activation by immunocompetent cells following "pathogen associated molecular patterns" (PAMPs) activation of toll-like receptors (TLRs) or nucleotide-binding oligomerization domain (NOD-like) receptors. Cytokine release activates inflammatory programs for pathogen removal and the subsequent induction of homoeostatic tissue repair mechanisms. Notably, in humans affected by various forms of pharmacoresistant epilepsy of differing etiolo-

gies (e.g. Rasmussen's (RE) and limbic encephalitis (LE), malformations of cortical development, and mesial temporal lobe epilepsy (mTLE)) increased inflammatory mediators are measured in epileptogenic foci in the absence of an identifiable active infectious process. However, it is also important to note that CNS infection, which is a common cause of TLE, can also result in a cytokine storm that affects excitability. In this context, evidence of HHV6 infected astrocytes and neurons has been reported in about 2/3 of patients with mTLE [108]. Moreover, recent work has shown the presence of Human Papilloma virus in human focal cortical dysplasia type II which might be responsible for focal epileptogenic malformations during fetal brain development in association with enhanced mTORC1 signaling [18].

The so-called *sterile inflammation* in the brain can be induced when TLRs are activated by endogenous molecules released by injured brain cells, named "danger signals" or "damage-associated molecular patterns" (DAMPs). In particular, the activation of TLR4, which can also be activated by gram-negative bacteria, is induced by the ubiquitous nuclear protein High Mobility Group Box 1 (HMGB1) which is released, upon its cytoplasmatic translocation, by neurons and glial cells. In concert with IL-1β released by glia, thereafter activating IL-1 receptor type 1 (IL-1R1), HMGB1 induces the transcriptional up-regulation of various inflammatory genes, therefore promoting the generation of the brain inflammatory cascade in glia and endothelial cells of the BBB (Fig. 14.1). In the context of malformations of cortical development, the inflammatory cascade is also induced in aberrant neuronal cells [3]. The activation of the IL-1R1/TLR4 signaling in neurons, which overexpress these receptors in pathologic conditions, in concert with pathways induced by other cytokines such as TNF-α, IL-6, the complement system and some prostaglandins, alters neuronal excitability by modifying either glutamate or GABA receptor subunit composition, or trafficking of receptors, or the function of voltage-gated ion channels via rapid onset post-translational mechanisms [118, 123]. Furthermore, initiation of the JAK/STAT and other signaling pathways through these mechanisms can also result in activation of

Table 14.2 Antagonism of IL-1R1/TLR4 in rodent models of seizures

Seizure reduction in rodents exposed to an acute challenge

Kainic acid (lesional model), bicuculline and febrile seizures (non lesional models) [28, 87, 114, 119]

   Status epilepticus [24, 64]

   Electrical rapid kindling [88, 5, 6]

Chronic recurrent seizures reduced in

   mTLE mouse model [66, 67]

   SWD in GAERS & WAG/Rij (absence seizures models) [1, 49]

Other inflammatory signaling contributing to seizures are mediated by

   TNF-α, IL-6, COX-2 & complement system (*reviewed in* [50, 115, 3])

glial cells, inducing a cascade of events that alters their structure and function in a variety of ways that can also contribute to aberrant excitability [99].

In animal models, pharmacological intervention to block or activate specific inflammatory pathways induced in human epilepsy brain specimens has shown that: (i) cytokines such as IL-1β, TNF-α, and IL-6, and danger signals such as HMGB1 and S100β, contribute to seizures in a receptor-dependent manner; (ii) the complement system contributes to seizure generation and cell loss; and (iii) PGE2 contributes to cell loss by activating EP2 receptors in neurons (Table 14.2). This set of evidence is corroborated by the assessment of susceptibility to seizures and cell loss in transgenic mouse models with impaired or overexpressed inflammatory signalings [118].

14.1.2 IL-1β, HMGB1 and the NMDA and GABA Receptors

IL-1β and HMGB1 both potentiate NMDA receptor function in cultured hippocampal neurons using post-translational mechanisms mediated by activation of IL-1R1 and TLR4, respectively [8, 53, 121]. In particular, these cytokines enhance NMDA-mediated Ca^{2+} influx by activating Src kinases-dependent NR2B phosphorylation (Fig. 14.2). This signaling has been demonstrated

to underlie the proictogenic and proneurotoxic properties of these cytokines [7, 8, 40, 121].

This rapid onset (within 2 min) mechanism is reminiscent of that induced by IL-1β in hypothalamic neurons, which underlies the initial rise in body temperature induced by this cytokine [23, 91, 105], and it involves MyD88-dependent and ceramide-mediated activation of Src kinases. IL-1β also down-regulates AMPA receptor expression and their phosphorylation state in a Ca^{2+}- and NMDA-dependent manner in hippocampal neurons [53]. Recent evidence shows that HMGB1 effects on neuronal excitability may also include a physical, receptor unrelated, interaction with presynaptic NMDA receptors resulting in enhanced Ca^{2+}-dependent glutamate release from presynaptic terminals evoked upon NMDAR stimulation [80]. Notably, HMGB1 per se can also induce glutamate release from hippocampal gliosome preparations implying that this molecule may increase gliotransmission [81]. While the effect of IL-1β and HMGB1 on NMDA-induced Ca^{2+}-influx in neuronal cell soma and dendrites mediates cell loss and increases seizures [7, 8, 121], whether the effect of HMGB1 on presynaptic or glial glutamate release results in pathologic outcomes has not been yet investigated.

Excitatory actions of IL-1β have been reported in hippocampal slices or cultured pyramidal neurons where the cytokine reduces synaptically-mediated GABA inhibition in CA3 hippocampal region via still unidentified kinases [123, 129], and increases CA1 neurons excitability by reducing NMDA-induced outward current. This latter action involves activation of cytoplasmatic P38 MAPK phosphorylating large-conductance Ca^{2+}-dependent K channels [131].

14.1.3 Cytokines, Synaptic Transmission/Plasticity and Seizures

Cytokine receptors are expressed by the same resident CNS cells that express their cognate cytokines, namely neurons, microglia, and astrocytes. Binding of ligands to these receptors set

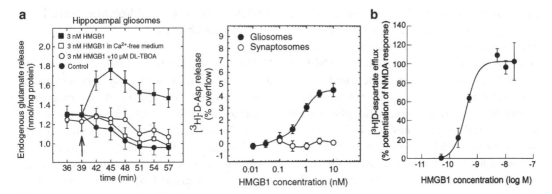

Fig. 14.2 *Presynaptic and postsynaptic effects of HMGB1 on glutamatergic transmission.* HMGB1 protein evokes (^{3}H)D-aspartate and glutamate release from re-sealed glial (*gliosomes*) and neuronal (*synaptosomes*) subcellular particles isolated from the mouse hippocampus (**a**). This protein per se augments the calcium-independent neurotransmitter outflow from gliosomes, but not from synaptosomes, in a concentration-dependent manner. This outflow is likely mediated by reversal of glutamate transporter (GLAST) since it is blocked by DL-threo-b-benzyloxyaspartate (TBOA) [81]. HMGB1 augments the NMDA-induced (^{3}H)D-aspartate calcium-dependent release from synaptosomes (**b**). This enhancing effect is mediated by increased intracellular calcium via the MK-801 sensitive channel. This HMGB1-NMDA receptor interaction involves the NR2B subunit [80]

into motion a variety of signaling pathways that activate glial cells and can also lead to enhanced excitability of neurons.

IL-1β. In the hippocampus, IL-1β was reported to induce rapid changes in synaptic transmission, and to inhibit LTP via activation of MAPK and PKC [12, 75, 84, 96]. Fast neuronal actions of IL-1β were described in the preoptic/anterior hypothalamic neurons involving A-type K$^+$ currents and the consequent reduced synaptic release of GABA [105].

TNF-α. Work by Stellwagen et al. demonstrated that TNF-α released by astrocytes binds to the TNF-α receptors (TNFR) on neurons and induces an increase in AMPA-type glutamate receptors and a concomitant decrease of GABA$_A$ receptors at synapses [102]. Specifically, TNF-α has been shown to increase trafficking of GluR2-lacking AMPA receptors to synaptic membranes in both hippocampal and motor neurons [11, 55, 56, 102, 103, 126]. In hippocampal neurons, this trafficking has been shown to depend on the PI3K–Akt pathway [102]. GluR2-lacking receptors are permeable to Ca^{2+} and activation of these receptors could dramatically alter synaptic strengths at these synapses or contribute to excitotoxicity. While TNFR knock out mice do not appear to have impaired long term potentiation (LTP) or long term depression (LTD), synaptic scaling may be modulated by TNF-α [101, 103]. While it is currently unclear what role TNF-α signaling may be playing in receptor trafficking in epilepsy, recent work using the Theiler's Murine Encephalomyelitis Virus (TMEV) model of TLE has demonstrated that there is over a 120-fold increase in whole brain TNF-α mRNA soon after infection in C57Bl/6 mice [47]. This dramatic increase in TNF-α expression is associated with acute seizures and changes in mEPSC amplitudes and decay times in hippocampal brain slices prepared from animals acutely infected with TMEV [57, 98, 104]. In addition, TNFR1 knockout mice are much less likely to exhibit seizures during the acute infection period. Taken together, the evidence suggests an important role of TNF-α in modulating excitatory circuits and excessive amounts of TNF-α may contribute to seizure activity. Accordingly, a proictogenic role of TNF-α mediated by TNFR1, and an opposite anti-ictogenic role of this cytokine mediated by TNFR2 have been reported in chemoconvulsant models of seizures [7–9, 124]. Molecular and functional interactions between TNFR and the glutamatergic system in the hippocampus appear to be implicated in the effect of this cytokine in seizure susceptibility [8].

In addition to modifying synaptic transmission, TNF-α is also known to stimulate the release of glutamate from microglia [17, 107] and astrocytes [92, 93], and these additional sources of extracellular glutamate likely contribute to excitotoxicity in injured brain regions. Activation of TNFR in cultured microglia results in an increased expression of glutaminase, which converts glutamine to glutamate. This excess intracellular glutamate is then released through connexin 36 hemi-channels and can be blocked by the gap junction inhibitor, carbenoxolone [107]. It is thought that this mechanism can contribute to neuronal cell death that often accompanies chronic or prolonged tissue inflammation.

IL-6. Recent work has demonstrated that IL-6, another cytokine that is increased in response to epileptogenic insults, decreases GABA and glycine-mediated inhibitory synaptic currents following bath application to spinal cord slices [46]. Such changes in synaptic neurotransmitter receptor function can result in tipping the balance of excitation and inhibition towards hyperexcitability. Binding of IL-6 to its receptor results in the activation of the JAK/STAT pathway and this pathway is known to regulate the expression of many different receptor gated ion channel subunits [60] and underlies NMDA-dependent LTD in the hippocampus [72]. Therefore, changes in IL-6 expression levels could dramatically influence excitability of neural circuits responsible for seizure generation. Recent work with the TMEV mouse model of TLE, demonstrated that IL-6 mRNA expression increases significantly during the acute infection period and this increase parallels the onset of seizures in this model. Furthermore, IL-6 receptor knockout mice have a reduced incidence of seizures following TMEV infection, suggesting that this cytokine, which is largely expressed in this animal model by infiltrating macrophages, contributes to lowering seizure thresholds [21, 47]. Finally, treatment of TMEV infected mice with either minocycline or wogonin, were both found to dramatically reduce concomitantly the number of infiltrating macrophages in the brain and seizure incidence [21]. These results suggest that IL-6 may be an important regulator, possibly through the JAK/STAT pathway, of synaptic plasticity and seizure activity.

14.1.4 Cytokines and Voltage-Gated Ion Channels

While cytokines have been extensively studied in neuropathic pain and in epilepsy, very few studies have examined the effects of the prominent cytokines on voltage gated ion channels (see [122]). Nevertheless, the limited available literature demonstrates that cytokines can modulate a variety of voltage gated ion channels through multiple mechanisms [95]. For example, TNF-α has been shown to increase expression of TTX resistant sodium channels in isolated dorsal root ganglion cells, increase Ca^{2+} currents in cultured hippocampal neurons and decrease inwardly rectifying K^+ currents in cultured cortical astrocytes [35, 44, 48]. IL-1β has been shown to decrease Ca^{2+} currents in cultured hippocampal and cortical neurons [83, 84, 132, 133] as well as Na^+ and K^+ currents in dissociated retinal ganglion cells [26].

The effect of cytokines on ion channel function is an area where clearly further work is necessary so as to inform hypotheses about the full range of activity of cytokines in epilepsy, particularly in view of the plethora of differing effects on neuronal functions that cytokines may have depending on their concentration, timing of tissue exposure, the type of neuronal cells expressing the relevant receptors, and the concomitant presence of other neuromodulatory molecules.

14.1.5 Prostaglandins, Synaptic Plasticity and Seizure Activity

Arachidonic acid (AA) is converted to prostanoids via activity of the enzyme cyclooxygenase (COX). COX-2 is constitutively active at low levels in the hippocampus, its expression rapidly increases as a consequence of neural activity, and is necessary for some forms of synaptic plasticity, such as LTP in the dentate gyrus [42]. Prostaglandin E2 (PGE2), one of the most common of the prostanoids to be formed in the hippocampus, binds to the G-protein coupled EP2 receptor on neurons, activates cAMP and mediates synaptic plasticity via the cAMP–protein kinase A (PKA)–cAMP-responsive element binding protein (CREB) pathway [42, 116]. Following status epilepticus (SE), COX-2 expression is increased in the hippocampus and prostaglandins, including PGE2, are also subsequently increased and hypothesized to be involved in mediating neurodegeneration that occurs in multiple brain regions following SE. This neurotoxic effect may be due to excessive stimulation of EP2 receptors expressed by microglia and the consequent activation of an alternative pathway, the cAMP-Epac signaling pathway promoting upregulation of various inflammatory mediators and oxidative stress [42]. Whereas pharmacological inhibition of COX-2 can be neuroprotective following CNS insults, this approach has not yielded great success in preventing the development of epilepsy following SE although disease-modifying effects have been reported [45, 51, 61, 85]. Depending on the drug used to inhibit COX-2 and the trigger of SE, adverse events have also been described in epileptic rats [39, 85]. Therefore, the search is on for drugs that can selectively interfere with downstream pathways of COX-2 in an effort to mitigate the detrimental inflammatory actions that can occur in the CNS following SE. Recently, Jiang et al. evaluated the ability of a novel small molecule and brain permeable EP2 antagonist, TG6-10-1, to confer neuroprotection and prevent the development of epilepsy in mice treated with pilocarpine [43]. Encouragingly, there was significant neuroprotection and decreased mortality following SE in the treated mice. However, there were no differences with vehicle-treated mice in spontaneous seizure frequency, suggesting that epileptogenesis was not interrupted with this treatment [43]. This suggests that adjunctive therapy with an EP2 antagonist may be important for attaining neuroprotection in patients experiencing SE, but additional approaches will be necessary to prevent the development of epilepsy. In this context, a recent study reported that co-treatment with IL-1 receptor antagonist (IL-1Ra, anakinra) and a COX-2 inhibitor given at the time of SE induction were required to reduce both cell loss and epileptogenesis in rats [52]. Similarly, combined treatment with IL-1Ra and VX-765, an inhibitor of IL-1β biosynthesis, given systemically to rats after 3 h of uninterrupted

SE, afforded significant neuroprotection although not inhibiting epilepsy development [74]. This evidence highlights the need of both early intervention and combined anti-inflammatory treatments for optimizing beneficial clinical outcomes.

Another strategy to be investigated is a combination of specific antiinflammatory drugs with classical antiepileptic drugs (AED) targeting complementary mechanisms. Indeed, some AEDs afford neuroprotection or decrease the severity of spontaneous seizures induced in SE models [71].

14.1.6 TLR4 and Neuronal Excitability

Out of 11 members of the TLRs family, TLR4 is the most extensively studied in CNS for its involvement in increasing brain excitability and cell loss, and for reducing neurogenesis.

Rat cortical application of lipopolysaccharide (LPS), a PAMP component of gram-negative bacteria wall and prototypical activator of TLR4, has been reported to rapidly increase the excitability of local neurons as assessed by measuring amplitudes of sensory evoked field potentials following rat forepaw stimulation and spontaneous activity [90]. A ten-fold higher LPS concentration could evoke epileptiform activity which was prevented by pre-application of IL-1Ra, implicating a role of IL-1β released from LPS-activated microglia [90].

We recently discovered that intracerebral LPS application reduces hyperpolarization-activated ion channel (HCN1) protein in hippocampal tissue, an effect associated with a reduction in Ih current as assessed in whole-cell patch recording of CA1 pyramidal neurons. This effect is long-lasting but reversible upon resolution of both microglia activation and induction of proinflammatory cytokines in these cells. The activation of IL-1R1/TLR4 signaling is responsible for this effect since it was precluded in TLR4 or IL-1R1 knock-out mice, and by pharmacological blockade of these receptors with selective antagonists (Bernard et al., 2013, personal communication).

The reported LTP and LTD impairment induced by TLR4 stimulation is compatible with neurological dysfunction and cognitive deficits induced by early life exposure to LPS which are associated with specific and persistent changes in NMDA receptor subunits expression in the cortex and hippocampus, predicting modifications in CNS excitability (for review see [89, 127]).

14.1.7 Inflammation-Induced Functional Changes in Astrocytes

Reactive astrogliosis occurs as a consequence of cytokine activation of the IL-1R/TLR and JAK/STAT pathway and other signaling pathways following CNS insults such as traumatic brain injury (TBI), SE, and infection [99]. Astrogliosis is a graded process and is characterized by hypertrophy of primary processes, dramatic increases in the expression of intermediate filament proteins such as glial fibrillary acidic protein (GFAP), a decrease and cell redistribution in glutamine synthetase [20, 29, 78, 125], an increase in expression of adenosine kinase, and, in some cases, a disruption in domain organization of glial processes [76, 99]. There is also a dramatic increase in gap junction coupling between astrocytes in animal models [106] and resected human tissue [19, 32, 70], and a number of specific subunits of kainate receptors (KAR) were recently found to be expressed in reactive astrocytes following chemoconvulsant-induced SE in rodents [112]. There are, therefore, a multitude of changes in astrocytes following seizure-inducing insults and these changes may have a dramatic impact on the circuit dynamics underlying seizure generation [25, 36].

As astrocytes are intricately involved in regulating neuronal activity at the tri-partite synapse (review [2]), some of the changes in glial function that are observed in rodent models and human epilepsy could easily lead to hyperexcitability in neural circuits and contribute to seizure generation. For example, decreases in the endogenous anticonvulsant adenosine as a consequence of increased expression of adenosine kinase can lead

to hyperexcitability and seizure activity [4, 15] and, while early after SE, glutamate uptake by astrocytes seems to be functioning well [106], there are numerous reports of cytokine-mediated decreases in glutamate transporter function in epilepsy and other disorders which could readily lead to excess excitation and cell death in vulnerable neurons [62, 68, 86, 94]. Reactive astrocytes have also been reported to have a decrease in the inward rectifier potassium channel (K_{IR}), namely Kv4.1, a critical ion channel that aids in the buffering of extracellular potassium concentrations, and this altered expression may be mediated by IL-1β [134]. Electrophysiological recordings in acute brain slices obtained from surgical specimens of patients with mTLE, have revealed a reduced K_{IR} conductance in reactive astrocytes [38]. However, we recently demonstrated that K_{IR} mediated currents were not altered in astrocytes during the latent period up to 2 weeks following SE in the KA-treated rat [106], and this is consistent with a recent report demonstrating that initial decreases in Kv4.1 mRNA and protein return to control levels by day 7 after SE [134]. Therefore, reactive astrocyte function may change over time as epilepsy develops.

While many of the observed changes in astrocytes that occur as a consequence of inflammation may actively contribute to network hyperexcitability, other components of reactive astrogliosis, such as increased gap junctional coupling, or increased neurotrophins may be critical compensatory mechanisms following injury, and may act to dampen excitability and protect neurons [36]. Thus, simply blocking the inflammatory response in glial cells may be too global an approach for disease modification during epileptogenesis, while targeting specific processes, such as maintaining K_{IR} function, might prove to be a more useful approach.

14.1.8 Cytokines Effects on BBB: Consequences for Neuronal Excitability

Evidence obtained using in vitro models of the BBB [31, 130] or epilepsy models [58, 77, 111, 116] demonstrated that cytokines and prostaglandins compromise the permeability properties of the BBB, and that such alteration in brain vessels is a common feature of drug-resistant epileptogenic foci in humans and experimental models. In particular, there is evidence of the presence of IL-1β in perivascular glia and astrocytic endfeet impinging on brain vessels in epilepsy tissue where the BBB is altered, as shown by the parenchymal extravasation of serum macromolecules such as albumin and IgG. One mechanism of BBB damage induced by cytokines involves breakdown of tight-junction proteins in brain vessels [58, 59, 69, 73] induced by activation of Src kinases. This evidence highlights that key molecular pathways activated by cytokines in epilepsy result in different outcomes depending on the target cell population (expressing the relevant receptors), i.e. BBB permeability function is compromised in vessels, hyperexcitability is induced in neurons, and astrocyte function is greatly modified.

BBB damage leads to albumin extravasation which induces TGF-β signaling in astrocytes by activating the TGF-β receptor type 2 [33]. This signaling mediates transcriptional up-regulation of IL-1β and other inflammatory genes in astrocytes [16, 34] while glutamate transporter and Kir4.1 channels are down-regulated. These pathologic changes have been shown to establish a hyperexcitable milieu in surrounding neurons due to increased extracellular K+ and glutamate [97] which decreases seizure threshold and may induce per se epileptiform activity [22, 34].

14.1.9 Leukocytes, Autoantibodies and Neuronal Excitability

There is evidence of adaptive immunity activation in rare disorders such as Rasmussen's encephalitis (RE), viral and limbic encephalitis and neurologic or systemic autoimmune disorders. These conditions are often associated with seizures and epilepsy development. In RE brain tissue, cytotoxic CD8+ T lymphocytes have been demonstrated in close apposition to neurons and astrocytes, then provoking their apoptosis by releasing granzyme

B [10, 79]. The presence of these cells, and more in general CD3+ leuckocytes, appears to be much less prominent in more common forms of epilepsy. For example, in focal cortical dysplasia (FCD) type 2, scattered lymphocytes have been described in brain tissue while this phenomenon occurs at a minor extent in FCD type 1, and is almost undetectable in mTLE [41, 65, 110]. Others have detected leukocytes in brain parenchyma surrounding brain vessels also in mTLE [30, 128]. In animal models of epilepsy the role of these cells is still uncertain since they were reported to mediate anti-epileptogenic and neuroprotective effects in KA-treated rats [128] whereas they contribute to the pathology in pilocarpine-treated mice [30]. Notably, in this latter instance the effects of leukocytes may be ascribed to the peculiar mechanisms mediating seizures caused by pilocarpine and which are not shared by other chemoconvulsants [64, 109, 117].

A recent randomized clinical study using tacrolimus, which impedes T cell proliferation and activation, in recent onset RE patients showed delayed deterioration of neurological deficits but the treatment did not ameliorate drug resistant seizures [13]. However, case reports have shown decreased seizure frequency in one RE patient treated with natalizumab, a blocker of T cell entry into the CNS [14] and in a patient with multiple sclerosis and refractory epilepsy [101]. The authors discussed that interpretation of data was limited by an additional coadministration of varying antiepileptic medications.

In limbic encephalitis and autoimmune disorders, circulating autoantibodies against various neuronal proteins have been detected (for review, see [120]). These antibodies recognizing membrane neuronal proteins may have a pathologic role, in addition to their diagnostic value. In particular, antibodies against NR1/NR2 subunits obtained from serum of affected patients can increase extracellular hippocampal glutamate levels when intracerebrally infused in rats. Increased sensitivity to AMPA receptor-mediated neuronal excitability and GABAergic dysfunction have also been reported [63]. Antibodies directed against voltage-gated K+ channel complex increase excitability of hippocampal CA3 pyramidal cells by reducing channel function at mossy fiber-CA3 synapses [54]. AMPA receptor antibodies alter synaptic receptor location and number by reducing those receptors containing the GLUR2 subunit, therefore increasing the relative abundance of Ca^{2+}-permeable receptors [53].

14.2 Conclusions

While understanding of the role of the innate immune system and the associated molecules with inflammatory properties in epilepsy and seizure threshold changes has advanced tremendously over the last decade, there are still a number of questions that yet remain open and require further investigation. For example, it is not yet clear which molecules and inflammatory pathways activated following epileptogenic brain insults will make the most appropriate targets for intervening to prevent seizure occurrence and/or the process of epileptogenesis. The complex network changes that occur in a number of cell types in the CNS, including neurons, microglia and astrocytes, in response to increases in a myriad of neuromodulatory and inflammatory molecules such as IL-1β, TNF-α, IL-6 and interferon-γ to name but a few, are difficult to decipher. Moreover, it has still to be determined which are the master regulators of the inflammatory cascade, and when and how to prevent the induction of brain inflammation or rather promote its resolution by implementing the effects of the endogenous antiinflammatory molecules, which are defective in epilepsy [82, 87].

Nevertheless, the increasing recognition that the innate immune system is tightly coupled to epileptogenesis and seizure threshold changes is encouraging as it opens up many potential novel molecular targets for therapeutics. Most AEDs are mainly antiseizure, symptomatic drugs that target neuronal proteins such as sodium channels or glutamate receptors. Their adverse effects on cognition and induction of sedation, coupled with the knowledge that nearly 30 % of patients with epilepsy do not have their seizures adequately controlled with current AEDs, suggest that targeting the neuromodulatory inflammatory pathways is a promising novel strategy with disease-modifying

potential. Considering that prolonged administration in epilepsy is likely to be required, and the constraints imposed by the BBB, both the efficacy and the safety of drugs that preclude or reverse the over-activation of specific innate immune mechanisms should be carefully considered. Importantly, some of these antiinflammatory drugs are already in clinical use showing therapeutic effects in peripheral inflammatory conditions [27, 37, 113]. These drugs might be considered to complement the symptomatic treatment provided by available AEDs for resolving the inflammatory processes in the brain, therefore raising seizure threshold and decreasing the likelihood of seizure recurrence. In this context, a phase 2 clinical study with VX765 has given promising results in adult patients with drug resistant partial onset seizures (http://clinicaltrials.gov/ct2/show/NCT01048255; www.epilepsy.com/files/Pipeline2012/6-7).

Acknowledgements A. Vezzani is very grateful to Phil Schwarzkroin for the intense and fruitful collaboration during the 8 years of shared editorial work for Epilepsia. During the time A.V. was serving as associated editor for basic science, she could fully appreciate P.S. extensive and deep scientific knowledge, his patience, constructive criticism, support and commitment. She thanks Phil, in particular, for his willingness to share responsibility and decisions as well as complaints and rewards, and for the very educational and formative time.

K.S. Wilcox would like to acknowledge the kind support Phil Schwartzkroin has demonstrated over the years. While she never had the opportunity to work with him, he was always interested in her work, encouraged her to speak out at meetings, to get involved in the American Epilepsy Society, and provided a welcoming environment for someone entering the field of epilepsy. He is a mentor to all.

Other Acknowledgements This work was supported by National Institutes of Health Grants NS 078331 (KSW), NS065434 (KSW), and the Margolis Foundation (KSW) and Ministero della salute Grant N. RF-2009-1506142 (AV).

References

1. Akin D, Ravizza T, Maroso M, Carcak N, Eryigit T, Vanzulli I et al (2011) IL-1beta is induced in reactive astrocytes in the somatosensory cortex of rats with genetic absence epilepsy at the onset of spike-and-wave discharges, and contributes to their occurrence. Neurobiol Dis 44(3):259–269

2. Araque A, Parpura V, Sanzgiri RP, Haydon PG (1999) Tripartite synapses: glia, the unacknowledged partner. Trends Neurosci 22(5):208–215

3. Aronica E, Ravizza T, Zurolo E, Vezzani A (2012) Astrocyte immune responses in epilepsy. Glia 60(8):1258–1268

4. Aronica E, Sandau US, Iyer A, Boison D (2013) Glial adenosine kinase – a neuropathological marker of the epileptic brain. Neurochem Int 63:688–695

5. Auvin S, Mazarati A, Shin D, Sankar R (2010) Inflammation enhances epileptogenesis in the developing rat brain. Neurobiol Dis 40(1):303–310

6. Auvin S, Shin D, Mazarati A, Sankar R (2010) Inflammation induced by LPS enhances epileptogenesis in immature rat and may be partially reversed by IL1RA. Epilepsia 51(Suppl 3):34–38

7. Balosso S, Maroso M, Sanchez-Alavez M, Ravizza T, Frasca A, Bartfai T et al (2008) A novel non-transcriptional pathway mediates the proconvulsive effects of interleukin-1beta. Brain 131(Pt 12):3256–3265

8. Balosso S, Ravizza T, Aronica E, Vezzani A (2013) The dual role of TNF-alpha and its receptors in seizures. Exp Neurol 247C:267–271

9. Balosso S, Ravizza T, Perego C, Peschon J, Campbell IL, De Simoni MG et al (2005) Tumor necrosis factor-alpha inhibits seizures in mice via p75 receptors. Ann Neurol 57(6):804–812

10. Bauer J, Vezzani A, Bien CG (2012) Epileptic encephalitis: the role of the innate and adaptive immune system. Brain Pathol 22(3):412–421

11. Beattie EC, Stellwagen D, Morishita W, Bresnahan JC, Ha BK, Von Zastrow M et al (2002) Control of synaptic strength by glial TNFalpha. Science 295(5563):2282–2285

12. Bellinger FP, Madamba S, Siggins GR (1993) Interleukin 1 beta inhibits synaptic strength and long-term potentiation in the rat CA1 hippocampus. Brain Res 628(1–2):227–234

13. Bien CG, Tiemeier H, Sassen R, Kuczaty S, Urbach H, von Lehe M et al (2013) Rasmussen encephalitis: incidence and course under randomized therapy with tacrolimus or intravenous immunoglobulins. Epilepsia 54(3):543–550

14. Bittner S, Simon OJ, Gobel K, Bien CG, Meuth SG, Wiendl H (2013) Rasmussen encephalitis treated with natalizumab. Neurology 81(4):395–397

15. Boison D (2013) Adenosine kinase: exploitation for therapeutic gain. Pharmacol Rev 65(3):906–943

16. Cacheaux LP, Ivens S, David Y, Lakhter AJ, Bar-Klein G, Shapira M et al (2009) Transcriptome profiling reveals TGF-beta signaling involvement in epileptogenesis. J Neurosci 29(28):8927–8935

17. Chen CJ, Ou YC, Chang CY, Pan HC, Liao SL, Chen SY et al (2012a) Glutamate released by Japanese encephalitis virus-infected microglia involves TNF-alpha signaling and contributes to neuronal death. Glia 60(3):487–501

18. Chen J, Tsai V, Parker WE, Aronica E, Baybis M, Crino PB (2012b) Detection of human papillo-mavirus in human focal cortical dysplasia type IIB. Ann Neurol 72(6):881–892

19. Collignon F, Wetjen NM, Cohen-Gadol AA, Cascino GD, Parisi J, Meyer FB et al (2006) Altered expression of connexin subtypes in mesial temporal lobe epilepsy in humans. J Neurosurg 105(1):77–87

20. Coulter DA, Eid T (2012) Astrocytic regulation of glutamate homeostasis in epilepsy. Glia 60(8):1215–1226

21. Cusick MF, Libbey JE, Patel DC, Doty DJ, Fujinami RS (2013) Infiltrating macrophages are key to the development of seizures following virus infection. J Virol 87(3):1849–1860

22. David Y, Cacheaux LP, Ivens S, Lapilover E, Heinemann U, Kaufer D et al (2009) Astrocytic dysfunction in epileptogenesis: consequence of altered potassium and glutamate homeostasis? J Neurosci 29(34):10588–10599

23. Davis CN, Tabarean I, Gaidarova S, Behrens MM, Bartfai T (2006) IL-1beta induces a MyD88-dependent and ceramide-mediated activation of Src in anterior hypothalamic neurons. J Neurochem 98(5):1379–1389

24. De Simoni MG, Perego C, Ravizza T, Moneta D, Conti M, Marchesi F et al (2000) Inflammatory cytokines and related genes are induced in the rat hippocampus by limbic status epilepticus. Eur J Neurosci 12(7):2623–2633

25. Devinsky O, Vezzani A, Najjar S, De Lanerolle NC, Rogawski MA (2013) Glia and epilepsy: excitability and inflammation. Trends Neurosci 36(3):174–184

26. Diem R, Hobom M, Grotsch P, Kramer B, Bahr M (2003) Interleukin-1 beta protects neurons via the interleukin-1 (IL-1) receptor-mediated Akt pathway and by IL-1 receptor-independent decrease of transmembrane currents in vivo. Mol Cell Neurosci 22(4):487–500

27. Dinarello CA, Simon A, van der Meer JW (2012) Treating inflammation by blocking interleukin-1 in a broad spectrum of diseases. Nat Rev Drug Discov 11(8):633–652

28. Dube C, Vezzani A, Behrens M, Bartfai T, Baram TZ (2005) Interleukin-1beta contributes to the generation of experimental febrile seizures. Ann Neurol 57(1):152–155

29. Eid T, Thomas MJ, Spencer DD, Runden-Pran E, Lai JC, Malthankar GV et al (2004) Loss of glutamine synthetase in the human epileptogenic hippocampus: possible mechanism for raised extracellular glutamate in mesial temporal lobe epilepsy. Lancet 363(9402):28–37

30. Fabene PF, Navarro Mora G, Martinello M, Rossi B, Merigo F, Ottoboni L et al (2008) A role for leukocyte-endothelial adhesion mechanisms in epilepsy. Nat Med 14(12):1377–1383

31. Ferrari CC, Depino AM, Prada F, Muraro N, Campbell S, Podhajcer O et al (2004) Reversible demyelination, blood–brain barrier breakdown, and pronounced neutrophil recruitment induced by chronic IL-1 expression in the brain. Am J Pathol 165(5):1827–1837

32. Fonseca CG, Green CR, Nicholson LF (2002) Upregulation in astrocytic connexin 43 gap junction levels may exacerbate generalized seizures in mesial temporal lobe epilepsy. Brain Res 929(1): 105–116

33. Friedman A, Kaufer D, Heinemann U (2009) Blood–brain barrier breakdown-inducing astrocytic transformation: novel targets for the prevention of epilepsy. Epilepsy Res 85(2–3):142–149

34. Frigerio F, Frasca A, Weissberg I, Parrella S, Friedman A, Vezzani A et al (2012) Long-lasting pro-ictogenic effects induced in vivo by rat brain exposure to serum albumin in the absence of concomitant pathology. Epilepsia 53(11):1887–1897

35. Furukawa K, Mattson MP (1998) The transcription factor NF-kappaB mediates increases in calcium currents and decreases in NMDA- and AMPA/kainate-induced currents induced by tumor necrosis factor-alpha in hippocampal neurons. J Neurochem 70(5):1876–1886

36. Gibbons MB, Smeal RM, Takahashi DK, Vargas JR, Wilcox KS (2013) Contributions of astrocytes to epileptogenesis following status epilepticus: opportunities for preventive therapy? Neurochem Int 63:660–669

37. Hennessy EJ, Parker AE, O'Neill LA (2010) Targeting Toll-like receptors: emerging therapeutics? Nat Rev Drug Discov 9(4):293–307

38. Hinterkeuser S, Schroder W, Hager G, Seifert G, Blumcke I, Elger CE et al (2000) Astrocytes in the hippocampus of patients with temporal lobe epilepsy display changes in potassium conductances. Eur J Neurosci 12(6):2087–2096

39. Holtman L, van Vliet EA, Edelbroek PM, Aronica E, Gorter JA (2010) Cox-2 inhibition can lead to adverse effects in a rat model for temporal lobe epilepsy. Epilepsy Res 91(1):49–56

40. Iori V, Maroso M, Rizzi M, Iyer AM, Vertemara R, Carli M et al (2013) Receptor for advanced glycation endproducts is upregulated in temporal lobe epilepsy and contributes to experimental seizures. Neurobiol Dis 58:102–114

41. Iyer A, Zurolo E, Spliet WG, van Rijen PC, Baayen JC, Gorter JA et al (2010) Evaluation of the innate and adaptive immunity in type I and type II focal cortical dysplasias. Epilepsia 51(9):1763–1773

42. Jiang J, Dingledine R (2013) Prostaglandin receptor EP2 in the crosshairs of anti-inflammation, anti-cancer, and neuroprotection. Trends Pharmacol Sci 34(7):413–423

43. Jiang J, Quan Y, Ganesh T, Pouliot WA, Dudek FE, Dingledine R (2013) Inhibition of the prostaglandin receptor EP2 following status epilepticus reduces delayed mortality and brain inflammation. Proc Natl Acad Sci U S A 110(9):3591–3596

44. Jin X, Gereau RW (2006) Acute p38-mediated modulation of tetrodotoxin-resistant sodium channels in mouse sensory neurons by tumor necrosis factor-alpha. J Neurosci 26(1):246–255

45. Jung KH, Chu K, Lee ST, Kim J, Sinn DI, Kim JM et al (2006) Cyclooxygenase-2 inhibitor, celecoxib, inhibits the altered hippocampal neurogenesis with attenuation of spontaneous recurrent seizures following pilocarpine-induced status epilepticus. Neurobiol Dis 23(2):237–246

46. Kawasaki Y, Zhang L, Cheng JK, Ji RR (2008) Cytokine mechanisms of central sensitization: distinct and overlapping role of interleukin-1beta, interleukin-6, and tumor necrosis factor-alpha in regulating synaptic and neuronal activity in the superficial spinal cord. J Neurosci 28(20): 5189–5194

47. Kirkman NJ, Libbey JE, Wilcox KS, White HS, Fujinami RS (2010) Innate but not adaptive immune responses contribute to behavioral seizures following viral infection. Epilepsia 51(3):454–464

48. Koller H, Allert N, Oel D, Stoll G, Siebler M (1998) TNF alpha induces a protein kinase C-dependent reduction in astroglial K+ conductance. Neuroreport 9(7):1375–1378

49. Kovacs Z, Czurko A, Kekesi KA, Juhasz G (2011) Intracerebroventricularly administered lipopolysaccharide enhances spike-wave discharges in freely moving WAG/Rij rats. Brain Res Bull 85(6): 410–416

50. Kulkarni SK, Dhir A (2009) Cyclooxygenase in epilepsy: from perception to application. Drugs Today (Barc) 45(2):135–154

51. Kunz T, Oliw EH (2001) The selective cyclooxygenase-2 inhibitor rofecoxib reduces kainate-induced cell death in the rat hippocampus. Eur J Neurosci 13(3):569–575

52. Kwon YS, Pineda E, Auvin S, Shin D, Mazarati A, Sankar R (2013) Neuroprotective and antiepileptogenic effects of combination of anti-inflammatory drugs in the immature brain. J Neuroinflammation 10:30

53. Lai AY, Swayze RD, El-Husseini A, Song C (2006) Interleukin-1 beta modulates AMPA receptor expression and phosphorylation in hippocampal neurons. J Neuroimmunol 175(1–2):97–106

54. Lalic T, Pettingill P, Vincent A, Capogna M (2011) Human limbic encephalitis serum enhances hippocampal mossy fiber-CA3 pyramidal cell synaptic transmission. Epilepsia 52(1):121–131

55. Leonoudakis D, Braithwaite SP, Beattie MS, Beattie EC (2004) TNFalpha-induced AMPA-receptor trafficking in CNS neurons; relevance to excitotoxicity? Neuron Glia Biol 1(3):263–273

56. Leonoudakis D, Zhao P, Beattie EC (2008) Rapid tumor necrosis factor alpha-induced exocytosis of glutamate receptor 2-lacking AMPA receptors to extrasynaptic plasma membrane potentiates excitotoxicity. J Neurosci 28(9):2119–2130

57. Libbey JE, Kirkman NJ, Smith MC, Tanaka T, Wilcox KS, White HS et al (2008) Seizures following picornavirus infection. Epilepsia 49(6): 1066–1074

58. Librizzi L, Noe F, Vezzani A, de Curtis M, Ravizza T (2012) Seizure-induced brain-borne inflammation sustains seizure recurrence and blood–brain barrier damage. Ann Neurol 72(1):82–90

59. Librizzi L, Regondi MC, Pastori C, Frigerio S, Frassoni C, de Curtis M (2007) Expression of adhesion factors induced by epileptiform activity in the endothelium of the isolated guinea pig brain in vitro. Epilepsia 48(4):743–751

60. Lund IV, Hu Y, Raol YH, Benham RS, Faris R, Russek SJ et al (2008) BDNF selectively regulates GABAA receptor transcription by activation of the JAK/STAT pathway. Sci Signal 1(41):ra9

61. Ma L, Cui XL, Wang Y, Li XW, Yang F, Wei D et al (2012) Aspirin attenuates spontaneous recurrent seizures and inhibits hippocampal neuronal loss, mossy fiber sprouting and aberrant neurogenesis following pilocarpine-induced status epilepticus in rats. Brain Res 1469:103–113

62. Mandolesi G, Musella A, Gentile A, Grasselli G, Haji N, Sepman H et al (2013) Interleukin-1beta alters glutamate transmission at purkinje cell synapses in a mouse model of multiple sclerosis. J Neurosci 33(29):12105–12121

63. Manto M, Dalmau J, Didelot A, Rogemond V, Honnorat J (2010) In vivo effects of antibodies from patients with anti-NMDA receptor encephalitis: further evidence of synaptic glutamatergic dysfunction. Orphanet J Rare Dis 5:31

64. Marchi N, Fan Q, Ghosh C, Fazio V, Bertolini F, Betto G et al (2009) Antagonism of peripheral inflammation reduces the severity of status epilepticus. Neurobiol Dis 33(2):171–181

65. Marchi N, Teng Q, Ghosh C, Fan Q, Nguyen MT, Desai NK et al (2010) Blood–brain barrier damage, but not parenchymal white blood cells, is a hallmark of seizure activity. Brain Res 1353:176–186

66. Maroso M, Balosso S, Ravizza T, Iori V, Wright CI, French J et al (2011) Interleukin-1beta biosynthesis inhibition reduces acute seizures and drug resistant chronic epileptic activity in mice. Neurotherapeutics 8(2):304–315

67. Maroso M, Balosso S, Ravizza T, Liu J, Aronica E, Iyer AM et al (2010) Toll-like receptor 4 and high-mobility group box-1 are involved in ictogenesis and can be targeted to reduce seizures. Nat Med 16(4):413–419

68. Mathern GW, Mendoza D, Lozada A, Pretorius JK, Dehnes Y, Danbolt NC et al (1999) Hippocampal GABA and glutamate transporter immunoreactivity in patients with temporal lobe epilepsy. Neurology 52(3):453–472

69. Morin-Brureau M, Lebrun A, Rousset MC, Fagni L, Bockaert J, de Bock F et al (2011) Epileptiform activity induces vascular remodeling and zonula occludens 1 downregulation in organotypic hippocampal cultures: role of VEGF signaling pathways. J Neurosci 31(29):10677–10688

70. Naus CC, Bechberger JF, Caveney S, Wilson JX (1991) Expression of gap junction genes in astrocytes and C6 glioma cells. Neurosci Lett 126(1):33–36

71. Nehlig A (2007) What is animal experimentation telling us about new drug treatments of status epilepticus? Epilepsia 48(Suppl 8):78–81

72. Nicolas CS, Peineau S, Amici M, Csaba Z, Fafouri A, Javalet C et al (2012) The Jak/STAT pathway is involved in synaptic plasticity. Neuron 73(2): 374–390

73. Nicoletti JN, Shah SK, McCloskey DP, Goodman JH, Elkady A, Atassi H et al (2008) Vascular endothelial growth factor is up-regulated after status epilepticus and protects against seizure-induced neuronal loss in hippocampus. Neuroscience 151(1):232–241

74. Noe FM, Polascheck N, Frigerio F, Bankstahl M, Ravizza T, Marchini S et al (2013) Pharmacological blockade of IL-1beta/IL-1 receptor type 1 axis during epileptogenesis provides neuroprotection in two rat models of temporal lobe epilepsy. Neurobiol Dis 59:183–193

75. O'Donnell E, Vereker E, Lynch MA (2000) Age-related impairment in LTP is accompanied by enhanced activity of stress-activated protein kinases: analysis of underlying mechanisms. Eur J Neurosci 12(1):345–352

76. Oberheim NA, Tian GF, Han X, Peng W, Takano T, Ransom B et al (2008) Loss of astrocytic domain organization in the epileptic brain. J Neurosci 28(13):3264–3276

77. Oby E, Janigro D (2006) The blood–brain barrier and epilepsy. Epilepsia 47(11):1761–1774

78. Ortinski PI, Dong J, Mungenast A, Yue C, Takano H, Watson DJ et al (2010) Selective induction of astrocytic gliosis generates deficits in neuronal inhibition. Nat Neurosci 13(5):584–591

79. Pardo CA, Vining EP, Guo L, Skolasky RL, Carson BS, Freeman JM (2004) The pathology of Rasmussen syndrome: stages of cortical involvement and neuropathological studies in 45 hemispherectomies. Epilepsia 45(5):516–526

80. Pedrazzi M, Averna M, Sparatore B, Patrone M, Salamino F, Marcoli M et al (2012) Potentiation of NMDA receptor-dependent cell responses by extracellular high mobility group box 1 protein. PLoS One 7(8):e44518

81. Pedrazzi M, Raiteri L, Bonanno G, Patrone M, Ledda S, Passalacqua M et al (2006) Stimulation of excitatory amino acid release from adult mouse brain glia subcellular particles by high mobility group box 1 protein. J Neurochem 99(3):827–838

82. Pernhorst K, Herms S, Hoffmann P, Cichon S, Schulz H, Sander T et al (2013) TLR4, ATF-3 and IL8 inflammation mediator expression correlates with seizure frequency in human epileptic brain tissue. Seizure 22(8):675–678

83. Plata-Salaman CR, Ffrench-Mullen JM (1992) Interleukin-1 beta depresses calcium currents in CA1 hippocampal neurons at pathophysiological concentrations. Brain Res Bull 29(2):221–223

84. Plata-Salaman CR, ffrench-Mullen JM (1994) Interleukin-1 beta inhibits Ca2+ channel currents in hippocampal neurons through protein kinase C. Eur J Pharmacol 266(1):1–10

85. Polascheck N, Bankstahl M, Loscher W (2010) The COX-2 inhibitor parecoxib is neuroprotective but not antiepileptogenic in the pilocarpine model of temporal lobe epilepsy. Exp Neurol 224(1):219–233

86. Proper EA, Hoogland G, Kappen SM, Jansen GH, Rensen MG, Schrama LH et al (2002) Distribution of glutamate transporters in the hippocampus of patients with pharmaco-resistant temporal lobe epilepsy. Brain 125(Pt 1):32–43

87. Ravizza T, Lucas SM, Balosso S, Bernardino L, Ku G, Noe F et al (2006) Inactivation of caspase-1 in rodent brain: a novel anticonvulsive strategy. Epilepsia 47(7):1160–1168

88. Ravizza T, Noe F, Zardoni D, Vaghi V, Sifringer M, Vezzani A (2008) Interleukin converting enzyme inhibition impairs kindling epileptogenesis in rats by blocking astrocytic IL-1beta production. Neurobiol Dis 31(3):327–333

89. Riazi K, Galic MA, Pittman QJ (2010) Contributions of peripheral inflammation to seizure susceptibility: cytokines and brain excitability. Epilepsy Res 89(1):34–42

90. Rodgers KM, Hutchinson MR, Northcutt A, Maier SF, Watkins LR, Barth DS (2009) The cortical innate immune response increases local neuronal excitability leading to seizures. Brain 132(Pt 9):2478–2486

91. Sanchez-Alavez M, Tabarean IV, Behrens MM, Bartfai T (2006) Ceramide mediates the rapid phase of febrile response to IL-1beta. Proc Natl Acad Sci U S A 103(8):2904–2908

92. Santello M, Bezzi P, Volterra A (2011) TNFalpha controls glutamatergic gliotransmission in the hippocampal dentate gyrus. Neuron 69(5):988–1001

93. Santello M, Volterra A (2012) TNFalpha in synaptic function: switching gears. Trends Neurosci 35(10):638–647

94. Sarac S, Afzal S, Broholm H, Madsen FF, Ploug T, Laursen H (2009) Excitatory amino acid transporters EAAT-1 and EAAT-2 in temporal lobe and hippocampus in intractable temporal lobe epilepsy. APMIS 117(4):291–301

95. Schafers M, Sorkin L (2008) Effect of cytokines on neuronal excitability. Neurosci Lett 437(3):188–193

96. Schneider H, Pitossi F, Balschun D, Wagner A, del Rey A, Besedovsky HO (1998) A neuromodulatory role of interleukin-1beta in the hippocampus. Proc Natl Acad Sci U S A 95(13):7778–7783

97. Seiffert E, Dreier JP, Ivens S, Bechmann I, Tomkins O, Heinemann U et al (2004) Lasting blood–brain barrier disruption induces epileptic focus in the rat somatosensory cortex. J Neurosci 24(36):7829–7836

98. Smeal RM, Stewart KA, Iacob E, Fujinami RS, White HS, Wilcox KS (2012) The activity within the CA3 excitatory network during Theiler's virus encephalitis is distinct from that observed during chronic epilepsy. J Neurovirol 18(1):30–44

99. Sofroniew MV (2009) Molecular dissection of reactive astrogliosis and glial scar formation. Trends Neurosci 32(12):638–647

100. Sotgiu S, Murrighile MR, Constantin G (2010) Treatment of refractory epilepsy with natalizumab in a patient with multiple sclerosis. Case report. BMC Neurol 10:84

101. Steinmetz CC, Turrigiano GG (2010) Tumor necrosis factor-alpha signaling maintains the ability of cortical synapses to express synaptic scaling. J Neurosci 30(44):14685–14690

102. Stellwagen D, Beattie EC, Seo JY, Malenka RC (2005) Differential regulation of AMPA receptor and GABA receptor trafficking by tumor necrosis factor-alpha. J Neurosci 25(12):3219–3228

103. Stellwagen D, Malenka RC (2006) Synaptic scaling mediated by glial TNF-alpha. Nature 440(7087): 1054–1059

104. Stewart KA, Wilcox KS, Fujinami RS, White HS (2010) Development of postinfection epilepsy after Theiler's virus infection of C57BL/6 mice. J Neuropathol Exp Neurol 69(12):1210–1219

105. Tabarean IV, Korn H, Bartfai T (2006) Interleukin-1beta induces hyperpolarization and modulates synaptic inhibition in preoptic and anterior hypothalamic neurons. Neuroscience 141(4):1685–1695

106. Takahashi DK, Vargas JR, Wilcox KS (2010) Increased coupling and altered glutamate transport currents in astrocytes following kainic-acid-induced status epilepticus. Neurobiol Dis 40(3):573–585

107. Takeuchi H, Jin S, Wang J, Zhang G, Kawanokuchi J, Kuno R et al (2006) Tumor necrosis factor-alpha induces neurotoxicity via glutamate release from hemichannels of activated microglia in an autocrine manner. J Biol Chem 281(30):21362–21368

108. Theodore WH, Epstein L, Gaillard WD, Shinnar S, Wainwright MS, Jacobson S (2008) Human herpes virus 6B: a possible role in epilepsy? Epilepsia 49(11):1828–1837

109. Uva L, Librizzi L, Marchi N, Noe F, Bongiovanni R, Vezzani A et al (2008) Acute induction of epileptiform discharges by pilocarpine in the in vitro isolated guinea-pig brain requires enhancement of blood–brain barrier permeability. Neuroscience 151(1):303–312

110. van Gassen KL, de Wit M, Koerkamp MJ, Rensen MG, van Rijen PC, Holstege FC et al (2008) Possible role of the innate immunity in temporal lobe epilepsy. Epilepsia 49(6):1055–1065

111. van Vliet EA, Zibell G, Pekcec A, Schlichtiger J, Edelbroek PM, Holtman L et al (2010) COX-2 inhibition controls P-glycoprotein expression and promotes brain delivery of phenytoin in chronic epileptic rats. Neuropharmacology 58(2):404–412

112. Vargas JR, Takahashi DK, Thomson KE, Wilcox KS (2013) The expression of kainate receptor subunits in hippocampal astrocytes after experimentally induced status epilepticus. J Neuropathol Exp Neurol 72(10):919–932

113. Vezzani A, Balosso S, Maroso M, Zardoni D, Noe F, Ravizza T (2010) ICE/caspase 1 inhibitors and IL-1beta receptor antagonists as potential therapeutics in epilepsy. Curr Opin Investig Drugs 11(1):43–50

114. Vezzani A, Conti M, De Luigi A, Ravizza T, Moneta D, Marchesi F et al (1999) Interleukin-1beta immunoreactivity and microglia are enhanced in the rat hippocampus by focal kainate application: functional evidence for enhancement of electrographic seizures. J Neurosci 19(12):5054–5065

115. Vezzani A, French J, Bartfai T, Baram TZ (2011) The role of inflammation in epilepsy. Nat Rev Neurol 7(1):31–40

116. Vezzani A, Friedman A, Dingledine RJ (2013) The role of inflammation in epileptogenesis. Neuropharmacology 69:16–24

117. Vezzani A, Janigro D (2009) Leukocyte-endothelial adhesion mechanisms in epilepsy: cheers and jeers. Epilepsy Curr 9(4):118–121

118. Vezzani A, Maroso M, Balosso S, Sanchez MA, Bartfai T (2011) IL-1 receptor/Toll-like receptor signaling in infection, inflammation, stress and neurodegeneration couples hyperexcitability and seizures. Brain Behav Immun 25(7):1281–1289

119. Vezzani A, Moneta D, Conti M, Richichi C, Ravizza T, De Luigi A et al (2000) Powerful anticonvulsant action of IL-1 receptor antagonist on intracerebral injection and astrocytic overexpression in mice. Proc Natl Acad Sci U S A 97(21):11534–11539

120. Vincent A, Irani SR, Lang B (2011) Potentially pathogenic autoantibodies associated with epilepsy and encephalitis in children and adults. Epilepsia 52(Suppl 8):8–11

121. Viviani B, Bartesaghi S, Gardoni F, Vezzani A, Behrens MM, Bartfai T et al (2003) Interleukin-1beta enhances NMDA receptor-mediated intracellular calcium increase through activation of the Src family of kinases. J Neurosci 23(25):8692–8700

122. Viviani B, Gardoni F, Marinovich M (2007) Cytokines and neuronal ion channels in health and disease. Int Rev Neurobiol 82:247–263

123. Wang S, Cheng Q, Malik S, Yang J (2000) Interleukin-1beta inhibits gamma-aminobutyric acid type A (GABA(A)) receptor current in cultured hippocampal neurons. J Pharmacol Exp Ther 292(2):497–504

124. Weinberg MS, Blake BL, McCown TJ (2013) Opposing actions of hippocampus TNFalpha receptors on limbic seizure susceptibility. Exp Neurol 247:429–437

125. Wilhelmsson U, Bushong EA, Price DL, Smarr BL, Phung V, Terada M et al (2006) Redefining the concept of reactive astrocytes as cells that remain within their unique domains upon reaction to injury. Proc Natl Acad Sci U S A 103(46):17513–17518

126. Yin HZ, Hsu CI, Yu S, Rao SD, Sorkin LS, Weiss JH (2012) TNF-alpha triggers rapid membrane insertion of Ca(2+) permeable AMPA receptors into adult motor neurons and enhances their susceptibility to slow excitotoxic injury. Exp Neurol 238(2):93–102

127. Yirmiya R, Goshen I (2011) Immune modulation of learning, memory, neural plasticity and neurogenesis. Brain Behav Immun 25(2):181–213

128. Zattoni M, Mura ML, Deprez F, Schwendener RA, Engelhardt B, Frei K et al (2011) Brain infiltration of leukocytes contributes to the pathophysiology of temporal lobe epilepsy. J Neurosci 31(11): 4037–4050

129. Zeise ML, Espinoza J, Morales P, Nalli A (1997) Interleukin-1beta does not increase synaptic inhibition in hippocampal CA3 pyramidal and dentate gyrus granule cells of the rat in vitro. Brain Res 768(1–2):341–344

130. Zhang J, Takahashi HK, Liu K, Wake H, Liu R, Maruo T et al (2011) Anti-high mobility group box-1 monoclonal antibody protects the blood–brain barrier from ischemia-induced disruption in rats. Stroke 42(5):1420–1428

131. Zhang R, Sun L, Hayashi Y, Liu X, Koyama S, Wu Z et al (2010) Acute p38-mediated inhibition of NMDA-induced outward currents in hippocampal CA1 neurons by interleukin-1beta. Neurobiol Dis 38(1):68–77

132. Zhou C, Tai C, Ye HH, Ren X, Chen JG, Wang SQ et al (2006a) Interleukin-1beta downregulates the L-type Ca2+ channel activity by depressing the expression of channel protein in cortical neurons. J Cell Physiol 206(3):799–806

133. Zhou C, Ye HH, Wang SQ, Chai Z (2006b) Interleukin-1beta regulation of N-type Ca2+ channels in cortical neurons. Neurosci Lett 403(1–2):181–185

134. Zurolo E, de Groot M, Iyer A, Anink J, van Vliet EA, Heimans JJ et al (2012) Regulation of Kir4.1 expression in astrocytes and astrocytic tumors: a role for interleukin-1 beta. J Neuroinflammation 9:280

Are Changes in Synaptic Function That Underlie Hyperexcitability Responsible for Seizure Activity?

15

John G.R. Jefferys

Abstract

The synaptic and intrinsic mechanisms responsible for epileptic seizures and briefer interictal epileptic discharges have been characterized in some detail. This chapter will outline some aspects of this work in the context of focal epilepsies, particularly in the temporal lobe, and will identify some of the major questions that remain. Early work, mainly using the actions of convulsant treatments on brain slices in vitro, revealed synaptic circuitry that could recruit populations of neurons into synchronous epileptic discharges. Subsequent investigations into cellular mechanisms of chronic experimental and clinical foci, again often in vitro, have revealed complex changes in synaptic properties, synaptic connectivity, intrinsic neuronal properties and selective losses of neurons: unraveling their roles in generating seizures, interictal discharges and interictal dysfunctions/comorbidities remains a significant challenge. In vivo recordings have revealed aspects of the pathophysiology of epileptic foci that have practical implications, for instance high-frequency oscillations, and potentially high-frequency hypersynchronous neuronal firing, which have been useful in localizing the epileptogenic zone for surgical resection.

Keywords

Temporal lobe epilepsy • Disease models • Cellular electrophysiology • Synaptic transmission • Chronic models • Hippocampus

15.1 Introduction

Phil Schwartzkroin pioneered cellular electrophysiology and basic research on epilepsy. I started my research career a couple of years after he did. His work impressed and inspired me from the start. His impact goes far wider than his considerable innovations and insights into the basic mechanisms

J.G.R. Jefferys (✉)
Departments of Pharmacology and Biochemistry,
University of Oxford, Oxford OX1 3QT, UK
e-mail: john.jefferys@pharm.ox.ac.uk

H.E. Scharfman and P.S. Buckmaster (eds.), *Issues in Clinical Epileptology: A View from the Bench*,
Advances in Experimental Medicine and Biology 813, DOI 10.1007/978-94-017-8914-1_15,
© Springer Science+Business Media Dordrecht 2014

of epilepsy, notably through his leadership roles in epilepsy societies and his development of influential monographs such as the "Encyclopedia of Basic Epilepsy Research" and "Models of Seizures and Epilepsy" [53, 62]. In summary Phil has made major contributions to the development of the basic science of epilepsy, which thoroughly justify this volume celebrating his career.

Seizures are the diagnostic feature of epilepsy and are classically considered as hypersynchronous electrophysiological activity [52]. The idea that seizures are hypersynchronous has been challenged recently, a point I will return to at the end of this chapter [36]. While the remit of this chapter is on the role of synaptic function in hyperexcitability and seizure generation, I will address broader issues on the pathophysiology of focal epilepsy. A quick definition of the terms of the title: hyperexcitability is a condition of neurons or neuronal networks in which they respond more intensely or more readily to normally innocuous activity, or may become spontaneously active; such responses or spontaneous activity may lead to the generation of seizures or other (briefer) epileptic discharges. Hyperexcitability can lead to seizures but the two are distinct concepts.

Normal brain tissue can generate seizures when exposed to convulsant conditions, as in acute models of epilepsy. Such acute models laid much of the groundwork for our understanding of basic mechanisms of clinical epilepsy, as well as being directly relevant to clinical symptomatic seizures. However, epilepsy is by definition a chronic condition where seizures occur spontaneously under physiological conditions; understanding why they do is a major challenge for both clinical and basic research.

Phil Schwartzkroin pioneered many of the models, preparations and concepts involved in understanding seizures and hyperexcitability, including: acute and chronic models of epilepsy, in vitro brain slices, synaptic properties, synaptic connectivity, intrinsic neuronal properties, glial properties ([39, 40]; for example: [60, 61, 63–68]). This chapter will outline work on basic mechanisms

of epilepsy, particularly focal epilepsy of the medial temporal lobe, which has, in large part, developed from these innovations.

15.2   Acute Epilepsy and Hyperexcitability

Some of the earliest epilepsy research used acute pharmacological block of inhibition (e.g. with penicillin, picrotoxin or bicuculline) to produce epileptic discharges, initially in vivo and then in hippocampal and neocortical slices in vitro. The introduction of brain slices in vitro into epilepsy research was a major step towards developing detailed cellular models of epileptic activity. Early advances included the discovery that the intrinsic electrical properties of central neurons were rather complex, and could look a lot like epileptic bursts [67, 75]. Another early discovery was that excitatory pyramidal cells were interconnected to form a recurrent excitatory network that was held in check by networks of inhibitory neurons [40, 46]. It is hard to think back to the state of the field 40+ years ago, but the idea that excitatory neurons within a brain structure made connections with each other seemed novel at that time. Changing that mindset is an example of how epilepsy research can make major contributions to our understanding of normal brain mechanisms [44]. Essentially the idea that emerged is that excitatory neurons in regions such as the hippocampus and neocortex form interconnected excitatory synaptic networks which present the risk of a chain reaction of positive feedback [77]. Normally negative feedback, mediated by some types of inhibitory neuron, prevents the build-up of a chain reaction, but blocking or depressing GABAergic transmission clearly disrupts this control. Other convulsant treatments include changes in extracellular ions (e.g. K^+ or Mg^{2+} or Cl^-) or channel blockers (e.g. 4-aminopyridine) have also been investigated in some depth (for review see [31]). Arguably all of these treatments make neurons and/or neuronal networks hyperexcitable.

In practice the ability of isolated hippocampal slices in vitro to generate synchronous epileptiform discharges depends on both synaptic and intrinsic neuronal properties [76]. Synaptic properties provide the most obvious mechanism for synchronization, although non-synaptic mechanisms can play a role [30], particularly when neurons are firing spontaneously as is discussed in the section on high-frequency oscillations. Intrinsic neuronal properties determine cellular excitability and amplify feedback excitation, as is the case with voltage-gated Ca^{2+} currents in CA3 pyramidal cells in disinhibited hippocampal slices [77]. Intrinsic neuronal properties also can shape the morphology of epileptic discharges, for instance where voltage-gated Ca^{2+} currents and Ca^{2+}-dependent K^+ currents generate rhythmic afterdischarges [76]. The importance of intrinsic neuronal properties is underlined by the many mutations in the genes for voltage-gated ion channels and their accessory subunits that have been associated with certain clinical epilepsies [5], and experimental evidence of "acquired channelopathies" following induction of epileptic foci in rodents [7, 54].

A great deal of progress was made on the cellular pathophysiology of acute epilepsy models in hippocampal and neocortical slices during the 1980s and 1990s. As I will outline in the next section, similar work on chronic models has developed rapidly since then. However, acute models in vitro continue to play important roles in epilepsy research. One example is the concept of clustering of neuronal firing during high-frequency oscillations, where subsets of neurons firing at lower rates combine to generate a collective high-frequency rhythm [8, 24, 35]. Another example comes from work on neocortical slices exposed to low concentrations of extracellular Mg^{2+}: bursts of rapid neuronal firing were spatially restricted and propagated relatively slowly across the slices. This study exploited particular advantages of brain slices in vitro for integrating optical and electrical recordings of neuronal activity, concluding that the rate of propagation was controlled by feed-forward or surround inhibition which balanced excitatory synaptic outputs from the discharging neurons [78].

15.3 Chronic Models and Hyperexcitability

Epilepsy is a chronic disease and the most realistic models also are chronic and characterized by spontaneous seizures. Again Phil Schwartzkroin made pioneering contributions, particularly with the alumina gel model and with human recordings in vitro [68]. The idea of using the precision of in vitro methods to investigate chronic models and clinical conditions is fundamentally important, particularly in epilepsy. I will outline some recent progress on chronic models, mainly of temporal lobe epilepsy.

15.3.1 Cellular Pathophysiology in Chronic Foci

Perhaps the most common methods to model medial temporal lobe epilepsy in rodents rely on inducing an initial status epilepticus, by injection of pilocarpine (usually systemically) [19], injection of kainic acid (systemic or intrahippocampal) [9, 85] or prolonged electrical stimulation [50, 80, 81]. Chapters on all these models can be found in a major monograph on epilepsy models which was co-edited by Phil Schwartzkroin [53]. Intrahippocampal tetanus toxin does not cause status epilepticus and causes little or no histopathology in the short term [32, 45], but can cause spatially limited hippocampal sclerosis in a minority of cases as well as a late loss of somatostatin-containing interneurons [47]. All these models have a "latent period" during which epileptogenesis transforms normal into epileptic brain tissue and results in spontaneous seizures, which are electrographically similar to those seen clinically.

One productive approach uses in vivo (or perhaps more accurately, ex vivo) brain slices to investigate cellular mechanisms of chronic epileptic foci, both experimental and clinical [17, 29, 42, 68]. Such studies have revealed diverse cellular changes, usually several coexisting in specific kinds of epileptic foci. These changes include synaptic, intrinsic neuronal, glial and structural. The problem gets more complex because while some changes increase excitability and probably promote epileptic activity, others may be adaptive, reducing excitability and tending to control epileptic activity, and some may even be epiphenomena with no direct consequence for generation of epileptic seizures.

Many synaptic changes have been found in chronic epileptic foci, e.g. affecting transmitter release, receptor expression and synaptic modulation (reviewed in [16]). I will outline a few examples here. Several of these epilepsy-related changes affect GABA-ergic inhibition. Selective losses of inhibitory interneurons provide an attractive mechanism for parallel increases in excitability and propensity to seizures. Histopathological evidence has been variable and a review of the substantial clinical and experimental evidence is beyond the scope of this chapter. Several studies have shown losses of somatostatin containing interneurons in chronic models of temporal lobe epilepsy, both in the hilus of the dentate area [15, 47] and in CA1 [18, 22]. There also is evidence of loss of axon- and soma-targeting interneurons in clinical [21] and experimental material [41, 58] although this is less consistent and is complicated by changes in expression of markers, including parvalbumin, used to identify specific classes of interneurons [71] although evidence on soma-targeting interneurons is contradictory [22, 71]. Survival of normal numbers of interneurons does not necessarily mean that inhibitory function remains intact. Inhibition may be dysfunctional, as in the original dormant basket cell hypothesis or other conditions where excitation of interneurons is impaired [69–71, 82, 88]. Inhibition can also be weakened indirectly by changes in chloride homeostasis in the postsynaptic neurons, as shown in subicular slices from humans with medically intractable epilepsy undergoing surgical resection [27]. Synchronous interictal discharges in this clinical tissue have substantial contribu-

tions from depolarizing GABAergic synaptic potentials, which appear to be associated with decreased expression of the KCC2 chloride transporter which maintains chloride equilibria hyperpolarized to rest.

Excitatory synapses also can be altered in epileptic foci. Receptors subunits can be modified, as is the case with AMPA receptors in more than one chronic model [55, 56]. Aberrant expression of different classes of receptor can affect synaptic function, as in the expression of kainic acid receptors in dentate gyrus, which prolong EPSPs and strengthen synaptic integration [2]. It has long been known that seizures induce changes in synaptic connectivity in the brain. The prototypical case is mossy fibre sprouting in the dentate gyrus [74], raising the prospect of increased recurrent excitation through this glutamatergic pathway. Recent studies suggest that aberrant postsynaptic receptors make the new synapses particularly effective [2], although it looks as though their effects may be controlled by inhibitory mechanisms, at least in vitro [51]. Despite the robust connection between chronic temporal lobe epilepsy and sprouting, it turns out that preventing sprouting with rapamycin after an episode of status epilepticus fails to prevent the development of chronic epilepsy [14, 26]. This is one example of the importance of testing the functional implications of cellular changes identified in epileptic foci: even plausible phenomena like sprouting of excitatory synaptic connections are not necessarily responsible for epileptogenesis.

Several neuropeptides change in chronic epilepsy, as reviewed in Casillas-Espinosa et al. [16], often in directions that suggest they may act as endogenous anticonvulsants. These effects have attracted attention for translational research and is providing leads for potential innovative treatments for epilepsies that are refractory to currently available drugs [72]. Finally, intrinsic properties due to voltage-gated ion channels change in both genetic and acquired epilepsies, with examples for sodium, potassium, calcium and HCN channels amongst others; this topic is beyond the scope of this chapter, but is reviewed in Poolos & Johnston [54].

This short overview covers a small part of the diverse cellular pathologies and pathophysiologies that have been found in chronic focal epilepsies. Even the most reproducible of models reveals multiple distinct cellular changes. Major questions remain on the roles each plays. It may be that in isolation none are sufficient to induce epileptic foci, but that several need to be present. It also is clear that some changes are antiepileptic and could provide leads for new treatments, as in the example of the work on peptides mentioned above. Of course some changes may be epiphenomena, perhaps induced by repeated seizures, but with no material impact on seizure susceptibility or generation. It is likely that different kinds of epileptic foci, both clinical and experimental, may have their individual combinations of cellular abnormalities. Finally, epilepsy is more than the seizures, with a range of comorbidities that can be detected between seizures [13], many of which will have underlying cellular mechanisms that may be identified by the kinds of investigations outlined above.

15.3.2 The Epileptogenic Zone, Hypersynchrony and High-Frequency Oscillations

Around one in three persons with epilepsy fail to gain adequate seizure control with currently available drugs, maybe even more for medial temporal lobe epilepsy [84]. Surgical resection of the tissue responsible for seizures can be remarkably effective as long as the correct tissue is removed [48]. The epileptogenic zone is defined as the area that is necessary and sufficient for resection to result in seizure freedom [57]. If the seizures stop after surgery then the resection must have been sufficient, but it is harder to be certain that all the resection was necessary. Presurgical work-up can include non-invasive imaging, scalp EEG, subdural and depth recordings. Here I will focus on the discovery that high-frequency oscillations may help define the epileptogenic zone in clinical and experimental foci [11].

High-frequency oscillations have frequencies greater than used to be recorded by routine EEG, and typically are considered as 80 Hz and above. Paper-based EEGs meant that high-frequency oscillations were missed in clinical electrophysiological investigations, but increasing computerization and improved amplifiers led to their discovery in clinical recordings during the 1990s [1, 23]. The early studies of high-frequency oscillations divided them between physiological ripples and pathophysiological "fast ripples", separated by a boundary at around 200–250 Hz [11, 12]. Fast ripples have been associated with neuronal loss and hippocampal sclerosis [24, 73]. However in an experimental model lacking status epilepticus and with minimal or no neuronal loss we found that fast ripples (>250 Hz) were reliably associated with the primary focus [37], which supports the idea that electrographic markers can extend surgery into more difficult cases.

It is well established that timing of gamma oscillations depends on inhibitory synaptic transmission [4, 83]. Inhibitory mechanisms also are important in physiological ripples [87]. However it is more difficult to see how fast ripples at >250 Hz, and reaching >500 Hz, can depend on synaptic mechanisms. Fast ripples appear to represent synchronous firing of excitatory pyramidal and granule neurons [8, 10, 35]. These neurons do not fire as fast as the fast ripple oscillation; rather they fire every few cycles. In vitro studies suggest that excitatory neurons are weakly but significantly synchronized in small fluctuating groups extending over distances of a few hundred microns [35]. The potential mechanisms of synchronization on a millisecond timescale are limited. Perhaps the most plausible is that groups of neurons which are close to threshold synchronize through electrical field (sometimes called ephaptic) interactions [30, 33]: this effect is relatively weak under physiological conditions but weak fields are sufficient to entrain neurons which are firing spontaneously [20].

As mentioned above, removal of the epileptogenic zone is "necessary and sufficient" for seizure freedom. Determining how much tissue needs to be removed is a difficult challenge. An interesting approach to this important problem comes from a clinical study which found that successful surgical outcome is associated with the proportion of tissue generating high-frequency oscillations that was resected [28].

It is increasingly clear that, while the distinction between ripples and fast ripples have been quite successful, the frequencies of oscillations are not sufficient to define their functional significance [34], either in terms of markers for epileptogenic tissue or in terms of contributions to seizure generation (e.g. by strengthening synaptic summation) [6]. The distinction between interictal pathological and normal physiological activities may depend on many factors: some cortical areas may differ from the hippocampus [43], not all epileptic foci are necessarily alike, and, from a practical point of view, electrode size can have an impact on recorded frequencies of high-frequency oscillations [86]. What does appear useful is finding recording sites with anomalous features, which may include faster activity than found in other sites in the same person [25, 38].

A distinctive approach to identifying epileptogenic zone used multichannel microelectrodes, the Utah or NeuroPort arrays, in people undergoing invasive ECoG recordings to find the regions in which neuronal hyperactivity first occurs at seizure onset. These arrays contain ~100 microelectrodes extending 1 mm from their bases which can be inserted into the cortex to record from neurons, probably located in layer 4–5. The big surprise was when Truccolo et al. [79] found that neurons recorded by their microelectrode arrays mostly stopped firing at electrographic (ECoG) seizure onset. This deviates substantially from experimental models of focal epilepsies, but does confirm earlier clinical studies with single microelectrodes [3]. Subsequently Schevon et al. [59] did find neuronal firing accelerating at seizure onset in some of their arrays, but the advancing wave of neuronal hyperactivity and hypersynchrony was much more restricted than the epileptic ECoG. It is not clear whether they were luckier or more careful in their microelectrode positioning, but it is reassuring that seizures are associated with accelerating neuronal firing which is phase linked to the simultaneously recorded epileptic ECoG [49]. The discrepancy

in localization was attributed to "inhibitory restraint" outlined above [59]: to recap, the idea is that feedforward inhibition constrains the advancing front of hyperactivity of excitatory neurons participating in the chain reaction of the epileptic seizure. This has been demonstrated explicitly in rodent neocortical slices in vitro [59, 78]. On this model the focal slowly-propagating population of hyperactive hypersynchronous excitatory neurons projects to more widespread regions where excitation is held in check by feedforward inhibition so that most of epileptic ECoG occurs in the absence of accelerating neuronal firing. This challenges the original description of epileptic EEGs as "hypersynchronous" [52], but it can be argued that the original use is a reasonable label for the large-amplitude relatively rhythmic EEG or ECoG, as long as it is clear that it does not necessarily mean hypersynchronous neuronal firing in the same area. These clinical investigations are difficult, but (ethics permitting) need repeating, ideally with critical experimental tests to determine whether the cellular interpretation of the role of feedforward inhibition derived from reductionist experiments really do apply to epileptic cortex in humans.

Perhaps the biggest challenges for this line of research are (a) whether the regions initiating hypersynchronous firing really do mark the epileptogenic zone, and if so, (b) how to exploit it for presurgical evaluation in preparation for resection of medically intractable epileptic foci. Microelectrode arrays would be very difficult to implement in most clinical settings: inserting microelectrode arrays into human cortex can be difficult and can damage the recorded tissue, while analyzing the resulting data needs the tools of cellular electrophysiology normally found in basic neuroscience laboratories. However it may be that other, technically more straightforward, markers can be found. Perhaps the most promising are the high-frequency oscillations discussed above. They represent coincident firing of principal neurons, and may provide markers for hyperactivity and hypersynchrony of neuronal firing. If they do, they would prove much more straightforward for clinical investigation than unit recordings from penetrating arrays of microelectrodes.

Future work needs to refine presurgical identification of the epileptogenic zone, using high-frequency oscillations and other biomarkers. The relationship between these biomarkers and pathological high-frequency neuronal firing detected by penetrating arrays may play a role in solving that clinical challenge. This relationship may also provide insights into the cellular mechanisms responsible for high-frequency oscillations, providing an in vivo approach to complement ex vivo (or in vitro) investigations to provide insights into the organization of the pathophysiological networks of epileptic foci.

15.4 Concluding Remarks

The last few decades have seen spectacular advances in our understanding of the cellular pathophysiology of epileptic activity in acute and chronic models and in clinical foci. Multiple cellular pathologies and pathophysiologies operate in chronic foci, whether clinical or experimental. One set of major challenges is to distinguish between those: responsible for generating seizures, helping control them, responsible for comorbidities, and functionally neutral epiphenomena. Progressive advances in chronic experimental investigations in vivo have started to help us understand epileptic foci in situ, and to refine our concepts of the epileptogenic zone which should expand the application of surgery in cases of pharmacologically intractable epilepsies.

Acknowledgements I am grateful for research support from Epilepsy Research UK (P1102) and the Medical Research Council of the UK (G082162).

References

1. Allen PJ, Fish DR, Smith SJM (1992) Very high-frequency rhythmic activity during SEEG suppression in frontal lobe epilepsy. Electroencephalogr Clin Neurophysiol 82:155–159
2. Artinian J, Peret A, Marti G, Epsztein J, Crepel V (2011) Synaptic kainate receptors in interplay with I-NaP shift the sparse firing of dentate granule cells to a sustained rhythmic mode in temporal lobe epilepsy. J Neurosci 31:10811–10818
3. Babb TL, Wilson CL, Isokawa-Akesson M (1987) Firing patterns of human limbic neurons during stereoencephalography (SEEG) and clinical temporal lobe seizures. Electroencephalogr Clin Neurophysiol 66:467–482
4. Bartos M, Vida I, Jonas P (2007) Synaptic mechanisms of synchronized gamma oscillations in inhibitory interneuron networks. Nat Rev Neurosci 8:45–56
5. Baulac S, Baulac M (2009) Advances on the genetics of mendelian idiopathic epilepsies. Neurol Clin 27:1041–1061
6. Berger T, Luscher HR (2003) Timing and precision of spike initiation in layer V pyramidal cells of the rat somatosensory cortex. Cereb Cortex 13:274–281
7. Bernard C, Anderson A, Becker A, Poolos NP, Beck H, Johnston D (2004) Acquired dendritic channelopathy in temporal lobe epilepsy. Science 305:532–535
8. Bikson M, Fox JE, Jefferys JGR (2003) Neuronal aggregate formation underlies spatiotemporal dynamics of nonsynaptic seizure initiation. J Neurophysiol 89:2330–2333
9. Bouilleret V, Ridoux V, Depaulis A, Marescaux C, Nehlig A, Le Gal LS (1999) Recurrent seizures and hippocampal sclerosis following intrahippocampal kainate injection in adult mice: electroencephalography, histopathology and synaptic reorganization similar to mesial temporal lobe epilepsy. Neuroscience 89:717–729
10. Bragin A, Benassi SK, Kheiri F, Engel J (2011) Further evidence that pathologic high-frequency oscillations are bursts of population spikes derived from recordings of identified cells in dentate gyrus. Epilepsia 52:45–52
11. Bragin A, Engel JJ, Wilson CL, Fried I, Mathern GW (1999) Hippocampal and entorhinal cortex high-frequency oscillations (100–500 Hz) in human epileptic brain and in kainic acid – treated rats with chronic seizures. Epilepsia 40:127–137
12. Bragin A, Engel JJ, Wilson CL, Vizentin E, Mathern GW (1999) Electrophysiologic analysis of a chronic seizure model after unilateral hippocampal KA injection. Epilepsia 40:1210–1221
13. Brooks-Kayal AR, Bath KG, Berg AT, Galanopoulou AS, Holmes GL, Jensen FE, Kanner AM, O'Brien TJ, Whittemore VH, Winawer MR, Patel M, Scharfman HE (2013) Issues related to symptomatic and disease-modifying treatments affecting cognitive and neuropsychiatric comorbidities of epilepsy. Epilepsia 54:44–60
14. Buckmaster PS, Ingram EA, Wen XL (2009) Inhibition of the mammalian target of rapamycin signaling pathway suppresses dentate granule cell axon sprouting in a rodent model of temporal lobe epilepsy. J Neurosci 29:8259–8269
15. Buckmaster PS, Jongen-Relo AL (1999) Highly specific neuron loss preserves lateral inhibitory circuits in the dentate gyrus of kainate-induced epileptic rats. J Neurosci 19:9519–9529
16. Casillas-Espinosa PM, Powell KL, O'Brien TJ (2012) Regulators of synaptic transmission: roles in the pathogenesis and treatment of epilepsy. Epilepsia 53:41–58

17. Cohen I, Navarro V, Clemenceau S, Baulac M, Miles R (2002) On the origin of interictal activity in human temporal lobe epilepsy in vitro. Science 298:1418–1421

18. Cossart R, Dinocourt C, Hirsch JC, Merchan-Perez A, De Felipe J, Ben Ari Y, Esclapez M, Bernard C (2001) Dendritic but not somatic GABAergic inhibition is decreased in experimental epilepsy. Nat Neurosci 4:52–62

19. Curia G, Longo D, Biagini G, Jones RSG, Avoli M (2008) The pilocarpine model of temporal lobe epilepsy. J Neurosci Methods 172:143–157

20. Deans JK, Bikson M, Fox JE, Jefferys JGR (2003) Effects of AC fields at powerline frequencies on gamma oscillations in vitro. Abstracts Viewer/Itinerary Planner, Program No. 258.1

21. DeFelipe J (1999) Chandelier cells and epilepsy. Brain 122:1807–1822

22. Dinocourt C, Petanjek Z, Freund TF, Ben Ari Y, Esclapez M (2003) Loss of interneurons innervating pyramidal cell dendrites and axon initial segments in the CA1 region of the hippocampus following pilocarpine-induced seizures. J Comp Neurol 459:407–425

23. Fisher RS, Webber WRS, Lesser RP, Arroyo S, Uematsu S (1992) High-frequency EEG activity at the start of seizures. J Clin Neurophysiol 9:441–448

24. Foffani G, Uzcategui YG, Gal B, Menendez de la Prida L (2007) Reduced spike-timing reliability correlates with the emercience of fast ripples in the rat epileptic hippocampus. Neuron 55:930–941

25. Gnatkovsky V, Francione S, Cardinale F, Mai R, Tassi L, Lo Russo G, de Curtis M (2011) Identification of reproducible ictal patterns based on quantified frequency analysis of intracranial EEG signals. Epilepsia 52:477–488

26. Heng K, Haney MM, Buckmaster PS (2013) High-dose rapamycin blocks mossy fiber sprouting but not seizures in a mouse model of temporal lobe epilepsy. Epilepsia 54:1535–1541

27. Huberfeld G, Wittner L, Clemenceau S, Baulac M, Kaila K, Miles R, Rivera C (2007) Perturbed chloride homeostasis and GABAergic signaling in human temporal lobe epilepsy. J Neurosci 27:9866–9873

28. Jacobs J, Zijlmans M, Zelmann R, Chatillon CE, Hall J, Olivier A, Dubeau F, Gotman J (2010) High-frequency electroencephalographic oscillations correlate with outcome of epilepsy surgery. Ann Neurol 67:209–220

29. Jefferys JGR (1989) Chronic epileptic foci *in vitro* in hippocampal slices from rats with the tetanus toxin epileptic syndrome. J Neurophysiol 62:458–468

30. Jefferys JGR (1995) Non-synaptic modulation of neuronal activity in the brain: electric currents and extracellular ions. Physiol Rev 75:689–723

31. Jefferys JGR (2007) Epilepsy in vitro: electrophysiology and computer modeling. In: Engel J Jr, Pedley TA, Aicardi J, Dichter MA, Moshé SL (eds) Epilepsy: a comprehensive textbook. Lippincott, Williams & Wilkins, Philadelphia

32. Jefferys JGR, Evans BJ, Hughes SA, Williams SF (1992) Neuropathology of the chronic epileptic syndrome induced by intrahippocampal tetanus toxin in the rat: preservation of pyramidal cells and incidence of dark cells. Neuropathol Appl Neurobiol 18:53–70

33. Jefferys JGR, Haas HL (1982) Synchronized bursting of CA1 pyramidal cells in the absence of synaptic transmission. Nature 300:448–450

34. Jefferys JGR, Menendez de la Prida L, Wendling F, Bragin A, Avoli M, Timofeev I, da Silva FHL (2012) Mechanisms of physiological and epileptic HFO generation. Prog Neurobiol 98:250–264

35. Jiruska P, Csicsvari J, Powell AD, Fox JE, Chang WC, Vreugdenhil M, Li X, Palus M, Bujan AF, Dearden RW, Jefferys JGR (2010) High-frequency network activity, global increase in neuronal activity and synchrony expansion precede epileptic seizures in vitro. J Neurosci 30:5690–5701

36. Jiruska P, De Curtis M, Jefferys JGR, Schevon CA, Schiff SJ, Schindler K (2013) Synchronization and desynchronization in epilepsy: controversies and hypotheses. J Physiol 591:787–797

37. Jiruska P, Finnerty GT, Powell AD, Lofti N, Cmejla R, Jefferys JGR (2010) High-frequency network activity in a model of non-lesional temporal lobe epilepsy. Brain 133:1380–1390

38. Jiruska P, Tomasek M, Netuka D, Otahal J, Jefferys JGR, Li X, Marusic P (2008) Clinical impact of high-frequency seizure onset zone in a case of bi-temporal epilepsy. Epileptic Disord 10:231–238

39. Knowles WD, Funch PG, Schwartzkroin PA (1982) Electrotonic and dye coupling in hippocampal CA1 pyramidal cells in vitro. Neuroscience 7:1713–1722

40. Knowles WD, Schwartzkroin PA (1981) Local circuit synaptic interactions in hippocampal brain slices. J Neurosci 1:318–322

41. Kobayashi M, Buckmaster PS (2003) Reduced inhibition of dentate granule cells in a model of temporal lobe epilepsy. J Neurosci 23:2440–2452

42. Köhling R, Lücke A, Straub H, Speckmann EJ, Tuxhorn I, Wolf P, Pannek H, Oppel F (1998) Spontaneous sharp waves in human neocortical slices excised from epileptic patients. Brain 121:1073–1087

43. Köhling R, Staley K (2011) Network mechanisms for fast ripple activity in epileptic tissue. Epilepsy Res 97:318–323

44. Lockard JS, Ward AA Jr (1980) Epilepsy: a window to brain mechanisms. Raven Press, New York

45. Mellanby J, George G, Robinson A, Thompson P (1977) Epileptiform syndrome in rats produced by injecting tetanus toxin into the hippocampus. J Neurol Neurosurg Psychiatry 40:404–414

46. Miles R, Wong RKS, Traub RD (1984) Synchronized afterdischarges in the hippocampus: contribution of local synaptic interactions. Neuroscience 12:1179–1189

47. Mitchell J, Gatherer M, Sundstrom LE (1995) Loss of hilar somatostatin neurons following tetanus toxin-induced seizures. Acta Neuropathol (Berl) 89:425–430

48. Mohammed HS, Kaufman CB, Limbrick DD, Steger-May K, Grubb RL, Rothman SM, Weisenberg JLZ, Munro R, Smyth MD (2012) Impact of epilepsy surgery on seizure control and quality of life: a 26-year follow-up study. Epilepsia 53:712–720

49. Mormann F, Jefferys JGR (2013) Neuronal firing in human epileptic cortex: the ins and outs of synchrony during seizures. Epilepsy Curr 13:100–102

50. Norwood BA, Bauer S, Wegner S, Hamer HM, Oertel WH, Sloviter RS, Rosenow F (2011) Electrical stimulation-induced seizures in rats: a "dose-response" study on resultant neurodegeneration. Epilepsia 52:E109–E112

51. Patrylo PR, Dudek FE (1998) Physiological unmasking of new glutamatergic pathways in the dentate gyrus of hippocampal slices from kainate-induced epileptic rats. J Neurophysiol 79:418–429

52. Penfield W, Jasper H (1954) Epilepsy and the functional anatomy of the human brain. Little Brown, Boston

53. Pitkänen A, Schwartzkroin PA, Moshé SL (2005) Models of seizures and epilepsy. Elsevier Academic, Amsterdam

54. Poolos NP, Johnston D (2012) Dendritic ion channelopathy in acquired epilepsy. Epilepsia 53(Suppl 9): 32–40. doi:10.1111/epi.12033

55. Rajasekaran K, Todorovic M, Kapur J (2012) Calcium-permeable AMPA receptors are expressed in a rodent model of status epilepticus. Ann Neurol 72:91–102

56. Rosa MLNM, Jefferys JGR, Sanders MW, Pearson RCA (1999) Expression of mRNAs encoding flip isoforms of GluR1 and GluR2 glutamate receptors is increased in rat hippocampus in epilepsy induced by tetanus toxin. Epilepsy Res 36:243–251

57. Rosenow F, Luders H (2001) Presurgical evaluation of epilepsy. Brain 124:1683–1700

58. Sayin U, Osting S, Hagen J, Rutecki P, Sutula T (2003) Spontaneous seizures and loss of axo-axonic and axo-somatic inhibition induced by repeated brief seizures in kindled rats. J Neurosci 23:2759–2768

59. Schevon CA, Weiss SA, McKhann G, Goodman RR, Yuste R, Emerson RG, Trevelyan AJ (2012) Evidence of an inhibitory restraint of seizure activity in humans. Nat Commun 3:1060

60. Schwartzkroin PA (1975) Characteristics of CA1 neurons recorded intracellularly in hippocampal in vitro slice preparation. Brain Res 85:423–436

61. Schwartzkroin PA (1986) Hippocampal slices in experimental and human epilepsy. Adv Neurol 44:991–1010

62. Schwartzkroin PA (2009) Encyclopedia of basic epilepsy research. Academic, London

63. Schwartzkroin PA, Futamachi KJ, Noebels JL, Prince DA (1975) Transcallosal effects of a cortical epileptiform focus. Brain Res 99:59–68

64. Schwartzkroin PA, Prince DA (1977) Penicillin-induced epileptiform activity in the hippocampal in vitro preparation. Ann Neurol 1:463–469

65. Schwartzkroin PA, Prince DA (1978) Cellular and field potential properties of epileptogenic hippocampal slices. Brain Res 147:117–130

66. Schwartzkroin PA, Prince DA (1979) Recordings from presumed glial cells in the hippocampal slice. Brain Res 161:533–538

67. Schwartzkroin PA, Slawsky M (1977) Probable calcium spikes in hippocampal neurons. Brain Res 133:157–161

68. Schwartzkroin PA, Turner DA, Knowles WD, Wyler AR (1983) Studies of human and monkey "epileptic" neocortex in the in vitro slice preparation. Ann Neurol 13:249–257

69. Sloviter RS (1991) Permanently altered hippocampal structure, excitability, and inhibition after experimental status epilepticus in the rat: the "dormant basket cell" hypothesis and its possible relevance to temporal lobe epilepsy. Hippocampus 1:41–66

70. Sloviter RS (1994) The functional organization of the hippocampal dentate gyrus and its relevance to the pathogenesis of temporal lobe epilepsy. Ann Neurol 35:640–654

71. Sloviter RS, Zappone CA, Harvey BD, Bumanglag AV, Bender RA, Frotscher M (2003) "Dormant basket cell" hypothesis revisited: relative vulnerabilities of dentate gyrus mossy cells and inhibitory interneurons after hippocampal status epilepticus in the rat. J Comp Neurol 459:44–76

72. Sorensen AT, Kokaia M (2013) Novel approaches to epilepsy treatment. Epilepsia 54:1–10

73. Staba RJ, Frighetto L, Behnke EJ, Mathern G, Fields T, Bragin A, Ogren J, Fried I, Wilson CL, Engel J (2007) Increased fast ripple to ripple ratios correlate with reduced hippocampal volumes and neuron loss in temporal lobe epilepsy patients. Epilepsia 48:2130–2138

74. Tauck DL, Nadler JV (1985) Evidence of functional mossy fiber sprouting in hippocampal formation of kainic acid-treated rats. J Neurosci 5:1016–1022

75. Traub RD, Jefferys JGR, Miles R, Whittington MA, Tóth K (1994) A branching dendritic model of a rodent CA3 pyramidal neurone. J Physiol 481:79–95

76. Traub RD, Miles R, Jefferys JGR (1993) Synaptic and intrinsic conductances shape picrotoxin-induced synchronized after-discharges in the guinea-pig hippocampal slice. J Physiol 461:525–547

77. Traub RD, Wong RKS (1982) Cellular mechanism of neuronal synchronization in epilepsy. Science 216:745–747

78. Trevelyan AJ, Sussillo D, Yuste R (2007) Feedforward inhibition contributes to the control of epileptiform propagation speed. J Neurosci 27:3383–3387

79. Truccolo W, Donoghue JA, Hochberg LR, Eskandar EN, Madsen JR, Anderson WS, Brown EN, Halgren E, Cash SS (2011) Single-neuron dynamics in human focal epilepsy. Nat Neurosci 14:635–641

80. Vicedomini JP, Nadler JV (1990) Stimulation-induced status epilepticus: role of the hippocampal mossy fibers in the seizures and associated neuropathology. Brain Res 512:70–74

81. Walker MC, Perry H, Scaravilli F, Patsalos PN, Shorvon SD, Jefferys JGR (1999) Halothane as a neuroprotectant during constant stimulation of the perforant path. Epilepsia 40:359–364

82. Whittington MA, Jefferys JGR (1994) Epileptic activity outlasts disinhibition after intrahippocampal tetanus toxin in the rat. J Physiol (Lond) 481:593–604

83. Whittington MA, Traub RD, Jefferys JGR (1995) Synchronized oscillations in interneuron networks driven by metabotropic glutamate receptor activation. Nature 373:612–615

84. Wiebe S, Jette N (2012) Pharmacoresistance and the role of surgery in difficult to treat epilepsy. Nat Rev Neurol 8:669–677

85. Williams PA, White AM, Clark S, Ferraro DJ, Swiercz W, Staley KJ, Dudek FE (2009) Development of spontaneous recurrent seizures after kainate-induced status epilepticus. J Neurosci 29:2103–2112

86. Worrell GA, Gardner AB, Stead SM, Hu SQ, Goerss S, Cascino GJ, Meyer FB, Marsh R, Litt B (2008) High-frequency oscillations in human temporal lobe: simultaneous microwire and clinical macroelectrode recordings. Brain 131:928–937

87. Ylinen A, Bragin A, Nádasdy Z, Jandó G, Szabo I, Sik A, Buzsáki G (1995) Sharp wave-associated high-frequency oscillation (200 Hz) in the intact hippocampus: network and intracellular mechanisms. J Neurosci 15:30–46

88. Zhang W, Buckmaster PS (2009) Dysfunction of the dentate basket cell circuit in a rat model of temporal lobe epilepsy. J Neurosci 29:7846–7856

Does Epilepsy Cause a Reversion to Immature Function?

16

Aristea S. Galanopoulou and Solomon L. Moshé

Abstract

Seizures have variable effects on brain. Numerous studies have examined the consequences of seizures, in light of the way that these may alter the susceptibility of the brain to seizures, promote epileptogenesis, or functionally alter brain leading to seizure-related comorbidities. In many –but not all- situations, seizures shift brain function towards a more immature state, promoting the birth of newborn neurons, altering the dendritic structure and neuronal connectivity, or changing neurotransmitter signaling towards more immature patterns. These effects depend upon many factors, including the seizure type, age of seizure occurrence, sex, and brain region studied. Here we discuss some of these findings proposing that these seizure-induced immature features do not simply represent rejuvenation of the brain but rather a de-synchronization of the homeostatic mechanisms that were in place to maintain normal physiology, which may contribute to epileptogenesis or the cognitive comorbidities.

Keywords

GABA receptor • Chloride cotransporter • Neurogenesis • mTOR • Dysplasia • Epileptogenesis

A.S. Galanopoulou, M.D., Ph.D. (✉)
Saul R. Korey Department of Neurology,
Dominick P. Purpura Department of Neuroscience,
The Laboratory of Developmental Epilepsy,
Comprehensive Einstein/Montefiore Epilepsy Center,
Albert Einstein College of Medicine,
1410 Pelham Parkway South, Kennedy Center Rm 306,
Bronx, NY 10461, USA
e-mail: aristea.galanopoulou@einstein.yu.edu

S.L. Moshé, M.D.
Saul R. Korey Department of Neurology,
Dominick P. Purpura Department of Neuroscience,
Department of Pediatrics, The Laboratory
of Developmental Epilepsy, Comprehensive Einstein/
Montefiore Epilepsy Center, Albert Einstein College
of Medicine, 1410 Pelham Parkway South,
Kennedy Center Rm 316, Bronx, NY 10461, USA
e-mail: solomon.moshe@einstein.yu.edu

H.E. Scharfman and P.S. Buckmaster (eds.), *Issues in Clinical Epileptology: A View from the Bench*,
Advances in Experimental Medicine and Biology 813, DOI 10.1007/978-94-017-8914-1_16,
© Springer Science+Business Media Dordrecht 2014

16.1   Introduction

Epilepsies have multiple causes and phenotypes, leading to different seizure and epilepsy syndromes. A variety of genetic, toxic/metabolic, or structural abnormalities have been causally associated with epilepsies. Epilepsy may occur as a "system disorder", attributed to dysfunction –but no overt structural pathology – of specific neuronal networks, as typically occurs in genetic generalized epilepsies, like absence epilepsy [4]. In other cases, specific pathologies, e.g., cortical malformations or hippocampal sclerosis, may lead to the generation of an epileptogenic focus.

Seizures and epilepsies may disrupt brain development. Often, these maldevelopmental consequences of seizures may manifest as age-inappropriate reversal to immature functions and developmental processes. For example, seizures may trigger the aberrant re-emergence of immature features of $GABA_A$ receptor ($GABA_AR$) signaling in neurons from adult animals or may cause morphological changes reminiscent of immature neurons. Immature features include the generation of new neuronal progenitor cells, functional alteration of selected signaling pathways or morphological changes. Many of these immature features have been documented in surgically resected epileptic tissues from individuals with drug-resistant epilepsies, like temporal lobe epilepsy (TLE), hypothalamic hamartomas, cortical dysplasias, or peritumoral epileptic tissue. Comparisons with nonepileptic post-mortem or surgically resected tissues have indicated that some of these changes are specific for the epileptic tissue [46, 47]. Yet, the appearance of these changes after seizures in animal models often depends upon a variety of factors. Here we will discuss the animal studies that have supported these observations and have provided insights on the complex interactions between the immature features of the epileptic focus and epilepsies, their etiologies and treatments and how these can be modified by age, sex, region-specific or other factors.

16.2   Neurogenesis in TLE

Perhaps the most classic argument for a reversal of normal age-specific functions with a re-emergence of patterns observed during development is the observation that there is an increased number of newborn cells in the dentate gyrus, in response to seizures or during the epileptic state [101]. Increased neurogenesis in the dentate gyrus of adult rats has been shown using post-SE models of epilepsy [48, 86, 100] or kindling [84, 105] or hyperthermic seizures [[61] and reviewed in [85, 101]] (Table 16.1). Newborn cells manifest many of the electrophysiological and morphological features of the granule cells, but also some distinctive characteristics. For example, they may be more dispersed [48, 100], have bipolar rather than polarized dendrites and they do not stain for Neuropeptide Y (NPY) or glutamic acid decarboxylase (GAD) immunoreactivity [100]. Furthermore, newborn cells may integrate abnormally into the hippocampus after seizures. Newborn neurons that migrate towards the CA3 pyramidal region may synchronize with CA3 neurons into epileptiform bursts [100]. Doublecortin-positive newborn neurons in the hilar dentate of epileptic rats exhibit long and recurrent basal dendrites directed towards the granule cell layer and also receive excitatory synaptic input which is unusual in seizure-naïve rats [95]. These seizure-induced changes may contribute to the excitability of the hippocampus. It has also been proposed that newborn neurons may not be capable to integrate normally in processes controlling cognitive processes, contributing therefore to cognitive deficits [30, 88].

The effects of seizures on neurogenesis at the dentate is age, sex, region, model specific and may depend on the number and type of seizures that the animal experiences (reviewed in Table 16.1). In brief, neonatal rats may respond instead with reduced or unaltered neurogenesis. Furthermore, aged rats may not respond as robustly with neurogenesis following seizures as younger adults do. Longitudinal studies may

Table 16.1 Effects of seizures on neurogenesis in the dentate gyrus

Animal characteristics	Model of seizures	Effect on neurogenesis in the dentate gyrus	Reference
PN0-4 Sprague–Dawley rats	PN0-4 flurothyl seizures (brief, repetitive)	1-5 brief flurothyl seizures had no effect on neurogenesis	[69]
		25 flurothyl seizures over 4 days *reduced* neurogenesis in the dentate	
PN1-7 Wistar rats	Recurrent pilocarpine SE (PN1, PN4, PN7)	*Reduced* neurogenesis on PN8, PN14	[124]
	BrdU (PN7, PN13, PN20, PN48)	*Increased* neurogenesis on PN49	
PN9 rats	PN9: kainic acid SE (2–3 h)	*Reduced* neurogenesis in the superior blade of the dentate	[59]
	BrdU: 3 h after kainic acid		
PN6-20 Sprague–Dawley rats	1–3 episodes of kainic acid SE between PN6-20	*Reduced* number of BrdU-positive neurons in rats with 3 SEs, assessed on PN13, PN20, PN30, but not at earlier timepoints	[63]
	BrdU after each seizure and 4 h prior to sacrifice		
PN10	PN10: Hyperthermic seizures (<30 min)	Normothermia-exposed males had more BrdU-positive cells than females	[61]
Sprague–Dawley male, female rats	PN11-16: Brdu injections	Hyperthermia had *no acute effect* on neurogenesis (assessed at PN17)	
		Following hyperthermic seizures, newborn neurons in males *survived better* till PN66 than in females	
PN15 Sprague–Dawley rats	PN15: flurothyl SE	*Increased* neurogenesis after SE	[78]
	PN17: BrdU injection	Further increased in malnourished animals	
PN21, PN35 Sprague–Dawley rats, both sexes	Lithium-pilocarpine SE	*Increased* neurogenesis in both age groups	[98]
	BrdU 3th–6th day after SE	No association with cell loss or subsequent probability for epilepsy	
Adult	Pilocarpine-SE (3–5 h)	*Increased* neurogenesis at 3, 6, and 13 days post-pilocarpine SE	[86]
Sprague–Dawley rats	BrdU: 1–27 days post-pilocarpine		
Adult	Pilocarpine or kainic acid induced SE (1 h)	Newborn neurons born after SE migrate into the CA3 layer, maintain many granule cell characteristics (electrophysiological, morphological). However, they are NPY or GAD negative, have bipolar dendrites, and integrate abnormally, firing synchronously to CA3 pyramidal neurons	[100]
Sprague–Dawley rats	BrdU: 4–11 or 26–30 days post SE		
Adult female mice (nestin-GFP transgenic mice)? (8 week old)	Kainic acid SE (2–3 h)	*Increased* neurogenesis post-SE seen with the doublecortin positive neurons but not with the nestin or calretinin positive neurons	[48]
	BrdU: 8 days post SE	Increased dispersion of newborn cells was seen with both doublecortin and calretinin positive neurons after SE	
Adult male Sprague–Dawley rats	Amygdala kindling	*Increased* neurogenesis after ≥9 stage 4–5 seizures but not after 4–6 seizures	[84]
	BrdU: 1 day after last kindled seizure or stimulation	Neurogenesis may not play a role in kindling development	

(continued)

Table 16.1 (continued)

Animal characteristics	Model of seizures	Effect on neurogenesis in the dentate gyrus	Reference
Adult male Wistar rats	Amygdala kindling	*Increased* neurogenesis seen only at the BrdU late group (after stage 5 seizures)	[105]
	BrdU early group: on the 2nd–4th stimulation days		
	BrdU late group: on the days of their 2nd–4th stage 5 seizure		
Adult C57BL/6J mice	Flurothyl kindling	*Increased* neurogenesis after:	[28, 29]
	BrdU injections 0–28 days after 1 or 8 flurothyl seizures	1–3 days following 1 seizure	
		0–7 days after 8 seizures	
		Greater degree of neurogenesis in dorsal than in ventral hippocampus, but the seizure induced increase in newborn cells was greater in the ventral hippocampus	
Adult F344 rats (4 months old)	Kainic acid i.c.v. or graded kainic acid SE (<6 h) i.p.	*16 days post-SE*: *Increased* number of doublecortin positive neurons in the dentate	[40]
		5 months after SE: *decreased* numbers of doublecortin positive neurons in the dentate	
Adult F344 rats (12 month old)	Kainic acid SE (i.p.)	*Increased* neurogenesis in the dentate, but to a less degree than in younger rats	[106]
	BrdU: day 0–12 after SE		

Seizures have age and model-specific effects on neurogenesis in the dentate gyrus

BrdU bromodeoxyuridine, *GAD* glutamic acid decarboxylase, *GFP* Green fluorescent protein, *i.c.v* intracerebroventricular, *i.p.* intraperitoneal, *NPY* neuropeptide Y, *PN* postnatal day, *SE* status epilepticus

reveal time-dependent changes in neurogenesis, which may be influenced also by the ability of these newborn cells to survive. For example, hyperthermic seizures caused the newborn neurons to survive longer in males than in females till adulthood, suggesting sex-specific factors controlling their function [61]. Few brief seizures may not be as sufficient to affect neurogenesis, as frequent or prolonged seizures do.

Investigations into whether aberrant neurogenesis may contribute to epileptogenesis have yielded variable results. Administration of antimitotics that prevent neurogenesis may decrease the frequency of spontaneous seizures in post-SE animals [51]. However, other treatments that reduce seizure-induced neurogenesis have resulted in either reduction [109] or no effect [88] on the frequency of spontaneous seizures. The developmental studies on the effects of SE in 2–3 week old rats which show increased SE-induced neurogenesis, even though neither cell loss nor epileptogenesis always ensue have also failed to associate the increase seizure-

induced neurogenesis with either of these consequences of SE [98]. Seizure-induced neurogenesis appears therefore to contribute to the excitability of the epileptic hippocampus and possibly to the associated cognitive dysfunction, but there is no definite evidence that it is required for or mediates the ensuing epileptogenesis. Future research into deciphering the mechanisms leading to seizure-induced neurogenesis and how these are modified by age or sex or seizure-specific factors would be needed.

16.3 Evidence for Immaturity of GABA$_A$ Receptor (GABA$_A$R) Signaling in Epilepsies

GABA$_A$R signaling is well known to undergo structural and functional changes through development. The subunit composition of the GABA$_A$R complexes changes to include subunits that will provide electrophysiologic and pharmacological properties more akin to mature neurons. A typical example is the developmental shift from alpha 2 or 3 (GABRA2 or GABRA3) to alpha 1 (GABRA1) subunits, which attribute faster kinetics of the inhibitory post-synaptic currents (IPSCs) and higher sensitivity to benzodiazepines [17, 47, 60]. In addition, GABA$_A$R signaling changes from depolarizing early in development to hyperpolarizing in more mature neurons, rendering GABA$_A$R-mediated inhibition more effective in older animals [70]. This is thought to be due to the developmental shift in the balance of the activity of cation/Cl⁻ cotransporters (CCCs) that control the intracellular Cl⁻ concentration to favor cotransporters that maintain high intracellular Cl⁻ (i.e., NKCC1) in immature neurons and low intracellular Cl⁻ in mature neurons (i.e., KCC2) [6, 26, 33, 90, 97]. The developmental increase in the expression and activity of KCC2, a Cl⁻ exporting transporter, and the parallel decrease in NKCC1 eventually reduce intracellular Cl⁻, permitting the appearance of hyperpolarizing GABA$_A$R signaling in more mature neurons.

The presence of depolarizing GABA$_A$R signaling is critical for normal development, as it promotes neuronal growth, differentiation and synaptogenesis, by controlling calcium-sensitive signaling processes. In parallel, KCC2 may also modify the development of glutamatergic synapses in dendritic spines via interactions with cytoskeletal proteins, like 4.1 N, independently of any effects on GABA$_A$R regulation [62]. The absence of depolarizing GABA$_A$R signaling early in life can either be incompatible with life or disrupt neuronal differentiation and communication [6, 16, 26, 33, 43, 118, 119]. Considering the neurotrophic effects of depolarizing GABA$_A$R signaling, it is not entirely surprising that depolarizing GABA$_A$Rs are also found in pathologic conditions that favor neuritic growth and differentiation so as to promote aberrant synaptogenesis, connectivity and re-wiring, as occurs in various forms of acquired, focal-onset epilepsies (Table 16.2). Indeed, depolarizing GABA$_A$R signaling can be facilitated by neurotrophins, like brain-derived neurotrophic growth factor (BDNF), which are released after seizures [96].

Abnormal shifts in the CCC activity towards an NKCC1-dominant state or depolarizing GABA$_A$R signaling have also been found in a number of pathological conditions predisposing to or leading to epilepsy, like trauma [11, 74], ischemia [45, 83], anoxia/glucose deprivation [36] as well as after kindling [80, 96] or during the latent or epileptic state in post-status epilepticus (SE) rodent models of epilepsy [7, 12, 13, 22, 87] (Table 16.2). Under such pathological conditions, the role of GABA$_A$R signaling is not just to promote the healing and re-wiring of the brain but may acquire a pathogenic role, by promoting neuronal excitability, due to the impairment in inhibition. In further support, KCC2 deficient mice manifest early life epilepsy and histopathologic alterations reminiscent of hippocampal sclerosis [122]. Pharmacologic inhibition of depolarizing GABA$_A$R signaling using the NKCC1 inhibitor bumetanide in combination with GABA$_A$R agonists has shown antiseizure effects in certain seizure models [18, 25, 65, 68, 75, 94, 103], although model-, region-, age-, or time-dependent differences have been reported [65, 66, 68, 117, 127]. Administration of bumetanide with phenobarbital prior to seizure onset in the kainic acid induced SE model significantly enhanced the antiseizure effect of phenobarbital, in an age-dependent manner, that was attributed to the developmental decrease in NKCC1 expression [25]. Similarly, bumetanide inhibited rapid kindling of PN11 Wistar rats when it was administered prior to kindling stimuli [68] or hypoxic seizures when given prior to hypoxia in PN10 rats, even though the brain levels of bumetanide are significantly low [18]. On the other hand, in vitro studies demonstrated variable results of bumetanide when given after seizure onset that followed model, age, and region dependent patterns [54, 117]. In addition, NKCC1-knockout mice show greater susceptibility to 4-aminopyridine than wild type animals [127]. It is therefore possible that bumetanide administration prior to seizure onset and younger ages may facilitate its ability to enhance the antiseizure effects of GABA$_A$R agonists. However it is also evident that model and region specific factors or other competing mechanisms may modify its effect.

Bumetanide has also been proposed to alleviate the febrile seizure-induced neurogenesis [56] and

Table 16.2 Epilepsies associated with depolarizing $GABA_AR$ signaling

Epilepsy type/animal model	Stage in epilepsy	Findings	Reference
Human epilepsies			
Human TLE	Following surgery for intractable epilepsy	*Subiculum*	[19, 42]
		Depolarizing $GABA_AR$ signaling	
		Bicuculline inhibits interictal epileptic discharges in vitro	
		Higher probability for depolarizing $GABA_AR$ in KCC2-negative neurons	
		Microinjections of hippocampal/temporal lobe extracts in Xenopus oocytes yield depolarizing $GABA_AR$ and high NKCC1 and low KCC2 mRNA expression	[81]
		Lower probability for NKCC1 to colocalize with KCC2 in epileptic subiculum / CA1	[72]
Human epilepsy due to hypothalamic hamartomas	Following surgery for intractable epilepsy	Depolarizing $GABA_AR$ signaling in hypothalamic hamartomas	[55]
Human epilepsy due to cortical dysplasias	Following surgery for intractable epilepsy	Reduced KCC2 expression in focal cortical dysplasias	[107]
		TSC, FCD type IIB	[110]
		Increased NKCC1, reduced KCC2	
		TSC (single case)	
		Depolarizing $GABA_AR$ signaling	
		FCD type IIA	
		Increased NKCC1 and KCC2	
		FCD type I or II	[47]
		Abnormal developmental changes in the expression of NKCC1, KCC2	
		Increase in NKCC1, altered subcellular expression of KCC2 in cortical malformations (FCD type IIB, hemimegalencephaly, gangliogliomas)	[3]
Tumor-associated human epilepsy	Peritumoral cells	Increased NKCC1 expression	[20]
Animal models of epilepsy			
Post-SE epileptic rats, pilocarpine model, male Wister rats	Latent phase, 3 weeks post-SE	Depolarizing $GABA_AR$ signaling in layer 5 entorhinal cortex but not in entorhinal layer 3, subiculum, dentate gyrus, or perirhinal cortex.	[13]
Post-SE epileptic rats, pilocarpine model, adult male Sprague–Dawley rats	Established epilepsy, 2–5 months after SE	Depolarizing $GABA_AR$ signaling in granule cells of the dentate gyrus, insular, subicular neurons or the deep layers of the piriform cortex	[7, 12, 22, 87]
		Reduction of KCC2 expression in the dentate gyrus, subiculum or the deep layers of the piriform cortex	

Abnormal shift to depolarizing $GABA_AR$ signaling and/or expression of cation chloride cotransporters KCC2 and NKCC1 have been described in both human tissue derived from epileptogenic areas of patients with epilepsies, as well as in animal models of epilepsy

FCD Focal cortical dysplasia, *SE* status epilepticus

the post-SE epilepsy-associated behavioral deficits [14], but has not been shown to have anti-epileptogenic effects in post-SE epilepsy or an in vitro model [14, 75]. Depolarizing GABA$_A$R signaling also renders the injured neurons dependent upon neurotrophic factors, like BDNF, for survival, by augmenting the expression of the pan-neurotrophin receptor p75^{NTR} [108]. Neurons with depolarizing GABA signaling are therefore more amenable to dying in injured areas, which are deprived of BDNF.

Epilepsy and seizures have also been associated with disruption in the normal developmental patterns of expression of the subunits of GABA$_A$Rs (see Table 16.3). In certain – but not all – cases these reflect a return to a more immature type of GABA$_A$R subunit composition, as in studies demonstrating a reduction in the α1

Table 16.3 Abnormalities in GABA$_A$R subunit expression in human epilepsies and animal models of SE or epilepsies

Epilepsy type/animal model	Stage in epilepsy	Findings	Reference
Human epilepsies			
Human TLE	Following surgery for intractable epilepsy	Decreased GABRA3 protein in temporal neocortex (layers I-III), no change in GABRA1 or GABRA2	[64]
		Increased GABRA3, GABRA5, GABRB1, GABRB2, GABRB3 mRNA in subiculum compared to neocortex	[82]
		Decreased GABRG2 mRNA in subiculum compared to neocortex	
		Decreased GABRA1, GABRA3, GABRB3, GABRG2 protein expression in sclerotic but not in nonsclerotic hippocampus (CA1)	[89]
		Increased GABRB1, GABRB2, GABRB3 protein expression in both sclerotic and nonsclerotic hippocampus	
Human mesial TLE	Following surgery for intractable epilepsy	No change in GABRA1, GABRB1, GABRB2 mRNA expression in the amygdala	[23]
Human epilepsy due to hypothalamic hamartomas	Following surgery for intractable epilepsy	No change in GABA$_A$R subunit mRNA	[123]
Human epilepsy due to cortical dysplasias	Following surgery for intractable epilepsy	*TSC, FCD type IIB*	[110]
		Decreased GABRA1 protein	
		FCD type IIA	
		Decreased GABRA4 protein	
		FCD type I or II	[47]
		Abnormal developmental changes in the expression of GABRA1, GABRA4, GABRG2 protein	
Animal models of SE or epilepsy			
Post-SE, pilocarpine model, PN10 rats	Nonepileptic (adult)	*In dentate gyrus granule cells*	[126]
		Increase in GABRA1 mRNA	
		No change in GABRA4, GABRD mRNA	
Post-SE rats, pilocarpine model, adult male Sprague–Dawley rats	1–8 days post-SE	*In CA1 pyramidal neurons*	[38]
		Decrease in GABRA4, GABRB2/3, GABRG2, gephyrin protein	

(continued)

Table 16.3 (continued)

Epilepsy type/animal model	Stage in epilepsy	Findings	Reference
Post-SE rats, pilocarpine model, adult rats	1–5 months post-SE	*In dentate gyrus granule cells (hippocampus)*	[15]
		Decrease in GABRA1, GABRB1 mRNA	
		Increase in GABRA4, GABRB3, GABRD, GABRE mRNA	
Post-SE rats, kainic acid, adult male Sprague–Dawley	1 month post-SE	*In dorsal hippocampus*	[104]
		Increase in GABRA1, GABRA2, GABRA4, GABRB2, GABRB3, GABRG2 protein	
		Decrease in GABRD	
Post-SE rats, kainic acid, adult male Sprague–Dawley rats	7–30 days post SE	*In dorsal hippocampus*	[113]
		Decrease in GABRA5 and GABRD mRNA	
Post-SE, Electrically induced, adult male Sprague–Dawley rats	7–30 days post SE	*In dorsal hippocampus*	[76]
		Increase in GABRA1, GABRA4, GABRB1, GABRB2, GABRB3 mRNA	
		Decrease in GABRD mRNA	
Post-SE rats, pilocarpine, adult male Sprague–Dawley rats	3–4 months post-SE	*In CA1, CA3 pyramidal neurons*	[41]
		Decrease in GABRA5 protein	
Post-SE rats, electrical stimulation of the amygdala, adult male Sprague–Dawley rats	Epileptic rats	*In hippocampus*	[58]
		Increase in GABRB3 mRNA (all regions)	
		Decrease in GABRA2 mRNA (CA3c) and GABRA4 mRNA (CA1)	
Post-SE, pilocarpine model, adult Wistar rats	Epileptic rats	*In cerebral cortex*	[67]
		Decrease in GABRA1, GABRG3, GABRD mRNA	
		Increase in GABRA5 mRNA	
Post-SE, electrical stimulation of amygdala, adult female Sprague–Dawley rats	Epileptic rats	*In hippocampus*	[8]
		Phenobarbital non-responders are more likely to have reduced GABRA1, GABRB2/3, GABRG2 protein expression in the hippocampus than responders	

SE and epilepsies have different effects upon the expression of $GABA_A R$ subunits (GABR). Their effects depend upon the type and/or model of SE or epilepsy, age at seizure occurrence, the region and timepoint after seizures when the study is conducted, and the specific subunit examined

subunit, whereas in others they indicate disrupted development [47]. Changes in $GABA_A R$ subunits may contribute to either drug refractoriness [15] or epileptogenesis [93] or the development of comorbidities.

Most of the above studies have been done in either adult animals or are derived from individuals with drug-resistant epilepsy that underwent surgical resection of the epileptogenic focus at ages when the brain is relatively more mature. Age-specific patterns of regulation by seizures have been extensively shown for the seizure-induced changes in $GABA_A R$ subunits [126].

Similarly, the effects of neonatal seizures on $GABA_A R$ signaling and CCCs are not only age [32, 53] but sex-specific as well [32]. Kainic acid induced SE in PN4-6 rats accelerated the switch to hyperpolarizing $GABA_A R$ signaling in the CA1 pyramidal neurons of males, due to an increase in KCC2 expression and decrease in NKCC1 activity [32]. In contrast, kainic acid induced SE in PN4-6 female rats, in which $GABA_A R$ signaling is not depolarizing, causes a transient return to the depolarizing signaling mode due to an increase in NKCC1 activity [32]. In this study, the sexually dimorphic response to neonatal seizures seemed to depend upon the earlier maturation of $GABA_AR$ signaling in the female hippocampus, attributed to a higher expression of KCC2 and lower NKCC1 activity in females [32]. Sex differences in the expression of KCC2 and NKCC1 or $GABA_AR$ signaling in the hippocampus have also been confirmed in other studies [73, 79]. In addition, brief kainic acid seizures augment the activity of KCC2 shortly after induction of seizures in neonatal male rats [53]. It should be noted however that these studies relate to the postictal – acute or sub-acute – stages of neonatal SE. During the acute ictal phase of the SE, there is plenty of evidence to support that $GABA_AR$ signaling becomes depolarizing [25, 52].

The consequences of these seizure effects on the direction of $GABA_AR$ signaling could impact upon the subsequent susceptibility of the animal to seizures, affect its ability to stop seizure propagation, or alter cognitive abilities. For example, activation of $GABA_AR$ signaling in the anterior substantia nigra pars reticulata (SNR) in rats has important age and sex specific role in controlling seizure propagation in the flurothyl model [114, 115]. Exposure of male and female PN4-6 rats to kainic acid induced SE, at the time when $GABA_AR$ signaling is depolarizing, causes a precocious appearance of hyperpolarizing $GABA_AR$ signaling due to increase in KCC2 expression [35] and disrupts the $GABA_AR$-sensitive anticonvulsant function of the anterior SNR in the flurothyl seizure model (unpublished data). It is possible that the early deprivation of the SNR of the neurotrophic effects of the depolarizing $GABA_AR$ signaling effects may impair its development, leading to these long-lasting deficits.

16.4 Other Immature or Dysmature Features Associated with Epilepsies

Seizures may cause long-lasting changes in other signaling pathways involved in neurodevelopmental plasticity. The mTOR pathway has attracted a lot of research interest recently because it is central to cellular differentiation and growth. The mTOR pathway may become dysregulated in several seizure and epilepsy models [34, 92, 111, 121, 125] even if not necessarily caused by genetic disruption of components of the mTOR pathway. The ability of rapamycin, an mTOR inhibitor, to suppress epilepsy in these models as well as prevent or reverse certain of the histopathological or cognitive abnormalities has supported its role as a potential epileptostatic and potentially disease-modifying treatment. We use the term "epileptostatic" (i.e., epilepsy is on hold) to indicate that inhibition of the expression of epilepsy and associated histopathological abnormalities occur only in the presence of mTOR inhibition but re-appear after the mTOR inhibitor is withdrawn. Other neurodevelopmental processes may also be affected, such as excitatory signaling or myelination. A neonatal brief kainic acid seizure may reduce the surface expression of the NMDA receptor (NR) subunit that normally emerges through developmental maturation, NR2A [21]. Seizures during the period of myelination can halt or impair myelination in both animal and human studies [24, 50, 91].

Loss of dendritic spines and less frequently shortening of dendritic length or abnormal dendritic branching patterns may be seen in patients with TLE or focal epilepsies [5, 10, 31, 44, 71, 102, 116]. Whether dendritic pathologies cause epilepsy is a matter open for investigation. Certainly many known etiologies of epilepsies demonstrate similar dendritic pathologies, including Rett syndrome [2] and tuberous sclerosis (TSC) [112] implicating the affected pathways (MeCP2, mTOR) in their pathogenesis. However the evidence that dendritic pathology causes epilepsy is currently lacking. Animal studies of seizures or epilepsy, in models like kindling, iron-induced cortical epilepsy, tetanus toxin model, or post-SE models of epilepsy have demonstrated similar dendritic abnormalities suggesting that seizures may impair dendritic architecture and spine development [1, 39, 49, 57, 77, 120, 125]. The lack of selectivity of the dendritic abnormalities for the epileptogenic focus, rather poses this feature as contributory to the overall neuronal dysfunction and seizure-associated comorbidities and to a lesser degree as causative of epilepsy.

In addition, dysplastic lesions may be encountered in pathological specimens from patients with TLE [9]. These can be found as clusters of granular neurons in layer 2 of the neocortex, nodular heterotopias in the temporal lobe, or heterotopic isolated neurons in the gray-white matter junction or deep subcortical white matter. It is currently unclear whether these dysplastic lesions are causative of or secondary to TLE. However, the possibility that such lesions may predispose to the development of TLE is supported by studies that demonstrate epileptogenic potential of these dysplastic lesions [27] as well as the animal studies demonstrating the pro-epileptogenic potential of pre-existing dysplastic lesions in two-hit seizure models [37, 99].

16.5 Conclusions

Seizures and several pathologies predisposing to focal-onset epilepsies may trigger the re-acquisition of immature features in mature neurons that are integrated in the epileptogenic focus. The appearance of these immature features is influenced by age and sex-specific factors, at least for certain of the events that precipitate epilepsies. We propose that this untimely re-acquisition of the immature features is not equivalent to rejuvenation of the brain but may rather represent a de-synchronization of the homeostatic mechanisms that were in place to maintain normal physiology. In other words the maladaptive interactions and integration of these immature components with otherwise appropriately functioning brain regions may contribute to the increased excitability and underlying pathological changes seen in the epileptic focus. Furthermore, such effects may disrupt normal brain development, leading to long-lasting impairments in networks that are critical for either seizure control, like the SNR, or for information processing leading to cognitive dysfunction.

A number of important unresolved questions arise. Under which conditions does the untimely presence of immature features and functions in the seizure-exposed brain promote epileptogenesis or cognitive decline? Conversely, what are the factors that can compensate and prevent disease progression? Are these functional changes different in epileptogenic foci than in regions that are secondarily affected by propagated seizures and why? What are the mechanisms leading to seizure-induced neurogenesis and how are these modified by age or sex or seizure-specific factors? Under which conditions might aberrant neurogenesis or abnormal $GABA_AR$ signaling have a pathogenic role in epileptogenesis or cognitive processes? What is the key switch mechanism that shifts depolarizing $GABA_AR$ signaling from promoting neurotrophic and healing processes in seizure-exposed or injured brains to facilitating excitability, seizure maintenance, and potentially epileptogenesis? Does the altered expression of $GABA_AR$ subunits in post-seizure or epileptic brain impair inhibition or could it, in certain situations, protect from the potentially excitatory effects of depolarizing GABA? It is evident from the examples presented in Tables 16.1, 16.2 and 16.3, that there is significant variability across studies, animal models, disease states, and regions suggesting that the answer may not be ubiquitous. Therefore, even if certain answers may be obtained in specific experimental paradigms, it is critical to be able to translate them into the human situation and, most specifically, to a specific individual in need of specific prognosis or treatment after a specific insult. Identifying markers that will enable us to detect and follow longitudinally, in vivo, the evolution of these changes and their functional alterations would be critical in both validating their significance and implementing individualized targeted treatments to prevent disease progression.

Acknowledgements Dr. Solomon Moshé would like to emphasize the role that Dr. Phil Schwartzkroin has had in developing the current state of affairs in the field of developmental epilepsy. Their interactions from very early on helped establish the need to investigate why the immature brain is more susceptible to seizures than perhaps other age groups, with different consequences that must be taken into account when novel interventions are designed. Dr Schwartzkroin, through a series of outstanding ground breaking studies and with his incisive remarks and keen eye for editing, has helped sculpture the current state of affairs to the field and in the years to come. Thank you Phil.

Other Acknowledgements Studies at Albert Einstein College of Medicine were supported by grants from NIH/NINDS NS 20253 and NS078333, CURE, Autism speaks, the Department of Defense, the Heffer Family Medical Foundation and the Siegel Family Foundation.

References

1. Ampuero E, Dagnino-Subiabre A, Sandoval R, Zepeda-Carreno R, Sandoval S, Viedma A, Aboitiz F, Orrego F, Wyneken U (2007) Status epilepticus induces region-specific changes in dendritic spines, dendritic length and TrkB protein content of rat brain cortex. Brain Res 1150:225–238
2. Armstrong DD (1992) The neuropathology of the Rett syndrome. Brain Dev 14(Suppl):S89–S98
3. Aronica E, Boer K, Redeker S, Spliet WG, van Rijen PC, Troost D, Gorter JA (2007) Differential expression patterns of chloride transporters, Na$^+$-K$^+$-2Cl$^-$ cotransporter and K$^+$-Cl$^-$ cotransporter, in epilepsy-associated malformations of cortical development. Neuroscience 145:185–196
4. Avanzini G, Manganotti P, Meletti S, Moshe SL, Panzica F, Wolf P, Capovilla G (2012) The system epilepsies: a pathophysiological hypothesis. Epilepsia 53:771–778
5. Belichenko PV, Dahlstrom A (1995) Studies on the 3-dimensional architecture of dendritic spines and varicosities in human cortex by confocal laser scanning microscopy and Lucifer yellow microinjections. J Neurosci Methods 57:55–61
6. Ben-Ari Y (2002) Excitatory actions of gaba during development: the nature of the nurture. Nat Rev Neurosci 3:728–739
7. Benini R, Avoli M (2006) Altered inhibition in lateral amygdala networks in a rat model of temporal lobe epilepsy. J Neurophysiol 95:2143–2154
8. Bethmann K, Fritschy JM, Brandt C, Loscher W (2008) Antiepileptic drug resistant rats differ from drug responsive rats in GABA A receptor subunit expression in a model of temporal lobe epilepsy. Neurobiol Dis 31:169–187
9. Blumcke I, Thom M, Aronica E, Armstrong DD, Bartolomei F, Bernasconi A, Bernasconi N, Bien CG, Cendes F, Coras R, Cross JH, Jacques TS, Kahane P, Mathern GW, Miyata H, Moshe SL, Oz B, Ozkara C, Perucca E, Sisodiya S, Wiebe S, Spreafico R (2013) International consensus classification of hippocampal sclerosis in temporal lobe epilepsy: a Task Force report from the ILAE Commission on Diagnostic Methods. Epilepsia 54:1315–1329
10. Blumcke I, Zuschratter W, Schewe JC, Suter B, Lie AA, Riederer BM, Meyer B, Schramm J, Elger CE, Wiestler OD (1999) Cellular pathology of hilar neu-

rons in Ammon's horn sclerosis. J Comp Neurol 414:437–453
11. Bonislawski DP, Schwarzbach EP, Cohen AS (2007) Brain injury impairs dentate gyrus inhibitory efficacy. Neurobiol Dis 25:163–169
12. Bortel A, Longo D, de Guzman P, Dubeau F, Biagini G, Avoli M (2010) Selective changes in inhibition as determinants for limited hyperexcitability in the insular cortex of epileptic rats. Eur J Neurosci 31:2014–2023
13. Bragin DE, Sanderson JL, Peterson S, Connor JA, Muller WS (2009) Development of epileptiform excitability in the deep entorhinal cortex after status epilepticus. Eur J Neurosci 30:611–624
14. Brandt C, Nozadze M, Heuchert N, Rattka M, Loscher W (2010) Disease-modifying effects of phenobarbital and the NKCC1 inhibitor bumetanide in the pilocarpine model of temporal lobe epilepsy. J Neurosci 30:8602–8612
15. Brooks-Kayal AR, Shumate MD, Jin H, Rikhter TY, Coulter DA (1998) Selective changes in single cell GABA(A) receptor subunit expression and function in temporal lobe epilepsy. Nat Med 4:1166–1172
16. Cancedda L, Fiumelli H, Chen K, Poo MM (2007) Excitatory GABA action is essential for morphological maturation of cortical neurons in vivo. J Neurosci 27:5224–5235
17. Chudomel O, Herman H, Nair K, Moshe SL, Galanopoulou AS (2009) Age- and gender-related differences in GABAA receptor-mediated postsynaptic currents in GABAergic neurons of the substantia nigra reticulata in the rat. Neuroscience 163:155–167
18. Cleary RT, Sun H, Huynh T, Manning SM, Li Y, Rotenberg A, Talos DM, Kahle KT, Jackson M, Rakhade SN, Berry G, Jensen FE (2013) Bumetanide enhances phenobarbital efficacy in a rat model of hypoxic neonatal seizures. PLoS One 8:e57148
19. Cohen I, Navarro V, Clemenceau S, Baulac M, Miles R (2002) On the origin of interictal activity in human temporal lobe epilepsy in vitro. Science 298:1418–1421
20. Conti L, Palma E, Roseti C, Lauro C, Cipriani R, de Groot M, Aronica E, Limatola C (2011) Anomalous levels of Cl- transporters cause a decrease of GABAergic inhibition in human peritumoral epileptic cortex. Epilepsia 52:1635–1644
21. Cornejo BJ, Mesches MH, Coultrap S, Browning MD, Benke TA (2007) A single episode of neonatal seizures permanently alters glutamatergic synapses. Ann Neurol 61:411–426
22. de Guzman P, Inaba Y, Biagini G, Baldelli E, Mollinari C, Merlo D, Avoli M (2006) Subiculum network excitability is increased in a rodent model of temporal lobe epilepsy. Hippocampus 16:843–860
23. de Moura JC, Tirapelli DP, Neder L, Saggioro FP, Sakamoto AC, Velasco TR, Panepucci RA, Leite JP, Assirati Junior JA, Colli BO, Carlotti Junior CG (2012) Amygdala gene expression of NMDA and GABA(A) receptors in patients with mesial temporal lobe epilepsy. Hippocampus 22:92–97

24. Dwyer BE, Wasterlain CG (1982) Electroconvulsive seizures in the immature rat adversely affect myelin accumulation. Exp Neurol 78:616–628
25. Dzhala VI, Talos DM, Sdrulla DA, Brumback AC, Mathews GC, Benke TA, Delpire E, Jensen FE, Staley KJ (2005) NKCC1 transporter facilitates seizures in the developing brain. Nat Med 11:1205–1213
26. Farrant M, Kaila K (2007) The cellular, molecular and ionic basis of GABA(A) receptor signalling. Prog Brain Res 160:59–87
27. Fauser S, Schulze-Bonhage A (2006) Epileptogenicity of cortical dysplasia in temporal lobe dual pathology: an electrophysiological study with invasive recordings. Brain 129:82–95
28. Ferland RJ, Gross RA, Applegate CD (2002) Differences in hippocampal mitotic activity within the dorsal and ventral hippocampus following flurothyl seizures in mice. Neurosci Lett 332:131–135
29. Ferland RJ, Gross RA, Applegate CD (2002) Increased mitotic activity in the dentate gyrus of the hippocampus of adult C57BL/6J mice exposed to the flurothyl kindling model of epileptogenesis. Neuroscience 115:669–683
30. Fournier NM, Botterill JJ, Marks WN, Guskjolen AJ, Kalynchuk LE (2013) Impaired recruitment of seizure-generated neurons into functional memory networks of the adult dentate gyrus following long-term amygdala kindling. Exp Neurol 244:96–104
31. Freiman TM, Eismann-Schweimler J, Frotscher M (2011) Granule cell dispersion in temporal lobe epilepsy is associated with changes in dendritic orientation and spine distribution. Exp Neurol 229:332–338
32. Galanopoulou AS (2008) Dissociated gender-specific effects of recurrent seizures on GABA signaling in CA1 pyramidal neurons: role of GABA(A) receptors. J Neurosci 28:1557–1567
33. Galanopoulou AS (2008) GABA(A) receptors in normal development and seizures: friends or foes? Curr Neuropharmacol 6:1–20
34. Galanopoulou AS, Gorter JA, Cepeda C (2012) Finding a better drug for epilepsy: the mTOR pathway as an antiepileptogenic target. Epilepsia 53:1119–1130
35. Galanopoulou AS, Kyrozis A, Moshé SL (eds) (2004) Kainate-induced status epilepticus upregulates KCC2 mRNA expression in the substantia nigra of immature rats. Epilepsia 45(Suppl 7):12
36. Galeffi F, Sah R, Pond BB, George A, Schwartz-Bloom RD (2004) Changes in intracellular chloride after oxygen-glucose deprivation of the adult hippocampal slice: effect of diazepam. J Neurosci 24:4478–4488
37. Germano IM, Sperber EF, Ahuja S, Moshe SL (1998) Evidence of enhanced kindling and hippocampal neuronal injury in immature rats with neuronal migration disorders. Epilepsia 39:1253–1260
38. Gonzalez MI, Cruz Del Angel Y, Brooks-Kayal A (2013) Down-regulation of gephyrin and GABAA receptor subunits during epileptogenesis in the CA1 region of hippocampus. Epilepsia 54:616–624
39. Gonzalez-Burgos I, Lopez-Vazquez MA, Beas-Zarate C (2004) Density, but not shape, of hippocampal dendritic spines varies after a seizure-inducing acute dose of monosodium glutamate in rats. Neurosci Lett 363:22–24
40. Hattiangady B, Rao MS, Shetty AK (2004) Chronic temporal lobe epilepsy is associated with severely declined dentate neurogenesis in the adult hippocampus. Neurobiol Dis 17:473–490
41. Houser CR, Esclapez M (2003) Downregulation of the alpha5 subunit of the GABA(A) receptor in the pilocarpine model of temporal lobe epilepsy. Hippocampus 13:633–645
42. Huberfeld G, Wittner L, Clemenceau S, Baulac M, Kaila K, Miles R, Rivera C (2007) Perturbed chloride homeostasis and GABAergic signaling in human temporal lobe epilepsy. J Neurosci 27:9866–9873
43. Hubner CA, Stein V, Hermans-Borgmeyer I, Meyer T, Ballanyi K, Jentsch TJ (2001) Disruption of KCC2 reveals an essential role of K-Cl cotransport already in early synaptic inhibition. Neuron 30:515–524
44. Isokawa M, Levesque MF (1991) Increased NMDA responses and dendritic degeneration in human epileptic hippocampal neurons in slices. Neurosci Lett 132:212–216
45. Jaenisch N, Witte OW, Frahm C (2010) Downregulation of potassium chloride cotransporter KCC2 after transient focal cerebral ischemia. Stroke 41:e151–e159
46. Jansen LA, Peugh LD, Ojemann JG (2008) GABA(A) receptor properties in catastrophic infantile epilepsy. Epilepsy Res 81:188–197
47. Jansen LA, Peugh LD, Roden WH, Ojemann JG (2010) Impaired maturation of cortical GABA(A) receptor expression in pediatric epilepsy. Epilepsia 51:1456–1467
48. Jessberger S, Romer B, Babu H, Kempermann G (2005) Seizures induce proliferation and dispersion of doublecortin-positive hippocampal progenitor cells. Exp Neurol 196:342–351
49. Jiang M, Lee CL, Smith KL, Swann JW (1998) Spine loss and other persistent alterations of hippocampal pyramidal cell dendrites in a model of early-onset epilepsy. J Neurosci 18:8356–8368
50. Jorgensen OS, Dwyer B, Wasterlain CG (1980) Synaptic proteins after electroconvulsive seizures in immature rats. J Neurochem 35:1235–1237
51. Jung KH, Chu K, Kim M, Jeong SW, Song YM, Lee ST, Kim JY, Lee SK, Roh JK (2004) Continuous cytosine-b-D-arabinofuranoside infusion reduces ectopic granule cells in adult rat hippocampus with attenuation of spontaneous recurrent seizures following pilocarpine-induced status epilepticus. Eur J Neurosci 19:3219–3226
52. Khalilov I, Holmes GL, Ben-Ari Y (2003) In vitro formation of a secondary epileptogenic mirror focus by interhippocampal propagation of seizures. Nat Neurosci 6:1079–1085

53. Khirug S, Ahmad F, Puskarjov M, Afzalov R, Kaila K, Blaesse P (2010) A single seizure episode leads to rapid functional activation of KCC2 in the neonatal rat hippocampus. J Neurosci 30:12028–12035

54. Kilb W, Sinning A, Luhmann HJ (2007) Model-specific effects of bumetanide on epileptiform activity in the in-vitro intact hippocampus of the newborn mouse. Neuropharmacology 53:524–533

55. Kim DY, Fenoglio KA, Kerrigan JF, Rho JM (2009) Bicarbonate contributes to GABAA receptor-mediated neuronal excitation in surgically resected human hypothalamic hamartomas. Epilepsy Res 83:89–93

56. Koyama R, Tao K, Sasaki T, Ichikawa J, Miyamoto D, Muramatsu R, Matsuki N, Ikegaya Y (2012) GABAergic excitation after febrile seizures induces ectopic granule cells and adult epilepsy. Nat Med 18:1271–1278

57. Kurz JE, Moore BJ, Henderson SC, Campbell JN, Churn SB (2008) A cellular mechanism for dendritic spine loss in the pilocarpine model of status epilepticus. Epilepsia 49:1696–1710

58. Lauren HB, Pitkanen A, Nissinen J, Soini SL, Korpi ER, Holopainen IE (2003) Selective changes in gamma-aminobutyric acid type A receptor subunits in the hippocampus in spontaneously seizing rats with chronic temporal lobe epilepsy. Neurosci Lett 349:58–62

59. Lauren HB, Ruohonen S, Kukko-Lukjanov TK, Virta JE, Gronman M, Lopez-Picon FR, Jarvela JT, Holopainen IE (2013) Status epilepticus alters neurogenesis and decreases the number of GABAergic neurons in the septal dentate gyrus of 9-day-old rats at the early phase of epileptogenesis. Brain Res 1516:33–44

60. Laurie DJ, Wisden W, Seeburg PH (1992) The distribution of thirteen GABAA receptor subunit mRNAs in the rat brain. III. Embryonic and postnatal development. J Neurosci 12:4151–4172

61. Lemmens EM, Lubbers T, Schijns OE, Beuls EA, Hoogland G (2005) Gender differences in febrile seizure-induced proliferation and survival in the rat dentate gyrus. Epilepsia 46:1603–1612

62. Li H, Khirug S, Cai C, Ludwig A, Blaesse P, Kolikova J, Afzalov R, Coleman SK, Lauri S, Airaksinen MS, Keinanen K, Khiroug L, Saarma M, Kaila K, Rivera C (2007) KCC2 interacts with the dendritic cytoskeleton to promote spine development. Neuron 56:1019–1033

63. Liu H, Kaur J, Dashtipour K, Kinyamu R, Ribak CE, Friedman LK (2003) Suppression of hippocampal neurogenesis is associated with developmental stage, number of perinatal seizure episodes, and glucocorticosteroid level. Exp Neurol 184:196–213

64. Loup F, Picard F, Andre VM, Kehrli P, Yonekawa Y, Wieser HG, Fritschy JM (2006) Altered expression of alpha3-containing GABAA receptors in the neocortex of patients with focal epilepsy. Brain 129:3277–3289

65. Mares P (2009) Age- and dose-specific anticonvulsant action of bumetanide in immature rats. Physiol Res 58:927–930

66. Margineanu DG, Klitgaard H (2006) Differential effects of cation-chloride co-transport-blocking diuretics in a rat hippocampal slice model of epilepsy. Epilepsy Res 69:93–99

67. Mathew J, Balakrishnan S, Antony S, Abraham PM, Paulose CS (2012) Decreased GABA receptor in the cerebral cortex of epileptic rats: effect of Bacopa monnieri and Bacoside-A. J Biomed Sci 19:25

68. Mazarati A, Shin D, Sankar R (2009) Bumetanide inhibits rapid kindling in neonatal rats. Epilepsia 50:2117–2122

69. McCabe BK, Silveira DC, Cilio MR, Cha BH, Liu X, Sogawa Y, Holmes GL (2001) Reduced neurogenesis after neonatal seizures. J Neurosci 21:2094–2103

70. Mueller AL, Chesnut RM, Schwartzkroin PA (1983) Actions of GABA in developing rabbit hippocampus: an in vitro study. Neurosci Lett 39:193–198

71. Multani P, Myers RH, Blume HW, Schomer DL, Sotrel A (1994) Neocortical dendritic pathology in human partial epilepsy: a quantitative Golgi study. Epilepsia 35:728–736

72. Munoz A, Mendez P, DeFelipe J, Alvarez-Leefmans FJ (2007) Cation-chloride cotransporters and GABA-ergic innervation in the human epileptic hippocampus. Epilepsia 48:663–673

73. Murguia-Castillo J, Beas-Zarate C, Rivera-Cervantes MC, Feria-Velasco AI, Urena-Guerrero ME (2013) NKCC1 and KCC2 protein expression is sexually dimorphic in the hippocampus and entorhinal cortex of neonatal rats. Neurosci Lett 552:52–57

74. Nabekura J, Ueno T, Okabe A, Furuta A, Iwaki T, Shimizu-Okabe C, Fukuda A, Akaike N (2002) Reduction of KCC2 expression and GABAA receptor-mediated excitation after in vivo axonal injury. J Neurosci 22:4412–4417

75. Nardou R, Ben-Ari Y, Khalilov I (2009) Bumetanide, an NKCC1 antagonist, does not prevent formation of epileptogenic focus but blocks epileptic focus seizures in immature rat hippocampus. J Neurophysiol 101:2878–2888

76. Nishimura T, Schwarzer C, Gasser E, Kato N, Vezzani A, Sperk G (2005) Altered expression of GABA(A) and GABA(B) receptor subunit mRNAs in the hippocampus after kindling and electrically induced status epilepticus. Neuroscience 134:691–704

77. Nishizuka M, Okada R, Seki K, Arai Y, Iizuka R (1991) Loss of dendritic synapses in the medial amygdala associated with kindling. Brain Res 552:351–355

78. Nunes ML, Liptakova S, Veliskova J, Sperber EF, Moshe SL (2000) Malnutrition increases dentate granule cell proliferation in immature rats after status epilepticus. Epilepsia 41(Suppl 6):S48–S52

79. Nunez JL, McCarthy MM (2007) Evidence for an extended duration of GABA-mediated excitation in the developing male versus female hippocampus. Dev Neurobiol 67:1879–1890

80. Okabe A, Ohno K, Toyoda H, Yokokura M, Sato K, Fukuda A (2002) Amygdala kindling induces upregulation of mRNA for NKCC1, a Na(+), K(+)-2Cl(–)

cotransporter, in the rat piriform cortex. Neurosci Res 44:225–229

81. Palma E, Amici M, Sobrero F, Spinelli G, Di Angelantonio S, Ragozzino D, Mascia A, Scoppetta C, Esposito V, Miledi R, Eusebi F (2006) Anomalous levels of Cl- transporters in the hippocampal subiculum from temporal lobe epilepsy patients make GABA excitatory. Proc Natl Acad Sci U S A 103:8465–8468

82. Palma E, Spinelli G, Torchia G, Martinez-Torres A, Ragozzino D, Miledi R, Eusebi F (2005) Abnormal GABAA receptors from the human epileptic hippocampal subiculum microtransplanted to Xenopus oocytes. Proc Natl Acad Sci U S A 102:2514–2518

83. Papp E, Rivera C, Kaila K, Freund TF (2008) Relationship between neuronal vulnerability and potassium-chloride cotransporter 2 immunoreactivity in hippocampus following transient forebrain ischemia. Neuroscience 154:677–689

84. Parent JM, Janumpalli S, McNamara JO, Lowenstein DH (1998) Increased dentate granule cell neurogenesis following amygdala kindling in the adult rat. Neurosci Lett 247:9–12

85. Parent JM, Jessberger S, Gage FH, Gong C (2007) Is neurogenesis reparative after status epilepticus? Epilepsia 48(Suppl 8):69–71

86. Parent JM, Yu TW, Leibowitz RT, Geschwind DH, Sloviter RS, Lowenstein DH (1997) Dentate granule cell neurogenesis is increased by seizures and contributes to aberrant network reorganization in the adult rat hippocampus. J Neurosci 17:3727–3738

87. Pathak HR, Weissinger F, Terunuma M, Carlson GC, Hsu FC, Moss SJ, Coulter DA (2007) Disrupted dentate granule cell chloride regulation enhances synaptic excitability during development of temporal lobe epilepsy. J Neurosci 27:14012–14022

88. Pekcec A, Fuest C, Muhlenhoff M, Gerardy-Schahn R, Potschka H (2008) Targeting epileptogenesis-associated induction of neurogenesis by enzymatic depolysialylation of NCAM counteracts spatial learning dysfunction but fails to impact epilepsy development. J Neurochem 105:389–400

89. Pirker S, Schwarzer C, Czech T, Baumgartner C, Pockberger H, Maier H, Hauer B, Sieghart W, Furtinger S, Sperk G (2003) Increased expression of GABA(A) receptor beta-subunits in the hippocampus of patients with temporal lobe epilepsy. J Neuropathol Exp Neurol 62:820–834

90. Plotkin MD, Snyder EY, Hebert SC, Delpire E (1997) Expression of the Na-K-2Cl cotransporter is developmentally regulated in postnatal rat brains: a possible mechanism underlying GABA's excitatory role in immature brain. J Neurobiol 33:781–795

91. Pulsipher DT, Seidenberg M, Morton JJ, Geary E, Parrish J, Hermann B (2007) MRI volume loss of subcortical structures in unilateral temporal lobe epilepsy. Epilepsy Behav 11:442–449

92. Raffo E, Coppola A, Ono T, Briggs SW, Galanopoulou AS (2011) A pulse rapamycin therapy for infantile spasms and associated cognitive decline. Neurobiol Dis 43:322–329

93. Raol YH, Lund IV, Bandyopadhyay S, Zhang G, Roberts DS, Wolfe JH, Russek SJ, Brooks-Kayal AR (2006) Enhancing GABA(A) receptor alpha 1 subunit levels in hippocampal dentate gyrus inhibits epilepsy development in an animal model of temporal lobe epilepsy. J Neurosci 26:11342–11346

94. Reid KH, Guo SZ, Iyer VG (2000) Agents which block potassium-chloride cotransport prevent sound-triggered seizures in post-ischemic audiogenic seizure-prone rats. Brain Res 864:134–137

95. Ribak CE, Shapiro LA, Yan XX, Dashtipour K, Nadler JV, Obenaus A, Spigelman I, Buckmaster PS (2012) Seizure-induced formation of basal dendrites on granule cells of the rodent dentate gyrus. In: Noebels JL, Avoli M, Rogawski MA, Olsen RW, Delgado-Escueta AV (eds) Jasper's basic mechanisms of the epilepsies, 4th edn. Oxford, New York, pp 484–493

96. Rivera C, Li H, Thomas-Crusells J, Lahtinen H, Viitanen T, Nanobashvili A, Kokaia Z, Airaksinen MS, Voipio J, Kaila K, Saarma M (2002) BDNF-induced TrkB activation down-regulates the K+-Cl- cotransporter KCC2 and impairs neuronal Cl- extrusion. J Cell Biol 159:747–752

97. Rivera C, Voipio J, Payne JA, Ruusuvuori E, Lahtinen H, Lamsa K, Pirvola U, Saarma M, Kaila K (1999) The K+/Cl- co-transporter KCC2 renders GABA hyperpolarizing during neuronal maturation. Nature 397:251–255

98. Sankar R, Shin D, Liu H, Katsumori H, Wasterlain CG (2000) Granule cell neurogenesis after status epilepticus in the immature rat brain. Epilepsia 41(Suppl 6):S53–S56

99. Scantlebury MH, Gibbs SA, Foadjo B, Lema P, Psarropoulou C, Carmant L (2005) Febrile seizures in the predisposed brain: a new model of temporal lobe epilepsy. Ann Neurol 58:41–49

100. Scharfman HE, Goodman JH, Sollas AL (2000) Granule-like neurons at the hilar/CA3 border after status epilepticus and their synchrony with area CA3 pyramidal cells: functional implications of seizure-induced neurogenesis. J Neurosci 20:6144–6158

101. Scharfman HE, Pierce JP (2012) New insights into the role of hilar ectopic granule cells in the dentate gyrus based on quantitative anatomic analysis and three-dimensional reconstruction. Epilepsia 53(Suppl 1):109–115

102. Scheibel ME, Crandall PH, Scheibel AB (1974) The hippocampal-dentate complex in temporal lobe epilepsy. A Golgi study. Epilepsia 15:55–80

103. Schwartzkroin PA, Baraban SC, Hochman DW (1998) Osmolarity, ionic flux, and changes in brain excitability. Epilepsy Res 32:275–285

104. Schwarzer C, Tsunashima K, Wanzenbock C, Fuchs K, Sieghart W, Sperk G (1997) GABA(A) receptor subunits in the rat hippocampus II: altered distribution in kainic acid-induced temporal lobe epilepsy. Neuroscience 80:1001–1017

105. Scott BW, Wang S, Burnham WM, De Boni U, Wojtowicz JM (1998) Kindling-induced neurogenesis in the dentate gyrus of the rat. Neurosci Lett 248:73–76

106. Shetty AK, Hattiangady B, Rao MS, Shuai B (2012) Neurogenesis response of middle-aged hippocampus to acute seizure activity. PLoS One 7:e43286

107. Shimizu-Okabe C, Tanaka M, Matsuda K, Mihara T, Okabe A, Sato K, Inoue Y, Fujiwara T, Yagi K, Fukuda A (2011) KCC2 was downregulated in small neurons localized in epileptogenic human focal cortical dysplasia. Epilepsy Res 93:177–184

108. Shulga A, Magalhaes AC, Autio H, Plantman S, di Lieto A, Nykjaer A, Carlstedt T, Risling M, Arumae U, Castren E, Rivera C (2012) The loop diuretic bumetanide blocks posttraumatic p75NTR upregulation and rescues injured neurons. J Neurosci 32:1757–1770

109. Sugaya Y, Maru E, Kudo K, Shibasaki T, Kato N (2010) Levetiracetam suppresses development of spontaneous EEG seizures and aberrant neurogenesis following kainate-induced status epilepticus. Brain Res 1352:187–199

110. Talos DM, Sun H, Kosaras B, Joseph A, Folkerth RD, Poduri A, Madsen JR, Black PM, Jensen FE (2012) Altered inhibition in tuberous sclerosis and type IIb cortical dysplasia. Ann Neurol 71:539–551

111. Talos DM, Sun H, Zhou X, Fitzgerald EC, Jackson MC, Klein PM, Lan VJ, Joseph A, Jensen FE (2012) The interaction between early life epilepsy and autistic-like behavioral consequences: a role for the mammalian target of rapamycin (mTOR) pathway. PLoS One 7:e35885

112. Tavazoie SF, Alvarez VA, Ridenour DA, Kwiatkowski DJ, Sabatini BL (2005) Regulation of neuronal morphology and function by the tumor suppressors Tsc1 and Tsc2. Nat Neurosci 8:1727–1734

113. Tsunashima K, Schwarzer C, Kirchmair E, Sieghart W, Sperk G (1997) GABA(A) receptor subunits in the rat hippocampus III: altered messenger RNA expression in kainic acid-induced epilepsy. Neuroscience 80:1019–1032

114. Veliskova J, Moshe SL (2001) Sexual dimorphism and developmental regulation of substantia nigra function. Ann Neurol 50:596–601

115. Veliskova J, Moshe SL (2006) Update on the role of substantia nigra pars reticulata in the regulation of seizures. Epilepsy Curr 6:83–87

116. von Campe G, Spencer DD, de Lanerolle NC (1997) Morphology of dentate granule cells in the human epileptogenic hippocampus. Hippocampus 7:472–488

117. Wahab A, Albus K, Heinemann U (2011) Age- and region-specific effects of anticonvulsants and bumetanide on 4-aminopyridine-induced seizure-like events in immature rat hippocampal-entorhinal cortex slices. Epilepsia 52:94–103

118. Wang DD, Kriegstein AR (2008) GABA regulates excitatory synapse formation in the neocortex via NMDA receptor activation. J Neurosci 28:5547–5558

119. Wang DD, Kriegstein AR (2011) Blocking early GABA depolarization with bumetanide results in permanent alterations in cortical circuits and sensorimotor gating deficits. Cereb Cortex 21:574–587

120. Willmore LJ, Ballinger WE Jr, Boggs W, Sypert GW, Rubin JJ (1980) Dendritic alterations in rat isocortex within an iron-induced chronic epileptic focus. Neurosurgery 7:142–146

121. Wong M, Crino PB (2012) mTOR and epileptogenesis in developmental brain malformations. In: Noebels JL, Avoli M, Rogawski MA, Olsen RW, Delgado-Escueta AV (eds) Jasper's basic mechanisms of the epilepsies, 4th edn. Oxford, New York, pp 835–844

122. Woo NS, Lu J, England R, McClellan R, Dufour S, Mount DB, Deutch AY, Lovinger DM, Delpire E (2002) Hyperexcitability and epilepsy associated with disruption of the mouse neuronal-specific K-Cl cotransporter gene. Hippocampus 12:258–268

123. Wu J, Chang Y, Li G, Xue F, DeChon J, Ellsworth K, Liu Q, Yang K, Bahadroani N, Zheng C, Zhang J, Rekate H, Rho JM, Kerrigan JF (2007) Electrophysiological properties and subunit composition of GABAA receptors in patients with gelastic seizures and hypothalamic hamartoma. J Neurophysiol 98:5–15

124. Xiu-Yu S, Ruo-Peng S, Ji-Wen W (2007) Consequences of pilocarpine-induced recurrent seizures in neonatal rats. Brain Dev 29:157–163

125. Zeng LH, Xu L, Rensing NR, Sinatra PM, Rothman SM, Wong M (2007) Kainate seizures cause acute dendritic injury and actin depolymerization in vivo. J Neurosci 27:11604–11613

126. Zhang G, Raol YH, Hsu FC, Coulter DA, Brooks-Kayal AR (2004) Effects of status epilepticus on hippocampal GABAA receptors are age-dependent. Neuroscience 125:299–303

127. Zhu L, Polley N, Mathews GC, Delpire E (2008) NKCC1 and KCC2 prevent hyperexcitability in the mouse hippocampus. Epilepsy Res 79:201–212

Are Alterations in Transmitter Receptor and Ion Channel Expression Responsible for Epilepsies?

17

Kim L. Powell, Katarzyna Lukasiuk, Terence J. O'Brien, and Asla Pitkänen

Abstract

Neuronal voltage-gated ion channels and ligand-gated synaptic receptors play a critical role in maintaining the delicate balance between neuronal excitation and inhibition within neuronal networks in the brain. Changes in expression of voltage-gated ion channels, in particular sodium, hyperpolarization-activated cyclic nucleotide-gated (HCN) and calcium channels, and ligand-gated synaptic receptors, in particular GABA and glutamate receptors, have been reported in many types of both genetic and acquired epilepsies, in animal models and in humans. In this chapter we review these and discuss the potential pathogenic role they may play in the epilepsies.

Keywords

Genetic generalized epilepsy • Acquired epilepsy • Voltage-gated ion channels • Ligand-gated ion channels • Animal models of epilepsy

K.L. Powell • T.J. O'Brien (✉)
Department of Medicine, The Royal Melbourne Hospital, The University of Melbourne,
4th Floor Clinical Sciences Building, Royal Parade, Parkville 3050, VIC, Australia
e-mail: kpowell@unimelb.edu.au;
obrientj@unimelb.edu.au

K. Lukasiuk
The Nencki Institute of Experimental Biology, Polish Academy of Sciences,
3 Pasteur St, 02 093 Warsaw, Poland
e-mail: k.lukasiuk@nencki.gov.pl

A. Pitkänen, M.D., Ph.D. (✉)
Department of Neurobiology, A.I. Virtanen Institute for Molecular Sciences,
University of Eastern Finland,
Neulaniementie 2, P.O. Box 1627,
FIN-70 211 Kuopio, Finland

Department of Neurology, Kuopio University Hospital, Kuopio, Finland
e-mail: asla.pitkanen@uef.fi

H.E. Scharfman and P.S. Buckmaster (eds.), *Issues in Clinical Epileptology: A View from the Bench*, Advances in Experimental Medicine and Biology 813, DOI 10.1007/978-94-017-8914-1_17, © Springer Science+Business Media Dordrecht 2014

17.1   Introduction

Neuronal voltage-gated ion channels and ligand-gated synaptic receptors play a critical role in maintaining the delicate balance between excitation and inhibition within neuronal networks in the brain that enables normal brain electrical function [10, 80]. The consequence of disturbing the expression or function of these, even relatively subtly, can render neuronal networks liable to fire in an inappropriate, hyper-synchronous, oscillatory manner which can be self-sustained, engage other neuronal networks, and result in a clinical epileptic seizure. Changes in expression of voltage-gated ion channels and/or ligand-gated synaptic receptors have been reported in many types of both genetic and acquired epilepsies. The causative relationship between these changes likely varies between different epilepsy syndromes. In some cases these changes are clearly causative of the epilepsy; in some they may represent susceptibility factors that render the brain more liable to become epileptic following a second insult (acquired or genetic); while in others they may be compensatory or even epiphenomena – related to the precipitating insult but not directly impacting on the epilepsy.

In this chapter we will outline some of the changes in expression of voltage-gated ion channels and ligand-gated synaptic receptors that have been reported in association with the development of both genetic and acquired epilepsies, in animals and humans, and discuss their potential pathogenic role.

17.2   Genetic Generalized Epilepsy (GGE)

17.2.1 Overview

The genetic generalised epilepsies (GGEs) represent approximately 20–30 % of epilepsy cases and have a particularly high prevalence among children and adolescents [29, 31]. Patients with GGE have seizures that arise synchronously in both hemispheres on the electroencephalogram

(EEG) without any identifiable structural brain abnormality [3]. Patients with GGE syndromes can experience a number of different seizure types, including generalised tonic-clonic seizures, myoclonic seizures and absence seizures. The GGEs are a complex group of disorders with the aetiology presumed to largely genetic [3]. The underlying pathophysiological basis of the GGEs is still incompletely understood, but it is generally believed that in most cases more than one genetic abnormality contributes to determine the epilepsy phenotype (i.e. polygenic).

There are many reports in the literature describing mutations in voltage-gated sodium [18], potassium [30], calcium [104], HCN channels [76] and GABA receptors [104] in patients with GGEs. However, because the epileptogenic networks that generate the seizures in GGE are bilaterally and diffusely distributed, patients do not undergo epilepsy surgery and therefore brain tissue to examine for protein expression are rarely available. As a result most studies investigating changes in expression of ion channels and receptors relevant to GGE come from animal models. Of particular importance has been the Genetic Absence Epilepsy Rat from Strasbourg (GAERS) and Wistar Albino Glaxo/Rij-rat (WAG/Rij), which are the two most validated polygenic rat models of GGE [12, 52]. GAERS and WAG/Rij rats were both independently derived from selective inbreeding of Wistar rat colonies that spontaneously expressed generalised spike-and-wave discharges (SWDs) on EEG recordings, to generate strains that expresses frequent and prominent spontaneous absence-like seizures accompanied by generalised SWDs that electrophysiologically resemble those seen in human GGE with absence seizures. In both strains, the rats usually begin to display seizures in the second and third month post-natal, becoming longer and more frequent as the animals mature, being fully manifest in most WAG/Rij and in all GAERS by 4 months of age [12, 53]. It is well established and accepted that GAERS and the WAG/Rij strains are excellent models of human GGE with absence seizures because they parallel many of the features seen in humans with GGE such as; seizure, behaviour, electrophysiology, pathophysiology and pharmacology (reviewed by [12, 15, 68]).

Many different ion channels and receptors, including GABA, glutamate, sodium, chloride, calcium and HCN channels, have been implicated in the pathogenesis of epilepsy. There are numerous reviews detailing mutations in these genes, however this review will focus on alterations in expression that occur in ion channels and receptors in rat polygenic models of GGE – GAERS and WAG/Rij.

17.2.2 Voltage-Gated Ion Channels

17.2.2.1 Hyperpolarization-Activated Cyclic Nucleotide-Gated (HCN) Channels

HCN channels the generate I_h current in the brain, which modulates pacemaker activity and cellular excitability [63, 79]. All four HCN isoforms, HCN1-4, are expressed with regional and developmental differences in the brain and are differentially modulated by cAMP [7, 101].

Within the thalamocortical circuit, which is critical in the generation of seizures in patients with GGE, HCN1 channels are abundantly expressed in the cortex and HCN2 and HCN4 channels are abundantly expressed in the thalamus [57]. Several studies have examined HCN channel expression in GAERS and WAG/Rij rats reporting similar results. Using *in situ* hybridization, HCN1 channel mRNA expression is increased in distinct thalamic nuclei in epileptic GAERS, namely the reticular nucleus and ventroposterior medial nucleus of the thalamus, with no changes in expression in the somatosensory cortex or of HCN2 and HCN4 channels [41]. In adult WAG/Rij rats a reduction in HCN1 protein expression has been reported in the neocortex, hippocampus and cerebellum [89] whereas HCN1 channel protein expression is increased in the dorsal lateral geniculate nucleus of the thalamus [9, 32]. In another study, HCN1 protein expression was reported to be down regulated by 33 % in the neocortex in 1 month old WAG/Rij rats that were not yet experiencing spontaneous absence seizures. The decrease in HCN1 expression became

more progressive by three (56 % reduction) and 6 months (68 % reduction) [40].

17.2.2.2 Voltage-Gated Calcium Channels

Low voltage-activated, *"T-type"*, calcium channels are recognized to play a key role in neuronal burst firing in neurons in the thalamus which are critical in generating the hypersynchronous thalamocortical oscillations that underlie generalized SWDs [68]. Therefore alterations in expression of T-type channels have significant potential to play a pathogenic role in GGE. The three T-type calcium channel subtypes, $Ca_V3.1$, $Ca_V3.2$ and $Ca_V3.3$, have unique biophysical, pharmacological and regulatory properties [67] with differential expression within the thalamocortical circuit [73]. Mutations in the human *CACNA1H*, which encodes $Ca_V3.2$, have been found in patients with different GGE syndromes [11, 31, 45]. Exogenous expression of mutant human $Ca_V3.2$ channels reveal a variety of biophysical changes [37, 64, 100]. Neurons from the thalamic reticular nucleus in GAERS, which plays a key role in regulation of the oscillatory thalamocortical network activity that underlies absence seizure-associated generalized SWDs in these animals, have been found to have a significant increase in T-type calcium currents compared to non-epileptic control rats [95]. In a developmental expression study in GAERS, $Ca_V3.2$ mRNA expression was found to be elevated in the reticular nucleus of the thalamus in young animals before the onset of spontaneous absence seizures and $Ca_V3.1$ and $Ca_V3.2$ mRNA expression was increased in the ventral posterior thalamic relay nuclei and reticular nucleus of the thalamus of adult epileptic animals respectively [93]. Complimentary findings have been documented in WAG/Rij rats. mRNA expression of all T-type calcium channels were found to be elevated in distinct thalamic nuclei in young WAG/Rij rats (P18-28) preceding seizure onset; $Ca_V3.1$ was shown to be increased in the lateral geniculate nucleus and centrolateral nucleus, $Ca_V3.2$ was increased only in the reticular nucleus and $Ca_V3.3$ was increased in the centrolateral nucleus and reticular nucleus [8].

17.2.2.3 Voltage-Gated Sodium Channels

Voltage-gated sodium channels are responsible for the initiation and propagation of action potentials. Functional channels consists of one α subunit (Nav1.1–Nav1.9) and a variable number of β (β1–β4) subunits. Subunits are composed of four domains. Each domain contains six transmembrane domains, as well as voltage sensor and pore forming domains. β subunits are smaller and contain one anchoring transmembrane domain and large extracellular domain [51]. In the somatosensory cortex of epileptic WAG/Rij animals, a significant upregulation of $Na_v1.1$ and $Na_v1.6$ mRNA and protein levels was reported with the changes being localised to layer II-IV cortical neurons with immunohistochemistry [39]. Interestingly, long term treatment of WAG/Rij rats with ethosuximide commencing prior to the onset of spontaneous absence seizures not only suppressed seizures but it also completely abolished the abnormal expression of $Na_v1.1$, $Na_v1.6$ and HCN1 when examined in 5 month old WAG/Rij rats [5].

17.2.3 Altered Expression of Ligand Gated Ion Channels

17.2.3.1 GABA Receptors

Fast responses to GABA are mediated by ligand-gated GABA_A receptors whereas slow responses are mediated by G-protein coupled GABA_B receptors. Homeostatic balance of GABAergic and glutamatergic neurotransmission is critical for the maintenance of neuronal excitability. In absence epilepsy, the neuronal hyperexcitability which underlies absence seizure generation in the thalamocortical circuit is hypothesised to be due to an imbalance between excitatory and inhibitory neurotransmission [15].

17.2.3.2 GABA_A Receptors

GABAA receptors are pentamers consisting of multiple subunit subtypes, including α (α1-α2), β (β1-β3), γ (γ1-γ3), δ, ε, π, θ, and σ (σ1-σ3) subunits. Properties of GABAA receptor strongly depend on subunit composition [25]. The most common subunit composition contains two α subunits, two β subunits and a γ subunit [50]. Expression of α

and β subunits is sufficient for the production of GABA-gated chloride channels, while the γ subunit is required for modulation by benzodiazepines. Alterations in GABAergic inhibitory neurotransmission can influence neuronal excitability and indeed alterations in expression of GABA receptors have been reported in GAERS and WAG/Rij. A study by Spreafico et al. [86] found decreased immunofluorescence for β2–β3 subunits of GABA_A receptors in the sensorimotor cortex and anterior thalamic areas of epileptic GAERS [86]. However, in WAG/Rij rats conflicting results have been reported. An increase in the expression of α4 and δ subunits of the GABA_A receptor was observed in the relay nuclei of adult epileptic WAG/Rij animals [70] whereas decreased immunoreactivity of α3 subunit of the GABA_A receptor was reported at inhibitory synapses in the reticular nucleus of the thalamus [47].

17.2.3.3 GABA_B Receptors

The GABA_B receptors are metabotropic transmembrane receptors that are linked to potassium channels via G-proteins, thus GABA_B receptors mediate GABAergic slow responses. They are composed of two subunits; GABAB1 and GABAB2 with two splice variants of GABAB1 [34]. The GABAB1a and GABAB1b subunits are thought to be the site of agonist binding, while the GABAB2 subunit activates the G-protein signalling pathway. Alterations in GABA_B receptor subunit expression and distribution have been reported in WAG/Rij rats with a marked reduction in $GABA_{B1b}$, $GABA_{B1ac}$, $GABA_{B1d}$ and $GABA_{B1bc}$ mRNA levels in the cortex, whereas $GABA_{B1a}$ and $GABA_{B2}$ mRNA levels were unchanged [55]. Alterations in GABA_B receptor expression in the thalamocortical circuit has been reported in one study on GAERS [74]. $GABA_{B1}$ mRNA expression was shown to be increased in the somatosensory cortex but decreased in the ventrobasal nucleus of the thalamus. However, protein expression showed a different pattern of expression. Both $GABA_{B1}$ and $GABA_{B2}$ receptors were shown to be increased in all regions of the thalamocortical circuit (somatosensory cortex, ventrobasal nucleus and reticular nucleus of the thalamus) [74]. Moreover, transgenic mice overexpressing either GABAB1 subunits show an epileptic phenotype characterized by spontaneous, recurrent atypical absence seizures [87, 103].

17.2.3.4 Ionotropic Glutamate Receptors

The ionotropic glutamate receptor (iGluR) family of excitatory synaptic receptors, are divided into four distinct subgroups based on their pharmacology and structural homology, including the AMPA receptors (GluA1–GluA4), kainate receptors (GluK1–GluK5), NMDA receptors (GluN1, GluN2A–GluN2D, GluN3A, and GluN3B), and δ receptors (GluD1 and GluD2) [16]. The iGluRs are tetramers with a binding site for glutamate on each subunit that assemble as dimers of dimers, and their composition can be homomeric or heteromeric [94].

AMPA receptors mediate fast glutamatergic neurotransmission, and GluA1 and GluA2 protein expression has been shown to be upregulated in adult epileptic GAERS in the cortical membrane fraction [35]. In conjunction with this increase, it was also shown that stargazin (γ2), a transmembrane AMPA receptor regulatory protein (TARP), was also increased specifically in the membrane of the somatosensory cortex. Juvenile pre-epileptic GAERS did not show any alterations in AMPA receptor or TARP expression [35]. The epileptic and ataxic phenotype of the *stargazer* mouse was found to be genetically determined by a mutation in the stargazin gene (*Cacng2*) resulting in decreased expression of stargazin in the brain [43]. WAG/Rij at 3 and 6 months of age show a reduction in the NMDA receptor GluN1 subunit and AMPA receptor GluA4 subunit immunoreactivity in the somatosensory cortex compared to control rats, which was especially evident in layers IV, V and VI [98]. Similarly, GluN2B protein expression has been shown to be decreased in layers III and V of the somatosensory cortex of 2 month and 6 month old WAG/Rij rats [33].

17.2.4 Metabotropic Glutamate Receptors

Metabotropic glutamate receptors (mGluR) constitute a family of eight G-protein-coupled receptor subtypes that can indirectly modulate ion channels via second messenger systems. The family of mGluRs is composed of eight receptor subtypes, grouped into three different families according to their amino acid homology, pharmacologic properties, and G-protein coupling [13]. Class I metabotropic glutamate receptors (mGluR1 and mGluR5) mediate an increase in neuronal excitability and their activation can induce seizures. Class II receptors (mGluR2 and mGluR3) and class III receptors (mGluR4, mGluR6-8) depress synaptic transmission.

Several mGluRs subtypes are localised at synapses of thalamocortical neurons and thus may play an important role in the generation of epileptic generalised SWD [61]. Indeed, mGluR1α subtype has been shown to be down regulated in the thalamus of 8 month old epileptic WAG/Rij rats but in young pre-symptomatic WAG/Rij rats this reduction was not observed indicating that this change in mGluR1α receptor is occurring as a consequence of the seizures [60]. Additionally, mGluR4 protein levels in the reticular nucleus and ventral posterolateral thalamic nuclei were significantly reduced in 2 month old pre-epileptic WAG/Rij, but in 8 month old epileptic WAG/Rij rats a significant increase in mGluR4 protein levels in the reticular nucleus of the thalamus was observed [59].

17.3 Acquired Epilepsies

17.3.1 Overview

Acquired epilepsies are caused by brain insult such as stroke, traumatic brain injury (TBI), brain inflammation, or status epilepticus. Consequently, the molecular and cellular pathology of acquired epilepsies is heterogeneous both in type, distribution, and temporal evolution [71, 72]. Previous studies in human tissue and animal models have shown that in addition to neurodegeneration, neurogenesis, vascular injury and angiogenesis, proliferation and activations of different types of glia, axonal/myelin injury and axonal sprouting, dendritic plasticity, and changes in the composition of extracellular matrix, also the composition of ligand and voltage-gated ion channels can change [72]. This is often referred as "acquired channelopathy" which can contribute to both epileptogenesis, evolution of comorbidities, and drug-refractoriness [77].

Previous global analyses of gene expression in epileptic tissue indicate that changes in the expression of mRNA, encoding for receptors and ion channels is not prominent. Some changes have, however, been observed in the level of mRNA for subunits of calcium channels or GABA receptors (e.g. [26, 49, 72, 102]). Moreover, the changes in mRNA levels often do not correlate with the protein level [1, 44, 75]. The level of the expressed protein is the key for the function of receptors or ion channels. Therefore, we focus on changes in protein expression of channels and receptors. We compare the findings in human tissue to that in animals undergoing epileptogenesis or already having established epilepsy (changes detected >7 days post-injury). The data are summarized in Table 17.1.

17.4 Altered Expression of Voltage-Gated Ion Channels

17.4.1 Voltage-Gated Sodium Channels

Our literature search did not reveal any data on expression of α subunits in human TLE. In a rodent model of TLE, in which epileptogenesis was induced by status epilepticus (SE), Hargus et al. [28] reported that $Na_V1.6$ was present in axon initiation segment and Nav1.2 in the soma of neurons located in layer II of the entorhinal cortex. In animals with epilepsy, the expression of both subunits was increased at 3 months post-SE. Instead, expression of Nav1.1 and Nav1.3 was low and did not differ between the control and epileptic animals. Authors concluded that changes in the expression of Nav1.2 and Nav1.6 participate in generation of hyperexcitability of layer II neurons [28].

There are few studies of expression of β subunits in human TLE. Navβ3 was found to be expressed in principal neurons of the hippocampus proper. In TLE patients without hippocampal

sclerosis, the expression of Navβ3 was reduced in the hippocampus as compared to that in TLE patients with hippocampal sclerosis [99]. In the normal hippocampus, Navβ1 subunit was expressed in neurons and a weak immunoreactivity was also observed in astrocytes. The astrocytic expression of Navβ1 showed a remarkable increase during epileptogenesis and epilepsy triggered by SE [27].

17.4.2 Voltage Gated Potassium Channels

Voltage gated potassium channels are six transmembrane proteins containing a pore consisting of two transmembrane fragments and a voltage sensor on N-terminal side. Usually channels are tetramers composed of identical subunits. Their function is to return the membrane potential to resting state after depolarization [14].

One of the most studied subunits in human TLE and acquired epilepsy models is Kv4.2 (KCND2) that is critical for mediating the A-currents crucial for regulation of neuronal excitability and control of threshold for action potential initiation. In TLE patients with hippocampal sclerosis, the level of Kv4.2 was increased in the somata of pyramidal cells and in activated astrocytes [1]. Immunoreactivity for its phosphorylated form, pKv4.2, was increased in granule cell and in molecular layers of the dentate gyrus as well as in the hippocampal CA3 principal cells. In areas of neurodegeneration, however, the dendritic immunoreactivity of Kv4.2 or pKv4.2 was reduced. In some pyramidal neurons pKv4.2 co-localized with postsynaptic markers.

Decrease in the expression of Kv4.2 in CA1 has also been observed in animal models of epilepsy, including epileptogenesis triggered by SE or by TBI [4, 56, 82]. Increased expression of pKv4.2, similar to that observed in human TLE, has been found in the CA1 after SE in rats [4]. These observations suggested that a decrease in Kv4.2 and an increase in pKv4.2 by ERK kinases can contribute to increased dendritic excitability, resulting in reduced seizure threshold after epileptogenic brain insults [1, 4].

Table 17.1 Changes in the expression in subunits of various ligand and voltage-gated ion channels after epileptogenic brain insults in animal models and in human temporal lobe epilepsy (TLE)

Subunit	Model	Brain area	Change	Reference
GABA receptors				
GABA-Aα1	Human TLE	Hippocampus	↑ in granule cell layer	[48]
		Hippocampal formation	↓ in subgranular region, CA2 and CA3	[69]
	Intraperitoneal KA in rat	Hippocampus	↓ in CA1	[85]
			↑ in DG molecular layer at 1 month	
			↓ in CA1-CA3	
	FPI in rat	Hippocampus	↓ at 7 days	[75, 23]
	Intrahippocampal KA in mice	Hippocampus	↑ in DG at 1 month	[6]
			↓ in sr and slm of CA1 and hilus at 1 month	
GABA-Aα2	Human TLE	Hippocampus	↑ in the DG granule cell layer and molecular layer	[48]
			↓ subgranular region	
	Intraperitoneal KA in rat	Hippocampus	↑ in the DG molecular layer and sr and so of CA3 at 1 month	[85]
	Intrahippocampal KA in mice	Hippocampus	↓ in the DG, ↓ in CA1 sr and slm and CA3 at 1 month	[6]
	Pilocarpine in rat	Hippocampus	↓ in CA3 at 6 weeks	[20]
GABA-Aα3	Human TLE	Hippocampus	↓ subgranular layer	[48]
		Hippocampal formation	↑ in DG and subiculum	[69]
			↓in CA1	
	Intraperitoneal KA in rat	Hippocampus	↑ in CA1-CA3 in pyramidal-shaped perikaryon at 1 month	[85]
	Intrahippocampal KA mice	Hippocampus	↓ in CA1 sr and slm and CA3 at 1 month	[6]
	Pilocarpine in rat	Hippocampus	↑ in DG	[20]
			↓ in CA3, hilus	
GABA-Aα4	Intraperitoneal KA in rat	Hippocampus	↑ in DG molecular layer at 30 days	[85]
	HC stimulation induced SE in rat	Hippocampus	↑ at inhibitory synapse	[91]
	Pilocarpine in mice	Dentate gyrus	↑ immunoreactivity at 14–60 days	[65]
	FPI in rat	Hippocampus	↓ at 7 days	[75]
GABA-Aα5	Intraperitoneal KA in rat	Hippocampus	↑ in DG interneurons	[85]
			↓ in CA1 at 1 month	
	Intrahippocampal KA in mice	Hippocampus	↑ in DG	[6]
			↓ in CA1 sr and slm and CA3 at 1 month	
	Pilocarpine in rat	Hippocampus	↓ in CA3	[20]
			↑in granule cell layer	

(continued)

Table 17.1 (continued)

Subunit	Model	Brain area	Change	Reference
GABA-Aβ1	Human TLE	Hippocampal formation	↑ in DG and subiculum	[69]
	Intraperitoneal KA in rat	Hippocampus	↑ in CA1–CA3 sr and so at 1 month	[85]
GABA-Aβ2	Human TLE	Hippocampus	↑ granule cell layer ↓ subgranular region	[48]
		Hippocampal formation	↑ in DG and subiculum	[69]
	Intraperitoneal KA in rat	Hippocampus	↑ in CA1 so and sr, ↑DG molecular layer at 1 month	[85]
	Pilocarpine in rat	Hippocampus	↑ in granule cell layer at 6 weeks ↓ in CA3 and hilus	[20]
GABA-Aβ3	Human TLE	Hippocampus	↑ granule cell layer ↓ subgranular region	[48]
		Hippocampal formation	↑ in DG and subiculum ↓ in CA1	[69]
	Pilocarpine in rat	Hippocampus	↑ in granule cell layer at 6 weeks ↓ in CA3 and hilus	[20]
GABA-Aγ1	Intraperitoneal KA in rat	Hippocampus	↑ in DG molecular layer at 1 month ↓ in whole hippocampus	[85]
GABA-Aγ2	Human TLE	Hippocampus	↓ subgranular region	[48]
		Hippocampal formation	↑ in DG and subiculum ↓ in CA1	[69]
	Intraperitoneal KA in rat	Hippocampus	↑ in DG molecular layer and hilar interneurneurons at 1 month	[85]
	Intrahippocampal KA in mice	Hippocampus	↑ in DG ↓ in CA1 slm, CA3 and at 1 month	[6]
	Pilocarpine in rat	Hippocampus	↑ in molecular layer at 6 weeks	[20]
	Pilocarpine in mice	Hippocampus	↑ translocation from synaptic to perisynaptic localization at 1 month	[105]
		Dentate gyrus	↑ immunoreactivity at 7–60 days	[65]
	CCI in rat	Hippocampus	↓ at 90 days	[36]
	FPI in rat	Hippocampus	↓ at 7 days	[75]
GABA-Aδ	Pilocarpine in mice	Dentate gyrus	↓ immunoreactivity at 7–60 days	[65]
		Hippocampus	↓ at symmetrical synapses at 1 month	[105]
	CCI in rat	Hippocampus	↑ at 3 months	[36]
	FPI in rat	Hippocampus	↓ at 7 days	[75]

GABA-B receptors

GABA-BR1	Human TLE	Hippocampal formation	↓ in CA1 and granule cells	[58]
	Intrahippocampal KA in mice	Hippocampus	↑ in DG at 1 and 3 months	[88]
GABA-BR2	Intrahippocampal KA in mice	Hippocampus	↑ in DG at 1 and 3 months	[88]
			↓ n CA1	
AMPA/Kainate receptors				
GluR1	Intraperitoneal KA in rat	Hippocampus	↓ in the hippocampus at 1 month	[83]
GluR5	Human TLE	Hippocampus and cortex	↑ in the hippocampus	[44]
	Intraperitoneal KA in rat	Hippocampus	↑ at 3 and 6 months	[97]
NMDA receptors				
NR1	HC stimulation induced SE in rat	Hippocampus	↓ at 1 month	[19]
NR2B	Human TLE	Hippocampus	↑	[54]
	CCI in rat	Hippocampus	↑ at 3 months	[36]
	Intraperitoneal KA in rat	Hippocampus	↓ at 4–6 weeks	[92]
	HC stimulation induced SE in rat	Hippocampus	↓ at 1 month with transient ↑ following seizures	[19]
			↑ in astrocytes at 1 month	
	HC stimulation induced SE in rat	Hippocampus	↑ at 3 months	[54]
	FPI at P19	Hippocampus, cortex	↓ at 7 days	[24]
Metabotropic glutamate receptors				
mGluR2/3	Pilocarpine in rat	Dentate gyrus	↑ increase in molecular layer at 2 months	[78]
	Angular bundle stimulation induced SE in rat	Hippocampus	↑ in reactive astrocytes up to 3 months	[2]
mGluR4	Human TLE	Hippocampus	↑ in the dentate gyrus	[46]
mGluR5	Intra-amygdalar KA in rat	Hippocampus	↑ in reactive non-neuronal cells at 2 months	[96]
	Angular bundle stimulation induced SE in rat	Hippocampus	↑ in reactive astrocytes up to 3 months	[2]
	Pilocarpine in rat	CA1	↓ in stratum radiatum 4–10 weeks	[38]
Voltage-gated Na+ channels				
Nav1.2	HC stimulation induced SE in rat	Medial entorhinal cortex layer II	↓ cell bodies at 3 months	[28]
Nav1.6	HC stimulation induced SE in rat	Medial entorhinal cortex layer II	↑ axon initial segment at 3 months	[28]

(continued)

Table 17.1 (continued)

Subunit	Model	Brain area	Change	Reference
β1	HC stimulation induced SE in rat		↑ in astrocytes at 7 days–3 months	[27]
β3	Human TLE	Hippocampus, cortex	↓ in the hippocampus but not cortex of non-HS patients when compared to HS patients	[99]
Voltage-gated K$^+$ channels				
Kv1.4	Pilocarpine in rats	Hippocampus	↑ in the inner molecular layer of DG and sl of CA3 at 1–12 weeks	[56]
Kv4.2	Human TLE with HC sclerosis	Hippocampus	↑ in neuronal somata and astrocytes; ↓ in neuropil	[1]
	Intraperitoneal KA in rat	CA1	↓ at 1 month	[82]
	Pilocarpine in rat	CA1	↑ in phosphorylation; ↓ protein levels	[4]
		Hippocampus	↓ in sr of CA1 translocation to outer molecular layer of the dentate gyrus at 1–12 weeks	[56]
	Pilocarpine in rat	Hippocampus	↓ CA1 and CA3 at 50 days	[90]
	CCI in mice	Hippocampus	↓ in ipsilateral CA1 and CA3 at 1 and 8 weeks	[42]
KV4.3	Pilocarpine in rat	Hippocampus	translocation to outer molecular layer at 1–12 weeks	[56]
Kv7.2 (KCNQ2)	HC stimulation induced SE in rat	amygdala	↑ in basolateral amygdala at 15 days	[66]

Abbreviation: *CCI* controlled cortical impact, *FPI* fluid percussion injury, *HC* hippocampus, *HS* hippocampal sclerosis, *KA* kainic acid, *SE* status epilepticus, *sl* stratum lucidum, *slm* stratum lacunosum moleculare, *so* stratum oriens, *sr* stratum radiatum, *TLE* temporal lobe epilepsy

The two other subunits, Kv4.3 and Kv1.4, are implicated in A-current. In SE model, Kv4.3 protein translocated within the molecular layer of the dentate gyrus, resulting in an increase in its concentration in the outer two thirds of the molecular layer [56]. Kv1.4 is localized in axons. Interestingly, in epileptic animals an increase in Kv1.4 was observed in stratum lucidum of the CA3 and in the inner molecular layer of the dentate gyrus, which are the areas of extensive axonal sprouting after epileptogenic brain insults [56].

Another voltage dependent potassium channel implicated in acquired epilepsy is Kv7.2 (KCNQ2). Kv7.2 contributes to M-current, controlling baseline excitability. Number of Kv7.2 immuno-positive neurons was increased in the basolateral amygdala in animals after SE. The increase was present only in animals with spontaneous seizures. An increase in Kv7.2 was proposed to decrease the baseline excitability of amygdaloid neurons [66].

17.5 Altered Expression of Ligand Gated Ion-Channels

17.5.1 GABA_A Receptors

In drug-resistant patients with TLE, studies using subunit-specific antibodies have revealed profound and complex alterations in the expression of GABAA receptor subunits in the hippocampus. In particular, decrease in immunoreactivity of $\alpha 1$, $\alpha 2$, $\alpha 3$, and $\gamma 2$ was observed in the CA1 in TLE patients with hippocampal sclerosis [21, 48]. This was probably related to the CA1 neurodegeneration. In the granule cells of dentate gyrus, however, the expression of $\alpha 1$ and $\alpha 2$ subunits was increased.

An increased expression of $\beta 2$ and $\beta 3$ subunits was observed in the granule cell layer of dentate gyrus of patients with TLE, while the data available on the expression of β subunits in the CA1-CA3 subfields of the hippocampus proper are conflicting [21, 48]. Interestingly, several β subunits increase their expression in the apical dendrites and decrease the expression in the basal dendrites of granule cells [48].

The literature reporting the changes in the expression of GABAA receptor subunits in animal models of acquired epilepsy is extensive (Table 17.1). In the normal rat hippocampus, the distribution, of GABAA receptor α subunits is topographically organized in different hippocampal subfields and layers [81]. Fritschy et al. [20] showed that at 6 weeks after SE in rats, the expression of $\alpha 1$ subunit was up-regulated in the granule cell and molecular layers of the dentate gyrus and down-regulated in the hippocampus proper. Interestingly, also the number of hilar $\alpha 1$ positive interneurons was reduced [20]. Accordingly, a decrease in $\alpha 1$ immunoreactivity in the CA1 subfield was reported at 1 month after SE. The decrease was accompanied by an increase in $\alpha 1$ immunoreactivity in the granule cell and molecular layers of the dentate gyrus TBI caused a decrease in $\alpha 1$ expression in hippocampal extracts at 1 week post-TBI. No such decrease in $\alpha 1$ expression was found when assessed at 90 days post-TBI [23, 36, 75].

SE in rats resulted in an increase in $\alpha 2$ subunit immunoreactivity in the molecular layer of the dentate gyrus which was accompanied with a decreased $\alpha 2$ expression in the CA1 [20, 85]. When kainic acid was injected directly into the hippocampus, a decrease in $\alpha 2$ immunoreactivity occurred in the ipsilateral CA1, CA3, and also in the dentate gyrus [6]. After TBI, $\alpha 2$ expression did not differ from that in controls [23, 75].

Immunoreactivity of GABAA receptor $\alpha 3$ subunit was decreased in the CA1 and CA3 subfields of the hippocampus after SE [6, 20]. In the dentate gyrus, however, $\alpha 3$ immunoreactivity was increased in rats with epilepsy [20].

Decrease in $\alpha 4$ subunits was observed in extracts from the whole rat hippocampus 1 week after TBI, but no changes was evident in rats 90 days after TBI nor in rats in which epilepsy was induced by SE [36, 75, 91].

Immunoreactivity of $\alpha 4$ subunit was increased in the molecular layer of the dentate gyrus at 30 days after SE in mice [65]. Moreover, Sun et al. [91] demonstrated that in epileptic animals the $\alpha 4$ subunits located on the somata and dendrites of the dentate granule cells were more commonly present within inhibitory synapses than extra-synaptically. This coincided with a diminished action of neurosteroids on synaptic current, possibly contributing to facilitation of seizures in epileptic animals [91].

Expression of α5 subunit of GABAA receptor decreased in the CA1 subfield of the hippocampus in rats that had experienced SE [20, 85]. Moreover, SE resulted in a slight increase in α5 subunit immunoreactivity in the dentate gyrus [20]. An increased expression of α5 in the dentate gyrus and a decrease in the CA1 were also observed at 1 month after SE [6]. However, no changes were observed in the expression of α5 expression at 7 days after TBI induced by FPI [23, 75].

In the normal rat brain, GABA$_A$ receptor β1 and β3 subunits are expressed in the dendritic areas of the hippocampus, including the stratum oriens and stratum radiatum of the CA1-CA3, and the molecular layer of the dentate gyrus. Staining for β2 subunit is light in pyramidal cell dendrites or in granule cells, but is present in hippocampal interneurons [84]. At 6 weeks after SE, immunoreactivity for β2 and β3 subunits was increased in the granule cell layer and decreased in the CA1-CA3 subfields of hippocampus proper as well as in the hilus of the dentate gyrus [20]. However, at 7 days post-TBI the hippocampal expression of β3 subunit remained unaltered [23].

Immunoreactivity of GABAA receptor γ1 and γ2 subunits is light in the normal hippocampus. γ1 is expressed in astrocyte-like profiles. γ2 subunit is highly expressed in the dendrites of CA1-CA3 neurons and in the molecular layer of the dentate gyrus as well as in perikarya of a subpopulation of hilar neurons. Expression of γ3 subunit is most remarkable in fibers [81]. After SE there is an increase in the immunoreactivity for γ2 subunit in the molecular layer of the dentate gyrus [6, 20, 65]. Zhang et al. [105] showed that in epileptic rats γ2 subunits are translocated to the perisynaptic location in the dendrites of granule cells. This resulted in a decrease in the expression of γ2 subunits at the synaptic region, and coincided with a decrease in phasic inhibition in the dendrites of granule cells [105]. Data on expression of γ2 subunit expression in the hippocampus proper are less consistent. After SE, Sperk et al. [85] observed an increase in γ2

immunoreactivity in stratum lacunosum moleculare and stratum radiatum of the CA3. However, γ2 immunoreactivity was decreased in the ipsilateral CA1 and CA3 [85]. A decrease in the hippocampal expression of γ2 subunit was observed also after TBI using Western blot [36, 75].

In the normal hippocampus, δ subunits are expressed in the molecular and granule cell layers and in interneurons of the dentate gyrus. Light immunoreactivity is also present in the CA1-CA3 subfields of hippocampus proper [81]. Chronically epileptic animals after SE showed a decrease in δ subunit immunoreactivity in the molecular layer of the dentate gyrus and an increase in interneurons. This was accompanied by an increase in excitability in hippocampal slices sectioned from epileptic animal [65]. As shown by Zhang et al. [105] the expression of δ subunit was decreased in the dendrites of dentate granule cells [105]. Unexpectedly, no impairment was observed in tonic inhibition, indicating that a reduction in the expression of δ subunit is compensated by other GABAA subunits [105]. In addition to SE models, a decrease in δ subunit was observed in the hippocampal extracts at 7 days following TBI [75].

In summary, the changes in the pattern of expression of different GABAA receptor subunits are complex and model specific. In several reports, the decrease in the expression of subunit protein correlated with the severity of neurodegeneration whereas the increases in the expression likely presented compensatory molecular plasticity in altered network. Undoubtedly, the reported alterations explain the impairment of GABAergic transmission tuning the network towards increased excitability.

17.5.2 GABA$_B$ Receptors

Contribution of the altered expression of metabotropic GABA$_B$ receptors to acquired epileptogenesis and ictogenesis are poorly understood as compared to that of GABA$_A$ receptors. In the normal human brain, neuronal expression of one of the GABA$_B$ receptors, GABABR1, has been reported in the hippocampus and entorhinal cortex. In TLE, the expression of GABABR1 was reduced in the dentate granule cells as well as in the hippocampus, particularly in areas of neurodegeneration. Interestingly, no compensatory change in the expression of GABABR1 was found in surviving neurons [58].

Straessle et al. [88] investigated the distribution of the two variants of GABABR1 receptor, GABABR1a and GABABR1b as well as GABABR2 in a mouse model of TLE. At 4–6 weeks or 3 month post-SE, ipsilateral CA1-CA3 showed a remarkable reduction in GABABR1a and GABA-BR1b as well as in GABABR2 immunoreactivities, which was associated with extensive hippocampal neurodegeneration. On the contrary, expression of GABABR1a, GABABR1b, and GABAR2 subunits was enhanced in the dentate granule cells. Moreover, temporary loss and then reappearance of interneurons stained for GABABR1a,b or GABABR1b was observed in the hilus and CA3, In contrast to GABAA receptor subunits, no changes GABABR1a, GABABR1b, or GABABR2 immunoreactivities were observed in the molecular layer of the dentate gyrus [88].

17.5.3 AMPA/Kainate Receptors

Studies investigating the expression of subunit proteins forming AMPA (GluR1-4) and KA (GluR5-7) receptors in epileptic tissue are meager, despite the fact that some of the non-NMDA glutamate receptors are targeted by antiepileptic drugs.

The expression of GluR1 and GluR2/3 was increased in the molecular layer and GluR2 also in the stratum radiatum, in TLE patients either with or without hippocampal sclerosis. An increase in GluR1 was also found in the CA3 principal cells as well as in hilar mossy cells [17, 54]. In hippocampal stimulation model of TLE, rats with epilepsy showed increased expression of GluR1 in the molecular layer of the dentate gyrus [54]. In kainate model, however, hippocampal expression of GluR1 expression was decreased at 1 month post-SE [83]. A decrease in GluR1 protein expression was also observed in the hippocampus at 3 months after TBI [36].

Some information is available on kainate receptor subunit GluR5. Li et al. [44] reported an increase in GluR5 protein level in the hippocampus, but not in the temporal neocortex of TLE patients [44]. An increase in GluR5, but not in GluR6 protein expression was also observed in rats at 3 or 6 months after SE in rats [97].

17.5.4 NMDA Receptors

NMDA receptors are implicated in synaptic plasticity, including LTP and LTD. This has created an interest whether they could play a role also in the development of aberrant synaptic plasticity found in nimal models and human TLE.

NMDA receptors are tetramers consisting of at least one NR1 subunit and NR2(A-D) or NR3 (A-B) subunits. Properties of NMDA receptor are determined by its subunit composition [22]. Changes in the expression of NMDA receptors have been studied mostly at mRNA level, and these studies have focused on early time points after epileptogenic insult [22]. Much less information is available on protein expression and on its localization in the epileptic tissue.

As NR1 subunit is an indispensable component of the NMDA receptor, its expression provides information on the presence and localization of all NMDA receptors. Frasca et al. [19] reported a decrease in the phosphorylated and non-phosphorylated forms of NR1 in animals with epilepsy after SE [19].

More information is available on NR2, a subunit that is critical for the localization of NMDA receptor. After SE, the expression of NR2B protein was decreased which was accompanied by a decrease in PSD-95 protein. Moreover, the decrease in NR2 correlated with behavioral deficits [92]. In another SE model, hippocampus showed a reduced expression of both NR2B as well as its phosphorylated form, p-NR2B. The decrease in p-NR2B in post-synaptic membranes was associated with its reduced interaction with postsynaptic density. Interestingly, spontaneous seizures in these animals caused a transient increase in p-NR2B. It was concluded that altered phosphorylation on NR2B leads to extra synaptic localization of NMDA receptors in epileptic animals [19]. A decrease in hippocampal NR2B protein was also found in the CCI model of TBI in rats [36].

17.5.5 Metabotropic Glutamate Receptors

To our knowledge, only mGluR4 protein expression has been studied in the human epileptic brain. In the normal human brain, almost no mGluR4 immunoreactivity was present. The hippocampus resected from patients with TLE, however, showed a strong mGluR4 immunoreactivity, particularly in the dentate gyrus. Interestingly, mGluR4 was localized in periphery of pre- and postsynaptic membranes [46].

mGluR5, a member of class I receptors has been studied only in animal models of epilepsy. In the normal hippocampus, mGluR5 protein is expressed in the dendritic fields of pyramidal cells [96]. After SE, the expression of mGluR5 was reduced in the ipsilateral hippocampus, which correlated with the severity of neurode-generation in CA1-CA3 pyramidal cells [96]. mGluR5 immunoreactivity was decreased also in CA1 following SE [38]. The decrease in mGluR5 immunoreactivity after SE occurred in neurons whereas astrocytes showed a strong immunola-beling [2, 96]. It was suggested that an increase in mGluR5 in astrocytes could associate with Ca2+ oscillations in astrocytes [2, 96] whereas a reduction in mGlur5 in the CA1 principal cells could associate with an impairment in LTD which is one of the post-SE functional consequences [38].

Similarly to mGlur5, expression of mGluR2/3 receptor proteins has been studied in models of TLE triggered by SE. When mGluR2/3 immuno-reactivity was analyzed at 1 week, 3 weeks, or 3 months after SE, its expression was increased in activated vimentin-positive astrocytes. It was proposed that this contributed to the propagation of calcium waves in astrocytic syncytium, resulting in generation of seizure focus [2]. Interestingly, at the 3 month time point the intensity of mGluR5 staining was reduced in the molecular layer and in stratum lacunosum moleculare, and these

changes coincided with neurodegeneration in the entorhinal cortex [2]. In another SE model, a decrease in mGluR2/3 immunoreactivity was detected in the stratum lacunosum moleculare of the CA1 and CA3 and in mossy fibers located in the CA3 and the hilus [62]. An increase in mGluR2/3 immunoreactivity was found in the molecular layer of the dentate gyrus [62, 78]. It remains to be studied whether the increase in mGluR2/3 also after SE occurs in astrocytes [78].

17.6   Conclusions

The spectrum of changes in expression of proteins forming ligand and voltage-gated ion channels in genetic and acquired epilepsies is wide and extends over different etiologies. Changes vary depending on the stage of epileptogenesis, brain area investigated, as well as the cell type and cellular compartment assessed. The overall picture of changes in receptors and ion channels is frag-mentary and their functional analysis is limited. However, considering the multiplicity of molecular and cellular changes present in the epileptogenic regions, it remains a viable hypothesis that acquired channelopathies form a specific compo-nent of the molecular fingerprint for epileptogen-esis, eventually leading to the development of epilepsy. To which extent they also contribute to the development of comorbidities and/or tissue recovery remains to be studied.

Acknowledgements This work was supported by a Polish Ministry of Science and Education grant DNP/N119/ESF-EuroEPINOMICS/2012 to K.L. and the Academy of Finland, the Sigrid Juselius Foundation, CURE grant to A.P.

References

1. Aronica E, Boer K, Doorn KJ, Zurolo E, Spliet WG, van Rijen PC, Baayen JC, Gorter JA, Jeromin A (2009) Expression and localization of voltage dependent potassium channel Kv4.2 in epilepsy associated focal lesions. Neurobiol Dis 36(1):81–95

2. Aronica E, van Vliet EA, Mayboroda OA, Troost D, da Silva FH, Gorter JA (2000) Upregulation of metabotropic glutamate receptor subtype mGluR3

and mGluR5 in reactive astrocytes in a rat model of mesial temporal lobe epilepsy. Eur J Neurosci 12(7):2333–2344

3. Berg AT, Berkovic SF, Brodie MJ, Buchhalter J, Cross JH, van Emde Boas W, Engel J, French J, Glauser TA, Mathern GW, Moshe SL, Nordli D, Plouin P, Scheffer IE (2010) Revised terminology and concepts for organization of seizures and epilepsies: report of the ILAE Commission on Classification and Terminology, 2005–2009. Epilepsia 51(4):676–685

4. Bernard C, Anderson A, Becker A, Poolos NP, Beck H, Johnston D (2004) Acquired dendritic channelopathy in temporal lobe epilepsy. Science 305(5683):532–535

5. Blumenfeld H, Klein JP, Schridde U, Vestal M, Rice T, Khera DS, Bashyal C, Giblin K, Paul-Laughinghouse C, Wang F, Phadke A, Mission J, Agarwal RK, Englot DJ, Motelow J, Nersesyan H, Waxman SG, Levin AR (2008) Early treatment suppresses the development of spike-wave epilepsy in a rat model. Epilepsia 49(3):400–409

6. Bouilleret V, Loup F, Kiener T, Marescaux C, Fritschy JM (2000) Early loss of interneurons and delayed subunit-specific changes in GABA(A)-receptor expression in a mouse model of mesial temporal lobe epilepsy. Hippocampus 10(3):305–324

7. Brewster AL, Chen Y, Bender RA, Yeh A, Shigemoto R, Baram TZ (2006) Quantitative analysis and subcellular distribution of mRNA and protein expression of the hyperpolarization-activated cyclic nucleotide-gated channels throughout development in rat hippocampus. Cereb Cortex 17:702–712

8. Broicher T, Kanyshkova T, Meuth P, Pape HC, Budde T (2008) Correlation of T-channel coding gene expression, IT, and the low threshold Ca2+ spike in the thalamus of a rat model of absence epilepsy. Mol Cell Neurosci 39(3):384–399

9. Budde T, Caputi L, Kanyshkova T, Staak R, Abrahamczik C, Munsch T, Pape HC (2005) Impaired regulation of thalamic pacemaker channels through an imbalance of subunit expression in absence epilepsy. J Neurosci 25(43):9871–9882

10. Casillas-Espinosa PM, Powell KL, O'Brien TJ (2012) Regulators of synaptic transmission: roles in the pathogenesis and treatment of epilepsy. Epilepsia 53(Suppl 9):41–58

11. Chen Y, Lu J, Pan H, Zhang Y, Wu H, Xu K, Liu X, Jiang Y, Bao X, Yao Z, Ding K, Lo WH, Qiang B, Chan P, Shen Y, Wu X (2003) Association between genetic variation of CACNA1H and childhood absence epilepsy. Ann Neurol 54(2):239–243

12. Coenen AM, Van Luijtelaar EL (2003) Genetic animal models for absence epilepsy: a review of the WAG/Rij strain of rats. Behav Genet 33(6):635–655

13. Conn PJ, Pin JP (1997) Pharmacology and functions of metabotropic glutamate receptors. Annu Rev Pharmacol Toxicol 37:205–237

14. Cooper EC (2012) Potassium channels (including KCNQ) and epilepsy. In: Noebels JL, Avoli M, Rogawski MA, Olsen RW, Delgado-Escueta AV (eds) Jasper's basic mechanisms of the epilepsies, 4th edn. Bethesda, pp 55–65

15. Danober L, Deransart C, Depaulis A, Vergnes M, Marescaux C (1998) Pathophysiological mechanisms of genetic absence epilepsy in the rat. Prog Neurobiol 55(1):27–57

16. Dingledine R, Borges K, Bowie D, Traynelis SF (1999) The glutamate receptor ion channels. Pharmacol Rev 51(1):7–61

17. Eid T, Kovacs I, Spencer DD, de Lanerolle NC (2002) Novel expression of AMPA-receptor subunit GluR1 on mossy cells and CA3 pyramidal neurons in the human epileptogenic hippocampus. Eur J Neurosci 15(3):517–527

18. Escayg A, Goldin AL (2010) Sodium channel SCN1A and epilepsy: mutations and mechanisms. Epilepsia 51(9):1650–1658

19. Frasca A, Aalbers M, Frigerio F, Fiordaliso F, Salio M, Gobbi M, Cagnotto A, Gardoni F, Battaglia GS, Hoogland G, Di Luca M, Vezzani A (2011) Misplaced NMDA receptors in epileptogenesis contribute to excitotoxicity. Neurobiol Dis 43(2):507–515

20. Fritschy JM, Kiener T, Bouilleret V, Loup F (1999) GABAergic neurons and GABA(A)-receptors in temporal lobe epilepsy. Neurochem Int 34(5):435–445

21. Furtinger S, Pirker S, Czech T, Baumgartner C, Sperk G (2003) Increased expression of gamma-aminobutyric acid type B receptors in the hippocampus of patients with temporal lobe epilepsy. Neurosci Lett 352(2):141–145

22. Ghasemi M, Schachter SC (2011) The NMDA receptor complex as a therapeutic target in epilepsy: a review. Epilepsy Behav 22(4):617–640

23. Gibson CJ, Meyer RC, Hamm RJ (2010) Traumatic brain injury and the effects of diazepam, diltiazem, and MK-801 on GABA-A receptor subunit expression in rat hippocampus. J Biomed Sci 17:38

24. Giza CC, Maria NS, Hovda DA (2006) N-methyl-D-aspartate receptor subunit changes after traumatic injury to the developing brain. J Neurotrauma 23(6):950–961

25. Gonzalez MI, Brooks-Kayal A (2011) Altered GABA(A) receptor expression during epileptogenesis. Neurosci Lett 497(3):218–222

26. Gorter JA, van Vliet EA, Aronica E, Breit T, Rauwerda H, Lopes da Silva FH, Wadman WJ (2006) Potential new antiepileptogenic targets indicated by microarray analysis in a rat model for temporal lobe epilepsy. J Neurosci 26(43):11083–11110

27. Gorter JA, van Vliet EA, Lopes da Silva FH, Isom LL, Aronica E (2002) Sodium channel beta1-subunit expression is increased in reactive astrocytes in a rat model for mesial temporal lobe epilepsy. Eur J Neurosci 16(2):360–364

28. Hargus NJ, Merrick EC, Nigam A, Kalmar CL, Baheti AR, Bertram EH 3rd, Patel MK (2011) Temporal lobe epilepsy induces intrinsic alterations in Na channel gating in layer II medial entorhinal cortex neurons. Neurobiol Dis 41(2):361–376

29. Hauser WA, Annegers JF, Kurland LT (1991) Prevalence of epilepsy in Rochester, Minnesota: 1940–1980. Epilepsia 32(4):429–445

30. Helbig I, Scheffer IE, Mulley JC, Berkovic SF (2008) Navigating the channels and beyond: unravelling the genetics of the epilepsies. Lancet Neurol 7(3):231–245

31. Heron SE, Khosravani H, Varela D, Bladen C, Williams TC, Newman MR, Scheffer IE, Berkovic SF, Mulley JC, Zamponi GW (2007) Extended spectrum of idiopathic generalized epilepsies associated with CACNA1H functional variants. Ann Neurol 62(6):560–568

32. Kanyshkova T, Meuth P, Bista P, Liu Z, Ehling P, Caputi L, Doengi M, Chetkovich DM, Pape HC, Budde T (2012) Differential regulation of HCN channel isoform expression in thalamic neurons of epileptic and non-epileptic rat strains. Neurobiol Dis 45(1):450–461

33. Karimzadeh F, Soleimani M, Mehdizadeh M, Jafarian M, Mohamadpour M, Kazemi H, Joghataei MT, Gorji A (2013) Diminution of the NMDA receptor NR subunit in cortical and subcortical areas of WAG/Rij rats. Synapse 67:839–846

34. Kaupmann K, Huggel K, Heid J, Flor PJ, Bischoff S, Mickel SJ, McMaster G, Angst C, Bittiger H, Froestl W, Bettler B (1997) Expression cloning of GABA(B) receptors uncovers similarity to metabotropic glutamate receptors. Nature 386(6622):239–246

35. Kennard JT, Barmanray R, Sampurno S, Ozturk E, Reid CA, Paradiso L, D'Abaco GM, Kaye AH, Foote SJ, O'Brien TJ, Powell KL (2011) Stargazin and AMPA receptor membrane expression is increased in the somatosensory cortex of Genetic Absence Epilepsy Rats from Strasbourg. Neurobiol Dis 42(1):48–54

36. Kharlamov EA, Lepsveridze E, Meparishvili M, Solomonia RO, Lu B, Miller ER, Kelly KM, Mtchedlishvili Z (2011) Alterations of GABA(A) and glutamate receptor subunits and heat shock protein in rat hippocampus following traumatic brain injury and in posttraumatic epilepsy. Epilepsy Res 95(1–2):20–34

37. Khosravani H, Altier C, Simms B, Hamming KS, Snutch TP, Mezeyova J, McRory JE, Zamponi GW (2004) Gating effects of mutations in the Cav3.2 T-type calcium channel associated with childhood absence epilepsy. J Biol Chem 279(11):9681–9684

38. Kirschstein T, Bauer M, Muller L, Ruschenschmidt C, Reitze M, Becker AJ, Schoch S, Beck H (2007) Loss of metabotropic glutamate receptor-dependent long-term depression via downregulation of mGluR5 after status epilepticus. J Neurosci 27(29):7696–7704

39. Klein JP, Khera DS, Nersesyan H, Kimchi EY, Waxman SG, Blumenfeld H (2004) Dysregulation of sodium channel expression in cortical neurons in a rodent model of absence epilepsy. Brain Res 1000(1–2):102–109

40. Kole MH, Brauer AU, Stuart GJ (2007) Inherited cortical HCN1 channel loss amplifies dendritic calcium electrogenesis and burst firing in a rat absence epilepsy model. J Physiol 578(Pt 2):507–525

41. Kuisle M, Wanaverbecq N, Brewster AL, Frere SG, Pinault D, Baram TZ, Luthi A (2006) Functional stabilization of weakened thalamic pacemaker channel regulation in rat absence epilepsy. J Physiol 575(Pt 1):83–100

42. Lei Z, Deng P, Li J, Xu ZC (2012) Alterations of A-type potassium channels in hippocampal neurons after traumatic brain injury. J Neurotrauma 29(2):235–245

43. Letts VA, Felix R, Biddlecome GH, Arikkath J, Mahaffey CL, Valenzuela A, Bartlett FS 2nd, Mori Y, Campbell KP, Frankel WN (1998) The mouse stargazer gene encodes a neuronal Ca^{2+}-channel gamma subunit. Nat Genet 19(4):340–347

44. Li JM, Zeng YJ, Peng F, Li L, Yang TH, Hong Z, Lei D, Chen Z, Zhou D (2010) Aberrant glutamate receptor 5 expression in temporal lobe epilepsy lesions. Brain Res 1311:166–174

45. Liang J, Zhang Y, Wang J, Pan H, Wu H, Xu K, Liu X, Jiang Y, Shen Y, Wu X (2006) New variants in the CACNA1H gene identified in childhood absence epilepsy. Neurosci Lett 406(1–2):27–32

46. Lie AA, Becker A, Behle K, Beck H, Malitschek B, Conn PJ, Kuhn R, Nitsch R, Plaschke M, Schramm J, Elger CE, Wiestler OD, Blumcke I (2000) Up-regulation of the metabotropic glutamate receptor mGluR4 in hippocampal neurons with reduced seizure vulnerability. Ann Neurol 47(1):26–35

47. Liu XB, Coble J, van Luijtelaar G, Jones EG (2007) Reticular nucleus-specific changes in alpha3 subunit protein at GABA synapses in genetically epilepsy-prone rats. Proc Natl Acad Sci U S A 104(30):12512–12517

48. Loup F, Wieser HG, Yonekawa Y, Aguzzi A, Fritschy JM (2000) Selective alterations in GABAA receptor subtypes in human temporal lobe epilepsy. J Neurosci 20(14):5401–5419

49. Lukasiuk K, Dabrowski M, Adach A, Pitkanen A (2006) Epileptogenesis-related genes revisited. Prog Brain Res 158:223–241

50. Macdonald RL, Olsen RW (1994) GABAA receptor channels. Annu Rev Neurosci 17:569–602

51. Mantegazza M, Catterall WA (2012) Voltage-gated Na+ channels:structure, function, and pathophysiology. In: Noebels JL, Avoli M, Rogawski MA, Olsen RW, Delgado-Escueta AV (eds) Jasper's basic mechanisms of the epilepsies, 4th edn. Bethesda, pp 41–54

52. Marescaux C, Vergnes M (1995) Genetic absence epilepsy in rats from Strasbourg (GAERS). Ital J Neurol Sci 16(1–2):113–118

53. Marescaux C, Vergnes M, Depaulis A (1992) Genetic absence epilepsy in rats from Strasbourg – a review. J Neural Transm Suppl 35:37–69

54. Mathern GW, Pretorius JK, Leite JP, Kornblum HI, Mendoza D, Lozada A, Bertram EH 3rd (1998) Hippocampal AMPA and NMDA mRNA levels and subunit immunoreactivity in human temporal lobe epilepsy patients and a rodent model of chronic mesial limbic epilepsy. Epilepsy Res 32(1–2): 154–171

55. Merlo D, Mollinari C, Inaba Y, Cardinale A, Rinaldi AM, D'Antuono M, D'Arcangelo G, Tancredi V, Ragsdale D, Avoli M (2007) Reduced GABAB receptor subunit expression and paired-pulse depression in a genetic model of absence seizures. Neurobiol Dis 25(3):631–641

56. Monaghan MM, Menegola M, Vacher H, Rhodes KJ, Trimmer JS (2008) Altered expression and localization of hippocampal A-type potassium channel subunits in the pilocarpine-induced model of temporal lobe epilepsy. Neuroscience 156(3):550–562

57. Monteggia LM, Eisch AJ, Tang MD, Kaczmarek LK, Nestler EJ (2000) Cloning and localization of the hyperpolarization-activated cyclic nucleotide-gated channel family in rat brain. Brain Res Mol Brain Res 81(1–2):129–139

58. Munoz A, Arellano JI, DeFelipe J (2002) GABABR1 receptor protein expression in human mesial temporal cortex: changes in temporal lobe epilepsy. J Comp Neurol 449(2):166–179

59. Ngomba RT, Ferraguti F, Badura A, Citraro R, Santolini I, Battaglia G, Bruno V, De Sarro G, Simonyi A, van Luijtelaar G, Nicoletti F (2008) Positive allosteric modulation of metabotropic glutamate 4 (mGlu4) receptors enhances spontaneous and evoked absence seizures. Neuropharmacology 54(2):344–354

60. Ngomba RT, Santolini I, Biagioni F, Molinaro G, Simonyi A, van Rijn CM, D'Amore V, Mastroiacovo F, Olivieri G, Gradini R, Ferraguti F, Battaglia G, Bruno V, Puliti A, van Luijtelaar G, Nicoletti F (2011a) Protective role for type-1 metabotropic glutamate receptors against spike and wave discharges in the WAG/Rij rat model of absence epilepsy. Neuropharmacology 60(7–8):1281–1291

61. Ngomba RT, Santolini I, Salt TE, Ferraguti F, Battaglia G, Nicoletti F, van Luijtelaar G (2011b) Metabotropic glutamate receptors in the thalamocortical network: strategic targets for the treatment of absence epilepsy. Epilepsia 52(7):1211–1222

62. Pacheco Otalora LF, Couoh J, Shigamoto R, Zarei MM, Garrido Sanabria ER (2006) Abnormal mGluR2/3 expression in the perforant path termination zones and mossy fibers of chronically epileptic rats. Brain Res 1098(1):170–185

63. Pape HC (1996) Queer current and pacemaker: the hyperpolarization-activated cation current in neurons. Annu Rev Physiol 58:299–327

64. Peloquin JB, Khosravani H, Barr W, Bladen C, Evans R, Mezeyova J, Parker D, Snutch TP, McRory JE, Zamponi GW (2006) Functional analysis of Ca3.2 T-type calcium channel mutations linked

to childhood absence epilepsy. Epilepsia 47(3): 655–658

65. Peng Z, Huang CS, Stell BM, Mody I, Houser CR (2004) Altered expression of the delta subunit of the GABAA receptor in a mouse model of temporal lobe epilepsy. J Neurosci 24(39):8629–8639

66. Penschuck S, Bastlund JF, Jensen HS, Stensbol TB, Egebjerg J, Watson WP (2005) Changes in KCNQ2 immunoreactivity in the amygdala in two rat models of temporal lobe epilepsy. Brain Res Mol Brain Res 141(1):66–73

67. Perez-Reyes E (2003) Molecular physiology of low-voltage-activated t-type calcium channels. Physiol Rev 83(1):117–161

68. Pinault D, O'Brien TJ (2007) Cellular and network mechanisms of genetically-determined absence seizures. Thalamus Relat Syst 3:181–203

69. Pirker S, Schwarzer C, Czech T, Baumgartner C, Pockberger H, Maier H, Hauer B, Sieghart W, Furtinger S, Sperk G (2003) Increased expression of GABA(A) receptor beta-subunits in the hippocampus of patients with temporal lobe epilepsy. J Neuropathol Exp Neurol 62(8):820–834

70. Pisu MG, Mostallino MC, Dore R, Mura ML, Maciocco E, Russo E, De Sarro G, Serra M (2008) Neuroactive steroids and GABAA receptor plasticity in the brain of the WAG/Rij rat, a model of absence epilepsy. J Neurochem 106(6): 2502–2514

71. Pitkanen A, Lukasiuk K (2009) Molecular and cellular basis of epileptogenesis in symptomatic epilepsy. Epilepsy Behav 14(Suppl 1):16–25

72. Pitkanen A, Lukasiuk K (2011) Mechanisms of epileptogenesis and potential treatment targets. Lancet Neurol 10(2):173–186

73. Powell KL, Cain SM, Snutch TP, O'Brien TJ (2013) Low threshold T-type calcium channels as targets for novel epilepsy treatments. Br J Clin Pharmacol. doi:10.1111/bcp.12205

74. Princivalle AP, Richards DA, Duncan JS, Spreafico R, Bowery NG (2003) Modification of GABA(B1) and GABA(B2) receptor subunits in the somatosensory cerebral cortex and thalamus of rats with absence seizures (GAERS). Epilepsy Res 55(1–2):39–51

75. Raible DJ, Frey LC, Cruz Del Angel Y, Russek SJ, Brooks-Kayal AR (2012) GABA(A) receptor regulation after experimental traumatic brain injury. J Neurotrauma 29(16):2548–2554

76. Reid CA, Phillips AM, Petrou S (2012) HCN channelopathies: pathophysiology in genetic epilepsy and therapeutic implications. Br J Pharmacol 165(1):49–56

77. Remy S, Gabriel S, Urban BW, Dietrich D, Lehmann TN, Elger CE, Heinemann U, Beck H (2003) A novel mechanism underlying drug resistance in chronic epilepsy. Ann Neurol 53(4):469–479

78. Rohde J, Kirschstein T, Wilkars W, Muller L, Tokay T, Porath K, Bender RA, Kohling R (2012)

Upregulation of presynaptic mGluR2, but not mGluR3 in the epileptic medial perforant path. Neuropharmacology 62(4):1867–1873

79. Santoro B, Tibbs GR (1999) The HCN gene family: molecular basis of the hyperpolarization-activated pacemaker channels. Ann N Y Acad Sci 868: 741–764

80. Schwartzkroin PA (2012) Cellular bases of focal and generalized epilepsies. Handb Clin Neurol 107:13–33

81. Schwarzer C, Tsunashima K, Wanzenbock C, Fuchs K, Sieghart W, Sperk G (1997) GABA(A) receptor subunits in the rat hippocampus II: altered distribution in kainic acid-induced temporal lobe epilepsy. Neuroscience 80(4):1001–1017

82. Shin M, Brager D, Jaramillo TC, Johnston D, Chetkovich DM (2008) Mislocalization of h channel subunits underlies h channelopathy in temporal lobe epilepsy. Neurobiol Dis 32(1):26–36

83. Solomonia R, Mikautadze E, Nozadze M, Kuchiashvili N, Lepsveridze E, Kiguradze T (2010) Myo-inositol treatment prevents biochemical changes triggered by kainate-induced status epilepticus. Neurosci Lett 468:277–281

84. Sperk G, Schwarzer C, Tsunashima K, Fuchs K, Sieghart W (1997) GABA(A) receptor subunits in the rat hippocampus I: immunocytochemical distribution of 13 subunits. Neuroscience 80(4):987–1000

85. Sperk G, Schwarzer C, Tsunashima K, Kandlhofer S (1998) Expression of GABA(A) receptor subunits in the hippocampus of the rat after kainic acid-induced seizures. Epilepsy Res 32(1–2):129–139

86. Spreafico R, Mennini T, Danober L, Cagnotto A, Regondi MC, Miari A, De Blas A, Vergnes M, Avanzini G (1993) GABAA receptor impairment in the genetic absence epilepsy rats from Strasbourg (GAERS): an immunocytochemical and receptor binding autoradiographic study. Epilepsy Res 15(3):229–238

87. Stewart LS, Wu Y, Eubanks JH, Han H, Leschenko Y, Perez Velazquez JL, Cortez MA, Snead OC 3rd (2009) Severity of atypical absence phenotype in GABAB transgenic mice is subunit specific. Epilepsy Behav 14(4):577–581

88. Straessle A, Loup F, Arabadzisz D, Ohning GV, Fritschy JM (2003) Rapid and long-term alterations of hippocampal GABAB receptors in a mouse model of temporal lobe epilepsy. Eur J Neurosci 18(8):2213–2226

89. Strauss U, Kole MH, Brauer AU, Pahnke J, Bajorat R, Rolfs A, Nitsch R, Deisz RA (2004) An impaired neocortical Ih is associated with enhanced excitability and absence epilepsy. Eur J Neurosci 19(11):3048–3058

90. Su T, Cong WD, Long YS, Luo AH, Sun WW, Deng WY, Liao WP (2008) Altered expression of voltage-gated potassium channel 4.2 and voltage-gated potassium channel 4-interacting protein, and changes in intracellular calcium levels following lithium-pilocarpine-induced status epilepticus. Neuroscience 157(3):566–576

91. Sun C, Mtchedlishvili Z, Erisir A, Kapur J (2007) Diminished neurosteroid sensitivity of synaptic inhibition and altered location of the alpha4 subunit of GABA(A) receptors in an animal model of epilepsy. J Neurosci 27(46):12641–12650

92. Sun QJ, Duan RS, Wang AH, Shang W, Zhang T, Zhang XQ, Chi ZF (2009) Alterations of NR2B and PSD-95 expression in hippocampus of kainic acid-exposed rats with behavioural deficits. Behav Brain Res 201(2):292–299

93. Talley EM, Solorzano G, Depaulis A, Perez-Reyes E, Bayliss DA (2000) Low-voltage-activated calcium channel subunit expression in a genetic model of absence epilepsy in the rat. Brain Res Mol Brain Res 75(1):159–165

94. Traynelis SF, Wollmuth LP, McBain CJ, Menniti FS, Vance KM, Ogden KK, Hansen KB, Yuan H, Myers SJ, Dingledine R (2010) Glutamate receptor ion channels: structure, regulation, and function. Pharmacol Rev 62(3):405–496

95. Tsakiridou E, Bertollini L, de Curtis M, Avanzini G, Pape HC (1995) Selective increase in T-type calcium conductance of reticular thalamic neurons in a rat model of absence epilepsy. J Neurosci 15(4): 3110–3117

96. Ulas J, Satou T, Ivins KJ, Kesslak JP, Cotman CW, Balazs R (2000) Expression of metabotropic glutamate receptor 5 is increased in astrocytes after kainate-induced epileptic seizures. Glia 30(4): 352–361

97. Ullal G, Fahnestock M, Racine R (2005) Time-dependent effect of kainate-induced seizures on glutamate receptor GluR5, GluR6, and GluR7 mRNA and protein expression in rat hippocampus. Epilepsia 46(5):616–623

98. van de Bovenkamp-Janssen MC, van der Kloet JC, van Luijtelaar G, Roubos EW (2006) NMDA-NR1 and AMPA-GluR4 receptor subunit immunoreactivities in the absence epileptic WAG/Rij rat. Epilepsy Res 69(2):119–128

99. van Gassen KL, de Wit M, van Kempen M, van der Hel WS, van Rijen PC, Jackson AP, Lindhout D, de Graan PN (2009) Hippocampal Nabeta3 expression in patients with temporal lobe epilepsy. Epilepsia 50(4):957–962

100. Vitko I, Chen Y, Arias JM, Shen Y, Wu XR, Perez-Reyes E (2005) Functional characterization and neuronal modeling of the effects of childhood absence epilepsy variants of CACNA1H, a T-type calcium channel. J Neurosci 25(19):4844–4855

101. Wainger BJ, DeGennaro M, Santoro B, Siegelbaum SA, Tibbs GR (2001) Molecular mechanism of cAMP modulation of HCN pacemaker channels. Nature 411(6839):805–810

102. Winden KD, Karsten SL, Bragin A, Kudo LC, Gehman L, Ruidera J, Geschwind DH, Engel J Jr (2011) A systems level, functional genomics analysis of chronic epilepsy. PLoS One 6(6):e20763

103. Wu Y, Chan KF, Eubanks JH, Guin Ting Wong C, Cortez MA, Shen L, Che Liu C, Perez Velazquez J, Tian Wang Y, Jia Z, Carter Snead O 3rd (2007) Transgenic mice over-expressing GABA(B)R1a

receptors acquire an atypical absence epilepsy-like phenotype. Neurobiol Dis 26(2):439–451

104. Yalcin O (2012) Genes and molecular mechanisms involved in the epileptogenesis of idiopathic absence epilepsies. Seizure 21(2):79–86

105. Zhang N, Wei W, Mody I, Houser CR (2007) Altered localization of GABA(A) receptor subunits on dentate granule cell dendrites influences tonic and phasic inhibition in a mouse model of epilepsy. J Neurosci 27(28):7520–7531

How Do We Make Models That Are Useful in Understanding Partial Epilepsies?

David A. Prince

Abstract

The goals of constructing epilepsy models are (1) to develop approaches to prophylaxis of epileptogenesis following cortical injury; (2) to devise selective treatments for established epilepsies based on underlying pathophysiological mechanisms; and (3) use of a disease (epilepsy) model to explore brain molecular, cellular and circuit properties. Modeling a particular epilepsy syndrome requires detailed knowledge of key clinical phenomenology and results of human experiments that can be addressed in critically designed laboratory protocols. Contributions to understanding mechanisms and treatment of neurological disorders has often come from research not focused on a specific disease-relevant issue. Much of the foundation for current research in epilepsy falls into this category. Too strict a definition of the relevance of an experimental model to progress in preventing or curing epilepsy may, in the long run, slow progress. Inadequate exploration of the experimental target and basic laboratory results in a given model can lead to a failed effort and false negative or positive results. Models should be chosen based on the specific issues to be addressed rather than on convenience of use. Multiple variables including maturational age, species and strain, lesion type, severity and location, latency from injury to experiment and genetic background will affect results. A number of key issues in clinical and basic research in partial epilepsies remain to be addressed including the mechanisms active during the latent period following injury, susceptibility factors that predispose to epileptogenesis, injury – induced adaptive versus maladaptive changes, mechanisms of pharmaco-resistance and strategies to deal with multiple pathophysiological processes occurring in parallel.

D.A. Prince, M.D. (✉)
Department of Neurology and Neurological Sciences,
Stanford University School of Medicine,
Stanford, CA, USA
e-mail: daprince@stanford.edu

H.E. Scharfman and P.S. Buckmaster (eds.), *Issues in Clinical Epileptology: A View from the Bench*,
Advances in Experimental Medicine and Biology 813, DOI 10.1007/978-94-017-8914-1_18,
© Springer Science+Business Media Dordrecht 2014

Keywords

Posttraumatic • Prophylaxis • Mechanisms • Latent period • Pathophysiology
• Translation • Maladaptive

18.1   Introduction

I have chosen to limit discussion here to models
of the partial or lesional epilepsies; however a
number of the issues are generic to understanding the relevance of other models to prevention
and treatment of clinical epilepsies. What follows is not meant to be a literature review, but
rather a discussion of unsolved issues and my
opinions relevant to the use of epilepsy models.
For additional discussion and references, the
reader is referred to Epilepsia, 54: Supplement 4,
1–74, 2013 and articles therein, generated by
participants of a joint AES/ILAE translational
workshop, and a number of recent reviews of
epilepsy models and mechanisms [10, 11, 24,
25, 28–30, 32, 37, 44–48]. Issues and difficulties
raised by these authors bear a remarkable resem-
blance to those highlighted in reviews more than
20 years ago (e.g. [9]), in spite of the introduction of a number of new models and antiepileptic
drugs.

   Making laboratory models "relevant" requires
several considerations. We should recognize that
important contributions to understanding the
mechanisms and treatment of neurological disorders has often come from "non-targeted"
research, not seemingly focused on a specific
disease-relevant issue. Much of the foundation
for current research in epilepsy falls into this category. For this reason, too strict a definition of
whether an experimental model is relevant or
non-relevant to progress in preventing or curing
epilepsy may, in the long run, slow progress.
How should one design a model relevant to our
clinical understanding and treatment? The first
step would be identification of specific key clinical issues that are roadblocks in preventing or
treating epilepsy, and would be feasible to
address in a critically- designed animal model.
Such issues can only be identified through
detailed observations of clinical phenomenology

and associated human research data, i.e. an
important "bedside to bench" approach. Extensive
basic research focused on one or more of these
key issues should follow. The third step would
use of data from the model to design a clinical
experiment or trial. Here is where further definition of the too-often-used term "translational"
becomes important. In literature, scholarly translation of a work requires intimate knowledge of
the vocabularies and nuances of two languages.
By analogy, application of data from a laboratory model to aspects of clinical disorder requires
detailed clinical and basic experimental data.
Inadequate definition of the experimental target and less than rigorous exploration of the
laboratory results in a given model can lead to
a failed effort, or the "Lost in Translation"
phenomenon.

18.2   Why Model at All? What Are the Long-Term Goals?

18.2.1 Prophylaxis of Seizure Development in Lesional Epilepsies

Reference to any classification of seizure disorders clearly reveals that the epilepsies are multifaceted and related to a large variety of etiologies.
Why should one expect that the same model of
epilepsy will be useful for research on prevention
of seizures resulting from a stroke versus a focal
tumor versus a traumatic brain injury? Although
each of these etiologies likely has a different
combination of underlying mechanisms that lead
to seizure generation, they may all involve some
common abnormalities that are sequelae of focal
injury, such as aberrant rewiring of cortical circuits or vulnerability of specific inhibitory interneuronal subtypes. Preventative treatments that
are selective for such specific subtypes of underlying pathophysiology might be effective in more

than one epilepsy syndrome and least likely to induce unwanted side effects. However, there are multiple mechanisms for epileptogenesis that occur in parallel for each subtype of lesional epilepsy. As a consequence, too focused an approach on one pathophysiology or pathway or gene may yield a false negative result, even though the target mechanism has been successfully affected. This may be particularly true in a model of epilepsy with a very high yield of seizures from diffuse brain lesions, such as some post-status epilepticus models. Therefore, progress may require a roadmap of pathophysiological mechanisms obtained from models of different types of epilepsies and even different models of a specific post-lesional epilepsy (e.g. [17]). Potential use of anti-epileptogenic cocktails containing more than one selective agent, treatments with single drugs that have multiple modes of action or treatments directed upstream to affect multiple pathways for epileptogenesis [11] would be a logical direction in studies of prophylaxis after injury.

18.2.2 Development of Selective Treatments for Established Epilepsies, Based on Underlying Pathophysiological Mechanisms

Unfortunately, as noted by many authors, in spite of the development of a number of new antiepileptic drugs, the proportion of individuals who have poorly controlled seizures remains the same, at about one third. It has been proposed that the reason for this is the use of the same models for initial drug screening over the years. There are a number of unknowns that should be considered in designing models for experiments to address this goal. The species and strain of the animal model selected for a given experiment will significantly affect the results [22, 43]. Susceptibility to seizures, and the efficacy and spectrum of toxicity of a given antiepileptic drug will also vary in individuals with different genetic backgrounds [15, 39]. Do such genetic differences extend also to the specific mechanisms underlying development of a particular epilepsy syndrome due to different etiologies, e.g., limbic

circuit epilepsy due to a head injury vs. following status epilepticus? Might these two etiologies for the same syndrome differ in their responses to a particular anti-seizure agent? Another important variable may be the temporal evolution of epileptogenesis after serious brain injury. This clearly varies markedly among individuals and may unfold over years [38, 41]. Is ongoing seizure activity responsible for the progressive loss of hippocampal volume seen in radiological studies? Do the mechanisms underlying seizures following an injury also vary over time, so that drugs might be selected on the basis of the duration of epilepsy in a particular model or patient? For example, early on after cortical injury, treatments that are directed against alterations in blood brain barrier, and immunological mechanisms or inflammation may be effective, however underlying mechanisms may shift over time so that later, formation of new synapses, recurrent excitatory circuits or disturbed inhibitory circuit function become important drug targets. A related question is whether emergence of drug resistance is in part due to shifts in underlying epileptogenic mechanisms over time? Are decreases in responsiveness related to progressive changes late after injury, such a increasing excitatory sprouting and/or death of neuronal subtypes, and what is the role the plastic changes in cortical circuits resulting from ongoing epileptiform activity in this process?

18.2.3 Disease as a Tool to Explore Brain Molecular, Cellular and Circuit Properties

"Epilepsy represents one of the most exquisite experiments of nature and its study may provide basic insight into fundamental functions of the brain." [20]. Epilepsy has long been used as a research tool to explore brain mechanisms such as circuit properties and connections, and mechanisms of synchronization within normal brain. Clementi [3] described reflex epilepsies in which a selective afferent input would trigger local seizure activity and could be used to assess connectivity. Much of the early information about

localization of sensory and motor functions in cerebral cortex was derived from experiments in which the sites for seizure activity were mapped in human brain (e.g. [33]). Epilepsy research has revealed normal brain mechanisms such as cortical "surround" inhibition [35] and aspects of dendritic function [31, 53]. Plastic changes in brain structure and function are key to many normal processes during development, as well as after injury [19]. Epileptogenesis is a striking example of such brain plasticity [18]. Issues such as sprouting of new connections, changes in receptor subunit composition, and alterations in intrinsic membrane properties that are characteristic of neural development are also found during epileptogenesis and following prolonged recurrent seizure activity. Clinical studies done with multiple implanted electrodes in patients with epilepsy, together with functional MRI have revealed sites of pathophysiological interaction, pathways for spread of activity and modifications of epileptic brain to experience and treatment.

18.3 Issues/Problems in Developing Models of Epilepsy

18.3.1 How Many Models Are Necessary for an Epilepsy with a Given Etiology?

As there is no perfect model of human partial epilepsy, models must be chosen on the basis of the long-term goal to be addressed. If the goal is to determine whether a specific therapeutic agent decreases the incidence of seizures either prophylactically or after epilepsy is established, a "high throughput" model with a relatively short latency between injury and seizure activity and a high proportion of animals developing behavioral seizures would be necessary to adequately power the experiment without exhausting available manpower or other resources. Lesional models with long latencies from injury and lower rates of occurrence of seizures would be impractical. The choice of models of chronic focal neocortical epilepsy for use in development of prophylactic or therapeutic strategies is particularly vexing

due to long latencies, relatively low incidence of clinical seizures and variability between laboratories or even in the same laboratory. These requirements have led to the predominant use of models of chronic limbic system epilepsy following status epilepticus induced by pilocarpine, kainic acid or repetitive electrical stimulation. This approach begs the issue of whether other models of epileptogenesis such as those following traumatic injury in temporal lobes or neocortex have the same distribution of underlying mechanisms, and whether post-status epileptogenesis is a common pathophysiology in man. In other words, are we putting "all of our eggs into one basket?" Obviously, all pathologies cannot be represented in a given model. In other CNS disorders, such as autism, schizophrenia and Alzheimer's disease, experiments have been done in a variety of models for a given condition, resulting in conclusions that multiple pathophysiologies may contribute to a given phenotype.

If, on the other hand, the experimental goal is to elucidate the basic cellular and synaptic mechanisms that may contribute to hyperexcitability and epileptogenesis, for example following focal cortical or hippocampal injury, a model in which hyperexcitability persists in vitro in a high proportion of cortical slices from a known focal area of injury would be preferred over the more diffuse or multifocal brain injuries that occur following status epilepticus or severe brain trauma. In this case, one might choose the partial cortical isolation model or epileptogenic focal areas in cortex due to infarction, controlled local cortical trauma, or experimentally induced focal inflammation/infection.

18.3.2 Multiple Pathophysiological Processes

Not only are there multiple abnormalities in any given model, but also these abnormalities do not occur in parallel over time. This has important implications for choice of therapy, be it prophylactic or after seizures have developed. For example, early on after injury, inflammation, alterations in the blood brain barrier, excessive release of glutamate from injured tissue and abnormalities in membrane properties or receptors of acutely damaged neurons may be most important as targets for whatever agents are chosen. Further, it may be unclear which of these processes or combination of them is epileptogenic, and underlies the later development of seizures. Inflammation and blood brain barrier disturbances are present following any cortical trauma, yet only a minority of mild to moderate injuries result in partial epilepsy. Likewise, only a small proportion of gray matter infarctions result in focal epileptogenesis, even though similar acute processes occur following most injuries. Over time other more indolent processes may occur such as progressive loss of nerve cells following repetitive seizures or slowly activating mechanisms that induce either adaptive or maladaptive circuitry (e.g. [23, 27, 42, 49]). The choice of a therapeutic agent would depend on which of these processes was ongoing at a given point in time; it might not be effective to treat an area of injury with an anti-inflammatory agent after epileptogenesis is well established. The best experimental strategy would be to attempt to isolate or control one or another of these potential epileptogenic processes and assess the end result in a preparation that is sensitive enough to detect small changes in whatever is being measured.

18.3.3   Variables That May Affect the Development of Epilepsy After Cortical Injury and the Interpretation of Results of Modeling Experiments

(i) Severity of injury and resulting epileptogenesis: It is clear that the severity of injury is a key prognostic factor in human posttraumatic epilepsy and one that may affect experimental results in a model [5]. Further, in models of severe traumatic injury or prolonged status epilepticus, multiple brain regions may be affected, making it difficult to determine site(s) of seizure origin. As discussed above, in experiments testing either prophylactic or therapeutic agents, it is desirable to use "high throughput" models in which there is frequent and intense seizure

activity. Under these circumstances, it is possible that a therapeutic trial would appear to be negative because of the intense epileptiform activity, even though the agent employed was altering its target, as hypothesized. Other epileptogenic mechanisms might be powerful enough to hide favorable actions, leading to a false-negative trial.

(ii) Site(s) and distribution of lesions (focal, multifocal, diffuse; hippocampus vs. neocortex) may influence results: Different cortical areas have varying susceptibilities to the development of epileptiform activity. Such differences in epileptogenic capacity from region to region with a given injury (e.g. [5]), or even within different laminae in the same cortical area [4, 36] may be due to variability in circuitry, receptors and intrinsic cell properties. Whatever the mechanisms, this variability makes it important to focus modeling experiments on specific neuronal types and structures comparable to those thought to be involved in clinical epileptogenesis. These intrinsic differences make it important to sample a given cell type or area, recognizing that there may be significant differences if experiments are carried out in another cortical region. There is marked variability in incidence, severity and frequency of seizures, even in the same posttraumatic model in the same laboratory [5]. This variability resembles that seen following human head injury, but also makes testing of antiepileptic or prophylactic strategies more difficult and raises questions about models in which almost all animals have frequent seizures.

(iii) The etiology as well as severity of a human cortical lesion may be a factor that determines the likelihood of epileptogenesis and success of a planned intervention. Penetrating injuries and those that induce intracerebral bleeding have a higher incidence of seizures than those resulting from infarction or closed head injury. There are also sometimes striking differences between incidence of seizures in different models in the same laboratory [17], and between different laboratories using the same model. Some of this variability may be due to errors in experimental design [21, 34].

(iv) Age and species: Assumptions regarding applicability of specific findings from models of epilepsy in one strain or species to another, or to human epilepsy, should be made with caution. These differences extend to transport of antiepileptic drugs [1]; induction of status epilepticus and its consequences [2, 22, 55]; seizure-induced cell injury or death [43]; kindling [12, 50], and effects of ischemia [54]. Susceptibility to epilepsy may be greater in the immature brain [16, 52], although some parameters that are thought to be important to epileptogenesis, such as the maturation of excitatory axonal arbors of cortical pyramidal cells, are slow to develop fully [40]. This makes results of experiments performed in models of epilepsy in immature in vitro slices difficult to generalize to mature cortical CNS structures.

18.4 What Are Some Key Issues for Clinical and Basic Research in Partial Epilepsies?

(a) Latent period between injury and seizures provides evidence for ongoing epileptogenic processes following cortical injury and an opportunity for prophylaxis.

There may be a critical period within the first few days after injury when therapeutic intervention will be effective, even though the latency to seizures is significantly longer (e.g. [6, 13, 14, 26]). Further analyses of pathophysiological events that occur during the critical period and are interrupted during such experiments may lead to new effective antiepileptogenic approaches.

(b) Non-epileptogenic vs. epileptogenic injury.

What are the genetic or other susceptibility factors that predispose an individual to epileptogenesis after injury (e.g. [8, 51])?

(c) Which changes in epileptogenic brain are adaptive vs. maladaptive?

Injury-induced axonal sprouting has been considered a key *maladaptive* epileptogenic mechanism in a variety of models, and in human cortical structures. (reviewed in [25,

37]). However, establishment of new connectivity may also be an important *adaptive* mechanism that underlies recovery from stroke and other injury [7, 23, 27]. Recent results show that excitatory synaptic connectivity and epileptogenesis can be significantly reduced in cortically injured rats by treatment in vivo with gabapentin, a drug that interferes with synaptogenesis induced by astrocytic thrombospondins [26]. Will such drugs also limit behavioral recovery from brain injury? Additional experiments are required in models of injury-induced epileptogenesis to determine whether maladaptive connectivity can be limited without affecting adaptive mechanisms that foster behavioral recovery.

(d) New (targeted) drug development; pharmaco-resistance.

Why have rates of seizure control not increased, in spite of introduction of multiple new anti-epileptic drugs?

(e) Mechanisms of interictal-ictal transitions. Why a seizure today? What starts it? How does it propagate? What ends it? *"Why does the relatively restricted sporadic discharge of chronically epileptic neurons become periodically enhanced and propagated to produce overt seizures?"* [20].

(f) What are the trigger mechanisms for sporadic seizures? Roles of "stress", sleep, fever, hormones, etc.

(g) Effects of epileptiform activity on neocortical and limbic structure/function.

(h) What are the long term impacts of epilepsy on the mature and developing brain?

(i) Co-morbidities as targets for research using animal models.

18.5 Conclusions

1. In assessing models of neocortical or temporal lobe epilepsy, it is important to first identify the specific issue to be addressed in the laboratory, derived from clinical observations and research, or results of previous experiments.

2. The long list of variables detailed above, that can influence results and conclusions, should be considered in advance and experimental and control groups and protocols planned accordingly.

3. Convenience is not the most important criterion for use of a given model. In the case of preclinical trials of agents for chronic partial epilepsy, the choices are quite limited, as seizure frequency sufficient to power the experiments is present predominantly in post-status models where widespread abnormalities are present and the analogy to the pathophysiology of spontaneously-occurring clinical partial epilepsy is unclear [29].

4. A major question for the model chosen will be whether expected results, based on known or expected variability of data, will yield an unambiguous answer to a specific issue, within practical limits of available resources.

5. A broader definition of the term "translation" is necessary, to include mechanisms by which defects at molecular, cellular and network levels are "translated" or evolve into the dysregulated cortical activities that generate epileptiform activity and behavioral events. Without knowledge of events at this level, the "Lost in Translation" phenomenon, i.e., failed clinical trials, false negative experiments, or collection of data irrelevant to the clinical issue, will be more likely.

6. *"A really complete understanding of epilepsy might require almost total knowledge of the central nervous system"* [20]. Much of our progress in epilepsy research derives from non-epilepsy related experiments in basic neuroscience. Too much emphasis on the "relevance" of a particular model or approach to epileptogenesis may limit discovery of major contributing mechanisms derived from less targeted experiments.

Acknowledgements I gratefully acknowledge the major contributions made by Phil Schwartzkroin to work in my laboratory from 1971 to 1979. He has been a valued colleague and friend over the years, and I admire the outstanding contributions he has made to many trainees and the epilepsy community since those exciting early days.

Other Acknowledgements Supported in part by grants NS06477, NS12151, NS39579, NS082664, and T32NS007280.

References

1. Baltes S, Gastens AM, Fedrowitz M, Potschka H, Kaever V, Löscher W (2007) Differences in the transport of the antiepileptic drugs phenytoin, levetiracetam and carbamazepine by human and mouse P-glycoprotein. Neuropharmacology 52(2):333–346

2. Bankstahl M, Bankstahl JP, Bloms-Funke P, Löscher W (2012) Striking differences in proconvulsant-induced alterations of seizure threshold in two rat models. Neurotoxicology 33(1):127–137

3. Clementi A (1959) Experimental epilepsy following simultaneous strychinization of centers pertaining to two heterologous zones (visual zone and sensory-motor zone) of the cerebral cortex of the dog. Boll Soc Ital Biol Sper 35:1435–1439

4. Connors BW (1984) Initiation of synchronized neuronal bursting in neocortex. Nature 310(5979):685–687

5. Curia G, Levitt M, Fender JS, Miller JW, Ojemann J, D'Ambrosio R (2011) Impact of injury location and severity on posttraumatic epilepsy in the rat: role of frontal neocortex. Cereb Cortex 21(7):1574–1592

6. D'Ambrosio R, Eastman CL, Darvas F, Fender JS, Verley DR, Farin FM, Wilkerson HW, Temkin NR, Miller JW, Ojemann J, Rothman SM, Smyth MD (2013) Mild passive focal cooling prevents epileptic seizures after head injury in rats. Ann Neurol 73(2): 199–209

7. Dancause N, Barbay S, Frost SB, Plautz EJ, Chen D, Zoubina EV, Stowe AM, Nudo RJ (2005) Extensive cortical rewiring after brain injury. J Neurosci 25(44):10167–10179

8. Darrah SD, Miller MA, Ren D, Hoh NZ, Scanlon JM, Conley YP, Wagner AK (2013) Genetic variability in glutamic acid decarboxylase genes: associations with post-traumatic seizures after severe TBI. Epilepsy Res 103(2–3):180–194

9. Fisher RS (1989) Animal models of the epilepsies. Brain Res Brain Res Rev 14(3):245–278

10. Galanopoulou AS, Kokaia M, Loeb JA, Nehlig A, Pitkänen A, Rogawski MA, Staley KJ, Whittemore VH, Edward Dudek F (2013) Epilepsy therapy development: technical and methodologic issues in studies with animal models. Epilepsia 54(Suppl 4):13–23

11. Goldberg EM, Coulter DA (2013) Mechanisms of epileptogenesis: a convergence on neural circuit dysfunction. Nat Rev Neurosci 14(5):337–349

12. Goldensohn ES (1984) The relevance of secondary epileptogenesis to the treatment of epilepsy: kindling and the mirror focus. Epilepsia 25(Suppl 2):S156–S173

13. Graber KD, Prince DA (1999) Tetrodotoxin prevents posttraumatic epileptogenesis in rats. Ann Neurol 46(2):234–242

14. Graber KD, Prince DA (2004) A critical period for prevention of neocortical post-traumatic epileptogenesis in rats. Ann Neurol 55:860–870

15. Helbig I, Lowenstein DH (2013) Genetics of the epilepsies: where are we and where are we going? Curr Opin Neurol 26(2):179–185

16. Holmes GL, Khazipov R, Ben-Ari Y (2002) Seizure-induced damage in the developing human: relevance of experimental models. Prog Brain Res 135:321–334

17. Huusko N, Römer C, Ndode-Ekane XE, Lukasiuk K, Pitkänen A (2013) Loss of hippocampal interneurons and epileptogenesis: a comparison of two animal models of acquired epilepsy. Brain Struct Funct [Epub ahead of print]

18. Isokawa M, Levesque MF, Babb TL, Engel J Jr (1993) Single mossy fiber axonal systems of human dentate granule cells studied in hippocampal slices from patients with temporal lobe epilepsy. J Neurosci 13(4):1511–1522

19. Jacobs KM, Graber KD, Kharazia VN, Parada I, Prince DA (2000) Postlesional epilepsy: the ultimate brain plasticity. Epilepsia 41:S153–S161

20. Jasper HH, Ward AA Jr, Pope A (1969) Basic mechanisms of the epilepsies. Little, Brown, Boston, pp ix–x

21. Landis SC, Amara SG, Asadullah K, Austin CP, Blumenstein R, Bradley EW, Crystal RG, Darnell RB, Ferrante RJ, Fillit H, Finkelstein R, Fisher M, Gendelman HE, Golub RM, Goudreau JL, Gross RA, Gubitz AK, Hesterlee SE, Howells DW, Huguenard J, Kelner K, Koroshetz W, Krainc D, Lazic SE, Levine MS, Macleod MR, McCall JM, Moxley RT 3rd, Narasimhan K, Noble LJ, Perrin S, Porter JD, Steward O, Unger E, Utz U, Silberberg SD (2012) A call for transparent reporting to optimize the predictive value of preclinical research. Nature 490(7419):187–191

22. Langer M, Brandt C, Löscher W (2011) Marked strain and substrain differences in induction of status epilepticus and subsequent development of neurodegeneration, epilepsy, and behavioral alterations in rats. [corrected]. Epilepsy Res 96(3):207–224

23. Lee JK, Kim JE, Sivula M, Strittmatter SM (2004) Nogo receptor antagonism promotes stroke recovery by enhancing axonal plasticity. J Neurosci 24(27): 6209–6217

24. Lerche H, Shah M, Beck H, Noebels J, Johnston D, Vincent A (2013) Ion channels in genetic and acquired forms of epilepsy. J Physiol 591(Pt 4):753–764

25. Li H, McDonald W, Parada I, Faria L, Graber K, Takahashi DK, Ma Y, Prince D (2011) Targets for preventing epilepsy following cortical injury. Neurosci Lett 497(3):172–176

26. Li H, Graber KD, Jin S, McDonald W, Barres BA, Prince DA (2012) Gabapentin decreases epileptiform discharges in a chronic model of neocortical trauma. Neurobiol Dis 48(3):429–438

27. Liauw J, Hoang S, Choi M, Eroglu C, Choi M, Sun GH, Percy M, Wildman-Tobriner B, Bliss T, Guzman RG, Barres BA, Steinberg GK (2008) Thrombospondins 1 and 2 are necessary for synaptic plasticity and functional recovery after stroke. J Cereb Blood Flow Metab 28(10):1722–1732

28. Löscher W, Brandt C (2010) Prevention or modification of epileptogenesis after brain insults: experimental approaches and translational research. Pharmacol Rev 62(4):668–700

29. Löscher W (2011) Critical review of current animal models of seizures and epilepsy used in the discovery and development of new antiepileptic drugs. Seizure 20(5):359–368

30. Löscher W, Schmidt D (2011) Modern antiepileptic drug development has failed to deliver: ways out of the current dilemma. Epilepsia 52(4):657–678

31. Masukawa LM, Prince DA (1984) Synaptic control of excitability in isolated dendrites of hippocampal neurons. J Neurosci 4(1):217–227

32. Pavlov I, Kaila K, Kullmann DM, Miles R (2013) Cortical inhibition, pH and cell excitability in epilepsy: what are optimal targets for antiepileptic interventions? J Physiol 591(Pt 4):765–774

33. Penfield W, Jasper HH (1954) Epilepsy and the functional anatomy of the human brain. Little, Brown, Boston

34. Pitkänen A, Nehlig A, Brooks-Kayal AR, Dudek FE, Friedman D, Galanopoulou AS, Jensen FE, Kaminski RM, Kapur J, Klitgaard H, Löscher W, Mody I, Schmidt D (2013) Issues related to development of antiepileptogenic therapies. Epilepsia 54(Suppl 4):35–43

35. Prince DA, Wilder BJ (1967) Control mechanisms in cortical epileptogenic foci. "Surround" inhibition. Arch Neurol 16(2):194–202

36. Prince DA, Tseng GF (1993) Epileptogenesis in chronically injured cortex: in vitro studies. J Neurophysiol 69(4):1276–1291

37. Prince DA, Parada I, Graber K (2012) Traumatic brain injury and posttraumatic epilepsy. In: Noebels JL, Avoli M, Rogawski MA, Olsen RW, Delgado-Escueta AV (eds) Jasper's basic mechanisms of the epilepsies, 4th edn. Oxford, New York, pp 315–330

38. Raymont V, Salazar AM, Lipsky R, Goldman D, Tasick G, Grafman J (2010) Correlates of posttraumatic epilepsy 35 years following combat brain injury. Neurology 75(3):224–229

39. Rogawski MA (2013) The intrinsic severity hypothesis of pharmacoresistance to antiepileptic drugs. Epilepsia 54(Suppl 2):33–40

40. Romand S, Wang Y, Toledo-Rodriguez M, Markram H (2011) Morphological development of thick-tufted layer v pyramidal cells in the rat somatosensory cortex. Front Neuroanat 5:5

41. Salazar AM, Jabbari B, Vance SC, Grafman J, Amin D, Dillon JD (1985) Epilepsy after penetrating head injury. I. Clinical correlates: a report of the Vietnam Head Injury Study. Neurology 35(10):1406–1414

42. Salin P, Tseng GF, Hoffman S, Parada I, Prince DA (1995) Axonal sprouting in layer V pyramidal neurons

of chronically injured cerebral cortex. J Neurosci 15(12):8234–8245

43. Schauwecker PE (2012) Strain differences in seizure-induced cell death following pilocarpine-induced status epilepticus. Neurobiol Dis 45(1):297–304

44. Schmidt D, Rogawski MA (2002) New strategies for the identification of drugs to prevent the development or progression of epilepsy. Epilepsy Res 50(1–2):71–78

45. Schwartzkroin PA (2012) Cellular bases of focal and generalized epilepsies. Handb Clin Neurol 107:13–33

46. Schwartzkroin PA (2012) Why – and how – do we approach basic epilepsy research? In: Noebels JL, Avoli M, Rogawski MA, Olsen RW, Delgado-Escueta AV (eds) Jasper's basic mechanisms of the epilepsies, 4th edn. Oxford, New York, pp 24–37

47. Simonato M, French JA, Galanopoulou AS, O'Brien TJ (2013) Issues for new antiepilepsy drug development. Curr Opin Neurol 26(2):195–200

48. Sloviter RS, Bumanglag AV (2013) Defining "epileptogenesis" and identifying "antiepileptogenic targets" in animal models of acquired temporal lobe epilepsy is not as simple as it might seem. Neuropharmacology 69:3–15

49. Tauck DL, Nadler JV (1985) Evidence of functional mossy fiber sprouting in hippocampal formation of kainic acid-treated rats. J Neurosci 4:1016–1022

50. Wada JA (1978) Kindling as a model of epilepsy. Electroencephalogr Clin Neurophysiol Suppl 34:309–316

51. Wagner AK, Miller MA, Scanlon J, Ren D, Kochanek PM, Conley YP (2010) Adenosine A1 receptor gene variants associated with post-traumatic seizures after severe TBI. Epilepsy Res 90(3):259–272

52. Wasterlain CG, Gloss DS, Niquet J, Wasterlain AS (2013) Epileptogenesis in the developing brain. Handb Clin Neurol 111:427–439

53. Wong RK, Prince DA (1979) Dendritic mechanisms underlying penicillin-induced epileptiform activity. Science 204(4398):1228–1231

54. Wölfer J, Bantel C, Köhling R, Speckmann EJ, Wassmann H, Greiner C (2006) Electrophysiology in ischemic neocortical brain slices: species differences vs. influences of anaesthesia and preparation. Eur J Neurosci 23(7):1795–1800

55. Xu B, McIntyre DC, Fahnestock M, Racine RJ (2004) Strain differences affect the induction of status epilepticus and seizure-induced morphological changes. Eur J Neurosci 20(2):403–418

Aligning Animal Models with Clinical Epilepsy: Where to Begin?

19

Stephen C. Harward and James O. McNamara

Abstract

Treatment of the epilepsies have benefitted immensely from study of animal models, most notably in the development of diverse anti-seizure medications in current clinical use. However, available drugs provide only symptomatic relief from seizures and are often ineffective. As a result, a critical need remains for developing improved symptomatic or disease-modifying therapies – or ideally, preventive therapies. Animal models will undoubtedly play a central role in such efforts. To ensure success moving forward, a critical question arises, namely "How does one make laboratory models relevant to our clinical understanding and treatment?" Our answer to this question: It all begins with a detailed understanding of the clinical phenotype one seeks to model. To make our case, we point to two examples – Fragile X syndrome and status epilepticus-induced mesial temporal lobe epilepsy – and examine how development of animal models for these distinct syndromes is based upon observations by astute clinicians and systematic study of the disorder. We conclude that the continuous and effective interaction of skilled clinicians and bench scientists is critical to the optimal design and study of animal models to facilitate insight into the nature of human disorders and enhance likelihood of improved therapies.

S.C. Harward
Department of Neurobiology, Duke University
Medical Center, Durham, NC 27710, USA
e-mail: sharward@neuro.duke.edu

J.O. McNamara (✉)
Department of Neurology, Medicine (Neurology),
Duke University Medical Center,
Durham, NC 27710, USA

Department of Pharmacology and Molecular Cancer
Biology, Duke University Medical Center,
Durham, NC 27710, USA
e-mail: jmc@neuro.duke.edu

H.E. Scharfman and P.S. Buckmaster (eds.), *Issues in Clinical Epileptology: A View from the Bench*,
Advances in Experimental Medicine and Biology 813, DOI 10.1007/978-94-017-8914-1_19,
© Springer Science+Business Media Dordrecht 2014

Keywords

Epilepsy • Animal models • Fragile X • Temporal lobe epilepsy

Animal models have played a critical role in epilepsy research dating back to 1937 when Putnam and Merritt [32] published the first animal model of seizures, the electroshock test in cats. Since that time, thousands of papers have been published that detail the development and utilization of animal models of seizures and epilepsy. Currently, there are over 100 different animal models employed in epilepsy research [33, 35]. These models utilize a wide array of species including drosophila, zebrafish, mice, rats, guinea pigs, cats, and even non-human primates. These models have provided insight into cellular and molecular mechanisms surrounding many aspects of the epilepsies. Moreover, some of these models have led to the development of novel therapies in the clinic.

That said, much work remains. Current pharmacologic treatments for epilepsy are "symptomatic" insofar as they suppress but do not prevent, modify, or cure the disorder. There is a critical need for new and improved treatment options that promote not only enhanced symptomatic therapy, but also (for the first time) provide disease-modifying or preventive therapy. Satisfying this need will require the use of appropriately designed and implemented animal models. To ensure success, we must address the following question: "How does one make laboratory models relevant to our clinical understanding and treatment?"

In our view, the answer to this question starts with a detailed understanding of the clinical phenotype one seeks to model. This understanding in turn guides design and analysis of the animal model. Here we choose two examples to illustrate our thinking. One consists of a monogenic disorder, Fragile X syndrome. The other is a subtype of the common, sporadic disorder temporal lobe epilepsy – namely the syndrome of mesial temporal lobe epilepsy emerging months to years after an episode of prolonged seizures (status epilepticus).

For each example, we will focus on a specific animal model that promises to inform clinical understanding and treatment.

19.1 Evaluating Clinical Relevance

Evaluating the clinical relevance of an animal model is not a question unique to epilepsy research. In fact, this question is seminal to preclinical investigation of most human diseases. One approach to considering the relevance of a model for a particular human disease involves model evaluation using three criteria: its construct validity, its face validity, and its predictive validity [5].

"Construct validity" refers to how closely an animal model recapitulates the causal mechanisms underlying the disease in humans [5]. Construct validity is most readily addressed with monogenic disorders in which clinical and molecular genetic analyses have elucidated the molecular etiology of the syndrome in humans. For example, in Dravet Syndrome, the underlying cause for most human cases consists of *de novo* mutations of the *SCN1A* gene that result in loss of function [7]. Consequently, approaches to developing an animal model of Dravet Syndrome with high construct validity would include engineering an experimental animal (e.g. fly or zebrafish or mouse) with a null mutation of *SCN1A* or by substituting the wild type gene with an actual mutation identified in a human, a strategy referred to as "knock-in".

"Face validity" refers to how closely an animal model recapitulates phenotypic characteristics of the human disease [5]. For example, patients with Rett Syndrome display several characteristic features, including cognitive impairment, breathing irregularities, and stereotypic hand movements [18, 29]. An animal model of Rett Syndrome with high face validity would reproduce most if not all of these phenotypic features.

Finally, "predictive validity" refers to how closely an animal model recapitulates treatment responsiveness observed in humans [5]. For example, absence seizures in humans are typically quite responsive to the pharmacologic agent ethosuximide [16]. An animal model for absence seizures that has high predictive validity would demonstrate a similar response to ethosuximide.

As we consider how to enhance the relevance of animal models to our clinical understanding and treatment, we will use these criteria as a framework. Ideally, animal models would have high validity for each of these three criteria. However, as we discuss below, this may not be possible, and – importantly – may not be necessary in order to inform clinical understanding and/or treatment of epilepsy. In fact, having high validity in only one criterion – construct, face, or predictive – may still provide a useful model for the appropriately selected question.

19.2   Mendelian Disorders of Epilepsy

Mendelian disorders of epilepsy are those in which clinical and molecular genetic evidence establishes the mutation of a single gene as the cause of the disorder. To date, mutations in over 100 genes comprising a wide range of proteins have been linked to human diseases in which epilepsy is one of the phenotypic manifestations [28, 34]. Collectively, these Mendelian epilepsies account for only a small fraction of all epilepsies [28]. That said, study of these disorders will hopefully benefit individuals affected with these mutations, and insights derived from such studies may also inform mechanisms of non-Mendelian epilepsies.

For these Mendelian epilepsies, identification of the causal mutant gene by clinical and molecular genetic studies creates the opportunity to engineer an animal model by introducing the mutant gene into the genome of an experimental animal (e.g. fly, zebrafish, mouse, etc.) and examining its phenotypic manifestations. Such models typically have high construct validity because scientists can incorporate the precise genetic abnormality seen in humans, whether this

abnormality is a point mutation, chromosomal translocation, a frameshift mutation, etc. This high construct validity commonly equates to high face validity – the models recapitulate the key phenotypic features of the human disease. In these situations, the high construct and face validity strengthen the likelihood that such models will have high predictive validity as well.

However, high construct validity does not assure high face validity. For example, cystic fibrosis is a disease characterized by multi-organ failure with recurrent and persistent pulmonary infections being quite prominent. A common cause of cystic fibrosis is a mutation of F508 in the *CFTR* gene [17]. Mouse models with this exact mutation in their endogenous *Cftr* gene do not reproduce the severe pulmonary phenotype seen in humans. There are numerous possible explanations for this disparity including differences in the genetic background, immune response, etc. In spite of such a disparity, the low face validity of these models does not preclude their usefulness for addressing important questions. In fact, these cystic fibrosis models have been used extensively and with good success to probe questions surrounding other aspects of the human disease. The key issue is that the question addressed in the animal model must be carefully aligned with a specific and important question arising in the human disorder.

To illustrate these considerations in greater detail, we consider Fragile X syndrome – a disorder in which epilepsy is a prominent manifestation and for which engineering genetically modified mice have produced useful models that promise to inform our clinical understanding and treatment of this disease.

19.2.1   Characterizing Fragile X and Developing an Animal Model

Fragile X syndrome is a genetic disorder occurring in 1:5,000 males [8]. Phenotypic manifestations include seizures, autism, cognitive impairment, hypersensitivity to sensory stimuli, motor incoordination, growth abnormalities, and various physical characteristics such as an elongated face, large protruding ears, and marcoorchidism [6, 9, 30]. Clinical and molecular genetic investigations have led to identification of the molecular etiology of this syndrome, namely, a mutation of a gene termed "Fragile X". Development of an animal model for Fragile X has been decades in the making. Its history began in 1943 when two clinicians (J. Purdon Martin and Julia Bell) described a family in England in which 11 males of two generations presented with mental retardation and social withdrawal [26]. Based on the pedigree, these clinicians hypothesized that this presentation of symptoms represented a novel, sex-linked recessive genetic disorder. Microscopic evaluation of chromosomal spreads isolated from these patients revealed the X-chromosome to be deformed or broken, thereby leading to the name "Fragile X" [24]. It took nearly 50 years, but the causative gene on the X-chromosome was finally identified to be *FMR1* [39]. The primary mutation within this gene leading to Fragile X was an expansion of the CGG trinucleotide repeat found within the 5′ untranslated region of *FMR1*. Investigators quickly demonstrated that this genomic expansion in turn leads to transcriptional silencing of the gene and thus a lack of the protein encoded by this gene – fragile X mental retardation protein (FMRP; [31]). With the gene, the mutation, and the effect on protein expression documented, scientists next set out to develop an animal model for Fragile X. By introducing a null mutation of the *Fmr1* gene into the genome of a mouse, an animal model was developed that recapitulated the loss of FMRP expression observed in humans [14]. By definition, this model does not exactly recapitulate the initial pathologic lesion underlying the human disorder, namely, the CGG trinucleotide expansion, and thus does not have *perfect* construct validity. However, it does recapitulate what is likely the primary consequence of the mutation, loss of FMRP expression.

The fact that the *Fmr1* knockout mouse is not identical to the human genetic abnormality, yet recapitulates the human protein abnormality, raises an interesting point regarding animal model development. Specifically, a model lacking *perfect* construct validity may nonetheless shed light on clinical understanding and treatment of a

disorder. Indeed the *Fmr1* knockout mouse does recapitulate many aspects of the human phenotype, thereby giving it high face validity. Similar to humans with Fragile X, these mice exhibit seizures, cognitive problems, hyperactivity, and macroorchidism [4, 14, 30]. The high face validity of this mouse model has led to an intense search for how the genotype causes the phenotype. Briefly, FMRP is highly expressed within neurons, especially at synapses [4, 10, 30]. Here, it binds many messenger RNAs and represses translation of these mRNAs into protein [2, 22, 36, 42]. Upstream of these events is the G-protein-coupled glutamate receptor, mGluR, the activation of which leads to protein translation [21, 40]. The loss of the repressive effects of FMRP in Fragile X allows for unopposed mGlu5 signaling, which in turn results in excessive protein translation (Fig. 19.1). It is this unopposed mGlu5 signaling that likely contributes to the phenotypic manifestations of Fragile X syndrome, because crossing FMRP mutant mice to mice in which one allele of mGlu5 has been eliminated reduces seizures and other abnormalities of the FMRP mutant mouse [11]. These findings have led to development of potent and selective inhibitors of the mGlu5 receptor. Continuous treatment of FMRP mutant mice with mGlu5 inhibitors commencing early in life eliminates seizures and other phenotypic abnormalities. Moreover, initiating treatment with mGluR5 inhibitors in adult FMRP mutant mice *after* the development of seizures and other abnormalities reduces these seizures and corrects these other abnormalities. Collectively, these findings have provided the foundation of a clinical trial for patients with Fragile X syndrome with an mGlu5 inhibitor, results of which will inform the predictive validity of this model.

19.2.2 *Fmr1* Knockout Mouse – A Model Facilitating Clinical Understanding and Treatment of Fragile X Syndrome

In sum, clinical recognition of the distinctive phenotype and its familial aggregation provided the foundation for discovery of the mutant gene

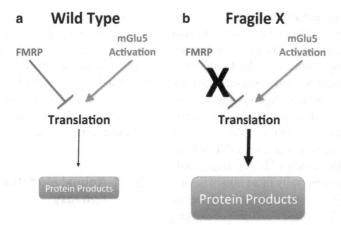

Fig. 19.1 FMRP and mGluR5 modulation of protein translation. (**a**) In wild type animals, FMRP inhibits protein translation while mGlu5 activation promotes translation. By balancing these opposing actions, the appropriate amount of protein products is generated. (**b**) However, in mouse models of Fragile X (*Fmrp* knockout animals), the loss of the repressive effects of FMRP leads to unopposed mGlu5 signaling and ultimately excessive protein synthesis

decades later. Insight into the nature of the causative mutations, in turn, enabled engineering a genetically modified mouse model. This model illustrates how discovery of the molecular mechanisms by which the genotype leads to the phenotype can give rise to identification of a target for development of small molecules that could be used as drugs to treat the disorder. Careful alignment of the animal model with the clinical phenotype, an alignment simplified by knowledge of the molecular etiology afforded by molecular genetics, has led to a sequence of discoveries that in turn have enabled design of a clinical trial based upon disease mechanism.

19.3 Non-Mendelian Disorders of Epilepsy

The vast majority of the epilepsies do not exhibit a Mendelian pattern of inheritance; instead they typically arise sporadically as a consequence of various cortical lesions including developmental abnormalities, neoplasms, traumatic brain injury, and vascular insults. In contrast to a Mendelian disorder in which development of an animal model is based upon a known molecular etiology (i.e., a mutant gene), here the model must rely on recapitulating some feature(s) of the clinical syndrome. Once again, however, a detailed characterization

of the syndrome in humans is of critical importance to both appropriate design and evaluation of the animal model. One non-Mendelian epilepsy syndrome that has been extensively characterized by clinicians is a form of temporal lobe epilepsy (TLE) arising long after an episode of prolonged seizure activity (status epilepticus – SE).

19.3.1 The Clinical Syndrome

TLE is the most common and also most devastating form of partial epilepsy in humans [33]. Broadly defined, TLE is an epilepsy in which seizures most commonly are initiated from the medial temporal lobe. A diversity of etiologies of TLE has been identified, implying that TLE comprises multiple disease subtypes. Despite such heterogeneity, some features are conserved – most notably the associated asymmetric pattern of hippocampal neuronal loss and gliosis, termed hippocampal or temporal lobe sclerosis [33].

One proposed subtype of TLE that presents with hippocampal sclerosis is that arising after an episode of SE [41]. Retrospective analysis of patients undergoing surgery for intractable TLE reveal that many of these patients experienced an episode of prolonged, focal, severe seizures (SE) many years prior to epilepsy development [15]. Most commonly, these severe seizures occurred in the context of complicated febrile seizures during infancy or childhood but similar observations have been made following afebrile status epilepticus arising *de novo* in adults. Longitudinal studies have confirmed these observations in that up to half of individuals experiencing *de novo* status epilepticus of either febrile or afebrile origin in childhood or adulthood develop recurrent seizures (epilepsy) after a seizure-free latent period of variable duration [1, 37]. Importantly, inducing SE experimentally in an otherwise normal animal is sufficient to trigger the subsequent development of TLE. Based on these converging lines of evidence, it seems likely that the occurrence of *de novo* SE during infancy or adulthood contributes to TLE development in humans.

One prominent feature of this syndrome is a structural abnormality referred to as Ammon's Horn or hippocampal sclerosis [43]. It has long been recognized that many patients with TLE have atrophic and damaged hippocampi as visualized on MRI or histopathologic examination [25, 43]. Animal studies provide convincing evidence that severe seizure activity is sufficient to induce hippocampal damage similar to hippocampal sclerosis observed in humans, namely, neuron loss predominantly in the hilus and CA1 as well as mossy fiber sprouting in the dentate gyrus [12]. However, the specific relationship between hippocampal sclerosis and epileptogenesis has been highly debated. In our view, it seems plausible that hippocampal sclerosis is both a consequence of SE and can contribute to development of TLE. In the context of this controversy, there emerged an important clinical observation: MRI evidence of acute hippocampal injury within days following complicated febrile seizures, an event followed months later by hippocampal atrophy [38]. This MRI abnormality is evident in a subset of children following an episode of febrile status epilepticus. The question arises as to whether the subsequent emergence of TLE years later occurs in the subset with hippocampal damage and not in those with normal hippocampi (as detected by MRI) following status epilepticus. Addressing this question is the objective of a multicenter, longitudinal study (the FEBSTAT study) of children undergoing

complicated febrile seizures, a study that will permit correlating the occurrence of acute hippocampal injury and subsequent hippocampal sclerosis with the later emergence of TLE [19]. Importantly, the detailed analysis of this syndrome will provide the information needed to design and characterize animal models properly aligned with the human disease.

19.3.2 Animal Models of SE-Induced Epilepsy

SE-induced TLE is a heterogeneous disorder. Variability in presentation can be seen at almost every aspect of the disease – SE etiology, latent period duration, epileptic seizure severity, histopathology, etc. As such, it is unlikely that one single model can perfectly recapitulate all of its many facets. For this reason, it is no surprise that many different models for SE-induced TLE have emerged. That said, there are several key features that an animal model of this syndrome is expected to reflect. First, the model should begin with a brief episode of SE that is followed by emergence of spontaneous recurrent seizures after a latent period. Second, the model should correlate SE with *unilateral* hippocampal sclerosis, consistent with the pathologic findings noted in humans on both MRI and histopathology. Third and finally, the model should produce adult-onset of TLE as a result of either SE in adulthood [37] or in infants and children [1].

One model that fulfills these three criteria is TLE arising following SE induced by microinfusion of the ionotropic glutamate receptor agonist, kainic acid (KA), into the amygdala. This model can be induced by infusion of KA into the amygdala of either young (P10) or adult rodents, the resulting SE leading to subsequent development of epilepsy. In adult mice and rats, microinjection of KA into the basolateral amygdala nucleus leads to almost immediate onset of status epilepticus [3, 27]. Typically, SE is allowed to continue for 40 min, at which point a benzodiazepine such as diazepam or lorazepam is administered to stop the seizure activity. Approximately 3 days after the initial

SE event, spontaneous recurrent seizures arise and appear to persist lifelong. In P10 rat pups, a similar approach has been utilized [13]. Kainic acid is microinjected into the basolateral amygdala nucleus leading to almost immediate onset of SE that lasts for several hours. Typically, SE is allowed to continue until its natural termination (as opposed to the pharmacologic intervention used in adult animals). When these animals are evaluated 4 months later, they exhibit both behavioral and electrographic seizure activity, demonstrating the emergence of TLE.

This model is one of several in which induction of SE in an otherwise normal rodent results in emergence of TLE. Other methods include systemic pilocarpine administration (a muscarinic agonist), systemic kainic acid administration, or focal electrical stimulation. These methods are effective and have been used extensively by many labs, with each of the models exhibiting advantages and disadvantages. Adapting the intra-amygdala KA model to mice [3] simplifies study of genetically modified animals, providing a powerful tool for elucidating molecular and cellular mechanisms of epilepsy. Additional advantages of the intra-amygdala KA model in the mouse include: 100 % of KA-injected mice develop SE; mortality is only 10–20 %; and 100 % of surviving animals become epileptic [3, 13, 27]. The efficiency together with low attrition provides important advantages, especially for studies of genetically modified animals with limited availability.

Importantly, a number of features of this model align with the clinical syndrome. To begin, this model mimics the initial pathologic insult observed in many patients, namely, SE. Furthermore, this model can induce epilepsy via SE in both young and adult animals, similarly to that observed in humans. However, the construct validity is not perfect in that in this model, SE is induced by a convulsant (KA) while in the majority of children, SE arises in the context of a febrile illness. That said, the fact that SE, whether induced by diverse chemical methods or electrical stimulation, causes TLE suggests that the key variable promoting epileptogenesis

in most instances is the occurrence of SE *per se*, not the cause of the SE.

In terms of face validity, this model does recapitulate many but not all components of the clinical syndrome. First, this model does mimic the temporal course of the disease in that SE leads to a latent period which in turn evolves into TLE. One area of debate with this model is the length of the latent period. In humans, the time between SE and TLE is on the order of months to years while in this model it is only a few days. Second, this model does yield unilateral hippocampal sclerosis following epilepsy onset that can be detected by both MRI and histopathologic analysis [13, 27]. One caveat is that the pattern of neuron loss within the hippocampus is different from that observed in humans. For most human specimens, neuron loss is most prominent in the hilus and CA1 regions [43]. In the intra-amygdala KA model, neuron loss is most prominent in CA3 and hilus, leaving CA1 relatively spared [13, 27]. Lastly, in humans with TLE, memory deficits and other comorbidities are common [25]. To date, there are no studies clearly documenting memory deficits following implementation of this animal model. However, recent work revealed the occurrence of anxiety-like behaviors in this model [23].

Regarding predictive validity, since there is currently no preventive therapy for TLE arising after SE in humans, it is not possible to assess the predictive validity of this model. However, the utility of this model has enabled discovery of two molecular targets that show promise for development of preventive therapy. First, David Henshall and colleagues reported that expression of the microRNA, miR-134, is increased following SE and that inhibiting miR-134 expression shortly after SE onset may be antiepileptogenic [20]. Second, work from our lab revealed that SE induced the enhanced activation of the BDNF receptor tyrosine kinase TrkB [23]. The utility of this model in the mouse enabled a powerful chemical-genetic approach using genetically modified mice. This approach led to the discovery that inhibition of the TrkB kinase activity, commencing following SE and continued for just 2 weeks, prevented development of TLE in more than 90 % of animals when they were tested a month later. These discoveries, made possible by adapting this model to the mouse (40), provide novel targets for development of preventive therapy for this particular syndrome. Whether similar molecular mechanisms underlie development of TLE induced by different causes (e.g. trauma, developmental abnormalities, etc.) is uncertain.

19.3.2.1 Aligning Animal Models with Human Disease

The epilepsies represent a collection of heterogeneous disorders for which only symptomatic treatment is currently available. The lack of efficacy, together with undesirable consequences of symptomatic therapy for many patients, underscores the need to develop preventive therapies. Development of preventive therapies based upon disease mechanism requires properly aligning the animal model with the clinical syndrome. This is a challenging task, one that must begin with a detailed characterization of the clinical disorder. Such information provides a context critical to design and study of animal models that recapitulate key features of the clinical disorder. This descriptive first step underscores the importance of continuous and effective interactions of clinicians and bench scientists to assure the optimal alignment of animal models with human diseases, thereby enhancing the likelihood that study of the models will contribute to understanding the mechanisms of the disease and improving treatment. In short, the most effective and efficient way to develop animal models is to start at the bedside, move to the bench, and with a lot of hard work and luck, return to the bedside with a novel therapy in hand.

Acknowledgements Dedicated to Philip A. Schwartzkroin. A dear friend and a wonderful scientist whose thought-provoking questions have shaped thinking in this field for at least three decades.

Other Acknowledgements This work was supported by NINDS grants RO1NS56217 (JOM), RO1NS060728 (JOM), and F31NS078847 (SCH).

References

1. Annegers JF, Hauser WA, Shirts SB, Kurland LT (1987) Factors prognostic of unprovoked seizures after febrile convulsions. N Engl J Med 316:493–498
2. Ashley C, Wilkinson K, Reines D, Warren S (1993) FMR1 protein: conserved RNP family domains and selective RNA binding. Science 262:563–566
3. Ben-Ari Y, Tremblay E, Ottersen OP (1980) Injections of kainic acid into the amygdaloid complex of the rat: an electrographic, clinical and histological study in relation to the pathology of epilepsy. Neuroscience 5:515–528
4. Bhakar AL, Dolen G, Bear MF (2012) The pathophysiology of fragile X (and what it teaches us about synapses). Annu Rev Neurosci 35:417–443
5. Chadman KK, Yang M, Crawley JN (2009) Criteria for validating mouse models of psychiatric diseases. Am J Med Genet B Neuropscyhiatr Genet 150B:1–11
6. Chudley AE, Hagerman RJ (1987) Fragile X syndrome. J Pediatr 110:821–831
7. Claes L, Del-Favero J, Ceulemans B, Lagae L, Van Broeckhoven C, De Jonghe P (2001) De novo mutations in the sodium-channel gene SCN1A causes severe myoclonic epilepsy of infancy. Am J Hum Genet 68:1327–1332
8. Coffee B, Keith K, Albizua I, Malone T, Mowrey J, Sherman SL, Warren ST (2009) Incidence of fragile X syndrome by newborn screening for methylated *FMR1* DNA. Am J Hum Genet 85:503–514
9. Cornish K, Sudhalter V, Turk J (2004) Attention and language in fragile X. Ment Retard Dev Disabil Res Rev 10:11–16
10. Devys D, Lutz Y, Rouyer N, Bellocq JP, Mandel JL (1993) The FMR-1 protein is cytoplasmic, most abundant in neurons and appears normal in carriers of a fragile X permutation. Nat Genet 4:335–340
11. Dolen G, Osterweil E, Rao BS, Smith GB, Auerbach BD, Chattarji S, Bear MF (2007) Correction of fragile X syndrome in mice. Neuron 56:955–962
12. Dudek EF, Clark S, Williams PA, Grabenstatter HL (2006) Kainate-induced status epilepticus: a chronic model of acquired epilepsy. In: Pitkanen A, Schwartzkroin PA, Moshe SL (eds) Models of seizures and epilepsy. Elsevier, Burlington, pp 415–432
13. Dunleavy M, Shinoda S, Schindler C, Ewart C, Dolan R, Gobbo OL, Kerskens CM, Henshall DC (2010) Experimental neonatal status epilepticus and the development of temporal lobe epilepsy with unilateral hippocampal sclerosis. Am J Pathol 176:330–342
14. Dutch-Belgian Fragile X Consort (1994) *Fmr1* knockout mice: a model to study fragile X mental retardation. Cell 78:23–33
15. French JA, Williamson PD, Thadani VM, Darey TM, Mattson RH, Spencer SS, Spencer DD (1993) Characteristics of medial temporal lobe epilepsy: I. Results of history and physical examination. Ann Neurol 34:774–780

16. Glauser TA, Cnaan A, Shinnar S, Hirtz DG, Dlugos D, Masur D, Clark PO, Capparelli EV, Adamson PC (2013) Ethosuximide, valproic acid, and lamotrigine in childhood absence epilepsy: initial monotherapy outcomes at 12 months. Epilepsia 54:141–155

17. Guilbault C, Saeed Z, Downey GP, Radzioch D (2007) Cystic fibrosis mouse models. Am J Respir Cell Mol Biol 36:1–7

18. Hagberg B, Aicardi J, Dias K, Ramos O (1983) A progressive syndrome of autism, dementia, ataxia, and loss of purposeful hand use in girls: Rett's syndrome: report of 35 cases. Ann Neurol 14:471–479

19. Hesdorffer DC, Shinnar S, Lewis DV, Moshe SL, Nordli DR Jr, Pellock JM, MacFall J, Shinnar RC, Masur D, Frank LM, Epstein LG, Litherland C, Seinfeld S, Bello JA, Chan S, Bagiella E, Sun S (2012) FEBSTAT study team. Design and phenomenology of the FEBSTAT study. Epilepsia 53:1471–1480

20. Jimenez-Mateos EM, Engel T, Merino-Serrais P, McKiernan RC, Tanaka K, Mouri G, Sano T, O'Tuathaigh C, Waddington JL, Prenter S, Delanty N, Farrell MA, O'Brien DF, Conroy RM, Stallings RL, DeFelipe J, Henshall DC (2012) Silencing microRNA-134 produces neuroprotective and prolonged seizure-suppressive effects. Nat Med 18:1087–1094

21. Krueger DD, Bear MF (2011) Toward fulfilling the promise of molecular medicine in fragile X syndrome. Annu Rev Med 62:411–429

22. Laggerbauer B, Ostareck D, Keidel EM, Ostareck-Lederer A, Fischer U (2001) Evidence that fragile X mental retardation protein is a negative regulator of translation. Hum Mol Genet 10:329–338

23. Liu G, Gu B, He XP, Joshi RB, Wackerle HD, Rodriguiz RM, Wetsel WC, McNamara JO (2013) Transient inhibition of TrkB kinase after status epilepticus prevents development of temporal lobe epilepsy. Neuron 79:31–38

24. Lubs HA (1969) A marker X chromosome. Am J Hum Genet 21:231–244

25. Luby M, Spencer DD, Kim JH, deLanerolle N, McCarthy G (1995) Hippocampal MRI volumetrics and temporal lobe substrates in medial temporal lobe epilepsy. Magn Reson Imaging 13:1065–1071

26. Martin JP, Bell J (1943) A pedigree of mental defect showing sex-linkage. J Neurol Psychiatry 6:154–157

27. Mouri G, Jimenez-Mateos E, Engel T, Dunleavy M, Hatazaki S, Paucard A, Matsushima S, Taki W, Henshall DC (2008) Unilateral hippocampal CA3-predominant damage and short latency epileptogenesis after intra-amygdala microinjection of kainic acid in mice. Brain Res 1213:140–151

28. Nance MA, Hauser WA, Anderson VE (1997) Genetic diseases associated with epilepsy. In: Engel J, Pedley TA (eds) Epilepsy: a comprehensive textbook, 1st edn. Lippincott-Raven, Philadelphia, pp 197–209

29. Neul JL, Kaufmann WE, Glaze DG, Chirstodoulou J, Clarke AJ, Bahi-Buisson N, Leonard H, Bailey ME, Schanen NC, Zappella M, Renieri A, Huppke P, Percy AK, RettSearch Consortium (2010) Rett syndrome: revised diagnostic criteria and nomenclature. Ann Neurol 68:944–950

30. Penagarikano O, Mulle JG, Warren ST (2007) The pathophysiology of fragile X syndrome. Annu Rev Genomics Hum Genet 8:109–129

31. Pieretti M, Zhang FP, Fu Y-H, Warren ST, Oostra BA, Caskey CT, Nelson DL (1991) Absence of expression of the FMR-1 gene in fragile X syndrome. Cell 66:817–822

32. Putnam TJ, Merritt HH (1937) Experimental determination of the anticonvulsant properties of some phenyl derivatives. Science 85:525–526

33. Raol YH, Brooks-Kayal AR (2012) Experimental models of seizures and epilepsies. Prog Mol Biol Transl Sci 105:57–82

34. Reid CA, Berkovic SF, Petrou S (2009) Mechanisms of inherited epilepsies. Prog Neurobiol 87:41–57

35. Sarkisian MR (2001) Overview of the current animal models for human seizure and epileptic disorders. Epilepsy Behav 2:201–216

36. Siomi H, Siomi MC, Nussbaum RL, Dreyfuss G (1993) The protein product of the fragile X gene, FMR1, has characteristics of an RNA-binding protein. Cell 74:291–298

37. Tsai MH, Chuang YC, Chang HW, Chang WN, Lai SL, Huang CR, Tsai NW, Wang HC, Lin YJ, Lu CH (2009) Factors predictive of outcome in patients with de novo status epilepticus. Q J Med 102:57–62

38. VanLandingham KE, Heinz ER, Cavazos JE, Lewis DV (1998) Magnetic resonance imaging evidence of hippocampal injury after prolonged focal febrile convulsions. Ann Neurol 43:413–426

39. Verkerk AJMH, Pieretti M, Sutcliffe JS, Fu Y-H, Kuhl DPA, Pizzuti A, Reiner O, Richards S, Victoria MF, Zhang FP, Eussen BE, van Ommen GJB, Blonden LAJ, Riggins GJ, Chastain JL, Kunst CB, Galjaard H, Caskey CT, Nelson DL, Oostra BA, Warren ST (1991) Identification of a gene (FMR-1) containing a CGG repeat coincident with a breakpoint cluster region exhibiting length variation in fragile X syndrome. Cell 65:905–914

40. Weiler IJ, Greenough WT (1993) Metabotropic glutamate receptors trigger postsynaptic protein synthesis. Proc Natl Acad Sci U S A 90:7168–7171

41. Wieser HG (2004) ILAE Commission report. Mesial temporal lobe epilepsy with hippocampal sclerosis. Epilepsia 45:695–714

42. Wilkinson KD, Keene JD, Darnell RB, Warren ST (2001) Microarray identification of FMRP-associated brain mRNAs and altered mRNA translational profiles in fragile X syndrome. Cell 107:477–488

43. Williamson PD, French JA, Thadani VM, Kim JM, Novelly RA, Spencer SS, Spencer DD, Mattson RH (1993) Characteristics of medial temporal lobe epilepsy: II. Interictal and ictal scalp electroencephalography, neuropsychological testing, neuroimaging, surgical results, and pathology. Ann Neurol 34:781–787

What Non-neuronal Mechanisms Should Be Studied to Understand Epileptic Seizures?

20

Damir Janigro and Matthew C. Walker

Abstract

While seizures ultimately result from aberrant firing of neuronal networks, several laboratories have embraced a non-neurocentric view of epilepsy to show that other cells in the brain also bear an etiologic impact in epilepsy. Astrocytes and brain endothelial cells are examples of controllers of neuronal homeostasis; failure of proper function of either cell type has been shown to have profound consequences on neurophysiology. Recently, an even more holistic view of the cellular and molecular mechanisms of epilepsy has emerged to include white blood cells, immunological synapses, the extracellular matrix and the neurovascular unit. This review will briefly summarize these findings and propose mechanisms and targets for future research efforts on non-neuronal features of neurological disorders including epilepsy.

Keywords

Anti-epileptic drugs • Cerebrovasculature • Drug resistance • Brain endothelium • Glia-neuronal interactions • Extracellular matrix

D. Janigro, Ph.D. (✉)
Department of Molecular Medicine, Cerebrovascular Research, Cleveland Clinic Lerner College of Medicine, Cleveland, OH, USA
e-mail: janigrd@ccf.org

M.C. Walker, Ph.D., FRCP
Department of Clinical and Experimental Epilepsy, UCL Institute of Neurology, University College London, London WC1N 3BG, UK
e-mail: m.walker@ucl.ac.uk

This book is devoted to one of the all-time leaders in epilepsy research, Philip Alan Schwartzkroin. Phil has not only changed the traditional understanding of mammalian neurophysiology but he also revolutionized the tools we employ to study the brain as one of the people to perfect the brain slice preparation [70, 71, 74–76]. Last but not least, Phil has edited many seminal books and papers, and incessantly contributed to the recruitment and scientific development of scores of young scientists. Under the shadow of this giant (and former mentor for one of us (DJ)) writing this review is a

H.E. Scharfman and P.S. Buckmaster (eds.), *Issues in Clinical Epileptology: A View from the Bench*,
Advances in Experimental Medicine and Biology 813, DOI 10.1007/978-94-017-8914-1_20,
© Springer Science+Business Media Dordrecht 2014

humbling experience; one way to start the process is to refer the reader to Phil's recent introduction to the field of epilepsy research [72].

20.1 Why Study Non-neuronal Mechanisms in Neurology or Neuroscience?

We study the brain for many reasons, not least of which is for its intrinsic interest (see Schwartzkroin's recent introduction to the field [72]). The fascination with neurons and neuronal circuitries is not surprising since neurons are the collectors and effectors of our daily experiences and actions. In the specific case of epilepsy research (clinical or basic/translational), the quest for "epileptic neurons" or "epileptic circuits" has produced remarkable results, leading to the discovery of viable anti-epileptic drug targets and to the multimodal definition of the "epileptic focus", an invaluable clinical tool for the neurosurgeon. However, as in other neurological disorders, a neurocentric approach has left certain questions unanswered and experimental opportunities remain. The most striking example of why neuroscience should become more "holistic" is embolic stroke, a disease stemming from cerebrovascular disease that has devastating consequences on brain function. After the NIH convened a Stroke Progress Review Group in 2001, stroke research shifted from a purely neurocentric focus to a more integrated view wherein dynamic interactions between all cell types contribute to function and dysfunction in the brain. In the field of epilepsy research and treatment, there is no pressing need for such a sharp re-direction, since the field is already characterized by the study of many cell types, and non-neuronal processes. For example:

1. Many neuronal molecular, morphological defects or functional abnormalities described in human epileptic brain are present throughout the cycle of interictal-to-ictal states that characterize the epileptic brain. The persistence of these neuronal abnormalities does not fully explain why at a given time point an interictal cortex develops a seizure. Other mechanisms, such as changes in cerebral blood flow or blood-brain barrier permeability have been proposed to mediate the interictal to ictal transition.

2. It has been proposed that the process of epileptogenesis is distinct from the process of ictogenesis. According to this hypothesis, what makes a brain epileptic (e.g., genetic mutations, acquired or inherited; malformations of brain development) does not directly cause seizures. In fact, seizures can occur in "non-epileptic" brain and people with epilepsy spend most of the time not having seizures, indeed many experience only a few seizures per year. Again, as in (1), non-neuronal mechanisms spanning from altered cerebral blood flow to glial dysfunction have been used to explain how an asymptomatic neurologic condition can suddenly develop into a seizure state or the fact that seizures can occur in non-epileptic brain (e.g., stroke).

3. Multiple drug resistance to anti-epileptic drugs affects over 20 % of patients with epilepsy. Multiple drug resistance cannot be fully explained in pharmacodynamic or neuronal terms, and great emphasis has been put on pharmacokinetic mechanisms that include the blood-brain barrier.

4. Analysis of resected or *post-mortem* epileptic brain reveals a number of pathophysiological changes in astrocytes and microglia. MRI studies show, in addition to persistent structural changes such as malformations of brain development, an array of transient changes that reflect post-ictal or interictal functional fluctuations in the extracellular space (increased FLAIR signal, perfusion changes *etc.*).

5. The analysis of molecular transcripts and changes in gene expression in patients with epilepsy reveal a surprising number of genes and proteins that are involved in astrocytic function, blood-brain barrier maintenance and transport, as well as immune signaling and extracellular matrix proteins.

The following paragraphs detail the rationale for new or corroborative experiments that will help understand the extent and nature of non-neuronal mechanisms of seizure disorders.

20.2 Identification of Important Problems

The translational nature of modern research affords the unique opportunity to use real life clinical problems and "translate" these into meaningful laboratory efforts. As beautifully illustrated by Phil in his summary of basic mechanisms [72], the tools used for research are not always the same used in clinical practice. In fact,

a substantial discrepancy in size and temporal resolution becomes evident when comparing clinical and laboratory-based approaches (Fig. 20.1). For this mini review, we will focus on three fundamental yet often neglected aspects of ictogenesis and epileptogenesis: the blood-brain barrier, glia (Fig. 20.2) and the extracellular matrix. The following paragraphs will summarize current understanding and knowledge gaps related to these cellular and molecular mechanisms of neuronal pathophysiology.

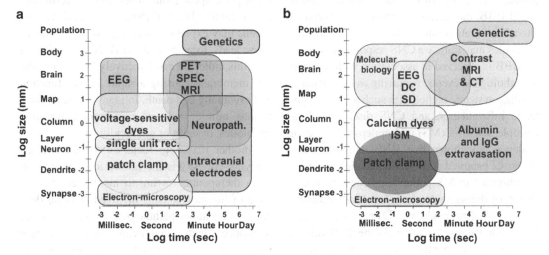

Fig. 20.1 Comparison of methods used in basic (**a**) or clinical (**b**) neuroscience. Note the partial overlap and significant differences

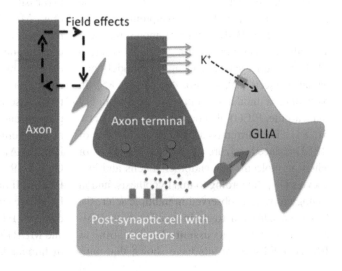

Fig. 20.2 Some mechanisms by which glia can affect seizure activity. Glia regulate the concentration of extracellular potassium, the size of extracellular space and so electrical field effects (ephaptic communication), and uptake of neurotransmitter

20.3   The Blood-Brain Barrier (BBB)

The BBB is the most important vascular barrier of the CNS. The BBB protects the brain from harmful substances of the blood stream, while supplying the brain with the nutrients required for proper function. The BBB strictly regulates the trafficking of cells of the immune system and pro-inflammatory cytokines from the blood into the brain. Recent findings indicate that neurovascular dysfunction is an integral part of many neurological disorders [35, 88]. In diseases with a compromised BBB, the microenvironment of neurons is altered; infiltration into the brain of cells, ions, or molecules may initiate a CNS response. Failure of the BBB is observed in association with a variety of pathological events, occurring as consequence of either systemic pathologies such as stroke, systemic inflammation and CNS disease such as multiple sclerosis (MS) and epilepsy. Increasing evidence has shown that BBB damage causes abnormal neuronal activity. For example, seizures are observed in MS patients, as consequence of stroke, or during systemic or local inflammation. As a proof-of-principle, we (DJ et al.) and others have demonstrated that failure of the BBB induced by "mechanical" means (such as osmotic shock) can play a key role in the onset of seizures [45].

In vitro and *in vivo* experiments on various models of neurological diseases have shown that blood-brain barrier damage accompanies the development of neurological symptoms; in contrast, managing BBB failure promotes recovery and affords neuroprotection. BBB disruption (BBBD) causes seizures in animal models and human subjects [19, 45, 46, 48, 50, 51, 85]. In particular, a model of temporal lobe epilepsy (pilocarpine, PILO) also depends on BBBD [19, 51, 84]. The currently accepted mechanism of BBBD-induced seizures predicts activation of adhesion molecules on endothelial cells and leukocytes [19]. According to this hypothesis, and in analogy to what is observed in multiple sclerosis, leukocyte adhesion to or interaction with BBB endothelial cells is an essential step leading to BBBD. Published results have shown that anti-inflammatory therapy (e.g., glucocorticosteroids) effectively reduce BBBD and associated symptoms [48]. The specific cell types involved in inflammation-promoted blood-brain barrier dysfunction are poorly understood but many leukocyte families have been shown to be involved, including natural killer cells and cytotoxic lymphocytes [4, 46, 50]. Attempts to curb the immune response, such as the extreme case of splenectomy, have been shown to decrease experimental seizures [50]. While BBBD-induced seizures were independent from the means used to obtain disruption (osmotic, pilocarpine, albumin), a specific molecular effector of pilocarpine-induced seizures, perforin, was only recently identified [50]. Perforin released by T cells may explain how activation of T lymphocytes leads to increased BBB permeability; in fact, this molecule can effectively "perforate" the cell membrane causing a rapid loss of function and eventually cell death. In many ways, perforin actions mimic those of membrane-permeating antibiotics, nystatin or gramicidin.

Another reason to focus on the BBB when studying epilepsy is the failure to generate new brain therapeutics owing to insufficient knowledge of the mechanisms involved in brain drug distribution under pathological conditions. Drug resistance affects a significant number of people with epilepsy; it is estimated that approximately 20–30 % of people with epilepsy fail to respond to available anti-epileptic drugs (AEDs) [4, 26, 30, 37, 48, 61, 67]. In the past decade the over-expression of multidrug transporter proteins (e.g., MDR1) at the blood-brain barrier (BBB) has been proposed as a mechanism that contributes to the failure of AEDs to penetrate into epileptic brain [1, 9, 16, 41–43, 47, 49, 59, 77]. In addition to multidrug transporters, it was shown that transcripts of P450 enzymes are elevated in primary endothelial cells (EC) isolated from drug resistant epileptic (DRE) patients; these enzymes include AED-metabolizers such as CYP3A4, CYP2C19, *etc.* [21]. In addition, transcripts for PHASE II metabolic enzymes are present in DRE EC; these enzymes are responsible for the metabolism of 1st and 2nd generation AEDs; CYP3A4 and MDR1 co-localize at the BBB (and neurons) in human DRE brain [22] and overexpression of CYP3A4 in DRE EC is associated with exaggerated carbamazepine (CBZ) metabolism. This new metabolic pathway produces the toxic CBZ metabolite quinolic acid (QA) leading to the paradoxical situation of an anti-epileptic drug being metabolized in the proximity of the epileptic focus to a seizure-promoting agent.

In summary, therapeutic considerations (use of anti-inflammatory therapy to treat seizures, BBB transporters in multiple drug resistance to anti-epileptic drugs) and etiologic factors (loss of BBB in seizures) suggest that the BBB is a viable and important target for studies aimed at the understanding and treatment of epilepsy. In addition to the role of the blood brain barrier, two other non-neuronal elements need to be considered – glia and brain extracellular matrix – both of which have been shown to have an increasing repertoire of roles in regulating network and brain excitability.

20.4 Neuroglia

"Glia" comes from the Greek meaning glue, and Virchow in his search for connective tissue in the brain, first coined the term neuroglia, considering them a sort of putty that supported the neurons [79]. Later, Golgi distinguished glia from neurons by the lack of an axon and ascribed to them a nutritive as well as supportive role. Ramon y Cajal determined that they were involved in the insulation of nerve cells and axons, a role later confirmed for oligodendroglia by a young Penfield who also established a role of glia in phagocytosis [24]. The repertoire of glia has, however, expanded in recent years from supportive tissue to playing an active role in determining network excitability, both modulating and responding to neuronal activity (Table 20.1).

20.4.1 Glia, Extracellular Space and Potassium Buffering

Glia play a critical part in the regulation of the size of the extracellular space, and extracellular ion homeostasis. In particular, they play a crucial role in the regulation of the concentration of

Table 20.1 Role of glia in the central nervous system

Roles of Glia
A supportive and protective role for neurons
A role in inflammation
Regulation of the size of the extracellular space
Maintenance of ion homeostasis in the extracellular space
Neurotransmitter uptake and synthesis
Providing neurons with energy
Detecting glutamate release from neurons and other glia
Release of neurotransmitters, and regulatory proteins
Synapse formation and regulation
Communication between neuronal activity and cerebral blood flow

potassium [73]. Glia express both aquaporins and potassium channels (inward rectifying and delayed rectifying) that play a role in this glial function through maintaining potassium and water homeostasis [5, 11, 18]. In addition, the connection of glia through gap junctions results in a glial syncytium, which facilitates not only water and potassium buffering but also glial communication [23]. Abnormalities of glial buffering of potassium result in potassium accumulation during neuronal activity. Such an increase in extracellular potassium will result in the depolarization of neurons and may therefore play a role in seizure initiation and spread [44, 73]. Reductions in the size of the extracellular space can affect neuronal communication through enhancement of ephaptic transmission (electrical interactions occurring though juxtaposed neuronal elements, which are lessened by increasing the conductive space between these elements), alterations in neurotransmitter "spill-over" and clearance, and changes in the regulation of extracellular ion concentrations. It is noteworthy that decreasing the extracellular space can promote seizure activity, whilst strategies aimed at increasing the extracellular space and decreasing glial and neuronal swelling can terminate seizure activity [31].

20.4.2 Glia and Neurotransmitter Concentrations

Glia also regulate the extracellular concentration of glutamate and GABA. They express the glutamate transporters GLAST (EAAT1) and GLT1 (EAAT2), which are responsible for most glutamate clearance [12]. These transporters determine the extracellular glutamate concentration, thus shaping the NMDA receptor response and the "spill-over" of glutamate following synaptic release onto other synapses (heterosynaptic activation) and extra-synaptic receptors [36]. Through this means, glial glutamate clearance plays a role in long-term synaptic plasticity. The expression of these transporters is regulated by an interaction between neurons and glia mediated by ephrins [55], which are extracellular proteins involved in neuronal development but which may be altered in injury and have been proposed to be involved in synaptic reorganisation following status epilepticus. Thus mechanisms that may play a part in synaptic reorganisation during epileptogenesis could also be involved in alterations in the expression of glutamate transporters. These possible roles of ephrins in epileptogenesis (see also below) have yet to be fully investigated.

The role of glia in the regulation of extracellular GABA is less clear since the glial GABA transporter (GAT3) seems to be mainly effective when the neuronal GABA transporters (predominantly GAT1) are blocked [34]. However, it is likely that GAT3 regulates a different pool of GABA that derives from non-vesicular sources. Further, GAT3 seems to play a greater part in regulating the extracellular GABA detected by interneurons than that detected by principal cells [80]. It has been proposed that GAT3 can reverse during periods of excessive activity, thus increasing extracellular GABA concentrations [28]. Finally, glia also are involved in the synthesis of neurotransmitters and in the glutamate-GABA shunt by which glutamate is converted to GABA [10]. Glutamate taken up by glia is converted to glutamine, which is then released into the extracellular space. Glutamine is taken up by neurons and converted to GABA. Inhibition of any of these processes results in a decrease in vesicular GABA content, GABA release and consequently GABAergic transmission [39]. Decreases in glutamate uptake that have been observed during

epileptogenesis could therefore not only increase extracellular glutamate but also decrease GABAergic transmission.

20.4.3 Glia and Metabolism

The uptake of glutamate by glia may have a further important role in neuronal energetics. Glutamate enters the Krebs cycle and therefore acts as an energy substrate. Glial glutamate uptake also activates the sodium-potassium ATPase, increasing glucose uptake and glycolysis [63]. Thus increases in extracellular glutamate during seizure activity can increase glial metabolism. Consequently, glia release lactate, which is taken up by neurons and used as an energy substrate, particularly during periods of excessive neuronal activity [6]. The role of glia in neuronal metabolism is probably even more extensive than this. Neurons lack pyruvate carboxylase [68], an enzyme that is crucial for replenishment of oxaloacetate in the Krebs cycle. As a result of this, the synthesis of GABA and glutamate can rapidly deplete Krebs cycle intermediaries in neurons. Replenishment of these intermediaries in neurons can, however, occur from direct transport of these intermediaries from glia to neurons. Glia are also a major producer of glutathione from glutamate, cysteine and glycine; glial glutathione production is necessary for protection of neurons from free radicals, which are produced during excessive neuronal activity [17]. Failure of glia to provide energy substrates for neurons could therefore promote neuronal death and disorders of neurotransmitter production and neuronal function. Indeed, glia play a crucial role in neurometabolism but how this is altered during and to what extent it plays a part in epileptogenesis are still unclear.

20.4.4 Glia, the Tripartite Synapse and Synaptic Plasticity

One of the main recent advances in our understanding of glia in modulating network activity has been the concept of the tripartite synapse in which glia in close proximity to synapses play a part in synaptic transmission, along with the presynaptic terminal and postsynaptic cell [2]. Vesicles and vesicle-associated proteins have been detected in astrocytes, often in close association to nerve terminals. Glia can detect glutamate via metabotropic glutamate receptors, which mediate a focal rise in astrocyte calcium, which has been proposed to mediate vesicular transmitter release. Calcium rises in one astrocyte can trigger a calcium wave through the glial syncytium, suggesting a mechanism by which focal activity can spread. Glia can also release neurotransmitter through reverse transport and membrane channels. Most of the studies in this area support glial release of glutamate, d-serine and ATP (which is converted to adenosine by extracellular ectonucleotidases) [27]. Glutamate released from glia can act at post-synaptic NMDA receptors and has been proposed to contribute to paroxysmal depolarizing shifts underlying epileptiform activity [82]. D-serine is a co-agonist at NMDA receptors and d-serine release from glia seems to be necessary for NMDA receptor mediated long term potentiation [29]. Lastly, increased adenosine levels through glial ATP release modulates presynaptic release of glutamate in a bimodal fashion through A1 (decreasing release probability) and A2 (increasing release probability) receptors [81]. Thus glia can alter network excitability over short time periods, and could play a role in both seizure initiation (glutamate/D-serine release) and termination (adenosine).

Glia can also play a longer term role in modulating synaptic transmission through the interaction of ephrins, specifically ephrin-A3 on astrocytes with the EphA4 receptor on dendrites [55]. This is a bidirectional interaction, which regulates the expression of glutamate transporters in glia and modulates spine and synapse formation in neurons. Such interactions are important in synaptic plasticity. In addition, glial ephrin signaling is important for neurogenesis, indicating a role for glia in modulating neuronal development and connectivity [55]. The role that this plays in epileptogenesis has yet to be explored.

20.4.5 Glia and Neurovascular Coupling

When neuronal activity increases in an area of the brain, there is a concomitant increase in cerebral blood flow to that area – a phenomenon termed "neurovascular coupling." There appear to be multiple mechanisms mediating this effect, but there is evidence that glutamate acting via metabotropic glutamate receptors and glutamate uptake by glia can affect the release of vasoactive compounds that directly affect cerebral vasculature [64]. One important consequence of this scenario is that neurovascular coupling may depend upon the release of glutamate rather than local neuronal firing. Indeed, there is accumulating evidence that, although neurovascular coupling correlates both with neuronal firing and local field potentials (i.e. post-synaptic receptor activation through glutamate release), the coupling with field potentials is stronger [40]. This increased blood flow is a critical component of seizure activity that can be detected with ictal SPECT or as an increase in the MRI blood oxygen level dependent (BOLD) signal.

20.4.6 Changes in Glia with Epileptogenesis

Brain injury and neuronal loss invariably leads to a reactive gliosis in which there is not only a proliferation of astrocytes but also changes in astrocytic morphology and gene expression [78]. Moreover, a reactive gliosis is observed in multiple pathologies associated with epileptogenesis, including traumatic brain injury, stroke, tumors, vascular lesions and hippocampal sclerosis. Abnormal glia are also found in tuberous sclerosis; specific knockout of the Tsc1 gene in glia results in seizures [83].

Reactive gliosis may alter regulation of the extracellular space and promote ephaptic transmission. Aquaporin expression in astrocytes changes from astrocyte end feet (i.e., their perivascular location) to a more diffuse expression [5]. This has been proposed to lead to abnormal water regulation, with perivascular water accumulation and increased water uptake by astrocytes resulting in astrocyte swelling and a decrease in the extracellular space. Breakdown of the blood brain barrier and accumulation of albumin within glia also leads to a reduction in glial inward rectifying potassium channel expression and so decreased buffering of potassium rises [13]. Moreover there is evidence in human epileptic tissue of a change in glial glutamate transporter expression and, from rodent studies of epileptogenesis, decreased efficacy of glutamate uptake [13, 65].

Glial metabolism also changes during epileptogenesis. There is an increase in the expression of adenosine kinase and along with astrocytosis, this leads to decreased adenosine levels with epileptogenesis [7]. There are decreased levels of glutamine synthetase, and a consequent decrease in the glutamate-GABA shunt, resulting in decreased inhibitory transmission [10]. Indeed, a specific reactive gliosis mediated by transfection with a viral vector had no effect on the intrinsic excitability of neighboring neurons, but selectively decreased inhibitory transmission, leading to an inhibitory deficit and increased propagation of excitatory transmission [60]. This is a clear demonstration that reactive gliosis alone is sufficient to promote hyperexcitability. Glial metabolism may also be affected by a reactive gliosis due to decreased glutamate uptake, although the role that changes in glial metabolism have on the development of epilepsy are unclear.

Although it is uncertain to what extent reactive gliosis affects the tripartite synapse, astrocyte calcium rises mediated by activation of metabotropic glutamate and purinergic receptors can promote the generation of seizure activity in vitro and *in vivo* [25]. Also glial metabotropic receptors are upregulated in epilepsy [3].

The critical role that glia play in the inflammatory process underlying epileptogenesis is discussed elsewhere in this book.

There has thus been growing evidence that glia can alter network excitability through multiple mechanisms. The possible roles of reactive gliosis and the part that it plays both in the development of epilepsy and the generation of seizures need to be further modeled and studied. The extensive role that glia play in many critical functions will need to be carefully dissected in order to target specific glia mediated processes during epileptogenesis (Fig. 20.2).

20.5 The Extracellular Matrix (ECM)

20.5.1 Physiological Role of the Extracellular Matrix

The extracellular matrix (ECM) consists of molecules that are secreted both by neurons and glia, and that aggregate in the extracellular space. About 20 % of the volume of the adult brain consists of extracellular matrix, and the extracellular matrix plays an essential role in determining the diffusion of small molecules [57]. In contrast to ECM elsewhere in the body, the brain ECM predominantly consists of proteoglycans, glycosaminoglycans (in particular hyaluronic acid), and glycoproteins of the tenascin family. There are also proteins that link the ECM to ECM and to molecules on neurons and glia [15].

The vast majority of the ECM is present in the extra-synaptic space. The ECM also makes up the basal lamina, which contributes to the blood-brain barrier. It has also been increasingly recognized that brain ECM consists of other well-defined components including peri-neuronal nets (mesh-like structures which surround cell bodies and proximal dendrites particularly of parvalbumin-expressing interneurons as a mesh-like structure), and specific components present at synapses which are linked to proteins at the post-synaptic and pre-synaptic membrane [15].

Peri-neuronal nets consist of proteoglycans of the lectican family which link with hyaluronic acid and tenascin-R [86]. Peri-neuronal nets are critical in development, closing critical periods and stabilizing synapses and neuronal plasticity. Digestion of proteoglycans associated with peri-neuronal nets or knockout of tenascin-R affect both synaptic plasticity and the excitability of interneurons. Peri-neuronal nets therefore play a crucial role in regulating network excitability and plasticity.

The extracellular matrix can undergo remodeling, which is dependent upon a series of serine proteases, such as plasminogen activators (in particular urokinase-type plasminogen activator), thrombin, metalloproteinase's, and reelin. All of these have been implicated in neuronal and network plasticity [14]. Alterations and remodeling of peri-neuronal nets permit neuronal reorganization following brain damage and seizures, and during development.

The interaction of the extracellular matrix with neurons can occur via specific receptors, integrins, which are transmembrane heterodimeric transmembrane glycoproteins composed of two of 26 subunits. Integrins bind to intracellular cytoskeleton and secondary messenger systems and extracellularly to other cells and the ECM [32]. They are closely associated with glutamate receptors and various ion channels. Integrins regulate multiple processes including synaptic plasticity, neuronal migration and development, axonal growth and synaptogenesis. They are also involved in angiogenesis.

20.5.2 Changes in the ECM in Epilepsy

There are persistent changes in multiple components of the ECM during the development of epilepsy. Peri-neuronal net components, including aggrecan, neurocan, hyaluronan, tenascin-R and some of the linking proteins, decrease during epileptogenesis; a progressive decrease in perineuronal nets is associated with a progressive decrease in inhibition and the occurrence of seizures (months after traumatic brain injury) [53, 62]. In addition, degradation of the ECM may permit aberrant neuronal and synaptic reorganisation. ECM remodelling and the increased secretion of proteases may also contribute to this process. There is robust evidence that expression of MMP-9 is increased during epileptogenesis, and that this increase may promote kindling [54]. Other serine proteases are also up-regulated in epilepsy including urokinase-type plasminogen activator (uPA) and its receptor (uPAR) [38]. Intriguingly, uPAR up-regulation may be

protective as uPAR knockouts develop a more severe epilepsy phenotype following status epilepticus [56]. This indicates that some of the changes of ECM during epileptogenesis may be adaptive rather than pathogenic.

In addition, mutations in the gene encoding SRPX2 (Sushi-repeat Protein, X-linked 2), one of the ligands of uPAR, results in bilateral perisylvian polymicrogyria and epilepsy in humans [66]. Integrin expression is also increased during epileptogenesis and in pathologies associated with the development of epilepsy [87].

Lastly, an extracellularly secreted molecule, leucine rich, glioma-inactivated 1 (LGI1) has been strongly associated with epilepsy [8, 20, 33, 58, 69]. LGI1 interconnects presynaptic disintegrin and metalloproteinase domain-containing protein 23 (ADAM23) to postsynaptic ADAM22 at the synaptic cleft. LGI1 is important for trafficking and kinetics of a presynaptic potassium channel, Kv1.1, and also for trafficking of post-synaptic AMPA receptors. In humans, mutations in LGI1 cause autosomal dominant lateral temporal epilepsy or autosomal dominant partial epilepsy with auditory features with onset in childhood/adolescence [58]. In addition, autoantibodies directed against LGI1 have been shown to underlie limbic encephalitis and temporal lobe seizures in humans [69].

Overall, there is growing evidence for the importance of the ECM in epileptogenesis, plasticity and determining network excitability. Further studies aimed at modeling disruption and reorganization of the ECM will be important for a greater understanding of the epileptogenic process. Moreover, the ECM provides an ideal target for therapies aimed at disrupting epileptogenesis and modifying established epilepsy, as it is extracellular and so easily accessible to drugs and has multiple downstream effects, regulating receptors, channels and synaptic transmission.

20.6 Conclusions

There is burgeoning evidence to support a critical role for non-neuronal mechanisms in epileptogenesis and the generation of seizures.

Both animal experiments and experiments of nature (gene mutations) indicate that pathology of non-neuronal elements are sufficient for epileptogenesis. However, most of our present therapies are neurocentric, indicating that there may be enormous undiscovered therapeutic potential in targeting these non-neuronal elements. Moreover, it is a concern that many of the large scale mathematical models of brain function (e.g., the blue brain project [52]) have thus far ignored the role of these non-neuronal constituents.

Acknowledgements This work was supported by the National Institutes of Health (R01NS078307, R01NS43284, R41MH093302, R21NS077236, R42MH093302, and R21HD057256 to DJ).

References

1. Abbott NJ, Khan EU, Rollinson CM, Reichel A, Janigro D, Dombrowski SM et al (2002) Drug resistance in epilepsy: the role of the blood-brain barrier. Novartis Found Symp 243:38–47
2. Araque A, Parpura V, Sanzgiri RP, Haydon PG (1999) Tripartite synapses: glia, the unacknowledged partner. Trends Neurosci 22:208–215
3. Aronica E, van Vliet EA, Mayboroda OA, Troost D, da Silva FH, Gorter JA (2000) Upregulation of metabotropic glutamate receptor subtype mGluR3 and mGluR5 in reactive astrocytes in a rat model of mesial temporal lobe epilepsy. Eur J Neurosci 12:2333–2344
4. Bauer S, Koller M, Cepok S, Todorova-Rudolph A, Nowak M, Nockher WA et al (2008) NK and CD4+ T cell changes in blood after seizures in temporal lobe epilepsy. Exp Neurol 211:370–377
5. Binder DK, Nagelhus EA, Ottersen OP (2012) Aquaporin-4 and epilepsy. Glia 60:1203–1214
6. Bittner CX, Valdebenito R, Ruminot I, Loaiza A, Larenas V, Sotelo-Hitschfeld T et al (2011) Fast and reversible stimulation of astrocytic glycolysis by K+ and a delayed and persistent effect of glutamate. J Neurosci 31:4709–4713
7. Boison D (2012) Adenosine dysfunction in epilepsy. Glia 60:1234–1243
8. Chabrol E, Navarro V, Provenzano G, Cohen I, Dinocourt C, Rivaud-Pechoux S et al (2010) Electroclinical characterization of epileptic seizures in leucine-rich, glioma-inactivated 1-deficient mice. Brain 133:2749–2762
9. Cornford EM, Oldendorf WH (1986) Epilepsy and the blood-brain barrier. Adv Neurol 44:787–812
10. Coulter DA, Eid T (2012) Astrocytic regulation of glutamate homeostasis in epilepsy. Glia 60:1215–1226
11. D'Ambrosio R, Wenzel J, Schwartzkroin PA, McKhann GM, Janigro D (1998) Functional specialization and topographic segregation of hippocampal astrocytes. J Neurosci 18:4425–4438
12. Danbolt NC (2001) Glutamate uptake. Prog Neurobiol 65:1–105
13. David Y, Cacheaux LP, Ivens S, Lapilover E, Heinemann U, Kaufer D et al (2009) Astrocytic dysfunction in epileptogenesis: consequence of altered potassium and glutamate homeostasis? J Neurosci 29:10588–10599
14. Dityatev A (2010) Remodeling of extracellular matrix and epileptogenesis. Epilepsia 51(Suppl 3):61–65
15. Dityatev A, Seidenbecher CI, Schachner M (2010) Compartmentalization from the outside: the extracellular matrix and functional microdomains in the brain. Trends Neurosci 33:503–512
16. Dombrowski SM, Desai SY, Marroni M, Cucullo L, Goodrich K, Bingaman W et al (2001) Overexpression of multiple drug resistance genes in endothelial cells from patients with refractory epilepsy. Epilepsia 42:1501–1506
17. Dringen R (2000) Metabolism and functions of glutathione in brain. Prog Neurobiol 62:649–671
18. Emmi A, Wenzel HJ, Schwartzkroin PA, Taglialatela M, Castaldo P, Bianchi L et al (2000) Do glia have heart? Expression and functional role for ether-a-go-go currents in hippocampal astrocytes. J Neurosci 20:3915–3925
19. Fabene PF, Navarro MG, Martinello M, Rossi B, Merigo F, Ottoboni L et al (2008) A role for leukocyte-endothelial adhesion mechanisms in epilepsy. Nat Med 14:1377–1383
20. Fukata Y, Lovero KL, Iwanaga T, Watanabe A, Yokoi N, Tabuchi K et al (2010) Disruption of LGI1-linked synaptic complex causes abnormal synaptic transmission and epilepsy. Proc Natl Acad Sci U S A 107:3799–3804
21. Ghosh C, Gonzalez-Martinez J, Hossain M, Cucullo L, Fazio V, Janigro D et al (2010) Pattern of P450 expression at the human blood-brain barrier: Roles of epileptic condition and laminar flow. Epilepsia 51:1408–1417
22. Ghosh C, Marchi N, Desai NK, Puvenna V, Hossain M, Gonzalez-Martinez J et al (2011) Cellular localization and functional significance of CYP3A4 in the human epileptic brain. Epilepsia 52:562–571
23. Giaume C, Koulakoff A, Roux L, Holcman D, Rouach N (2010) Astroglial networks: a step further in neuro-glial and gliovascular interactions. Nat Rev Neurosci 11:87–99
24. Gill AS, Binder DK (2007) Wilder Penfield, Pio del Rio-Hortega, and the discovery of oligodendroglia. Neurosurgery 60:940–948
25. Gomez-Gonzalo M, Losi G, Chiavegato A, Zonta M, Cammarota M, Brondi M et al (2010) An excitatory loop with astrocytes contributes to drive neurons to seizure threshold. PLoS Biol 8:e1000352
26. Granata T, Marchi N, Carlton E, Ghosh C, Gonzalez-Martinez J, Alexopoulos AV et al (2009) Management

of the patient with medically refractory epilepsy. Expert Rev Neurother 9:1791–1802

27. Halassa MM, Haydon PG (2010) Integrated brain circuits: astrocytic networks modulate neuronal activity and behavior. Annu Rev Physiol 72:335–355

28. Heja L, Nyitrai G, Kekesi O, Dobolyi A, Szabo P, Fiath R et al (2012) Astrocytes convert network excitation to tonic inhibition of neurons. BMC Biol 10:26

29. Henneberger C, Papouin T, Oliet SH, Rusakov DA (2010) Long-term potentiation depends on release of D-serine from astrocytes. Nature 463:232–236

30. Hitiris N, Mohanraj R, Norrie J, Sills GJ, Brodie MJ (2007) Predictors of pharmacoresistant epilepsy. Epilepsy Res 75:192–196

31. Hochman DW, Baraban SC, Owens JW, Schwartzkroin PA (1995) Dissociation of synchronization and excitability in furosemide blockade of epileptiform activity. Science 270:99–102

32. Hynes RO (2004) The emergence of integrins: a personal and historical perspective. Matrix Biol 23:333–340

33. Irani SR, Alexander S, Waters P, Kleopa KA, Pettingill P, Zuliani L et al (2010) Antibodies to Kv1 potassium channel-complex proteins leucine-rich, glioma inactivated 1 protein and contactin-associated protein-2 in limbic encephalitis, Morvan's syndrome and acquired neuromyotonia. Brain 133:2734–2748

34. Kersante F, Rowley SC, Pavlov I, Gutierrez-Mecinas M, Semyanov A, Reul JM et al (2013) A functional role for both -aminobutyric acid (GABA) transporter-1 and GABA transporter-3 in the modulation of extracellular GABA and GABAergic tonic conductances in the rat hippocampus. J Physiol 591:2429–2441

35. Krizanac-Bengez L, Mayberg MR, Janigro D (2004) The cerebral vasculature as a therapeutic target for neurological disorders and the role of shear stress in vascular homeostatis and pathophysiology. Neurol Res 26:846–853

36. Kullmann DM (2000) Spillover and synaptic cross talk mediated by glutamate and GABA in the mammalian brain. Prog Brain Res 125:339–351

37. Kwan P, Arzimanoglou A, Berg AT, Brodie MJ, Allen HW, Mathern G et al (2010) Definition of drug resistant epilepsy: consensus proposal by the ad hoc Task Force of the ILAE Commission on Therapeutic Strategies. Epilepsia 51:1069–1077

38. Lahtinen L, Huusko N, Myohanen H, Lehtivarjo AK, Pellinen R, Turunen MP et al (2009) Expression of urokinase-type plasminogen activator receptor is increased during epileptogenesis in the rat hippocampus. Neuroscience 163:316–328

39. Liang SL, Carlson GC, Coulter DA (2006) Dynamic regulation of synaptic GABA release by the glutamate-glutamine cycle in hippocampal area CA1. J Neurosci 26:8537–8548

40. Logothetis NK, Pauls J, Augath M, Trinath T, Oeltermann A (2001) Neurophysiological investigation of the basis of the fMRI signal. Nature 412:150–157

41. Loscher W (2007) Mechanisms of drug resistance in status epilepticus. Epilepsia 48(Suppl 8):74–77

42. Loscher W, Potschka H (2005) Drug resistance in brain diseases and the role of drug efflux transporters. Nat Rev Neurosci 6:591–602

43. Loscher W, Sills GJ (2007) Drug resistance in epilepsy: why is a simple explanation not enough? Epilepsia 48:2370–2372

44. Lux HD, Heinemann U, Dietzel I (1986) Ionic changes and alterations in the size of the extracellular space during epileptic activity. Adv Neurol 44:619–639

45. Marchi N, Angelov L, Masaryk T, Fazio V, Granata T, Hernandez N et al (2007) Seizure-promoting effect of blood-brain barrier disruption. Epilepsia 48(4):732–742

46. Marchi N, Fan QY, Ghosh C, Fazio V, Bertolini F, Betto G et al (2009) Antagonism of peripheral inflammation reduces the severity of status epilepticus. Neurobiol Dis 33:171–181

47. Marchi N, Gonzalez-Martinez J, Nguyen MT, Granata T, Janigro D (2010) Transporters in drug-refractory epilepsy: clinical significance. Clin Pharmacol Ther 87:13–15

48. Marchi N, Granata T, Freri E, Ciusani E, Puvenna V, Teng Q et al (2011) Efficacy of anti-inflammatory therapy in a model of acute seizures and in a population of pediatric drug resistant epileptics. PLoS One 6:e18200

49. Marchi N, Hallene KL, Kight KM, Cucullo L, Moddel G, Bingaman W et al (2004) Significance of MDR1 and multiple drug resistance in refractory human epileptic brain. BMC Med 2:37

50. Marchi N, Johnson A, Puvenna V, Tierney W, Ghosh C, Cucullo L et al (2011) Modulation of peripheral cytotoxic cells and ictogenesis in a model of seizures. Epilepsia 52:1627–1634

51. Marchi N, Oby E, Fernandez N, Uva L, de Curtis M, Batra A et al (2007) *In vivo* and *in vitro* effects of pilocarpine: relevance to epileptogenesis. Epilepsia 48(10):1934–1946

52. Markram H (2006) The blue brain project. Nat Rev Neurosci 7:153–160

53. McRae PA, Baranov E, Rogers SL, Porter BE (2012) Persistent decrease in multiple components of the perineuronal net following status epilepticus. Eur J Neurosci 36:3471–3482

54. Mizoguchi H, Nakade J, Tachibana M, Ibi D, Someya E, Koike H et al (2011) Matrix metalloproteinase-9 contributes to kindled seizure development in pentylenetetrazole-treated mice by converting pro-BDNF to mature BDNF in the hippocampus. J Neurosci 31:12963–12971

55. Murai KK, Pasquale EB (2011) Eph receptors and ephrins in neuron-astrocyte communication at synapses. Glia 59:1567–1578

56. Ndode-Ekane XE, Pitkanen A (2013) Urokinase-type plasminogen activator receptor modulates epileptogenesis in mouse model of temporal lobe epilepsy. Mol Neurobiol 47:914–937

57. Nicholson C, Sykova E (1998) Extracellular space structure revealed by diffusion analysis. Trends Neurosci 21:207–215

58. Nobile C, Michelucci R, Andreazza S, Pasini E, Tosatto SC, Striano P (2009) LGI1 mutations in autosomal dominant and sporadic lateral temporal epilepsy. Hum Mutat 30:530–536

59. Oby E, Janigro D (2006) The blood-brain barrier and epilepsy. Epilepsia 47:1761–1774

60. Ortinski PI, Dong J, Mungenast A, Yue C, Takano H, Watson DJ et al (2010) Selective induction of astrocytic gliosis generates deficits in neuronal inhibition. Nat Neurosci 13:584–591

61. Patsalos PN, Berry DJ, Bourgeois BF, Cloyd JC, Glauser TA, Johannessen SI et al (2008) Antiepileptic drugs – best practice guidelines for therapeutic drug monitoring: a position paper by the subcommission on therapeutic drug monitoring. ILAE Commission on Therapeutic Strategies. Epilepsia 49:1239–1276

62. Pavlov I, Huusko N, Drexel M, Kirchmair E, Sperk G, Pitkanen A et al (2011) Progressive loss of phasic, but not tonic, GABAA receptor-mediated inhibition in dentate granule cells in a model of post-traumatic epilepsy in rats. Neuroscience 194:208–219

63. Pellerin L, Magistretti PJ (1994) Glutamate uptake into astrocytes stimulates aerobic glycolysis: a mechanism coupling neuronal activity to glucose utilization. Proc Natl Acad Sci U S A 91:10625–10629

64. Petzold GC, Murthy VN (2011) Role of astrocytes in neurovascular coupling. Neuron 71:782–797

65. Proper EA, Hoogland G, Kappen SM, Jansen GH, Rensen MG, Schrama LH et al (2002) Distribution of glutamate transporters in the hippocampus of patients with pharmaco-resistant temporal lobe epilepsy. Brain 125:32–43

66. Roll P, Rudolf G, Pereira S, Royer B, Scheffer IE, Massacrier A et al (2006) SRPX2 mutations in disorders of language cortex and cognition. Hum Mol Genet 15:1195–1207

67. Sanchez Alvarez JC, Serrano Castro PJ, Serratosa Fernandez JM (2007) Clinical implications of mechanisms of resistance to antiepileptic drugs. Neurologist 13:S38–S46

68. Schousboe A, Bak LK, Waagepetersen HS (2013) Astrocytic control of biosynthesis and turnover of the neurotransmitters glutamate and GABA. Front Endocrinol (Lausanne) 4:102

69. Schulte U, Thumfart JO, Klocker N, Sailer CA, Bildl W, Biniossek M et al (2006) The epilepsy-linked Lgi1 protein assembles into presynaptic Kv1 channels and inhibits inactivation by Kvbeta1. Neuron 49:697–706

70. Schwartzkroin PA (1975) Characteristics of CA1 neurons recorded intracellularly in the hippocampal in vitro slice preparation. Brain Res 85:423–436

71. Schwartzkroin PA (1977) Further characteristics of hippocampal CA1 cells in vitro. Brain Res 128:53–68

72. Schwartzkroin PA (2012) Why – and how – do we approach basic epilepsy research? In: Noebels JL, Avoli M, Rogawski MA et al (eds) Jasper's basic mechanisms of the epilepsies, 4th edn. Oxford, New York, pp 24–37

73. Schwartzkroin PA, Baraban SC, Hochman DW (1998) Osmolarity, ionic flux, and changes in brain excitability. Epilepsy Res 32:275–285

74. Schwartzkroin PA, Mathers LH (1978) Physiological and morphological identification of a nonpyramidal hippocampal cell type. Brain Res 157:1–10

75. Schwartzkroin PA, Prince DA (1978) Cellular and field potential properties of epileptogenic hippocampal slices. Brain Res 147:117–130

76. Schwartzkroin PA, Prince DA (1979) Recordings from presumed glial cells in the hippocampal slice. Brain Res 161:533–538

77. Sisodiya SM, Goldstein DB (2007) Drug resistance in epilepsy: more twists in the tale. Epilepsia 48:2369–2370

78. Sofroniew MV, Vinters HV (2010) Astrocytes: biology and pathology. Acta Neuropathol 119:7–35

79. Somjen GG (1988) Nervenkitt: notes on the history of the concept of neuroglia. Glia 1:2–9

80. Song I, Volynski K, Brenner T, Ushkaryov Y, Walker M, Semyanov A (2013) Different transporter systems regulate extracellular GABA from vesicular and non-vesicular sources. Front Cell Neurosci 7:23

81. Stone TW, Ceruti S, Abbracchio MP (2009) Adenosine receptors and neurological disease: neuroprotection and neurodegeneration. Handb Exp Pharmacol 193:535–587

82. Tian GF, Azmi H, Takano T, Xu Q, Peng W, Lin J et al (2005) An astrocytic basis of epilepsy. Nat Med 11:973–981

83. Uhlmann EJ, Wong M, Baldwin RL, Bajenaru ML, Onda H, Kwiatkowski DJ et al (2002) Astrocyte-specific TSC1 conditional knockout mice exhibit abnormal neuronal organization and seizures. Ann Neurol 52:285–296

84. Uva L, Librizzi L, Marchi N, Noe F, Bongiovanni R, Vezzani A et al (2008) Acute induction of epileptiform discharges by pilocarpine in the in vitro isolated guinea-pig brain requires enhancement of blood-brain barrier permeability. Neuroscience 151:303–312

85. van Vliet EA, da Costa AS, Redeker S, van Schaik R, Aronica E, Gorter JA (2007) Blood-brain barrier leakage may lead to progression of temporal lobe epilepsy. Brain 130:521–534

86. Wang D, Fawcett J (2012) The perineuronal net and the control of CNS plasticity. Cell Tissue Res 349:147–160

87. Wu X, Reddy DS (2012) Integrins as receptor targets for neurological disorders. Pharmacol Ther 134:68–81

88. Zlokovic BV (2008) The blood-brain barrier in health and chronic neurodegenerative disorders. Neuron 57:178–201

What Epilepsy Comorbidities Are Important to Model in the Laboratory? Clinical Perspectives

21

Simon Shorvon

Abstract

In recent years, there has been a focus on studies of comorbidity in epilepsy. The concept of epilepsy comorbidity is complex. This is partly because epilepsy is essentially a symptom for which there are many underlying causes, with multiple genetic and environmental influences. These causal conditions themselves carry comorbidities which vary from condition to condition. The fact that some psychiatric comorbidities are 'bidirectional' complicates this further. These issues reduce the usefulness of any unitary study of 'epilepsy comorbidity'. Epilepsy comorbidities can be divided into direct/indirect and somatic/psychiatric categories. Only some aspects are susceptible to experimental modeling. This chapter briefly reviews the clinical studies of cause, frequency, epidemiology and mortality of comorbidities, and their use as biomarkers for epilepsy.

Keywords

Epilepsy • Epileptic seizures • Somatic comorbidity • Psychiatric comorbidity

21.1 Definitions and Divisions of Epilepsy Comorbidities

The term comorbidity has been said to have been first coined by Feinstein [7] to define the co-existence of different diseases or conditions. The original studies of epilepsy comorbidity emphasized migraine, psychiatric disorders and vascular disease, but since the early 2000s there

S. Shorvon (✉)
UCL Institute of Neurology. Box 5, National Hospital for Neurology and Neurosurgery,
Queen Square, London WC1N 3BG, UK
e-mail: s.shorvon@ucl.ac.uk

H.E. Scharfman and P.S. Buckmaster (eds.), *Issues in Clinical Epileptology: A View from the Bench*,
Advances in Experimental Medicine and Biology 813, DOI 10.1007/978-94-017-8914-1_21,
© Springer Science+Business Media Dordrecht 2014

has been a greater focus on this problem and more recent studies have demonstrated a wider range of comorbid disorders and have attempted to define their extent [10, 14, 15, 18] and underlying mechanisms [3, 11, 22, 31]. Some of the comorbid conditions are susceptible to experimental modeling and others are not. In broad terms, experimental or animal models are most appropriately employed to investigate the mechanisms and the causes of comorbidity.

The concept of 'comorbidity' of a condition such as epilepsy is complicated. As a rider to any discussion, it should of course be realized that epilepsy is essentially a symptom for which there are many underlying causes, with multiple genetic and environmental influences [25, 28]. These causal conditions themselves carry comorbidities, which vary from condition to condition, thus complicating any broad study of 'epilepsy comorbidity'. Studies in epilepsy have divided and defined the range of comorbidities in a number of different ways.

(i) Direct/indirect: The direct comorbidities are those that are due to epilepsy. The indirect comorbidities are those that are due to underlying causes of the epilepsies or to risk factors which are shared with epilepsy.

(ii) Psychiatric/somatic: The psychiatric comorbidities refer to the primary psychiatric diseases and the somatic comorbidities to systemic and neurological disease.

It is probably not surprising to know that these divisions are artificial and there are in each system grey areas where conditions overlap or are not easy to pigeon-hole. Understanding comorbidity is important for various reasons:

(i) The comorbidities may have an important influence on prognosis of epilepsy (including mortality) and indeed often have a greater influence than the epilepsy itself.

(ii) The therapy of epilepsy may be influenced by their presence (as well as the fact that some comorbidity is due to therapy)

(iii) The comorbidities may have diagnostic implications in some situations

(iv) Doctors dealing with epilepsy should be alert to the risk of comorbidities as these too may require treatment

(v) The comorbidities of epilepsy may in many instances cause more distress and dysfunction than the epilepsy itself.

There is also often a two-way relationship between comorbidity and epilepsy (often known as a 'bidirectional' relationship; discussed further below). Comorbidity can affect the course of the epilepsy directly (via organic effects on the brain) or indirectly (via chronic ill health, side-effects of treatment, secondary psychiatric effects). Comorbidities also affect health care utilization, and all the outcomes of epilepsy including mortality.

21.2 Causes of Comorbidity in Epilepsy

'Direct comorbidity' is that due to the epilepsy or the effects of seizures themselves. Examples of seizure-related comorbidity are fractures due to falls in seizures, or memory disturbance due to cerebral damage. Laboratory studies offer the opportunity for prevention and especially neuroprotection and these are topics which can be modeled experimentally. Other direct morbidity is due to the secondary handicap of epilepsy which includes chronic ill health, psychiatric problems, social drift and other pressures. These are topics which cannot be studied in laboratory models.

The indirect comorbidities may be: (a) due to the underlying causes themselves, such as stroke or cerebral tumour, which cause epilepsy and also other effects; (b) due to shared risk factors which have been shown to predispose to epilepsy and also to other medical condition, examples include vascular disease which predisposes to stroke, or alcoholism which predisposes to head trauma. Sometimes the risk factors are genetic (discussed further below); (c) due to the treatment of epilepsy, examples include hepatic or bone disease, or interactions between medications; (d) conditions where the mechanisms underlying the association are quite unknown for instance associations with asthma, bowel disease or thyroid disease. Many of these aspects can be studied experimentally.

Psychiatric comorbidity (which can be both direct and indirect) is particularly complex with genetic, environmental, shared underlying causes and also treatment and direct cerebral damage all potentially contributing to the epilepsy and the comorbidities.

21.3 Epidemiology and Frequency of Comorbidities of Epilepsy

There have been a number of large scale surveys of comorbidity, based on National Health Service statistics.

21.3.1 Somatic Comorbidity

The first database to be mined was the UK General Practice Research Database (GPRD) which covered a period between 1995 and 1998 [11]. Data were based on a population of 1.3 million, in which all ICD codes (codes defined by the International Classification of Disease (ICD) were recorded and of these 1,041,643 adults were studied. 5,834 persons with epilepsy were identified. The most common somatic conditions in adults with epilepsy were: fractures (10 %), with highest rates in women older than 64 years (17 %); asthma (9 % with the highest rates (11 %) in younger women); and migraine (8 %). Amongst the oldest patients, the most common somatic comorbidities were diabetes (9 %), transient ischaemic attacks (18 %), ischaemic heart disease (14 %), heart failure (12 %), neoplasia (7 %), and osteoarthritis (12 %). The most common neurologic disorders in this age group were brain degenerative diseases (14 %) and Parkinson's disease (4 %). The importance of environment is also shown by the study of Babu et al. [2] in India which showed increased rates of neurocysticercosis, sleep disorders, and tuberculosis compared with controls.

Téllez-Zenteno et al. [32] have used data obtained through two door-to-door Canadian health surveys, the National Population Health Survey (NPHS, N = 49,000) and the Community Health Survey (CHS, N = 130,882), covering 98 % of the Canadian population. They found that those with epilepsy had a statistically significant higher prevalence of many chronic conditions when compared to the general population; those conditions which occurred twice as often or more were (proportional risk): stomach/intestinal ulcers (CHS 2.5, NPHS 2.7), stroke (CHS 3.9, NPHS 4.7), urinary incontinence (CHS 3.2, NPHS 4.4), bowel disorders (CHS 2.0, NPHS 3.3), migraine (PR, CHS 2.0, NPHS 2.6), Alzheimer's disease (NPHS 4.3), and chronic fatigue (CHS 4.1). Of course several of these conditions are causal conditions of epilepsy (stroke, Alzheimers disease) and so it is not at all surprising that they cluster with epilepsy in population surveys, but the others were more surprising. It was postulated by the authors that gastro-intestinal diseases may be due drug therapy or autonomic ictal effects, although both explanations seem unconvincing.

21.3.2 Psychiatric Comorbidity

The commonest comorbidities of epilepsy are psychiatric. There are a number of epidemiological studies of comorbidities looking at this association. In the study mentioned above, Gaitatzis et al. [11] found the commonest psychiatric conditions in adults with epilepsy were: depression (18 %), anxiety (11 %) and psychosis (9 %). Overall, 41 % of patients with epilepsy received a psychiatric diagnosis at some point during the 3-year study period. Téllez-Zenteno et al. [32] used data from the Canadian Community Health Survey (CCHS) to compare the rates of psychiatric disease in those with and without epilepsy. The CCHS included 36,984 subjects. Those with a history of epilepsy reported higher lifetime anxiety disorders (odds ratio (OR) 2.4, 95 % confidence intervals (CI) = 1.5–3.8) or suicidal thoughts (OR 2.2 (1.4–3.3)). Surprisingly, the risk of major depressive disease or of panic disorder/agoraphobia were not greater in those with epilepsy (and may throw some doubt upon the methodology of this study).

There are also a number of case control studies, looking both at the frequency of epilepsy in psychiatric populations and vice versa (the so-called bidirectional relationship). A variety of mental disorders, alcoholism and dementia are found more commonly in patients with epilepsy than in non-epileptic controls. The strongest associations of epilepsy are with major depression, bipolar disease, and schizophrenia. Major depressive episodes are more common in patients with epilepsy than in the general population, with prevalence ranging from 11 to 62 %, compared with 3.7–6.7 % for the general population [6, 8, 17, 23, 32]. There is an even stronger association with psychosis; the prevalence of the interictal psychosis of epilepsy ranges (in different studies) between 4.3 and 44 % and in a recent review, rates of 19.4 % and 15.2 % in generalized epilepsies and temporal lobe epilepsy groups are recorded.

The association of neurological and psychiatric disorders to epilepsy is complex. The fact that there is an association was fully recognized in the late nineteenth century and the concept of the 'Neurological Trait' was universally accepted [19, 26, 27]. According to this concept, epilepsy was an essentially inherited condition and inherited together with other neurological and psychiatric disorders. It was accepted that within a family the same inherited tendency might manifest in one person as epilepsy and in other family members as other conditions, but all reflected the same underlying inherited influence (of course, "genes" were not recognized, nor were Mendelian principles widely known at this time). Although different authorities included different conditions within the inherited tendency, at the core were mental disturbances such as insanity, mental retardation, behavioral aberrations, alcoholism – and epilepsy. Gowers, for instance, in 1881 wrote: "There are few diseases in the production of which inheritance has great influence…. It is well known that the neuropathic tendency does not always manifest itself in the same form…. The chief other morbid states (besides epilepsy), in which the neuropathic tendency is manifest are insanity, and, to a much smaller degree, chorea, hysteria, and some forms of disease of the spinal cord. Intemperance is probably also due, in many cases, to a neuropathic disposition" [13]. In Gowers' personal series of 1,218 epilepsy cases, he found that 42 %

"presented evidence of neurotic inheritance." In the nineteenth century, the concept was also linked to that of 'degeneration' and it was widely believed that the manifestations of the trait worsened as it was inherited from generation to generation.

Another topic of current interest is the "bidirectional nature" of the comorbidity epilepsy with various neuropsychiatric conditions. The association is often considered to be due to such factors as recurrent epileptic seizures, social stigma, adverse effects of drug treatment or the underlying structural or metabolic brain injury. However, recent studies have shown that the 'bidirectionality' may in fact predate the development of epilepsy [1, 4, 24] and be due to shared genetic propensities. Qin et al. [23] found a family history of epilepsy to be a risk factor for schizophrenia or schizophrenia-like psychosis, even after adjusting for personal history of epilepsy. Similarly, adults with new-onset epilepsy are seven times more likely to have a prior history of depression. Adults and children with newly diagnosed epilepsy have been noted to have a prior history of attempted suicide which is five times that of the general population. One development in the field was the finding that copy number variants (CNVs) underpin the pathogenesis of some neuro-developmental disease. Several studies have demonstrated that the same large CNVs underpin epilepsy, autism, schizophrenia, mental retardation and attention deficit hyperactivity disorder [1, 4, 5, 16, 20, 21, 24, 30, 33, 34].

If there are shared genetic influences, both the epilepsy and the neuropsychiatric conditions are frequently 'neurodevelopmental' in origin [19]. *Functional annotation analysis* is one attempt to understand shared pathogenic mechanisms, and the effect of the dimension of time is another factor which complicates analysis and renders simple 'gene hunts' unlikely to be very revealing. The reasons for this are the differing gene expression at different times, the effect of development of the activation of functional genetic pathways and the strong effect of environmental factors and chance in development (see 20 for further discussion of this point). The genetic mechanisms of these shared propensities (which has eerie echoes of the concept of the "neurological trait") are the subject of study and certainly can be modeled experimentally – this could be an area of promising future research.

21.4   Mortality Associated with Comorbidity

The risk of death in epilepsy is elevated even when the epilepsy is in remission, In our own recently published 25-year follow-up study of people with newly diagnosed epilepsy the risk of premature death was twice that of the general population [29]. The underlying causes of epilepsy (stroke, brain tumour etc.) have an obviously increased rate of mortality. This not surprisingly increases the risk of premature mortality amongst those with epilepsy. Of more interest from the point of view of studies of comorbidities, are the 'external' causes of mortality in epilepsy (ie not underlying causes of epilepsy such as brain tumours or strokes) and the risk of premature mortality due to such causes.

An outstanding study in the field was recently published examining the relationship of psychiatric comorbidity to premature death [9]. This is the gold standard study in the area, and outshines all the others in terms of its comprehensive nature and intelligence. Data were obtained from all individuals born in Sweden between 1954 and 2009, via a variety of nationwide population registers in Sweden which were then linked: the Patient Register, the Censuses from 1970 to 1990, the Multi-Generation Register, and the Cause-of-Death Register. Epilepsy was identified through the National Patient Register, which includes individuals hospitalized or having outpatient appointments with specialist physicians in Sweden who had received a diagnosis of epilepsy (n=69,995). Data for causes of death were retrieved for all individuals who died between 1969 and 2009 from the Cause of Death register based on death certificates, which covers over 99 % of all deaths. Patients were compared with age-matched and sex-matched controls (n=660,869) from the general population as well as unaffected siblings (n=81,396). 6,155 (8.8 %)

people with epilepsy died during follow-up. The study had extensive sensitivity testing and the comparison with unaffected siblings was important for exploring interfamilial confounding.

The study found a very substantially elevated risk of premature death in epilepsy. The odds ratio for premature mortality was 11.1 [95 % CI=10.6–11.6] compared with general population controls, and 11.4 [10.4–12.5] compared with unaffected siblings. 15.8 % of the deaths were due to external causes. The external causes with the highest odd ratios were non-vehicle accidents (OR 5.5, 95 % CI 4.7–6.5) and suicide (3.7, 3.3–4.2). Of those who died from external causes, 75.2 % had comorbid psychiatric disorders, with the strongest associations being with depression (13.0, 10.3–16.6) and substance misuse (22.4, 18.3–27.3). This link between premature morbidity and psychiatric disease is of course of fundamental importance in clinical practice. Epilepsy was found in this study to be an independent risk factor for all-cause and external causes of death, a finding which was most clearly shown by the comparison of patients with their unaffected siblings, with the rate of mortality increased by 2.9× for suicide and 3.6× for accidents. Another important point recognized was that despite the high relative risks (odds ratios), the absolute rates of premature mortality from external causes was only 1.4 %. However, about a third of the epilepsy patients had at least one comorbid psychiatric diagnosis and about 10 % exhibited substance misuse [9].

21.5   Comorbidity as Biomarker for Epilepsy

If comorbidities are common and easily measured, they can be used as biomarkers for epilepsy. Examples might include genetic markers or even physiological changes. The study of epilepsy biomarkers is not yet well developed, but biomarkers potentially have great importance for diagnostic purposes and also for prognosis and for studying the effects of therapy. Galanopoulou and Moshe [12] divided the search for biomarkers into four categories. Some are not likely to be susceptible to experimental study, but others are. Two of the categories defined were (in their own words):

Biomarker of epileptogenicity: The desired features of these biomarkers include:

- Specificity in differentiating the epileptic state from reactive changes resulting from an initial precipitating event or the first seizure, and from developmental processes that have not yet reached maturity;
- Sensitivity in diagnosing epilepsy at the pre-clinical or early symptomatic stages, when clinical diagnosis has not yet been established;
- Ability to detect the reversal of epileptogenicity, to prevent unnecessary continuation of treatments.

Biomarkers of treatment implementation, tolerability or toxicity: Many antiepilepsy drugs have side effects, which result in comorbidity, the mechanisms for which offer the possibility of biomarker studies:

- Provide target identification for treatment selection, distinguishing it from age-specific relevant processes;
- Define the timing and therapeutic window of treatment administration, based on age- and sex-adapted criteria;
- Distinguish the treatment-responsive from the resistant patient populations early;
- Provide early risk identification and monitoring of treatment-related toxicities, based on age- and sex-adapted criteria, with sufficient specificity for the administered treatment;
- Have the ability to localize the epileptogenic focus accurately and facilitate more effective ablative treatments, if medical treatments are not curative.

21.6 Experimental and Animal Models of Comorbidity

As emphasized above, only some aspects of epilepsy comorbidity are susceptible to modeling in the experimental laboratory. Experimental studies which are most likely to be successful are those directed at the causal molecular, physiological and/or genetic mechanisms of the

relationship of epilepsy and its comorbid conditions The relationship, especially for the psychiatric (and other brain-related) comorbidities are likely to be complex and have developmental and time-sensitive dimensions. Those comorbidities that have priority are, in the author's opinion, in the following areas:

(a) Studies of the adverse effects of epilepsy on brain function, with experimental studies that focus on the mechanisms of brain damage and the role of neuroprotection
(b) Studies of the adverse effects of epilepsy on somatic function, with experimental studies that focus on the molecular mechanisms of these effects and ways of blocking these (the role of osteoporosis for instance in fractures).
(c) Studies of the underlying mechanisms of psychiatric comorbidities, with experimental studies focusing on the genetic and molecular basis, the bidirectionality of the relationship between epilepsy and comorbidities, and on shared pathways.
(d) Studies of the role of comorbidity as biomarker of either epileptogenicity or of the adverse effects of treatment. These two may have a developmental or age-related effects, with different vulnerabilities at different ages.

Acknowledgements The author acknowledges the many discussions he has had with Professor Philip Schwartzkroin, when both were co-Editors-in-Chief of *Epilepsia* between 2005 and 2013, which have enriched his knowledge and understanding of all aspects of epilepsy.

Other Acknowledgements This work was undertaken at UCLH/UCL which receives a proportion of funding from the Department of Health's NIHR Biomedical Research Centres funding scheme. The author has no conflicts of interest to declare in relation to this chapter.

References

1. Adachi N, Akanuma N, Ito M et al (2010) Epileptic, organic, and genetic vulnerabilities for timing of the development of interictal psychosis. Br J Psychiatry 196:212–216
2. Babu CS, Satishchandra P, Sinha S, Subbakrishna DK (2009) Co-morbidities in people living with epilepsy: hospital based case-control study from a resource-poor setting. Epilepsy Res 86:146–152

3. Boro A, Haut S (2003) Medical comorbidities in the treatment of epilepsy. Epilepsy Behav 4(Suppl 2):S2–S12

4. Craddock N, Owen MJ (2010) Molecular genetics and the relationship between epilepsy and psychosis. Br J Psychiatry 197:75–76

5. De Kovel CGF, Trucks H, Helbig I et al (2010) Recurrent microdeletions at 15q11.2 and 16p13.11 predispose to idiopathic generalized epilepsies. Brain 133:23–32

6. Elliott B, O'Donavan J (2011) Psychiatric disorders. In: Shorvon S, Andermann F, Guerrini R (eds) The causes of epilepsy. Cambridge University Press, Cambridge, pp 593–606

7. Feinstein AR (1970) The pretherapeutic classification of comorbidity in chronic disease. J Chronic Dis 23:455–468

8. Filho GM, Rosa VP, Lin K, Caboclo LO, Sakamoto AC, Yacubian EM (2008) Psychiatric comorbidity in epilepsy: a study comparing patients with mesial temporal sclerosis and juvenile myoclonic epilepsy. Epilepsy Behav 13:196–201

9. Fazel S, Wolf A, Långström N, Newton CR, Lichtenstein P (2013) Premature mortality in epilepsy and the role of psychiatric comorbidity: a total population study. Lancet 382(9905):1646–1654

10. Forsgren L (1992) Prevalence of epilepsy in adults in northern Sweden. Epilepsia 33:450–458

11. Gaitatzis A, Carroll K, Majeed A et al (2004) The epidemiology of the comorbidity of epilepsy in the general population. Epilepsia 45:1613–1622

12. Galanopoulou A, Moshe S (2011) In search of epilepsy biomarkers in the immature brain: goals, challenges and strategies. Biomark Med 5:615–628

13. Gowers W (1881) Epilepsy and other chronic convulsive disorders. Churchill, London

14. Gudmundsson G (1966) Epilepsy in Iceland. A clinical and epidemiological investigation. Acta Neurol Scand 43(Suppl 25):1–124

15. Hackett R, Hackett L, Bhakta P (1998) Psychiatric disorder and cognitive function in children with epilepsy in Kerala, South India. Seizure 7:321–324

16. Heinzen EL, Radtke RA, Urban TJ et al (2010) Rare deletions at 16p13.11 predispose to a diverse spectrum of sporadic epilepsy syndromes. Am J Hum Genet 86:707–718

17. Hesdorffer DC, Hauser WA, Annegers JF, Cascino G (2000) Major depression is a risk factor for seizures in older adults. Ann Neurol 47:246–249

18. Jalava M, Sillanpaa M (1996) Concurrent illnesses in adults with childhood-onset epilepsy: a population-based 35-year follow-up study. Epilepsia 37:1155–1163

19. Johnson MR, Shorvon SD (2011) Heredity in epilepsy: neurodevelopment, comorbidity, and the neurological trait. Epilepsy Behav 22:421–427

20. Marshall CR, Noor A, Vincent JB et al (2008) Structural variation of chromosomes in autism spectrum disorder. Am J Hum Genet 82:477–488

21. Mefford HC, Muhle H, Ostertag PC et al (2010) Genome-wide copy number variation in epilepsy: novel susceptibility loci in idiopathic generalized and focal epilepsies. PLoS Genet 6:e1000962

22. Mensah SA, Beavis JM, Thapar AK, Kerr M (2006) The presence and clinical implications of depression in a community population of adults with epilepsy. Epilepsy Behav 8:213–219

23. Qin P, Xu H, Laursen TM, Vestergaard M, Mortensen PB (2005) Risk for schizophrenia and schizophrenia-like psychosis among patients with epilepsy: population based cohort study. Br Med J 331:23–25

24. Sebat J, Lakshimi B, Malhotra D et al (2007) Strong association of de novo copy number mutations with autism. Science 316:445–449

25. Shorvon S (2011) The causes of epilepsy: changing concepts of etiology of epilepsy over the past 150 years. Epilepsia 52:1033–1044

26. Shorvon S (2011) Historical introduction: the causes of epilepsy in the pre-molecular era (1860–1960). In: Shorvon SD, Andermann F, Guerrini R (eds) The causes of epilepsy. Common and uncommon causes in adults and children. Cambridge University Press, Cambridge, pp 1–20

27. Shorvon S (2011) Heredity in epilepsy – an historical overview. Neurol Asia 16:5–8

28. Shorvon S, Andermann F, Guerrini R (2011) The causes of epilepsy: common and uncommon causes in adults and children. Cambridge University Press, Cambridge

29. Shorvon S, Goodridge D (2013) Longitudinal cohort studies of the prognosis of epilepsy: contribution of the National General Practice Study of Epilepsy and other studies. Brain 136:3497–3510

30. Stefansson H, Rujescu D, Cichon S et al (2008) Large recurrent microdeletions associated with schizophrenia. Nature 455:232–236

31. Tellez-Zenteno JF, Matijevic S, Wiebe S (2005) Somatic comorbidity of epilepsy in the general population in Canada. Epilepsia 46:1955–1962

32. Téllez-Zenteno JF, Patten SB, Jetté N, Williams J, Wiebe S (2007) Psychiatric comorbidity in epilepsy: a population-based analysis. Epilepsia 48:2336–2344

33. Walsh T, McClellan JM, McCarthy SE et al (2008) Rare structural variants disrupt multiple genes in neurodevelopmental pathways in schizophrenia. Science 320:539–543

34. Williams NM, Zaharieva I, Martin A et al (2010) Rare chromosomal deletions and duplications in attention-deficit hyperactivity disorder: a genome-wide analysis. Lancet 376:1401–1408

Epilepsy Comorbidities: How Can Animal Models Help?

22

Carl E. Stafstrom

Abstract

An epilepsy comorbidity is a condition or disorder that occurs at a frequency greater than chance in a person with epilepsy. Examples of common epilepsy comorbidities are depression, anxiety, and intellectual disability. Epilepsy comorbidities can be quite disabling, sometimes affecting a patient's quality of life to a greater extent than seizures. Animal models offer the opportunity to explore shared pathophysiological mechanisms, therapeutic options, and consequences of both the epilepsy syndrome and a given comorbidity. In this chapter, depression is used as an example of how animal models can inform translational questions about epilepsy comorbidities.

Keywords

Epilepsy • Comorbidity • Depression • Animal models

22.1 What Are Epilepsy Comorbidities and Can Animal Models Help?

While epilepsy is primarily considered to be a condition of recurrent, unprovoked seizures, it is increasingly evident that epilepsy involves a lot more than seizures. Epilepsy comorbidities, defined as medical or psychiatric disorders that occur at a frequency greater than chance in a patient with epilepsy, play a crucial role in the quality of life and treatment effectiveness in patients with epilepsy. Epilepsy comorbidities include disorders of cognition, mood, and behavior [3, 14, 26], as well as a variety of medical and neurological disorders [10]. Specific examples include depression, anxiety, intellectual impairment, autism, sleep disorders, migraine, and many others (Table 22.1). In some individuals, comorbidities can be more impairing than the seizures themselves [11]. Many patients have more than one comorbidity, underscoring the need to understand the roles played by single and multiple

C.E. Stafstrom, M.D., Ph.D. (✉)
Departments of Neurology and Pediatrics, University of Wisconsin, Centennial Building 7176, 1685 Highland Avenue, Madison, WI 53705, USA
e-mail: stafstrom@neurology.wisc.edu

Table 22.1 Examples of epilepsy comorbidities

Anxiety disorder
Autism spectrum disorder
Cardiovascular disease/stroke
Dementia/Alzheimer disease
Depression
Intellectual disability/cognitive impairment
Migraine
Sleep disorders
Suicidality

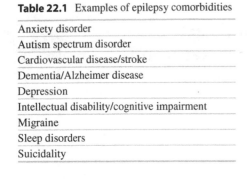

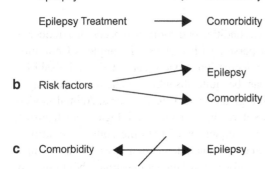

Fig. 22.1 Possible relationships between epilepsy and comorbidity. (**a**) A causal relationship might exist whereby epilepsy or an epilepsy treatment might cause a comorbidity, or a comorbidity might lead to epilepsy. (**b**) Common risk factors might exist (e.g., environmental, genetic, structural) that lead to both epilepsy and a comorbidity. (**c**) Comorbidity and epilepsy might be independent, unrelated associations

comorbidities in epilepsy, epileptogenesis, and quality of life in persons with epilepsy.

The concept of epilepsy comorbidity has been under-recognized but is not new. Recent attention has been focused on epilepsy comorbidities with the addition of comorbidities as a separate NIH Epilepsy Benchmark [23]. In addition, recognition of a comorbidity as a major cause of impaired quality of life of persons with epilepsy is elaborated in the recent Institute of Medicine report [5].

There are several possible relationships between epilepsy and a comorbidity (Fig. 22.1) [3, 10]. First, the relationship can be causal, with one disorder causing the other or making the other disorder more likely (Fig. 22.1a). That is, a comorbid condition can lead to epilepsy, or conversely,

the comorbid condition occurs as a result of the epilepsy or its treatment. Several examples will clarify this concept. The etiologies of most symptomatic epilepsies (for example, traumatic brain injury, stroke) correspond to this cause-and-effect model [10]. Conversely, epilepsy itself can lead to a comorbidity such as anxiety disorder in predisposed individuals [34]. Finally, numerous examples of epilepsy treatments leading to behavioral comorbidities can be cited, such as the association of phenobarbital with hyperactivity in children and the association of levetiracetam with altered mood [35].

Second, shared risk factors, which can be genetic, metabolic, structural or environmental, can lead to the development of both epilepsy and a comorbidity (Fig. 22.1b). An example is the structural brain damage caused by perinatal hypoxia-ischemia that leads to both epilepsy and comorbid cerebral palsy [3]. This type of relationship also includes the comorbidities that are considered "bidirectional", that is, common underlying mechanisms could facilitate the development of both epilepsy and the comorbidity. Depression is a common and critically important example of an epilepsy comorbidity and is discussed in detail below. Third, the relationship between epilepsy and a comorbidity could be incidental or even spurious (Fig. 22.1c).

Since so many people with epilepsy harbor one or more comorbidity, it is important to elucidate these relationships. For example, there could be shared pathophysiological mechanisms between epilepsy and a comorbidity, with the possibility that one or both conditions is amenable to a treatment or disease modification that exploits these common mechanisms. Of note, no specific therapy exists for a comorbidity in the context of epilepsy. That is, if a patient with epilepsy is diagnosed with a comorbidity such as anxiety or depression, treatment choice is limited to medications used to treat anxiety or depression, irrespective of the concurrent epilepsy. Novel treatments are needed that take into account the specific pathogenic mechanisms of both epilepsy and the comorbidity.

Given the prevalence of epilepsy comorbidities and the lack of understanding of their mechanisms,

Table 22.2 Factors to consider in animal models of epilepsy comorbidities

Age of onset (of seizures and comorbid symptoms)
Brain region and neurotransmitter system underlying comorbidity
Environmental factors (e.g., cage size and density, light/dark cycle)
Food intake (e.g., may be decreased in depressed animals)
Gender of animal
Handling by laboratory personnel
Species/strain/genetic background
Symptoms versus syndrome (i.e., concurrent additional comorbidities)

Table 22.3 What can be learned from studying epilepsy comorbidities in animal models?

Mechanisms of shared pathophysiology
Potential avenues for therapy and disease modification (e.g., relative roles of antidepressant and anticonvulsant medications on both epilepsy and depression)
Correlations between behavioral phenotype of the comorbidity and features of the epilepsy syndrome (seizure type, frequency, temporal relationship with comorbid symptoms, etc.)
Role of the comorbidity in epilepsy progression and epileptogenesis

the question arises as to whether animal models can provide useful information about pathogenesis or treatment [18]. The purpose of this chapter is to provide an overview of some of the theoretical issues in modeling epilepsy comorbidities in animals, followed by an example of how understanding one specific epilepsy comorbidity – depression – might enhance understanding of the pathophysiology of both disorders and could help to identify treatment targets. Comprehensive reviews of comorbidities in animal models of epilepsy already exist [3], as do detailed guidelines for testing specific cognitive functions in animal models of epilepsy [45].

The first question to consider is how closely an animal model resembles the human condition. This question applies to epilepsy as well as to the comorbidity, and when trying to model both conditions in one animal, obvious challenges arise (Table 22.2). Species differences are usually obvious, but not trivial. While at first glance, it

might seem implausible that a rodent could exhibit depression similar to that experienced by a patient. However, a burgeoning literature supports the idea that there are shared features and pathophysiological mechanisms between depression in animals and humans (discussed in greater detail below). Second, for any comorbidity under consideration, the experimenter must evaluate how the testing paradigm itself might contribute to the animal's performance; that is, does the test itself elicit stress or another set of behaviors that confound the original intention? Third, it is critical that longitudinal observations be employed – it is insufficient to test an animal only once in a behavioral paradigm since both epilepsy and most comorbidities are chronic (and often evolving) conditions (Table 22.3).

22.2 Depression as an Example of an Epilepsy Comorbidity

Depression is extremely common in the general population, but even more so among people with epilepsy [22]. In population-based studies, it has been estimated that approximately 25–35 % of individuals with epilepsy suffer from depression (even higher if the epilepsy is not well controlled) and that people with depression have a 3- to 7-times greater risk of developing epilepsy than the general population [15, 20, 47]. Depression also affects 8–26 % of children with epilepsy [9, 37]. These percentages far exceed those expected in the general population and may well underestimate the actual prevalence of depression in persons with epilepsy. A history of depression is a reliable predictor of worse epilepsy severity [20]. The bidirectional relationship of epilepsy and depression (epilepsy is more likely in people with depression, and depression is more likely among people with epilepsy) is validated by neurobiological data of several types, including neurotransmitter analyses, MRI and positron emission tomography studies of temporal or frontal lobe function, and investigations of hypothalamic-pituitary-adrenal (HPA) axis dysfunction [21]. The bidirectional relationship suggests that there may exist one or more common neurobiological mechanisms and that these mechanisms might be exploited for therapeutic advantage.

Depression is a heterogeneous disorder with several distinct subtypes classifiable using the Diagnostic and Statistical Manual of Mental Disorders (5th edition, DSM-V [1]). It is important to recognize that the DSM is based on expert consensus not validated biomarkers. DSM-V criteria for the diagnosis of depression include despair, anhedonia (inability to experience pleasure), vegetative symptoms (weight loss, appetite decrease or increase, decreased energy, insomnia), feelings of worthlessness and guilt, decreased focus/attention span, and suicidal ideation. It is uncertain whether depression in persons with epilepsy is identical to depression in persons without epilepsy. Data suggests that many "atypical" features that do not adhere to the strict DSM criteria typify depression in individuals with epilepsy [20]. Atypical features include a greater degree of anxiety, irritability, and mood lability. Importantly, the timing of depressive episodes may relate to seizure occurrence; a bout of depression may precede a seizure (interictal episode) or occur around the same time as a seizure (peri-ictal episode) [22]. Despite their frequent co-occurrence, the severity of depression, at least in temporal lobe epilepsy, is not proportionate to the number of seizures [12]. The treatment goal is reduction of both seizures and depressive symptoms, although seizure control does not always correlate with improvement in depression [13]. Ideally, this goal would be achieved using monotherapy, with one drug improving both seizure control and depression. Specific data about the impact of antidepressants on depression in epilepsy are scarce but much needed.

The effects of antidepressants on epilepsy and antiepileptic agents on depression are complex. Some antiepileptic drugs are well known for their mood stabilizing properties (e.g., carbamazepine, valproate, lamotrigine). Likewise, antidepressants have been shown to exert anticonvulsant effects in both patients and animals – selective serotonin reuptake inhibitors (SSRIs), serotonin-norepinephrine reuptake inhibitors (SNRIs), and tricyclic antidepressants (TCAs) can increase brain monoamines such as serotonin, norepinephrine, and dopamine, favoring an anticonvulsant action [16]. In addition, depression can be ameliorated by alterations of the primary neurotransmitter systems of the brain – glutamate receptor antagonists (e.g., dizocilpine, ketamine) or γ-amino-butyric acid (GABA)-receptor agonists [33, 40]. The multifaceted effects of these and other novel agents in epilepsy and depression are poised for study in animal models [28].

22.3 Evaluating Depression in Animal Models of Epilepsy

Depression is an exemplary disorder in which to explore the opportunities and challenges between epilepsy and a comorbidity using animal models [8]. Obviously, many subjective symptoms of depression cannot easily be extrapolated to animals, but an approximation of some of the symptoms makes the study of this comorbidity in animals quite tenable. To that end, a set of modified criteria for depression in rodents has been proposed [3]. Of the depression criteria listed in DSM-V, despair and anhedonia are most readily testable in animals, with validated laboratory tests available for those symptoms. The forced swim test (FST) is a measure of despair in rodents, while anhedonia, the failure to experience pleasure, is assessed by the taste preference test (TPT). These tests have been widely used to screen potential antidepressant compounds. It is important to recognize that a single administration of those or any other experimental measure of depression in animals represents only a single point in time, whereas a comorbidity typically evolves over time, necessitating serial assessments.

The FST is performed by placing an animal in a water-filled chamber with smooth sides, from which it cannot escape. Initially, the animal typically swims around frantically, trying to escape by climbing the walls (active escape phase). Eventually, the animal seems to give up this futile effort and becomes immobile, simply floating in the water, striving to keep its head above water to prevent drowning (immobility phase). These two phases are easily quantified, with the immobility

a Forced swim test

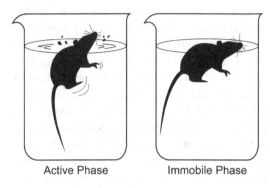

Active Phase   Immobile Phase

b Taste preference test

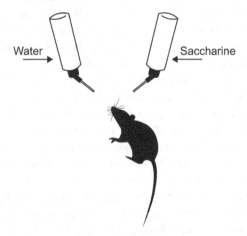

Water   Saccharine

Fig. 22.2 Laboratory tests of depression in rodents. (a) The forced swim test is a measure of despair, one of the core symptoms of depression. *Left*, active escape phase. *Right*, immobility phase, considered to represent despair. (b) The taste preference test is a measure of anhedonia, or loss of ability to experience pleasure. A normal animal prefers the sweetened liquid, while a depressed animal does not express this preference

phase comprising a validated measure of despair (Fig. 22.2a). In models of depression, animals that are depressed have shorter active escape phases and enter the immobile phase sooner. Importantly, the FST is itself a stressor for an animal. Antidepressant drugs increase the active escape phase duration, supporting the contention that the immobility phase represents despair. The FST has been used for antidepressant drug discovery in animal models of depression, but pharmacologic studies of antidepressants in epileptic animals have emerged only recently.

The TPT compares a rodent's preference for drinking a solution sweetened with saccharine (or sucrose) over plain water (Fig. 22.2b) [32]. Ordinarily, rodents prefer to drink the sweet solution. Sugar consumption stimulates dopaminergic fibers projecting from the ventral tegmental area to the nucleus accumbens, where the amount of dopamine released correlates with motivational aspects of reward [39]. In depressed animals, intake of sweet liquids such as sucrose or saccharine is decreased and there is no difference in rodents' consumption of the sweetened versus plain water, suggesting that they have less interest in the flavored fluid.

Other tests have also been employed for comorbidities in epilepsy research, some applicable to depression and others more reflective of anxiety, cognitive function, memory, or learning. Comprehensive lists of such tests (Table 1 in [3] and Table 1 in [8]) reveal that many are in need of validation in animals with epilepsy. As well, there is an urgent need for multi-dimensional behavioral tests to simultaneously assess concurrent comorbidities in the same subject – depression, anxiety, sleep dysfunction, etc.

22.4 Examples of Comorbid Epilepsy and Depression in Animal Models

To illustrate some of the insights that can be gained from animal models, examples are now provided that examine various aspects of the relationship between epilepsy and depression. These examples include both acquired and genetic etiologies. Space precludes detailed discussion of other relevant examples such as GAERs (genetic absence epilepsy rats from Strasbourg) [19] and genetically epilepsy prone rats (GEPRs) [17].

Chemoconvulsant models of temporal lobe epilepsy (TLE) in rats using either kainic acid (KA) or lithium/pilocarpine (LiP) allow detailed study of the relationship between seizures (number, frequency, duration and timing of spontaneous recurrent seizures) and the occurrence of behavioral and cognitive abnormalities. KA is a glutamate receptor agonist; pilocarpine is an agonist of muscarinic acetylcholine receptors. Both forms of chemoconvulsant-induced epilepsy mimic limbic epilepsy, with initial status epilepticus followed weeks-to-months later by spontaneous recurrent seizures and behavioral and cognitive abnormalities. In both KA- and LiP-epilepsy, evidence of depression has been documented on the FST and TPT [24, 36]. Following KA-induced status epilepticus, rats had shorter latencies to the immobile phase on the FST and longer duration of immobility, suggesting that these rats were depressed (increased despair) [24]. Using micro-array analysis, it was shown that depressed rats had a reduction in expression of the gene for serotonin receptor 5B. Most strikingly, environmental enrichment prevented both FST abnormalities and the underlying gene expression changes, suggesting that environmental factors play a crucial role in the development of depression as an epilepsy comorbidity. Investigation of structural brain injury and the roles of antidepressant and anticonvulsant drugs in this model would further clarify these relationships.

In the other chemoconvulsant model, intra-peritoneal injection of LiP causes limbic status epilepticus, followed in subsequent weeks by behavioral deficits such as learning and memory impairment and a depression phenotype. Compared to naïve rats, LiP-treated rats demonstrated increased immobility time in the FST and loss of taste preference in the TPT [29], supporting the depression phenotype of despair and anhedonia. These behavioral deficits were rescued by treatment with a blocker of the serotonin 5HT1a receptor, but there was no effect of selective serotonin reuptake inhibitors (SSRIs) [29], suggesting that depression in this model does not respond to medications typically used to treat clinical depression. These observations support the conclusion that depression, at least in some epilepsy disorders, represents an atypical form of the condition. This model provides the opportunity to dissect contributions of the multiple serotonin receptors involved in various depression subtypes [27]. The effects of standard anticonvulsants on depression and antidepressants on epilepsy have not yet been reported in this model.

To investigate the mechanism linking depression to epilepsy in this model, the authors noted that dysregulation of the HPA axis is a marker of depression, with increased levels of plasma glucocorticoid (cortisol) due to loss of negative feedback of cortisol on corticotrophin releasing hormone and adrenocorticotrophic hormone release [25]. LiP-treated rats had elevated cortisol levels, supporting the depression phenotype [31]. After status epilepticus in these animals, there was reduced serotonergic innervation from brainstem raphe nuclei to the hippocampus due to upregulation of raphe 5-HT1A autoreceptors, as found in some human depression [4]. Furthermore, a blocker of 5-HT1A receptors, WAY-100635, improved performance on the FST, forming a link between abnormal serotonergic function, depression, and behavior [30].

Further studies showed that increased hippocampal interleukin 1-β (IL1β) signaling might mediate both depressive symptoms and heightened hippocampal excitability leading to spontaneous seizures in this model. The authors proposed a scheme whereby epilepsy leads to depression by increasing IL1β signaling, which upregulates raphe 5-HT1A autoreceptors, compromising raphe-to-hippocampus serotonergic neurotransmission. These findings raise the possibility of a link between mechanisms of epilepsy, depression, stress, and the inflammatory response [49]. Potential loci for intervention might include blockade of glucocorticoid action, downregulation of raphe 5HT1A autoreceptors, or anti-inflammatory agents. This model can also be utilized to further characterize the mechanisms of neuronal excitability underlying epilepsy and depression.

The next example is rats bred for susceptibility or resistance to depression-like behaviors during swimming in the FST (named SwLo and SwHi, respectively). SwLo rats display increased immobility in the FST and anhedonic tendencies. Importantly, SwLo rats also have increased predisposition to limbic seizures induced by kainic acid or pilocarpine, providing an excellent opportunity to examine the joint mechanisms of depression and epilepsy, with particular relevance to temporal lobe epilepsy [7, 46]. Chronic antidepressant treatment reverses the FST deficits in

SwLo rats [50], substantiating the validity of this model in depression. In addition, the existence of the converse model – SwHi rats that are resistant to depression – provides a unique opportunity to examine whether this strain is also relatively resistant to seizure development. To date, there are no data regarding the effects of anticonvulsants on either depression or seizure development in this model. Finally, this model provides further evidence for the interaction of environment and genetics in the expression of both depression and epilepsy, as aerobic exercise was found to improve both FST performance and seizure resistance in SwLo rats compared to SwHi rats [6].

Lastly, a genetic model of absence epilepsy has revealed a number of important relationships between epilepsy predisposition and psychiatric comorbidities. The inbred WAG/Rij (Wistar Albino Glaxo/Rijswijk) rat strain develops absence seizures at approximately 2–3 months of age, in parallel with the onset of depression and anxiety phenotypes [41]. Therefore, this model is ideal to investigate the age-related onset and causal relationship between depression and epilepsy with spike-wave discharges. WAG/Rij rats have deficiencies in the FST and TPT, as well as anxiety-related behaviors in the open field test [44]. The depressive symptoms in this model can be rescued by chronic treatment with the TCA, imipramine (but the effect of imipramine on seizures is unknown). Chronic treatment of WAG/Rij rats with the anti-absence drug ethosuximide from 3 weeks to 5 months of age led to persistent seizure suppression many months after discontinuation of treatment [2]. Chronic ethosuximide treatment also reduced immobility time on the FST, suggesting that this anticonvulsant exerted both antiepileptic and antidepressant effects [48]. The authors concluded that there is a causal relationship between the development of the epileptic phenotype and depressive symptoms in this model [43]. Prominent involvement of the dopaminergic system in these behavioral dysfunctions is supported by acute treatment with a dopamine receptor D2/3 antagonist, raclopride, which exacerbated FST deficiencies, and a D2/3 receptor agonist, parlodel, which exerted antidepressant effects [42]. Recent work also implicates involvement of the mTOR pathway in both epileptogenesis and

depression in WAG/Rij rats [38]. Blockade of the mTOR pathway with rapamycin for either 7 days ("sub-chronic") or 17 weeks ("chronic") ameliorated absence seizures but had an opposite effect on depression using the FST and TPT – subchronic treatment with rapamycin had an antidepressant effect while chronic treatment produced a prodepressant effect. These results could form the basis of a novel treatment strategy for epilepsy and depression (mTOR inhibition), while raising the interesting caveat that the same agent (rapamycin) can exert different effects on depression, depending on the specific administration protocol. Taken together, data from more than three decades of study of the WAG/Rij rat absence epilepsy model strongly support a close interrelationship between seizures and psychiatric comorbidities, especially depression, and provide an excellent model in which to investigate correlations between seizure occurrence, cognitive dysfunction, and treatment parameters.

22.5 Conclusion

Potential pathophysiological overlaps between epilepsy and epilepsy comorbidities are eminently amenable to study in the laboratory using animal models. While acknowledging species differences and other inherent limitations of animal models of epilepsy and psychiatric diseases, the shared pathophysiology between epilepsy and depression, anxiety, and other comorbidities are readily amenable to laboratory investigation and could yield insights into the pathophysiological mechanisms in one or both conditions, as well as potential therapeutic modalities. This rigorous approach to translational neurobiology has been typified by laboratory models championed by Dr. Philip Schwartzkroin and his colleagues.

References

1. American Psychiatric Association (2013) Diagnostic and statistical manual of mental disorders, 5th edn. American Psychiatric Association, Arlington
2. Blumenfeld H, Klein JP, Schridde U, Vestal M, Rice T, Khera DS, Bashyal C, Giblin K, Paul-Laughinghouse C, Wang F, Phadke A, Mission J, Agarwal RK, Englot

DJ, Motelow J, Nerseyan H, Waxman SG, Levin AR (2008) Early treatment suppresses the development of spike-wave epilepsy in a rat model. Epilepsia 49:400–409

3. Brooks-Kayal AR, Bath KG, Berg AT, Galanopoulou AS, Holmes GL, Jensen FE, Kanner AM, O'Brien TJ, Whittemore VH, Winawer MR, Patel M, Scharfman HE (2013) Issues related to symptomatic and disease-modifying treatments affecting cognitive and neuropsychiatric comorbidities of epilepsy. Epilepsia 54(Suppl 4):44–60

4. Drevets WC, Thase ME, Moses-Kolko EL, Price J, Frank E, Kupfer DJ, Mathis C (2007) Serotonin-1A receptor imaging in recurrent depression: replication and literature review. Nucl Med Biol 34:865–877

5. England MJ, Liverman CT, Schultz AM, Strawbridge LM (2012) Epilepsy across the spectrum: promoting health and understanding. Institute of Medicine (US) Committee on the Public Health Dimensions of the Epilepsies. National Academies Press, Washington, DC

6. Epps SA, Kahn AB, Holmes PV, Boss-Williams KA, Weiss JM, Weinshenker D (2013) Antidepressant and anticonvulsant effects of exercise in a rat model of epilepsy and depression comorbidity. Epilepsy Behav 29:47–52

7. Epps SA, Tabb KD, Lin SJ, Kahn AB, Javors MA, Boss-Williams KA, Weiss JM, Weinshenker D (2012) Seizure susceptibility and epileptogenesis in a rat model of epilepsy and depression co-morbidity. Neuropsychopharmacology 37:2756–2763

8. Epps SA, Weinshenker D (2013) Rhythm and blues: animal models of epilepsy and depression comorbidity. Biochem Pharmacol 85:135–146

9. Ettinger AB, Weisbrot DM, Nolan EE, Gadow KD, Vitale SA, Andriola MR, Lenn NJ, Novak GP, Hermann BP (1998) Symptoms of depression and anxiety in pediatric epilepsy patients. Epilepsia 39:595–599

10. Gaitatzis AS, Sisodiya SM, Sander JW (2012) The somatic comorbidity of epilepsy: a weighty but often unrecognized burden. Epilepsia 53:1282–1293

11. Gilliam F, Hecimovic H, Sheline Y (2003) Psychiatric comorbidity, health, and function in epilepsy. Epilepsy Behav 4(Suppl 4):S26–S30

12. Gilliam FG, Maton BM, Martin RC, Sawrie SM, Faught RE, Hugg JW, Viikinsalo M, Kuzniecky RI (2007) Hippocampal 1H-MRSI correlates with severity of depression symptoms in temporal lobe epilepsy. Neurology 68:364–368

13. Hamid H, Liu H, Cong X, Devinsky O, Berg AT, Vickery BG et al (2011) Long-term association between seizure outcome and depression after resective epilepsy surgery. Neurology 77:1972–1976

14. Hamiwka LD, Wirrell EC (2009) Comorbidities in pediatric epilepsy: beyond "just" treating the seizures. J Child Neurol 24:734–742

15. Hesdorffer DC, Hauser WA, Olafsson E, Ludvigsson P, Kjartansson O (2006) Depression and suicide attempt as risk factors for incident unprovoked seizures. Ann Neurol 59:35–41

16. Igelström KM (2012) Preclinical antiepileptic actions of selective serotonin reuptake inhibitors – implications for clinical trial design. Epilepsia 53:596–605

17. Jobe PC (2003) Common pathogenic mechanisms between depression and epilepsy: an experimental perspective. Epilepsy Behav 4(Suppl 3):S14–S24

18. Jones NC, O'Brien TJ (2013) Stress, epilepsy, and psychiatric comorbidity: how can animal models inform the clinic? Epilepsy Behav 26:363–369

19. Jones NC, Salzberg MR, Kumar G, Couper A, Morris MJ, O'Brien TJ (2008) Elevated anxiety and depressive-like behavior in a rat model of genetic generalized epilepsy suggesting common causation. Exp Neurol 209:254–260

20. Kanner AM (2006) Depression and epilepsy: a new perspective on two closely related disorders. Epilepsy Curr 6:141–146

21. Kanner AM (2012) Can neurobiological pathogenic mechanisms of depression facilitate the development of seizure disorders? Lancet Neurol 11:1093–1102

22. Kanner AM (2013) The treatment of depressive disorders in epilepsy: what all neurologists should know. Epilepsia 54(Suppl 1):3–12

23. Kelley MS, Jacobs MP, Lowenstein DH (2009) The NINDS epilepsy research benchmarks. Epilepsia 50:579–582

24. Koh SR, Magid R, Chung H, Stine CD, Wilson DN (2007) Depressive behavior and selective downregulation of serotonin receptor expression after early-life seizures: reversal by environmental enrichment. Epilepsy Behav 10:26–31

25. Kondziella D, Alvestad S, Vaaler A, Sonnewald U (2007) Which clinical and experimental data link temporal lobe epilepsy with depression? J Neurochem 103:2136–2152

26. Lin JJ, Mula M, Hermann BP (2012) Uncovering the neurobehavioural comorbidities of epilepsy over the lifespan. Lancet 380:1180–1192

27. Lutz P-E (2013) Multiple serotonergic paths to antidepressant efficacy. J Neurophysiol 109:2245–2249

28. Machado-Vieira R, Salvadore G, Diaz Granados N, Ibrahim L, Latov D, Wheeler-Castillo C, Baumann J, Henter ID, Zarate CA Jr (2010) New therapeutic targets for mood disorders. ScientificWorldJournal 10:713–726

29. Mazarati A, Siddarth P, Baldwin RA, Shin D, Caplan R, Sankar R (2008) Depression after status epilepticus: behavioural and biochemical deficits and effects of fluoxetine. Brain 131(Pt 8):2071–2083

30. Mazarati AM, Pineda E, Shin D, Tio D, Taylor AN, Sankar R (2010) Comorbidity between epilepsy and depression: role of hippocampal interleukin-1beta. Neurobiol Dis 37:461–467

31. Mazarati AM, Shin D, Kwon YS, Bragin A, Pineda E, Tio D, Taylor AN, Sankar R (2009) Elevated plasma corticosterone level and depressive behavior in experimental temporal lobe epilepsy. Neurobiol Dis 34:457–461

32. McCaughey SA (2008) The taste of sugars. Neurosci Biobehav Rev 32:1024–1043

33. Möhler H (2012) The GABA system in anxiety and depression and its therapeutic potential. Neuropharmacology 62:42–53

34. Mula M (2013) Treatment of anxiety disorders in epilepsy: an evidence-based approach. Epilepsia 54(Suppl 1):13–18

35. Piedad J, Rickards H, Besag FM, Cavanna AE (2012) Beneficial and adverse psychotropic effects of antiepileptic drugs in patients with epilepsy: a summary of prevalence, underlying mechanisms and data limitations. CNS Drugs 26:319–335

36. Pineda E, Shin D, Sankar R, Mazarati AM (2010) Comorbidity between epilepsy and depression: experimental evidence for the involvement of serotonergic, glucocorticoid, and neuroinflammatory mechanisms. Epilepsia 51(Suppl 3):110–114

37. Russ SA, Larson K, Halfon N (2012) A national profile of childhood epilepsy and seizure disorder. Pediatrics 129:256–264

38. Russo E, Citraro R, Donato G, Camastra C, Iuliano R, Cuzzocrea S, Constanti A, De Sarro G (2013) mTOR inhibition modulates epileptogenesis, seizures and depressive behavior in a genetic rat model of absence epilepsy. Neuropharmacology 69:25–36

39. Salamone JD, Correa M (2012) The mysterious motivational functions of mesolimbic dopamine. Neuron 76:470–485

40. Sanacora G, Treccani G, Popoli M (2012) Towards a glutamate hypothesis of depression: an emerging frontier of neuropsychopharmacology for mood disorders. Neuropharmacology 62:63–77

41. Sarkisova K, van Luijtelaar G (2011) The WAG/Rij strain: a genetic animal model of absence epilepsy with comorbidity of depression. Prog Neuropsychopharmacol Biol Psychiatry 35:854–876

42. Sarkisova KY, Kulikov MA, Midzianovskaia IS, Folomkina AA (2008) Dopamine-dependent nature of depression-like behavior in WAG/Rij rats with genetic absence epilepsy. Neurosci Behav Physiol 38:119–128

43. Sarkisova KY, Kuznetsova GD, Kulikov MA, van Luijtelaar G (2010) Spike-wave discharges are necessary for the expression of behavioral depression-like symptoms. Epilepsia 51:146–160

44. Sarkisova KY, Midzianovskaia IS, Kulikov MA (2003) Depressive-like behavioral alterations and c-fos expression in the dopaminergic brain regions in WAG/Rij rats with genetic absence epilepsy. Behav Brain Res 144:211–226

45. Stafstrom CE (2006) Behavioral and cognitive testing procedures in animal models of epilepsy. In: Pitkänen A, Schwartzkroin PA, Moshé SL (eds) Models of seizures and epilepsy. Elsevier Academic Press, Amsterdam, pp 613–628

46. Tabb K, Boss-Williams KA, Weiss JM, Weinshenker D (2007) Rats bred for susceptibility to depression-like phenotypes have higher kainic acid-induced seizure mortality than their depression-resistant counterparts. Epilepsy Res 74:140–146

47. Tellez-Zenteno JF, Patten SB, Jetté N, Williams J, Wiebe S (2007) Psychiatric comorbidity in epilepsy: a population-based analysis. Epilepsia 48:2336–2344

48. van Luijtelaar G, Mishra AM, Edelbroek P, Coman D, Frankenmolen N, Schaapsmeerders P, Covolato G, Danielson N, Niermann H, Janeczko K, Kiemeneij A, Burinov J, Bashyal C, Coquillette M, Lüttjohann A, Hyder F, Blumenfeld H, van Rijn CM (2013) Antiepileptogenesis: electrophysiology, diffusion tensor imaging and behavior in a genetic absence model. Neurobiol Dis 60:126–138

49. Vezzani A, Friedman A, Dingledine RJ (2013) The role of inflammation in epileptogenesis. Neuropharmacology 69:16–24

50. West CH, Weiss JM (1998) Effects of antidepressant drugs on rats bred for low activity in the swim test. Pharmacol Biochem Behav 61:67–79

What New Modeling Approaches Will Help Us Identify Promising Drug Treatments?

23

Scott C. Baraban and Wolfgang Löscher

Abstract

Despite the development of numerous novel antiepileptic drugs (AEDs) in recent years, several unmet clinical needs remain, including resistance to AEDs in about 30 % of patients with epilepsy, adverse effects of AEDs that can reduce quality of life, and the lack of treatments that can prevent development of epilepsy in patients at risk. Animal models of seizures and epilepsy have been instrumental in the discovery and preclinical development of novel AEDs, but obviously the previously used models have failed to identify drugs that address unmet medical needs. Thus, we urgently need fresh ideas for improving preclinical AED development. In this review, a number of promising models will be described, including the use of simple vertebrates such as zebrafish (*Danio rerio*), large animal models such as the dog and newly characterized rodent models of pharmacoresistant epilepsy. While these strategies, like any animal model approach also have their limitations, they offer hope that new more effective AEDs will be identified in the coming years.

Keywords

Zebrafish • Epileptic dogs • Epileptic rodents • Pharmacoresistant epilepsy • Antiepileptic drugs • Epilepsy syndromes

W. Löscher (✉)
Department of Pharmacology,
Toxicology and Pharmacy,
University of Veterinary Medicine,
Hannover 30559, Germany

Center for Systems Neuroscience,
Hannover 30559, Germany
e-mail: Wolfgang.Loescher@tiho-hannover.de

S.C. Baraban (✉)
Epilepsy Research Laboratory,
Department of Neurological Surgery,
University of California,
San Francisco, CA 94143, USA
e-mail: Scott.Baraban@ucsf.edu

23.1 Introduction

Rodent models of seizures and epilepsy have played a fundamental role in advancing our understanding of basic mechanisms underlying ictogenesis and epileptogenesis. They have also been instrumental in the discovery and preclinical development of novel antiepileptic drugs (AEDs) [12]. Indeed, animal models with a similarly high predictive value do not exist for other neurological disorders, such as bipolar disease or migraine [62]. Despite the availability of predictive rodent models, at least 30 % of epilepsy patients are not controlled by currently available AEDs. One reason is that, with few exceptions, most AED candidates were identified in simple evoked seizure models in otherwise healthy rodents such as the maximal electroshock seizure (MES) or acute pentylenetetrazole (PTZ; metrazol) tests [48]. In these traditional models, in use since the 1940s, successful AED treatments suppress acute seizure events, but effects on drug-resistant seizure events or chronic spontaneous seizures are not routinely evaluated. Thus, we urgently need fresh ideas for improving preclinical AED development. Here, a number of promising models will be described, including the use of simple vertebrates such as zebrafish (*Danio rerio*), large animal models such as the dog and newly characterized rodent models of pharmacoresistant epilepsy. We will not discuss *in vitro* brain slice models or neurons derived from patients using induced pluripotent stem cell technology, because the network complexity of the brain and its alterations by seizure activity are difficult to recapitulate in the dish.

23.2 Zebrafish-Based Approaches to Epilepsy and Drug Discovery

Traditionally used as a model organism to study vertebrate development and embryogenesis, zebrafish only recently emerged as an important model for epilepsy research [5, 17, 27, 29, 53, 65, 70]. The rapid ex vivo development, genetic tractability and transparency of larval zebrafish make them ideally suited to these types of studies (Fig. 23.1). Because zebrafish are vertebrates with a fairly complex nervous system [2, 21, 61] recording electroencephalographic activity is also possible [7], and with exposure to standard convulsant manipulations (e.g., PTZ, pilocarpine, 4-aminopyridine, heat) abnormal electrical discharge with brief high-frequency small amplitude (interictal-like) and longer duration, complex multi-spike large amplitude (ictal-like) events can be readily observed. Sophisticated imaging approaches, taking advantage of the transparency of larval zebrafish and genetic modification to express calcium or bioluminescence indicators, provide additional evidence that central nervous system (CNS)-generated seizure-like activity is robust in response to PTZ. This is an important advantage of zebrafish as a model organism for epilepsy research as CNS-generated abnormal electrical events are often considered a hallmark feature of this disease. In the original description of the acute PTZ seizure model in wild-type zebrafish at 6 or 7 days post-fertilization (dpf), Baraban et al. [5] provided a framework for characterizing epilepsy in zebrafish: (i) evidence for seizure-induced gene (*c-Fos*) expression, (ii) a scoring system for seizure-like behaviours, (iii) electrophysiological examples of abnormal electrographic burst discharge and (iv) sensitivity to common AEDs (valproate, ethosuximide, carbamazepine, phenytoin, phenobarbital and diazepam). As expected from similar PTZ testing in rodents [71], valproate and diazepam were the most effective at inhibiting electrographic seizure events with approximate ED_{50}s of 1 mM and 5 µM, respectively. Using this same model, Berghmans et al. [11] extended this dataset to include 14 standard AEDs. These follow-up experiments used an assay where wild-type larvae were "incubated" in a test compound for 24 h prior to acute PTZ administration and monitoring of seizure-like behaviour exclusively in a locomotion-based tracking assay. These studies confirmed the results of Baraban et al. [5] but also highlight the limitations of a behaviour-only assay as two drugs that failed to alter electrographic burst discharge amplitude (ethosuximide

Fig. 23.1 Schematic illustration of the zebrafish assay

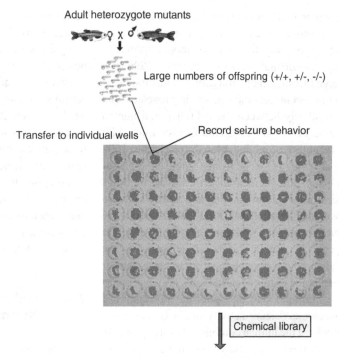

Adult heterozygote mutants

Large numbers of offspring (+/+, +/-, -/-)

Transfer to individual wells   Record seizure behavior

Chemical library

Phenotype-based screen e.g., inhibition of seizure behavior

and carbamazepine) were identified as "anticonvulsant" as measured by a reduction in swim activity. A likely explanation is that overnight exposure to these AEDs was either toxic or sedative to developing zebrafish, as both possibilities would appear as suppressed locomotion in motion-based tracking assay. More recently, Afrikanova et al. [1] revisited this overnight exposure-PTZ challenge assay and evaluated a similar list of 13 AEDs using a combination of locomotion tracking followed by electrophysiology on agar-immobilized larvae. These latter studies aligned most closely with the original PTZ findings, identifying valproate and diazepam, while also showing that ethosuximide altered burst frequency but not amplitude. Maximum-tolerated drug concentrations were studied in both papers highlighting an additional advantage of the zebrafish platform for simultaneous in vivo evaluation of drug toxicities e.g., one of the primary reasons that most compounds identified in preclinical trials ultimately fail to reach the clinic. In a recent paper by Baxendale et al. [10] also using PTZ, a high-throughput screen of a ~2,000 bioactive

small molecule library was reported. These studies used a first-pass assay based on increased *c-Fos* mRNA expression (as measured by in wholemount situ hybridization) following PTZ exposure at two dpf and a secondary locomotion-based assay at four dpf for additional concentration-response studies. Unfortunately, it is unclear whether the 46 compounds identified using this approach are antiepileptic as previous studies indicate the earliest possible developmental stage where confirmed electrographic seizures could be observed in zebrafish larvae is three dpf [6, 27]. Before this age, larvae are still in chorion and do not swim freely. Furthermore, these non-physiological assays should be interpreted with caution as the Baxendale et al. [10] study identified several candidate compounds with known neurotoxicity profiles e.g., lindane, rotenonic acid, deguelin, endrin and propanil.

Although seizures can be easily induced, drug discovery using acute seizure models, even in zebrafish, are prone to the same limitations as in rodents. Namely, these approaches use healthy animals, the seizure-events are acute and evoked using potentially non-physiological stimuli such as a stimulation electrode or convulsants, and most importantly they do not model spontaneously occurring seizure events. Zebrafish diverged from humans roughly 450 million years ago but recent genome sequencing revealed that the similarity between the zebrafish and human genome is ~70 % [28]. This fact, coupled with the fecundity of adult zebrafish (producing 100–200 offspring per week from a single adult breeding pair), the permeability of larvae to drugs placed in the bathing media, and ability to thrive in volumes as small as 100 μl make zebrafish an attractive model for a drug discovery program targeted to genetic forms of epilepsy. In the Baraban laboratory, we have focused on zebrafish designed to mimic monogenic epilepsy disorders of childhood as they offer the advantages of spontaneous seizure activity and a genetic basis mimicking the human condition. In this approach, one can model specific forms of pediatric epilepsy – Type I Lissencephaly (*Lis1*), Angelman syndrome (*Ube3A*), Tuberous Sclerosis Complex (*Tsc*) or Dravet syndrome for example (*Scn1a*) – then design drug screening programs targeted to that patient population. In some cases these are stable mutations carried in the zebrafish germline, where other models involve acute antisense knockdown of gene expression in immature zebrafish. Thus, a form of "personalized medicine" aimed at identifying new therapeutic options for relatively rare, but catastrophic, forms of epilepsy. Our recent studies are based on a two-stage screening process. First, zebrafish mutants are placed in individual wells and behaviour (locomotion) is tracked using a 96-well format. Once a baseline level of spontaneous seizure activity is established a test compound is added, and then a second locomotion assay is performed to evaluate the effect on seizure behaviour (with distance travelled and mean velocity of swim movement used as surrogate markers) [5, 16]. As freely behaving larvae can simultaneously be observed for heart rate, edoema or touch-sensitivity, in vivo toxicity is also determined with this strategy. Using a 96-well format it is relatively easy to power this research for statistical analysis and multiple drug concentrations can

be assessed in a given plate. The same fish can subsequently be used for electrophysiological analysis, which allows a determination of "false positives" in the locomotion assay that are lethal, sedative or paralyzing. With even a modest zebrafish facility, this approach can easily be used to screen 20–50 drugs per week. The disadvantage of this strategy is that it is not well-suited to acquired forms of epilepsy that develop more slowly over time or in the adult nervous system, or compounds that are not easily dissolved in embryo media. It is also difficult to directly translate concentrations that are effective via bath application in larval zebrafish to those that may be useful clinically in humans.

23.3   Rodent Models of Pharmacoresistant Seizures

The concept of developing rodent seizure or epilepsy models that do not respond to clinically approved AEDs and then using such models for the discovery of novel more effective AEDs is not new but, to our knowledge, was first proposed by Löscher in 1986 [38]. Since then, several models of pharmacoresistant seizures have been developed, including the phenytoin-resistant kindled rat [40], the lamotrigine-resistant kindled rat [68], and the phenobarbital-resistant epileptic rat [14]. In all these models, resistance to one AED extends to other AEDs (cf., [49]), thus fulfilling the criterion of pharmacoresistant epilepsy [32]. By using two of these models, Löscher and colleagues described several factors that differentiated AED-resistant from AED-responsive rats, including the extent of neurodegeneration in the hippocampus, genetic factors, AED target alterations, alterations in drug efflux transporters, and intrinsic severity of the epilepsy as a determinant of AED refractoriness [49]. Similar factors have been described for AED-resistant human epilepsy, so that the rat models obviously reflect clinically important mechanisms of refractoriness. The next logical step was to use such models for new treatment discovery. One example here is that inhibiting the drug efflux transporter P-glycoprotein (Pgp), which is increased at

Table 23.1 A comparison of elimination half-lives of antiepileptic drugs in humans, dogs and rats

AED	Half-life (h)		
	Human	Dog	Rat
Carbamazepine	25–50[a,b]	1–2[a,b]	1.2–3.5[a]
Clobazam	16–50	~1.5	1
Clonazepam	18–50	1–3	?
Diazepam	24–72[a] (DMD=40–130)	1–5[a] (DMD=4)	1.4[a] (DMD=1.1)
Ethosuximide	40–60	11–25	10–16
Felbamate	14–22	4–8	2–17[c]
Gabapentin	5–7	3–4	2–3
Lacosamide	13	2–2.5	3
Lamotrigine	21–50	2–5	12 to >30
Levetiracetam	6–11	4–5	2–3
Oxcarbazepine	1–2.5[a] (MHD=8–14)	~4[a] (MHD=3–4)	?[a] (MHD=0.7–4)
Perampanel	70	5	2
Phenobarbital	70–100[b]	25–90[b]	9–20[b]
Phenytoin	15–20[b,c]	2–6[b,c]	~1–8[b,c]
Potassium bromide	~300	~600	72–192
Pregabalin	6	6–7	2.5
Primidone	6–12[a] (PB=70–100)	4–12[a,b] (PB=25–90)	5[a] (PB=9–20)
Tiagabin	5–8	1–2	1
Topiramate	20–30	3–4	2–5
Valproate	8–15[a]	1–3[a]	~1–5[a,c]
Vigabatrin	5–7[d]	?[d]	~1[d]
Zonisamide	60–70	~15	8

Data are from previous reviews of Löscher [44, 46] and have been revised and updated for the present study. Note that rats and dogs eliminate most AEDs more rapidly than humans, which has to be considered when using such drugs for chronic studies in experimental animals

DMD desmethyldiazepam, *MHD* monohydroxy derivative, *PB* phenobarbital, ? indicates that no published data were found

[a]Active metabolites; [b]shortens on continuing exposure to the drug (because of enzyme induction); [c]non-linear kinetics (half-life increases with dose); [d]duration of action independent of half-life because of irreversible inhibition of GABA degradation

the blood–brain barrier of AED-resistant rats, counteracted resistance to phenobarbital in epileptic rats [15]. The increased Pgp functionality in epileptic rats can be visualized in vivo by positron emission tomography [4]. By using Pgp imaging, Feldmann et al. [19] demonstrated that about 40 % of AED-resistant patients exhibit increased brain functionality of Pgp and could potentially benefit from Pgp inhibition. This example illustrates that chronic rodent models of pharmacoresistant seizures are helpful to discover new strategies for treatment of medically intractable epilepsy.

The disadvantage of the described chronic epilepsy models is that they are not suited for large-scale testing of novel compounds but rather

for evaluation of selected treatment strategies as illustrated by the example of Pgp inhibition. Kindling models such as the phenytoin-resistant kindled rat [40] or the lamotrigine-resistant kindled rat [68] have the advantage that seizures can be induced at will, so that chronic drug administration is not needed, whereas models with spontaneous recurrent seizures (SRS) such as the phenobarbital-resistant epileptic rat [14] necessitate continuous (24/7) EEG/video recording for assessing drug efficacy. When testing drug effects on SRS in such rat models, the rapid elimination of most drugs, including AEDs, in rats (Table 23.1) necessitates the use of an adequate dosing regimen during prolonged drug administration to avoid false negative results [46]. The same is true when administering potential antiepileptogenic drugs in the latent period following epileptogenic brain insults in rats [46]. Mice developing SRS after intrahippocampal injection of kainate have been proposed as a model of pharmacoresistant seizures; these mice have the advantage that the frequency of SRS is so high that drug efficacy can be determined after single dose administration [54, 66]. However, as yet this model has only rarely been used for investigating the antiepileptic efficacy of novel compounds [54].

Based on the logistical problems associated with drug testing in chronic models, models such as the zebrafish or acute rodent seizure models are indispensable when testing large numbers of investigational compounds before evaluating the most interesting compounds in chronic models. One of these acute seizure models, the 6-Hz model of partial seizures in mice, was initially proposed to provide a useful model of therapy-resistant limbic seizures [9], but more recent studies have not confirmed this idea [49]. Rather, the 6-Hz model is a valuable part of a preclinical test battery to further differentiate compounds. Also, a more recent genetic mouse model of Dravet syndrome, in which clinical symptoms of this syndrome occur after Scn1a heterozygous knockout, may be an interesting possibility for testing drugs or drug combinations for treatment of as yet pharmacoresistant types of seizures [59, 60]. Furthermore, a zebrafish Scn1a mutant, such as the one recently described by the Baraban laboratory [8] would be an efficient first pass high-throughput approach to identify potential candidate compounds that can be further investigated in chronic rodent models of pharmacoresistant seizures.

23.4 Naturally Occurring Epilepsy in Dogs as a Translational Model

The dog is an important large animal model in various fields of biomedical research and fills a crucial step in the translation of basic research to new treatment regimens. For instance, because of the relative large body size of dogs and many similarities in physiology and pharmacology between dogs and humans, scaling doses from dogs to humans is much easier than using rodents in selecting doses for clinical trials in humans. To our knowledge, Löscher et al. [37] were the first to propose naturally occurring canine epilepsy as a translational model of human epilepsy. The prevalence and phenomenology of epilepsy in dogs are very similar to human epilepsy. Indeed, epilepsy is the most common chronic neurological disease in dogs, affecting about 0.6–1 % of the dog population [64, 69]. Furthermore, causes of canine epilepsy are similar to those in humans (Fig. 23.2) except that cerebrovascular disease does not play any significant role, because it is rare in dogs [69]. About 50 % of dogs with partial and generalized convulsive seizures are not controlled by treatment with AEDs, so that epileptic dogs have been proposed as a valuable model of pharmacoresistant epilepsy that can be used to unravel mechanisms of resistance and evaluate new strategies for treatment [44, 64]. However, clinical trials on new AEDs in epileptic dogs are as laborious and time-consuming as clinal trials in human patients, necessitating randomized trial designs in which the new drug is compared with either placebo or a standard comparator [57, 58]. Recently, different treatments, including AEDs, vagal stimulation, and ketogenic diet were compared with placebo in epileptic dogs, and an unexpectedly high placebo rate was found, which was similar to that known from controlled clinical trials in humans with epilepsy [57, 58]. In contrast to humans, the placebo effect has been largely disregarded in veterinary medicine. In humans, a placebo response seems to require a recognition by the patient of the intent of treatment efforts. Because it is generally presumed that animals lack certain cognitive capacities, e.g. the ability to comprehend the intent of the veterinarian's manipulations, the power of suggestion, and expectations of recovery and healing, the existence of a placebo effect in animals seems counterintuitive [55]. However, in veterinary studies, the placebo response may be a result of expectations of the pet owner regarding treatment in studies as those conducted by Munana et al.

Fig. 23.2 A comparison of the presumed causes of recurrent epileptic seizures in humans and dogs. The graph on humans illustrate the proportion of incidence cases of epilepsy by etiology in Rochester, Minnesota, U.S.A., 1935–1984 [24]; a similar graph was initially shown by Lowenstein [35]. The graph on dogs illustrates data from a recent epidemiologic study on canine epilepsy [69]

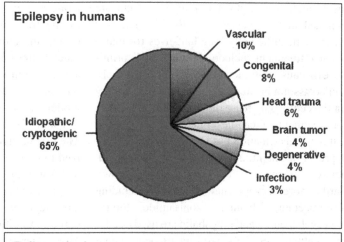

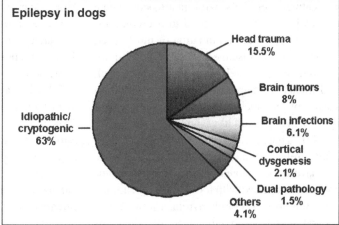

[57, 58] in epileptic dogs, where the owners are responsible for administration of treatment and outcome measures (i.e., seizure frequency) are derived solely from owner observations. Other factors that may be included in placebo responses in veterinary studies include regression to the mean, investigator bias, client bias, the potential for a higher level of care during the study, and improved adherence to treatment with active medication that is being administered in addition to the placebo during the study (for details see [57]). Furthermore, the placebo response can be because of effects of placebo administration on the animal, which is well documented in laboratory animals and may involve conditioned responses among others [55]. As a consequence, studies on new treatments in laboratory animals (or pets) should always include a "placebo" group receiv-

ing all manipulations (e.g., handling, injections, electrode implantation, seizure recording etc.) that are used for the new treatment.

In addition to chronic epilepsy, naturally occurring canine status epilepticus (SE) has been proposed as a translational platform for evaluating investigational compounds for eventual use in human trials [34] and a controlled study on i.v. levetiracetam for treatment of SE in dogs has been published recently [23].

One important caveat that has to be considered when using dogs for long-term studies on AEDs is that dogs, similar to rodents, eliminate many drugs, including most AEDs, much more rapidly than humans (Table 23.1). Thus, when using AEDs such as phenytoin, carbamazepine or valproate with too low half-lifes for maintenance treatment in epileptic dogs, no sufficient drug levels and, hence, no antiepileptic effects are obtained in this species [20, 36, 37]. The few AEDs with sufficiently long half-lives for maintenance treatment include phenobarbital, primidone (because of its metabolism to phenobarbital), and potassium bromide, which is the reason why until recently only these old drugs were approved for treatment of canine epilepsy in the US or Europe. This situation has changed by the recent approval of imepitoin for treatment of dogs with newly diagnosed epilepsy (see below). Furthermore, several newer AEDs, including levetiracetam, felbamate, zonisamide, topiramate, gabapentin, and pregabalin are used as add-on treatment in dogs with pharmacoresistant seizures [64]. It has been tried to overcome the problem of too rapid elimination of most AEDs by dogs by using sustained-release formulations; however, sustained-release preparations developed for use in humans are not suited for dogs because of the much higher gastrointestinal passage rate in dogs (~24 h) vs. humans (~65–100 h) [36, 44]. Thus, AED formulations that exhibit retarded release of the drug in the gastrointestinal tract have to be adapted to the dog to overcome problems associated with too rapid drug elimination in this species. For phenytoin, a slow-release preparation has been developed for dogs, by which therapeutic plasma levels could be maintained despite the rapid elimination of this drug in dogs [18], but, to our knowledge, no clinical experience with this preparation has been published. Vigabatrin has been evaluated for control of epilepsy in dogs, because its mechanism of action (irreversible inhibition of GABA degradation) allows an effective treatment which should be independent of species differences in drug elimination. Vigabatrin proved to be effective in epileptic dogs with phenobarbital-resistant seizures, but at least in part vigabatrin had to be withdrawn because of development of severe adverse effects, such as haemolytic anaemia [67].

Löscher's group has used dogs as a translational model over the recent 25 years in the development of a new category of AEDs, i.e., drugs that act as partial agonists at the benzodiazepine (BZD) site of the $GABA_A$ receptor. Such drugs have the wide spectrum of antiepileptic activity against diverse types of seizures as the traditional full BZD agonists such as diazepam, clonazepam or clobazam, but are much better tolerated and lack the tolerance and abuse liability of the full agonists [22, 41]. In our studies, we either used a canine seizure model, in which seizures are induced by i.v. infusion of pentylenetetrazole, or epileptic dogs. The first partial BZD agonist that was characterized in dogs (and compared with full BZD agonists) was the β-carboline abecarnil, providing proof-of-concept that partial BZD agonists are advantageous for treatment of seizures compared to traditional, full-agonist BZDs [39, 41]. More recently, the low-affinity partial BZD agonist imepitoin, an imidazolin derivative, was evaluated in the dog seizure model and epileptic dogs and reported to provide efficacious antiepileptic activity without the known disadvantages of full BZD agonists [45, 51]. Based on several randomized controlled clinical trials in epileptic dogs, imepitoin was recently approved in Europe for treatment of canine epilepsy [13, 51]. That imepitoin is an effective and safe AED in epileptic dogs indicates that low-affinity partial BZD agonists may offer a new mechanistic category of useful AEDs.

23.5 Network Approaches for Development of Novel Treatments

Several of the models described in this review may be particularly interesting for evaluating a novel strategy of AED development, the network approach [3, 26, 50]. One of the dominant strategies in drug discovery is designing maximally selective ligands to act on individual drug targets [26]. However, many effective drugs act via modulation of multiple targets rather than single proteins. Furthermore, most epilepsies develop not from alterations of a single target but rather from complex alterations resulting in an epileptic network in the brain. The only existing cure of epilepsy is resective surgery in which the regional epileptic network or part of this network is removed. Thus, treatments focusing exclusively on a single protein or individual biochemical pathway may be less effective than treatments targeting different proteins or pathways involved in the network. The latter approach has been recently termed "network pharmacology" and relates to principles of systems biology [3, 26]. The principle of network pharmacology is to develop combinations of existing drugs, which regulate activity via different targets within a biological network, for diseases that do not sufficiently respond to single drug treatment or for which no treatment exists. Integrating network biology and polypharmacology holds the promise of expanding the current opportunity space for druggable targets [26]. However, the rational design of polypharmacology faces considerable challenges in the need for new methods to validate target combinations and optimize multiple structure-activity relationships while maintaining drug-like properties. The advances in zebrafish chemical screening technologies may allow rapid identification of the most interesting drug combinations resulting from network approaches, followed by evaluating these combinations in chronic models of epilepsy.

Some examples for interesting network approaches include combinations of glutamate receptor antagonists that target different glutamate receptor subtypes. We reported that extremely low doses of the NMDA (N-methyl-D-aspartate) receptor antagonist MK-801 (dizocilpine) markedly potentiated the anticonvulsant effect the AMPA (alpha-amino-3-hydroxy-5-methyl-4-isoxazolepropionic acid) receptor antagonist NBQX (2,3-dihydroxy-6-nitro-7-sulfamoylbenzo(F)quinoxaline) without increasing its adverse effects [42]. Similar over-additive effects were seen when NBQX was combined with the competitive NMDA antagonist CGP39551 (the carboxyethylester of DL-(E)-2-amino-4-methyl-5-phosphono-3-pentenoic acid) or the low-affinity, rapidly channel blocking NMDA receptor antagonist memantine [42, 43]. We are currently evaluating combinations of clinically approved NMDA antagonists (ketamine, memantine) and the novel AMPA antagonist perampanel in models of difficult-to-treat seizures. Another interesting example is the combination

of phenobarbital with the diuretic bumetanide, which is currently evaluated clinically following promising preclinical data [31, 52]. The biologically plausible idea behind this combination is that a shift from inhibitory to excitatory GABA may be involved in difficult-to-treat neonatal and adult seizures [30, 56]. GABA-mediated excitation has been observed when expression of the chloride importer NKCC1 is higher than expression of the chloride exporter KCC2; e.g., early during development and in the hippocampus of adults with temporal lobe epilepsy [30, 56]. Bumetanide inhibits the neuronal chloride cotransporter NKCC1, thereby reverts the GABA shift and enables GABAmimetic drugs such as phenobarbital to potentiate inhibitory GABAergic transmission [52]. This recent work builds on an earlier demonstration from the Schwartzkroin laboratory that furosemide, another chloride cotransporter inhibitor, exhibits powerful anticonvulsant activity across a range of *in vitro* and in vivo seizure models [25]. Further examples for interesting network approaches include combined targeting of different inflammatory pathways, which are involved in seizure generation [33]. These examples strongly indicate that combinatorial treatment strategies offer new options for epilepsy therapy.

23.6   Conclusions

Models for the discovery of drugs with antiepileptic activity have traditionally relied on a relatively small number of acute seizure models employed in otherwise healthy rodents. While useful in the discovery of most drugs currently available in the clinic, more resistant types of epilepsies including temporal lobe epilepsy patients who are unresponsive to available AEDs and catastrophic, often genetically-based, types of epilepsies seen in children necessitate alternative drug discovery strategies. Zebrafish, canine and novel rodent approaches are described here and offer several unique advantages over these traditional models. While these strategies, like any animal model approach also have their limitations, they offer hope that new classes of

AEDs will be identified in the coming years. Furthermore, animal models in which epilepsy develops after brain insults or gene mutations are essential in the search for novel antiepileptogenic treatments that prevent or modify the development of epilepsy in patients at risk [47, 63]. Previously, this field was dominated by studies in SE models in rats, although SE is only rarely a cause of symptomatic epilepsy [47]. Thus, models of more common causes of acquired epilepsy, such as traumatic brain injury, and models in which epilepsy develops after gene mutations should be used more extensively in research on antiepileptogenesis. We have started to use the zebrafish and canine approaches to identify molecular pathways that may be involved in the epileptogenic process and may offer new targets for antiepileptogenic treatments.

Acknowledgements With gratitude and special thanks to Scott Baraban's postdoctoral mentor Phil Schwartzkroin. Scott's years in Seattle were rich in scientific interactions and opportunities. The environment created by Phil and Scott's fellow trainees (Daryl Hochman, Jim Owens, Catherine Woolley and Jurgen Wenzel) was conducive to open discussion, lively debate and exciting discoveries. Phil's scholarly approach to science and passion for epilepsy research was a guiding force in Scott's career. With the laboratory at UCSF, Scott strives to carry on some of these same principles. Wolfgang Löscher acknowledges the many thoughtful and constructive discussions with Phil that he had as an author of invited reviews in Epilepsia during the many years that Phil acted as a Managing Editor for this journal.

Other Acknowledgements Scott Baraban acknowledges funding from the National Institutes of Health, Citizens United for Research in Epilepsy and Dravet Syndrome Foundation, and Wolfgang Löscher funding from the German Research Foundation, the FP7 program of the European Commission and the National Institutes of Health.

References

1. Afrikanova T, Serruys AS, Buenafe OE, Clinckers R, Smolders I, de Witte PA et al (2013) Validation of the zebrafish pentylenetetrazol seizure model: locomotor versus electrographic responses to antiepileptic drugs. PLoS One 8(1):e54166
2. Ahrens MB, Li JM, Orger MB, Robson DN, Schier AF, Engert F et al (2012) Brain-wide neuronal dynamics during motor adaptation in zebrafish. Nature 485(7399):471–477
3. Ainsworth C (2011) Networking for new drugs. Nat Med 17(10):1166–1168
4. Bankstahl JP, Bankstahl M, Kuntner C, Stanek J, Wanek T, Meier M et al (2011) A novel positron emission tomography imaging protocol identifies seizure-induced regional overactivity of P-glycoprotein at the blood-brain barrier. J Neurosci 31(24):8803–8811
5. Baraban SC, Taylor MR, Castro PA, Baier H (2005) Pentylenetetrazole induced changes in zebrafish behavior, neural activity and c-fos expression. Neuroscience 131(3):759–768
6. Baraban SC, Dinday MT, Castro PA, Chege S, Guyenet S, Taylor MR (2007) A large-scale mutagenesis screen to identify seizure-resistant zebrafish. Epilepsia 48(6):1151–1157
7. Baraban SC (2013) Forebrain electrophysiological recording in larval zebrafish. J Vis Exp (71). pii: 50104
8. Baraban SC, Dinday MT, Hortopan GA (2013) Drug screening and transcriptomic analysis in Scn1a zebrafish mutants identifies potential lead compound for Dravet Syndrome. Nat Commun 4:2410
9. Barton ME, Klein BD, Wolf HH, White HS (2001) Pharmacological characterization of the 6 Hz psychomotor seizure model of partial epilepsy. Epilepsy Res 47:217–228
10. Baxendale S, Holdsworth CJ, Meza Santoscoy PL, Harrison MR, Fox J, Parkin CA et al (2012) Identification of compounds with anti-convulsant properties in a zebrafish model of epileptic seizures. Dis Model Mech 5(6):773–784
11. Berghmans S, Hunt J, Roach A, Goldsmith P (2007) Zebrafish offer the potential for a primary screen to identify a wide variety of potential anticonvulsants. Epilepsy Res 75(1):18–28
12. Bialer M, White HS (2010) Key factors in the discovery and development of new antiepileptic drugs. Nat Rev Drug Discov 9(1):68–82
13. Bialer M, Johannessen SI, Levy RH, Perucca E, Tomson T, White HS (2013) Progress report on new antiepileptic drugs: a summary of the Eleventh Eilat Conference (EILAT XI). Epilepsy Res 103(1):2–30
14. Brandt C, Volk HA, Löscher W (2004) Striking differences in individual anticonvulsant response to phenobarbital in rats with spontaneous seizures after status epilepticus. Epilepsia 45:1488–1497
15. Brandt C, Bethmann K, Gastens AM, Löscher W (2006) The multidrug transporter hypothesis of drug resistance in epilepsy: proof-of-principle in a rat model of temporal lobe epilepsy. Neurobiol Dis 24:202–211
16. Cario CL, Farrell TC, Milanese C, Burton EA (2011) Automated measurement of zebrafish larval movement. J Physiol 589(Pt 15):3703–3708
17. Chege SW, Hortopan GA, Dinday T, Baraban SC (2012) Expression and function of KCNQ channels in larval zebrafish. Dev Neurobiol 72(2):186–198
18. Derkx-Overduin LM (1994) Slow-release phenytoin in canine epilepsy. Thesis, Faculty of Veterinary Medicine, Utrecht, the Netherlands

19. Feldmann M, Asselin MC, Liu J, Wang S, McMahon A, Anton-Rodriguez J et al (2013) P-glycoprotein expression and function in patients with temporal lobe epilepsy: a case-control study. Lancet Neurol 12:777–785

20. Frey H-H, Löscher W (1985) Pharmacokinetics of anti-epileptic drugs in the dog: a review. J Vet Pharmacol Ther 8:219–233

21. Friedrich RW, Genoud C, Wanner AA (2013) Analyzing the structure and function of neuronal circuits in zebrafish. Front Neural Circuits 7:71

22. Haefely W, Facklam M, Schoch P, Martin JR, Bonetti EP, Moreau JL et al (1992) Partial agonists of benzodiazepine receptors for the treatment of epilepsy, sleep, and anxiety disorders. Adv Biochem Psychopharmacol 47:379–394

23. Hardy BT, Patterson EE, Cloyd JM, Hardy RM, Leppik IE (2012) Double-masked, placebo-controlled study of intravenous levetiracetam for the treatment of status epilepticus and acute repetitive seizures in dogs. J Vet Intern Med 26:334–340

24. Hauser WA, Annegers JF, Kurland LT (1993) Incidence of epilepsy and unprovoked seizures in Rochester, Minnesota: 1935–1984. Epilepsia 34:453–468

25. Hochman DW, Baraban SC, Owens JW, Schwartzkroin PA (1995) Dissociation of synchronization and excitability in furosemide blockade of epileptiform activity. Science 270:99–102

26. Hopkins AL (2008) Network pharmacology: the next paradigm in drug discovery. Nat Chem Biol 4(11): 682–690

27. Hortopan GA, Dinday MT, Baraban SC (2010) Spontaneous seizures and altered gene expression in GABA signaling pathways in a mind bomb mutant zebrafish. J Neurosci 30(41):13718–13728

28. Howe K, Clark MD, Torroja CF, Torrance J, Berthelot C, Muffato M et al (2013) The zebrafish reference genome sequence and its relationship to the human genome. Nature 496(7446):498–503

29. Hunt RF, Hortopan GA, Gillespie A, Baraban SC (2012) A novel zebrafish model of hyperthermia-induced seizures reveals a role for TRPV4 channels and NMDA-type glutamate receptors. Exp Neurol 237(1):199–206

30. Kahle KT, Staley KJ, Nahed BV, Gamba G, Hebert SC, Lifton RP et al (2008) Roles of the cation-chloride cotransporters in neurological disease. Nat Clin Pract Neurol 4(9):490–503

31. Kahle KT, Staley KJ (2008) The bumetanide-sensitive Na-K-2Cl cotransporter NKCC1 as a potential target of a novel mechanism-based treatment strategy for neonatal seizures. Neurosurg Focus 25(3):1–8

32. Kwan P, Arzimanoglou A, Berg AT, Brodie MJ, Allen HW, Mathern G et al (2010) Definition of drug resistant epilepsy: consensus proposal by the ad hoc Task Force of the ILAE Commission on Therapeutic Strategies. Epilepsia 51(6):1069–1077

33. Kwon YS, Pineda E, Auvin S, Shin D, Mazarati A, Sankar R (2013) Neuroprotective and antiepileptogenic effects of combination of anti-inflammatory drugs in the immature brain. J Neuroinflammation 10:30

34. Leppik IE, Patterson EN, Coles LD, Craft EM, Cloyd JC (2011) Canine status epilepticus: a translational platform for human therapeutic trials. Epilepsia 52(Suppl 8):31–34

35. Lowenstein DH (2009) Epilepsy after head injury: an overview. Epilepsia 50(Suppl 2):4–9

36. Löscher W (1981) Plasma levels of valproic acid and its metabolites during continued treatment in dogs. J Vet Pharmacol Ther 4:111–119

37. Löscher W, Schwartz-Porsche D, Frey H-H, Schmidt D (1985) Evaluation of epileptic dogs as an animal model of human epilepsy. Arzneim-Forsch (Drug Res) 35:82–87

38. Löscher W (1986) Experimental models for intractable epilepsy in nonprimate animal species. In: Schmidt D, Morselli PL (eds) Intractable epilepsy: experimental and clinical aspects. Raven Press, New York, pp 25–37

39. Löscher W, Hönack D, Scherkl R, Hashem A, Frey H-H (1990) Pharmacokinetics, anticonvulsant efficacy and adverse effects of the β-carboline abecarnil, a novel ligand for benzodiazepine receptors, after acute and chronic administration in dogs. J Pharmacol Exp Ther 255:541–548

40. Löscher W, Rundfeldt C (1991) Kindling as a model of drug-resistant partial epilepsy: selection of phenytoin-resistant and nonresistant rats. J Pharmacol Exp Ther 258:483–489

41. Löscher W (1993) Abecarnil shows reduced tolerance development and dependence potential in comparison to diazepam: animal studies. In: Stephens DN (ed) Anxiolytic β-carbolines. From molecular biology to the clinic. Springer, Berlin, pp 96–112

42. Löscher W, Rundfeldt C, Hönack D (1993) Low doses of NMDA receptor antagonists synergistically increase the anticonvulsant effect of the AMPA receptor antagonist NBQX in the kindling model of epilepsy. Eur J Neurosci 5:1545–1550

43. Löscher W, Hönack D (1994) Over-additive anticonvulsant effect of memantine and NBQX in kindled rats. Eur J Pharmacol 259:R3–R5

44. Löscher W (1997) Animal models of intractable epilepsy. Prog Neurobiol 53:239–258

45. Löscher W, Potschka H, Rieck S, Tipold A, Rundfeldt C (2004) Anticonvulsant efficacy of the low-affinity partial benzodiazepine receptor agonist ELB 138 in a dog seizure model and in epileptic dogs with spontaneously recurrent seizures. Epilepsia 45(10):1228–1239

46. Löscher W (2007) The pharmacokinetics of antiepileptic drugs in rats: consequences for maintaining effective drug levels during prolonged drug administration in rat models of epilepsy. Epilepsia 48:1245–1258

47. Löscher W, Brandt C (2010) Prevention or modification of epileptogenesis after brain insults: experimental approaches and translational research. Pharmacol Rev 62:668–700

48. Löscher W, Schmidt D (2011) Modern antiepileptic drug development has failed to deliver: ways out of the current dilemma. Epilepsia 52(4):657–678

49. Löscher W (2011) Critical review of current animal models of seizures and epilepsy used in the discovery and development of new antiepileptic drugs. Seizure 20:359–368

50. Löscher W, Klitgaard H, Twyman RE, Schmidt D (2013) New avenues for antiepileptic drug discovery and development. Nat Rev Drug Discov 12:757–776

51. Löscher W, Hoffmann K, Twele F, Potschka H, Töllner K (2013) The novel antiepileptic drug imepitoin compares favourably to other GABA-mimetic drugs in a seizure threshold model in mice and dogs. Pharmacol Res 77:39–46

52. Löscher W, Puskarjov M, Kaila K (2013) Cation-chloride cotransporters NKCC1 and KCC2 as potential targets for novel antiepileptic and antiepileptogenic treatments. Neuropharmacology 69:62–74

53. Mahmood F, Mozere M, Zdebik AA, Stanescu HC, Tobin J, Beales PL et al (2013) Generation and validation of a zebrafish model of EAST (epilepsy, ataxia, sensorineural deafness and tubulopathy) syndrome. Dis Model Mech 6(3):652–660

54. Maroso M, Balosso S, Ravizza T, Iori V, Wright CI, French J et al (2011) Interleukin-1beta biosynthesis inhibition reduces acute seizures and drug resistant chronic epileptic activity in mice. Neurotherapeutics 8(2):304–315

55. McMillan FD (1999) The placebo effect in animals. J Am Vet Med Assoc 215(7):992–999

56. Miles R, Blaesse P, Huberfeld G, Wittner L, Kaila K (2012) Chloride homeostasis and GABA signaling in temporal lobe epilepsy. In: Noebels JL, Avoli M, Rogawski MA, Olsen RW, Delgado-Escueta AV (eds) Jasper's basic mechanisms of the epilepsies, 4th edn. Oxford University Press, New York, pp 581–590

57. Munana KR, Zhang D, Patterson EE (2010) Placebo effect in canine epilepsy trials. J Vet Intern Med 24(1):166–170

58. Munana KR, Thomas WB, Inzana KD, Nettifee-Osborne JA, McLucas KJ, Olby NJ et al (2012) Evaluation of levetiracetam as adjunctive treatment for refractory canine epilepsy: a randomized, placebo-controlled, crossover trial. J Vet Intern Med 26:341–348

59. Oakley JC, Kalume F, Catterall WA (2011) Insights into pathophysiology and therapy from a mouse model of Dravet syndrome. Epilepsia 52(Suppl 2):59–61

60. Oakley JC, Cho AR, Cheah CS, Scheuer T, Catterall WA (2013) Synergistic GABA-enhancing therapy against seizures in a mouse model of Dravet syndrome. J Pharmacol Exp Ther 345(2):215–224

61. Panier T, Romano SA, Olive R, Pietri T, Sumbre G, Candelier R et al (2013) Fast functional imaging of multiple brain regions in intact zebrafish larvae using Selective Plane Illumination Microscopy. Front Neural Circuits 7:65

62. Perucca E, French J, Bialer M (2007) Development of new antiepileptic drugs: challenges, incentives, and recent advances. Lancet Neurol 6(9):793–804

63. Pitkänen A, Lukasiuk K (2011) Mechanisms of epileptogenesis and potential treatment targets. Lancet Neurol 10(2):173–186

64. Potschka H, Fischer A, von Rüden EL, Hülsmeyer V, Baumgärtner W (2013) Canine epilepsy as a translational model? Epilepsia 54(4):571–579

65. Ramirez IB, Pietka G, Jones DR, Divecha N, Alia A, Baraban SC et al (2012) Impaired neural development in a zebrafish model for Lowe syndrome. Hum Mol Genet 21(8):1744–1759

66. Riban V, Bouilleret V, Pham L, Fritschy JM, Marescaux C, Depaulis A (2002) Evolution of hippocampal epileptic activity during the development of hippocampal sclerosis in a mouse model of temporal lobe epilepsy. Neuroscience 112(1):101–111

67. Speciale J, Dayrell-Hart B, Steinberg SA (1991) Clinical evaluation of gamma-vinyl-gamma-aminobutyric acid for control of epilepsy in dogs. J Am Vet Med Assoc 198:995–1000

68. Srivastava AK, White HS (2013) Carbamazepine, but not valproate, displays pharmacoresistance in lamotrigine-resistant amygdala kindled rats. Epilepsy Res 104:26–34

69. Steinmetz S, Tipold A, Löscher W (2013) Epilepsy after head injury in dogs: a natural model of posttraumatic epilepsy. Epilepsia 54(4):580–588

70. Teng Y, Xie X, Walker S, Saxena M, Kozlowski DJ, Mumm JS et al (2011) Loss of zebrafish lgi1b leads to hydrocephalus and sensitization to pentylenetetrazol induced seizure-like behavior. PLoS One 6(9):e24596

71. Watanabe Y, Takechi K, Fujiwara A, Kamei C (2010) Effects of antiepileptics on behavioral and electroencephalographic seizure induced by pentetrazol in mice. J Pharmacol Sci 112(3):282–289

What Are the Arguments For and Against Rational Therapy for Epilepsy?

24

Melissa Barker-Haliski, Graeme J. Sills, and H. Steve White

Abstract

Although more than a dozen new anti-seizure drugs (ASDs) have entered the market since 1993, a substantial proportion of patients (~30 %) remain refractory to current treatments. Thus, a concerted effort to identify and develop new therapies that will help these patients continues. Until this effort succeeds, it is reasonable to re-assess the use of currently available therapies and to consider how these therapies might be utilized in a more efficacious manner. This applies to the selection of monotherapies in newly-diagnosed epilepsy, but perhaps, more importantly, to the choice of combination treatments in otherwise drug-refractory epilepsy. Rational polytherapy is a concept that is predicated on the combination of drugs with complementary mechanisms of action (MoAs) that work synergistically to maximize efficacy and minimize the potential for adverse events. Furthermore, rational polytherapy requires a detailed understanding of the MoA subclasses amongst available ASDs and an appreciation of the empirical evidence that supports the use of specific combinations. The majority of ASDs can be loosely categorized into those that target neurotransmission and network hyperexcitability, modulate intrinsic neuronal properties through ion channels, or possess broad-spectrum efficacy as a result of multiple mechanisms. Within each of these categories, there are discrete pharmacological profiles that differentiate individual ASDs. This chapter will consider how knowledge of MoA can help guide therapy in a

M. Barker-Haliski, Ph.D. • H.S. White, Ph.D. (✉)
Anticonvulsant Drug Development Program,
Department of Pharmacology and Toxicology,
University of Utah, 417 Wakara Way, Suite 3211,
Salt Lake City, UT 84108, USA
e-mail: melissa.barker@utah.edu;
steve.white@hsc.utah.edu

G.J. Sills, Ph.D.
Department of Molecular and Clinical Pharmacology,
University of Liverpool, Crown Street,
Liverpool L69 3BX, UK
e-mail: G.Sills@liverpool.ac.uk

H.E. Scharfman and P.S. Buckmaster (eds.), *Issues in Clinical Epileptology: A View from the Bench*,
Advances in Experimental Medicine and Biology 813, DOI 10.1007/978-94-017-8914-1_24,
© Springer Science+Business Media Dordrecht 2014

rational manner, both in the selection of monotherapies for specific seizure types and syndromes, but also in the choice of drug combinations for patients whose epilepsy is not optimally controlled with a single ASD.

Keywords

Mechanism of action • Anti-seizure drugs • Monotherapy • Polytherapy • Drug-refractory epilepsy

24.1 Introduction

Approximately 50 million people worldwide suffer from epilepsy. While more than 20 anti-seizure drugs (ASDs[1]) are currently available and many patients can be successfully treated with just one drug, there remains a substantial population (up to 40 % of all newly diagnosed patients) whose seizures are unresponsive to monotherapy [48, 57]. Most, if not all, of these patients will receive combination therapy at some point in the clinical management of their epilepsy [86]. Many will be exposed to newer ASDs, which are invariably brought to market as "adjunctive treatments" to those ASDs that are currently approved.

Emerging evidence suggests that combining ASDs with different mechanisms of action (MoAs) may be the most effective means to successfully manage difficult-to-control epilepsy [11]. Since many patients with refractory epilepsy may take three or more ASDs concurrently, it is essential that clinicians select drugs with the greatest potential for synergism and the lowest risk for adverse effects [11]. This can be considered "rational" polytherapy. Ultimately, this approach may offer the greatest potential to effectively manage seizures in patients with pharmacoresistant epilepsy: maximizing benefit and minimizing harm. Such an approach may also highlight novel

pathways or targets that might be exploited in future drug development efforts.

24.2 Does Mechanism of Action Really Matter?

It is logical to suggest that MoA should be considered at a number of steps in the treatment spectrum: when choosing an initial monotherapy for some primary generalized epilepsies; when considering a switch to a new ASD after a previous monotherapy has failed; or when adding a second or even third ASD in the therapy-resistant patient. Unfortunately, the absence of important clinical data from appropriate double-blind randomized clinical trials, which attempt to compare mechanistically distinct ASDs in discrete patient populations, prohibits such a logical therapeutic approach. Designing and delivering such a trial would be an enormous undertaking and one that is unlikely ever to be fully realized, on both logistical and financial grounds. Moreover, much of the evidence that is available to the patient with epilepsy and his or her clinician has been derived from clinical observation and often as a result of the desire to avoid poor outcomes, rather than to optimize the likelihood of good ones. This is most evident in the case of seizure worsening, where knowledge of MoA, the syndromic diagnosis and, in some cases, the underlying etiology can be beneficial. For example, clinical experience has demonstrated that GABAergic agonists and sodium channel blockers can worsen generalized spike-wave seizures in absence epilepsy. Similarly, sodium channel blockers, but

---

[1] Anti-seizure drugs (ASDs) is a new descriptive term considered by some to better reflect the effects of current therapies for epilepsy, in that they prevent only one of many sequelae of the disorder, i.e. the seizures, but not other comorbidities associated with epilepsy [14, 45].

not GABAergic agonists, can worsen seizures in patients with Dravet's syndrome or severe myoclonic epilepsy of infancy (SMEI). Time will tell whether ongoing improvements in our understanding of the underlying molecular etiology of the epilepsies will direct the choice of treatment in other seizure types.

Robust empirical evidence to support mechanism-driven therapy may be lacking, but proof-of-principle can be derived through post-hoc analysis of clinical trial data. This is unfortunately not done with sufficient regularity. Results from head-to-head monotherapy studies (where available) can be scrutinized for any evidence of preferential efficacy of a specific MoA within a specific seizure type [10, 37]. Likewise, add-on clinical trial studies of new ASDs can be interrogated for evidence of preferred combinations of ASDs, as was done with post-hoc analysis of lacosamide trial data [76]. In this analysis, lacosamide appeared to possess less efficacy and to be associated with more adverse effects when added to existing treatment regimens that contained at least one "traditional" voltage-gated sodium channel (VGSC) blocking ASD (*i.e.* phenytoin, carbamazepine, lamotrigine, oxcarbazepine) than when added to regimens that were devoid of sodium channel blockers [76]. Although the power of such post-hoc analyses is questionable and the original studies on which they are based are both heterogeneous and not necessarily reflective of real life, the results are important to direct future rational therapy decisions.

Such insight allows for some generalizations, not least of which is that, for newly-diagnosed focal epilepsies, MoA is mostly irrelevant. The majority of these patients will respond to a modest dose of whichever drug is chosen [9], with choice more often dictated not by MoA, but by clinical and demographic characteristics. In this population, MoA becomes more relevant when patients start to fail ASDs due to a lack of adequate seizure control at a therapeutic dose. Under those circumstances, for example, it would not make sense to replace one VGSC blocker with another. Failure due to adverse effects is different and it would be reasonable to replace carbamazepine with lamotrigine if carbamazepine

was effective, but not well tolerated. Arguably, MoA becomes most important in this population when the decision is made that monotherapy is not sufficient and that polytherapy is required. Under those circumstances, the best outcomes are often seen with drugs that work in different ways. For the drug refractory patient, the question then becomes: what is meant by "different"? Are lamotrigine and lacosamide different? Are benzodiazepines and barbiturates different? Is it enough to consider the class into which the drug might be arbitrarily placed, or is discrete consideration of the pharmacological minutiae more important? That remains unclear. A related issue is the supposed promiscuity of the majority of ASDs in terms of their cellular effects, resulting in negative perceptions of the efficacy of the drug in that particular circumstance. This unfortunate attitude often undermines efforts to explore and to implement rational treatment strategies for therapy-resistant epilepsy on the basis of MoA.

The understanding of how ASDs exert their effects at the cellular level has improved immeasurably in the past 25 years [52]. This advance will only further optimize treatment outcomes in epilepsy. Admittedly for some ASDs, the precise MoA remains frustratingly elusive, but for most, the primary cellular effects are now well described [93]. In the remainder of this chapter, we describe current understanding of ASD MoAs, categorized by target type (Table 24.1), and thereafter discuss the clinical implications of those actions and how therapeutic management may develop in future years from such observations.

24.3   Compounds That Target Neurotransmission and Network Synchronization

Epilepsy, in its broadest sense, is generally considered to arise due to an imbalance in, or abnormal synchronization of, inhibitory and excitatory signaling within neuronal networks [2, 80]. As such, it is not surprising that most currently available ASDs target ion channels or receptors involved in excitatory and/or inhibitory neurotransmission (Table 24.1) [49, 93]. Similarly, it

Table 24.1 Mechanism of action of approved anti-seizure drugs

Mechanism of action	Anti-seizure drug(s)
Neurotransmission and network synchronization	
Inhibitory neurotransmission: GABA system modulation	Barbiturates, Benzodiazepines, Felbamate, Stiripentol, Tiagabine, Topiramate, Valproate, Vigabatrin
Excitatory neurotransmission: Glutamate receptor modulation	Felbamate, Perampanel, Topiramate
Synaptic vesicle modulation: SV2A protein binding	Levetiracetam
Neuronal voltage-dependent ion channels	
Sodium (Na+) channels	
Fast-inactivated	Carbamazepine, Eslicarbazepine, Lamotrigine, Oxcarbazepine, Phenytoin, Rufinamide, Topiramate, Zonisamide
Slow-inactivated	Lacosamide
Calcium (Ca^{2+}) channels[a]	Ethosuximide, Gabapentin, Lamotrigine, Pregabalin, Topiramate, Valproate, Zonisamide
Potassium (K+) channels: Kv7.2/7.3 selective	Ezogabine

[a]As noted in the text, ASD effects on voltage-gated calcium channels have to be differentiated on the basis of whether they modify low or high voltage-gated calcium channels; e.g., ethosuximide, valproate and zonisamide have all been reported to modify the low voltage-gated T-type calcium current

makes sense that ASDs which display broad mechanistic profiles, *i.e.*, those that target multiple processes and pathways that are known to contribute to abnormal network synchronization, such as valproate, felbamate, topiramate, or zonisamide, often display broad spectrum clinical utility [51, 62]. Thus, understanding the specific MoAs of various ASDs may improve treatment outcomes when drug combinations are selected that display the most promising synergistic interactions at both inhibitory and excitatory synapses while conferring the least risk for adverse events.

Curbing excitatory neuronal activity can be achieved through GABAergic neuromodulation [5]. Some of the earliest marketed ASDs, including barbiturates and benzodiazepines, directly target the GABAergic system (Table 24.1) and although the MoA of valproate remains to be definitively identified, one of its many pharmacological effects is to increase synaptic GABA turnover [49, 51]. Of the newer ASDs, two were specifically designed to enhance synaptic GABAergic inhibitory neurotransmission. Tiagabine blocks synaptic GABA reuptake [66, 87] thereby prolonging the inhibitory action of GABA at GABAergic synapses, whereas vigabatrin selectively inhibits GABA transaminase [49, 77], an action that prevents the catabolism of GABA and increases readily releasable GABA within presynaptic terminals [49]. Topiramate enhances GABAergic neurotransmission through non-benzodiazepine site effects on the GABA$_A$ receptor [82, 83]. More recently, stiripentol has been approved for the treatment of Dravet's syndrome. Stiripentol is a positive allosteric modulator of α3-β3-γ2-containing GABA$_A$ receptors, increasing GABAergic neurotransmission in neuronal circuits where this receptor subtype is expressed. Preference for stiripentol over non-selective GABA$_A$ receptor drugs in Dravet's syndrome suggests that pursuing subunit selective agents in drug development may provide improved seizure control or tolerability in other epilepsies [18]. With multiple ASDs that target the GABAergic system (Table 24.1), pharmacological enhancement of inhibitory neurotransmission can be considered a well-proven strategy for seizure control.

Until recently, efforts to target glutamate-mediated excitatory neurotransmission have met with disappointment. Within the brain, excitatory synaptic transmission is mediated predominantly by AMPA- and NMDA-type glutamate receptors [20, 28]. Early preclinical evidence suggested that modulating glutamatergic signaling could effectively control or suppress seizures [78]. However, efforts to develop NMDA-receptor selective antagonists for the clinical management of epilepsy met with difficulty due to significant adverse behavioral effects [90]. To date, only felbamate possesses any substantial effects on NMDA-type glutamate receptors [41, 46, 73, 92]. Conversely, the modulation of AMPA-type glutamate receptors holds more clinical promise [74]. Modulation of AMPA, but not NMDA [42], receptor signaling exerts fewer effects on synaptic plasticity [38] and has greater potential to modulate network hyperexcitability [74], an effect largely attributable to activity-dependent AMPA receptor localization dynamics that underlie fast synaptic excitatory neurotransmission [1]. To this point, perampanel is the first glutamatergic system-selective ASD that acts as a noncompetitive AMPA receptor antagonist [72], decreasing neuronal excitability and synchronization [74]. Amongst many other proposed MoAs, topiramate has also been shown to suppress excitatory neurotransmission by blocking non-NMDA type-glutamate receptors [36] and can reduce high basal concentrations of extracellular glutamate in the hippocampi of spontaneously epileptic rats [44]. Taken together, the effect of felbamate on NMDA receptors, the AMPA-selective effects of perampanel, and the non-NMDA effects of topiramate further demonstrate that glutamatergic modulation can efficiently suppress epileptic activity.

Excitatory neurotransmission may also be influenced by the binding of the ASD levetiracetam to the SV2A protein (Table 24.1), a membrane glycoprotein of synaptic vesicles [3]. The specific role of SV2A protein is still under active investigation. It is currently hypothesized that SV2A contributes to excitatory neurotransmission by participating in synaptic vesicle exo- and endocytotic processes in response to calcium-triggered vesicle fusion [24, 97]. Interestingly, SV2A knockout mice develop severe seizures [22] and resected brain tissues from patients with temporal lobe epilepsy show decreased immunoreactivity for SV2A protein [89], suggesting a possible role of reduced levels of SV2A in epileptogenicity. Levetiracetam is effective in the 6 Hz model of psychomotor seizures, but ineffective in other "traditional" animal models of epilepsy, further highlighting its unique pharmacological profile [47]. The availability of a drug like levetiracetam might be considered advantageous from a rational therapy perspective. The unique and novel MoA may be effectively combined with other ASDs with diverse MoAs to mitigate seizure frequency and susceptibility. More importantly, where it is possible to use selective combinations of such diverse MoAs to enhance efficacy, it may be possible to minimize the likelihood of adverse events by decreasing the total exposure burden of the ASDs. Indeed, the fact that levetiracetam possesses a unique mode of action could explain why it has been so successful clinically, as both monotherapy and adjunctive treatment.

24.4 Compounds That Modulate Intrinsic Neuronal Properties

The above-described drug mechanisms modulate seizure susceptibility by selectively regulating excitatory and inhibitory neurotransmission and thereby suppressing aberrant neuronal network synchronization. However, many of the currently available treatments for epilepsy can also modulate the intrinsic excitability of individual neurons by targeting ion channels (Table 24.1). Multiple ASDs, such as carbamazepine, phenytoin, lamotrigine and oxcarbazepine modulate the fast-inactivated state of VGSC, whereas lacosamide is thought to have a preferential effect on the slow-inactivated state of the channel. In contrast, the gabapentinoids (gabapentin and pregabalin) and ezogabine decrease neuronal firing by selectively targeting calcium and potassium channels, respectively. It is likely that ion channel-selective mechanisms have naturally emerged amongst ASDs because ion channel dysfunction is so heavily implicated in the pathophysiology of many idiopathic epilepsies [55]. Recent evidence also demonstrates a link between genetic channelopathies and acquired epilepsies and supports the further development of ion channel modulators for the management of seizure disorders in general [68].

Ion channel-targeting drugs modulate depolarization and action potential generation and propagation. Many ASDs, including carbamazepine, phenytoin, lamotrigine and oxcarbazepine, bind VGSCs, with preferential affinity for the channel protein when in the fast-inactivated state. This leads to a prolongation of recovery following transient depolarizations [54], thereby limiting repetitive action potential firing. This effect is also both use- and frequency-dependent, meaning that it is enhanced during periods of high-frequency neuronal firing, as during epileptic discharges. Topiramate may also exert some antiseizure effects through blockade of use-dependent VGSCs, although this effect appears to be different from traditional VGSC-blocking ASDs [56]. Conversely, lacosamide also targets VGSCs but with preferential effects on the slow-inactivated state of the channel which predominates during sustained depolarization [23]. A selective action on the slow-inactivated rather than fast-inactivated state of VGSCs promotes the stabilization of hyperexcitable neuronal membranes, and suggests that lacosamide is pharmacologically distinct from traditional ASDs that target the fast-inactivated state (Table 24.1) [23]. The characteristic use- and frequency-dependence of sodium channel block is the only example of selectivity for a disease-related mechanism amongst current ASDs and explains why these drugs can interfere with a fundamental neurophysiological mechanism without significantly affecting normal neuronal activity.

In addition to blocking sodium channels, several ASDs act via blockade of voltage-gated calcium channels (Table 24.1), an action that effectively decreases intracellular calcium ion concentration. In the dendrites and cell soma, elevated intracellular calcium can promote destabilization of VGSCs and increase cellular excitability and the likelihood of action potential firing [69]. In pre-synaptic nerve terminals, elevated intracellular calcium is the trigger for neurotransmitter release. ASDs that selectively target high voltage-activated calcium channels have found success in the management of epilepsy and also neuropathic pain [85]. The gabapentinoids (gabapentin and pregabalin) selectively bind to the accessory subunit $\alpha_2\delta$-1 of voltage-gated calcium channels [32] to block P/Q-type calcium currents at nerve terminals, reducing the calcium-dependent release of glutamate [31]. Lamotrigine has also been shown to target P/Q-type, N-type and R-type channels [30, 91], all of which are expressed on pre-synaptic nerve terminals. Topiramate and felbamate are reported to have similar effects, although the channel subtypes are less well defined. However, a different action is seen with ethosuximide. This ASD interacts with the low voltage-activated T-type calcium channel that is predominantly expressed on thalamocortical relay neurons [21], which have in turn been implicated in the generation of the hypersyn-chronous discharges that underlie generalized absence epilepsy. Blockade of T-type channels by ethosuximide almost certainly explains its efficacy in this regard, and may also explain the anti-absence effects of both valproate and zonisamide [13, 88]. Thus, several currently available ASDs modulate high and low voltage-activated calcium channels; an effect that indirectly reduces excitatory neurotransmission at glutamatergic synapses and limits neuronal synchronization.

Rather than targeting cellular excitability by limiting depolarization, it is also possible to promote hyperpolarization via a facilitatory effect on potassium currents. This is the primary MoA of ezogabine, which is a positive allosteric modulator of K_v7.2/7.3 voltage-gated potassium channels that carry the so-called M-current [96]. The M-current is a non-inactivating potassium conductance, which exerts a hyperpolarizing influence on the resting membrane potential [53]. It serves as a natural brake on excitability in regions prone to synchronous network activity, with enhancement of the M-current by ezogabine suppressing epileptiform activity by prevention of spike bursting [75, 98, 99]. The role of potassium channels in modulating neuronal excitability is further underscored by the finding that mutations in K_v7.2/7.3 potassium channel genes provide the basis for seizures in benign familial neonatal convulsions (BFNC), a rare form of epilepsy that arises due to mutations in these channels [7, 8, 16]. Thus, ezogabine provides another example of how targeting intrinsic neuronal properties can attenuate epileptic activity.

24.5 Compounds That Reduce Seizure Susceptibility Through Multiple Mechanisms

As outlined above, currently available ASDs exert their effects through multiple mechanisms, including suppression of neuronal excitability, reduction in the propensity to fire an action potential, disruption of neurotransmission through interactions with synaptic vesicles, or an increased inhibitory tone. Some ASDs, namely valproate, felbamate, topiramate and zonisamide, are reported to possess two or more of these effects (Table 24.1). These ASDs are effective across multiple epilepsy indications [51], suggesting that one way to treat a heterogeneous disorder like epilepsy is to broadly target ion channels and neurotransmitter receptors. Indeed, several clinical and preclinical studies provide strong proof-of-concept for such a treatment strategy [29, 76], with some authors suggesting that combining a drug with a single MoA with a drug with multiple MoAs may improve seizure control [11]. However, this approach might also result in mechanistic redundancy (where a specific mechanism is unnecessary or unhelpful) or reinforcement (where a specific mechanism is duplicated) – either scenario could potentially elevate the risk of adverse effects without necessarily enhancing seizure control. This can be considered an inherent limitation of drugs with multiple MoAs, that in some patients not all of those mechanisms will be beneficial and some may indeed be detrimental. This would explain why these compounds are often considered to be powerful drugs with proven broad-spectrum efficacy but which are occasionally not well-tolerated.

Broad-spectrum ASDs pose an interesting pharmacological conundrum: whether a single drug with multiple MoAs (*i.e.* polypharmacology) is equivalent, superior, or inferior to multiple drugs, each with single MoAs (*i.e.* polypharmacy)? Would it be better to use a combination of phenytoin, ethosuximide and acetazolamide, which target sodium channels, T-type calcium channels, and carbonic anhydrase respectively, or zonisamide, which targets all of these mechanisms simultaneously? Rational polytherapy would suggest that the single drug should behave in exactly the same way, in terms of both efficacy and tolerability, as the three-drug combination, assuming dose equivalents can be found and drug interactions can be compensated for. This is a puzzle that will probably never be solved because there are likely to be few prescribers who would choose the combination therapy under these circumstances. Most would opt instead for zonisamide on the grounds of ease of use, but

also to limit adverse effects that are perceived to hinder polypharmacy approaches.

24.6 Preclinical Evidence of Potential for Polytherapy

Preclinical data suggest that the greatest potential for synergistic effects with polytherapy arises when ASDs with multiple MoAs are combined with ASDs with single, distinct MoAs [34, 43]. Preclinical isobolographic studies in rodent models of epilepsy have played an important role in teasing out either favorable or synergistic interactions from those that may be negative or antagonistic [25, 64, 79]. This preclinical evidence supports the concept of rational polypharmacy to control seizures using ASDs with diverse MoAs, although critics would argue that data from studies involving experimental animals are far more frequently positive than is observed clinically. With the possible exception of valproate combined with lamotrigine, synergistic combinations identified in animal models do not appear to reliably extend to the clinic.

Such discrepancies between preclinical and clinical observations can be explained, at least in part, by the inherent limitations in the preclinical studies. First, they are almost always conducted using high-throughput models of acute seizures in non-epileptic rodents. Given that the epileptic brain is undoubtedly remodeled relative to the normal brain [50], there are likely to be changes in the pharmacological responsiveness. Second, these studies are invariably conducted following acute drug dosing and with efficacy determined at or around the time of peak effect. This clearly does not reflect the clinical situation, and ignores any pharmacokinetic interaction that may exist between the compounds being tested or any tolerance that might develop from repeated administration. Such effects would also bypass the role of hepatic induction and/or inhibition seen when two or more ASDs are chronically combined in the patient population. This does not imply that these types of preclinical studies of ASD combinations lack value, but simply that the results should not be automatically assumed to translate to the clinic. The apparent discrepancy between the results of preclinical and clinical combination studies could just as easily be explained by the fact that clinical studies have never been systematically explored. It is possible that the same combinations are synergistic in both animal models and human patients; it is just that we do not yet possess the clinical evidence to prove it.

If nothing else, preclinical evaluation can provide important insights into potentially synergistic and antagonistic combinations, which can be taken forward for more detailed clinical investigation. In this regard, such an approach can help to triage the myriad of possible combinations and allow clinical researchers to focus on those likely to be most beneficial (or least detrimental). With the introduction of more etiologically relevant animal models, future studies can be designed to examine combinations using chronic dosing in animals with therapy-resistant epilepsy. Such studies should more clearly define the true clinical potential for synergism. However, as the models become more elaborate and the treatment schedules more demanding, the likelihood of undertaking in-depth isobolographic studies of every possible ASD combination diminishes, not least because of the time and cost involved. A compromise may be required such that combinations are initially identified using acute seizure models, later confirmed using chronic treatment in models of drug-resistant epilepsy, and only then advanced to clinical validation studies.

24.7 Clinical Evaluation of Polytherapy

Validating the findings of preclinical studies and thus establishing a basis for rational polytherapy requires the formal clinical evaluation of combinations of ASDs. This is a complex and challenging task that is unlikely to ever be systematically completed. It is more likely that clinical validation will be reserved for only the most robust combinations and in circumstances where evidence is considered absolutely essential to clinical implementation.

Clinical combination studies are notoriously difficult to undertake. They require the investigation of efficacy and tolerability of both single drugs and combinations in relatively homogeneous populations of patients using a design that is sufficiently powered to separate synergism from additivity alone, as well as adjusting combinations to balance overall drug load. Not surprisingly, such a study has never been attempted in epilepsy and it is debatable whether one ever will be. In the meantime, we are largely dependent on small proof-of-principle studies and anecdotal observations for evidence of effective polytherapy regimens. These include the classic and unexpected observations suggesting synergism with valproate and lamotrigine [12], which were later proven to hold true [67]. These studies provided validation of an ASD combination that was already in widespread use and probably considered useful by many investigators but for which there was no specific evidence of benefit. However, it is debatable whether there will ever be sufficient imperative or resources to pursue such a validation in the future.

The more applicable alternative strategy is the utilization of existing resources to search for at least indirect evidence of synergism. In this regard, post-hoc analysis of Phase III regulatory trial data provides a potential opportunity. The study by Sake and colleagues used the reanalysis of Phase III clinical trial data to demonstrate efficacy and tolerability with add-on lacosamide stratified by background therapy (*i.e.* whether it contained sodium channel blockers or not) [76]. Although the trials were never designed for this purpose and the analysis was arguably under-powered, some interesting findings were reported. Not least of these was the observation that lacosamide, which targets the slow-inactivation state of the VGSC, showed reduced efficacy and enhanced adverse effects when combined with traditional sodium blockers (which target the fast-inactivated state) than when combined with non-sodium channel blocking drugs [76]. If a similar approach were undertaken with all newly licensed ASDs, we could rapidly develop a picture of which mechanisms work best together and over time, with sufficient numbers of studies, it may be possible to start to investigate individual drug combinations. Making this a mandatory requirement in the regulatory approval process would expedite the generation of such data and insisting on its release for independent meta-analysis would add further validity.

24.8 Other Considerations for Rational Therapy

Consideration of multiple factors including MoA, route of metabolism and excretion must be made when determining the therapeutic treatment strategy. While it is generally considered best to combine therapies with different MoAs to maximize clinical effect, the risk for drug-drug interactions due to convergent induction or inhibition of cytochrome P450 enzymes poses a risk for extraneous adverse events in the context of polytherapy if such considerations are not accounted for in advance. Such risk/benefit assessments should be based on metabolic pathway and pharmacokinetics. The newer generation ASDs, at least those that do not undergo hepatic metabolism, are thought to possess the least potential for pharmacokinetic interactions as they are not primarily metabolized in the liver [35]. In contrast, phenobarbital, phenytoin, carbamazepine, primidone, valproate, lamotrigine, felbamate, rufinamide, and to some extent topiramate and oxcarbazepine are all associated with a risk of drug-drug interactions due to their route of metabolism [61, 63, 95]. When these drugs are combined, pharmacokinetic interactions may confound apparent pharmacodynamic effects and mask any possible synergism (or antagonism). That said, pharmacokinetic interactions can also be beneficial in a therapeutic sense, allowing a more rapid attainment of steady state with long half-life drugs in enzyme-induced patients (*i.e.* with zonisamide; [81]), or combining a beneficial pharmacokinetic interaction with a pharmacodynamic one (*i.e.*, valproate with lamotrigine). Such understanding of pharmacokinetics and drug interactions is essential when evaluating combination therapy and when reporting potential synergism between

ASDs, which will invariably be interpreted as being pharmacodynamic in nature.

Of additional consideration is the influence of genetics on therapeutic response. Siblings with epilepsy show similar responses to monotherapy or polytherapy [84]. There is, thus, strong evidence for a genetic contribution to the pharmacological management of epilepsy although the individual variants that predispose to treatment success or failure in general remain to be identified. However, there is now substantial evidence for specific genetic mutations in certain epilepsies; e.g., greater than 25 "epilepsy genes" have been identified [71], many of which encode the voltage- and ligand-gated ion channels that are also the predominant targets of ASDs [33, 70]. This is most clearly illustrated in Dravet's syndrome, which is associated with loss-of-function mutations in one allele of the $Na_v1.1$ channel [17]. $Na_v1.1$ channels are the predominant sodium channels in inhibitory interneurons [58]. Importantly, this information provides a better understanding of why the sodium channel blocker, lamotrigine, can exacerbate seizures in patients with $Na_v1.1$ mutations by possibly inhibiting the remaining functional sodium channels on inhibitory interneurons. Of course, this very elegant scientific explanation came long after clinical experience had already taught us to avoid sodium blockers in patients with a Dravet's phenotype [39]. In the future, however, it may be possible to predict likely treatment response on the basis of drug MoA and the underlying molecular etiology of the epilepsy. In this regard, Dravet's syndrome provides us with a clear example of the importance of translation (both forward and back) in directing a rational therapeutic approach. In the context of this chapter, it is also interesting that combination therapy with stiripentol, valproate and clobazam appears to be the most effective treatment in patients with Dravet's syndrome [19]. Finding an effective therapeutic approach with ASDs that do not specifically target the mutations in VGSCs in Dravet's patients thus illustrates one of the best examples to date of rational polytherapy for genetic epilepsy.

Indeed, several other genetic disorders present with seizures, which, unlike Dravet's syndrome, may be effectively managed with therapies that specifically target the mutated protein or pathway [26, 94, 100]. For example, tuberous sclerosis (TSC) arises as a result of a mutation in one of two proteins in the mammalian target of rapamycin (mTOR) proliferation pathway (TSC1 or TSC2) and which has recurrent seizures as a characteristic phenotype [65]. The mTOR inhibitor rapamycin effectively suppresses aberrant cellular proliferation in TSC [27] and has been proposed to be disease-modifying, although whether adjunct rapamycin will effectively reverse the epileptogenic process and protect against seizures remains to be determined [65]. A similar situation applies to Fragile X Syndrome (FXS), an autism spectrum disorder in which approximately 14 % of patients present with mild seizures [6]. FXS arises due to a triplet repeat in the *FMR1* gene that leads to loss of the RNA-binding protein, Fragile X Mental Retardation Protein (FMRP) [4]. This protein interacts with machinery essential to synaptic plasticity processes mediated by group I metabotropic glutamate receptors (mGluRs), including mGluR5 [4]. Clinical trials are currently ongoing to examine the use of mGluR5 antagonists in the targeted treatment of FXS [40]. Additionally, FXS patients may benefit from a rational polytherapy approach as preclinical studies in FMR1 knockout mice, which display audiogenic seizures [59], suggest that acute, combined treatment with an mGluR5 inverse agonist and a GABA_B receptor agonist can synergistically suppress seizures better than either treatment alone [60]. Furthermore, studies are underway to determine whether treatment for FXS could translate into effective means to suppress network hypersynchronization and changes in synaptic plasticity that arise in epilepsy in general.

Our limited experience from Dravet's syndrome, TSC, and FXS suggests that understanding the molecular etiology of epilepsy can promote rational therapeutic approaches by identifying pathways for targeted intervention or those that should be avoided. With current large-scale efforts to unravel the genetic contribution to epilepsy, including those coordinated by Epi4K, EpiPGX, CENet, and the ILAE Genetics Consortium, it is probable that opportunities for rational therapy guided by the underlying etiology of the disorder will expand considerably. For example, emergent evidence suggests that *de novo* mutations of ion channel-encoding genes are prevalent in severe childhood epileptic encephalopathies [15, 33]; this information will then likely be informative to direct personalized treatment strategies for patients with similar mutations. Obviously, the hope of such collaborative research endeavors is that the emerging data will eventually inform clinical practice and could play an important role in individualized therapy. Such observations will further illustrate the need for critical evaluation of the disease characteristics and genetic associations *a priori* before deciding on a rational mono- or polytherapeutic approach. We may not be able to hit every target in every patient, but a better understanding of the disorder from a molecular perspective can only be an improvement over current practice in which most patients are treated from a position of blissful ignorance regarding the cause of their epilepsy.

24.9 Summary and Conclusions

MoA is an important criterion in the selection of ASDs for individuals with epilepsy, particularly in the avoidance of seizure worsening in generalized epilepsies, in the replacement of ineffective monotherapies, and when instituting or adjusting polypharmacy regimens. In all of these cases, therapy (whether mono- or poly-) can be said to be rational when MoA is considered. As detailed above, currently available ASDs often share similar features in their MoA, allowing for selective application of ASDs in certain epilepsy patients. These MoAs dictate how individual drugs behave clinically, in the control of specific seizure types and in their propensity to elicit specific adverse effects, and also how ASDs perform within polytherapy regimens. For patients in whom monotherapy has proved inadequate, current evidence supports the combination of a drug with a single, selective MoA with one that possesses multiple cellular effects. Future effort to understand how drug combinations work in certain patient populations is clearly of critical importance. In most cases, however, clinical validation of combinations identified in experimental models is lacking and greater efforts should be made to conduct post-hoc analysis of clinical trial data, which may provide essential information to direct basic research efforts, and vice-versa. At present, rational therapy for epilepsy describes the use of existing medications to treat seizure types and syndromes of mostly unknown cause using knowledge of how those medications work and interact at the cellular level. In some ways, it is not surprising that this approach is sub-optimal. Future advances in our understanding of the underlying molecular etiologies of the epilepsies, driven at least in part by current global genomics efforts, are likely to improve rational therapy of epilepsy immeasurably. These rationally applied strategies to mono- and polytherapeutic management will thus be critical to future efforts to better treat the refractory epilepsy patient, as well as the newly diagnosed patient.

Acknowledgements This work was prepared in honor of Dr. Phil Schwartzkroin, whose professional and scientific dedication has provided invaluable hope of therapeutic progress for the patient living with epilepsy.

References

1. Anggono V, Huganir RL (2012) Regulation of AMPA receptor trafficking and synaptic plasticity. Curr Opin Neurobiol 22:461–469
2. Avoli M, de Curtis M (2011) GABAergic synchronization in the limbic system and its role in the generation of epileptiform activity. Prog Neurobiol 95:104–132
3. Bajjalieh SM, Peterson K, Linial M, Scheller RH (1993) Brain contains two forms of synaptic vesicle protein 2. Proc Natl Acad Sci U S A 90:2150–2154
4. Bardoni B, Schenck A, Mandel JL (2001) The fragile X mental retardation protein. Brain Res Bull 56:375–382
5. Bernard C, Cossart R, Hirsch JC, Esclapez M, Ben-Ari Y (2000) What is GABAergic inhibition? How is it modified in epilepsy? Epilepsia 41(Suppl 6):S90–S95
6. Berry-Kravis E, Raspa M, Loggin-Hester L, Bishop E, Holiday D, Bailey DB (2010) Seizures in fragile X syndrome: characteristics and comorbid diagnoses. Am J Intellect Dev Disabil 115:461–472

7. Biervert C, Schroeder BC, Kubisch C, Berkovic SF, Propping P, Jentsch TJ, Steinlein OK (1998) A potassium channel mutation in neonatal human epilepsy. Science 279:403–406
8. Biervert C, Steinlein OK (1999) Structural and mutational analysis of KCNQ2, the major gene locus for benign familial neonatal convulsions. Hum Genet 104:234–240
9. Brodie MJ, Barry SJ, Bamagous GA, Norrie JD, Kwan P (2012) Patterns of treatment response in newly diagnosed epilepsy. Neurology 78: 1548–1554
10. Brodie MJ, Perucca E, Ryvlin P, Ben-Menachem E, Meencke HJ, Levetiracetam Monotherapy Study Group (2007) Comparison of levetiracetam and controlled-release carbamazepine in newly diagnosed epilepsy. Neurology 68:402–408
11. Brodie MJ, Sills GJ (2011) Combining antiepileptic drugs – rational polytherapy? Seizure 20:369–375
12. Brodie MJ, Yuen AW (1997) Lamotrigine substitution study: evidence for synergism with sodium valproate? 105 Study Group. Epilepsy Res 26:423–432
13. Broicher T, Seidenbecher T, Meuth P, Munsch T, Meuth SG, Kanyshkova T, Pape HC, Budde T (2007) T-current related effects of antiepileptic drugs and a Ca2+ channel antagonist on thalamic relay and local circuit interneurons in a rat model of absence epilepsy. Neuropharmacology 53:431–446
14. Brooks-Kayal AR, Bath KG, Berg AT et al (2013) Issues related to symptomatic and disease-modifying treatments affecting cognitive and neuropsychiatric comorbidities of epilepsy. Epilepsia 54(Suppl 4):44–60
15. Carvill GL, Heavin SB, Yendle SC et al (2013) Targeted resequencing in epileptic encephalopathies identifies de novo mutations in CHD2 and SYNGAP1. Nat Genet 45:825–830
16. Castaldo P, del Giudice EM, Coppola G, Pascotto A, Annunziato L, Taglialatela M (2002) Benign familial neonatal convulsions caused by altered gating of KCNQ2/KCNQ3 potassium channels. J Neurosci 22:RC199
17. Catterall WA, Kalume F, Oakley JC (2010) NaV1.1 channels and epilepsy. J Physiol 588:1849–1859
18. Chiron C (2005) Stiripentol. Expert Opin Investig Drugs 14:905–911
19. Chiron C, Marchand MC, Tran A, Rey E, d'Athis P, Vincent J, Dulac O, Pons G (2000) Stiripentol in severe myoclonic epilepsy in infancy: a randomised placebo-controlled syndrome-dedicated trial. STICLO study group. Lancet 356:1638–1642
20. Collingridge GL, Lester RA (1989) Excitatory amino acid receptors in the vertebrate central nervous system. Pharmacol Rev 41:143–210
21. Coulter DA, Huguenard JR, Prince DA (1989) Characterization of ethosuximide reduction of low-threshold calcium current in thalamic neurons. Ann Neurol 25:582–593
22. Crowder KM, Gunther JM, Jones TA, Hale BD, Zhang HZ, Peterson MR, Scheller RH, Chavkin C,

Bajjalieh SM (1999) Abnormal neurotransmission in mice lacking synaptic vesicle protein 2A (SV2A). Proc Natl Acad Sci U S A 96:15268–15273

23. Curia G, Biagini G, Perucca E, Avoli M (2009) Lacosamide: a new approach to target voltage-gated sodium currents in epileptic disorders. CNS Drugs 23:555–568

24. Custer KL, Austin NS, Sullivan JM, Bajjalieh SM (2006) Synaptic vesicle protein 2 enhances release probability at quiescent synapses. J Neurosci 26:1303–1313

25. Czuczwar SJ, Kaplanski J, Swiderska-Dziewit G, Gergont A, Kroczka S, Kacinski M (2009) Pharmacodynamic interactions between antiepileptic drugs: preclinical data based on isobolography. Expert Opin Drug Metab Toxicol 5:131–136

26. D'Hulst C, Kooy RF (2009) Fragile X syndrome: from molecular genetics to therapy. J Med Genet 46:577–584

27. Dabora SL, Franz DN, Ashwal S et al (2011) Multicenter phase 2 trial of sirolimus for tuberous sclerosis: kidney angiomyolipomas and other tumors regress and VEGF- D levels decrease. PLoS One 6:e23379

28. Davies SN, Collingridge GL (1989) Role of excitatory amino acid receptors in synaptic transmission in area CA1 of rat hippocampus. Proc R Soc Lond B Biol Sci 236:373–384

29. Deckers CL, Czuczwar SJ, Hekster YA, Keyser A, Kubova H, Meinardi H, Patsalos PN, Renier WO, Van Rijn CM (2000) Selection of antiepileptic drug polytherapy based on mechanisms of action: the evidence reviewed. Epilepsia 41:1364–1374

30. Dibue M, Kamp MA, Alpdogan S, Tevoufouet EE, Neiss WF, Hescheler J, Schneider T (2013) Ca 2.3 (R-type) calcium channels are critical for mediating anticonvulsive and neuroprotective properties of lamotrigine in vivo. Epilepsia 54:1542–1550

31. Dooley DJ, Mieske CA, Borosky SA (2000) Inhibition of K(+)-evoked glutamate release from rat neocortical and hippocampal slices by gabapentin. Neurosci Lett 280:107–110

32. Dooley DJ, Taylor CP, Donevan S, Feltner D (2007) Ca2+ channel alpha2delta ligands: novel modulators of neurotransmission. Trends Pharmacol Sci 28:75–82

33. Epi4K Consortium, Epilepsy Phenome/Genome Project, Allen AS et al (2013) De novo mutations in epileptic encephalopathies. Nature 501:217–221

34. French JA, Faught E (2009) Rational polytherapy. Epilepsia 50(Suppl 8):63–68

35. French JA, Gidal BE (2000) Antiepileptic drug interactions. Epilepsia 41(Suppl 8):S30–S36

36. Gibbs JW 3rd, Sombati S, DeLorenzo RJ, Coulter DA (2000) Cellular actions of topiramate: blockade of kainate-evoked inward currents in cultured hippocampal neurons. Epilepsia 41(Suppl 1):S10–S16

37. Glauser TA, Cnaan A, Shinnar S et al (2010) Ethosuximide, valproic acid, and lamotrigine in childhood absence epilepsy. N Engl J Med 362:790–799

38. Goda Y, Stevens CF (1996) Synaptic plasticity: the basis of particular types of learning. Curr Biol 6:375–378

39. Guerrini R, Dravet C, Genton P, Belmonte A, Kaminska A, Dulac O (1998) Lamotrigine and seizure aggravation in severe myoclonic epilepsy. Epilepsia 39:508–512

40. Hagerman RJ, Berry-Kravis E, Kaufmann WE et al (2009) Advances in the treatment of fragile X syndrome. Pediatrics 123:378–390

41. Harty TP, Rogawski MA (2000) Felbamate block of recombinant N-methyl-D-aspartate receptors: selectivity for the NR2B subunit. Epilepsy Res 39:47–55

42. Hunt DL, Castillo PE (2012) Synaptic plasticity of NMDA receptors: mechanisms and functional implications. Curr Opin Neurobiol 22:496–508

43. Kaminski RM, Matagne A, Patsalos PN, Klitgaard H (2009) Benefit of combination therapy in epilepsy: a review of the preclinical evidence with levetiracetam. Epilepsia 50:387–397

44. Kanda T, Kurokawa M, Tamura S et al (1996) Topiramate reduces abnormally high extracellular levels of glutamate and aspartate in the hippocampus of spontaneously epileptic rats (SER). Life Sci 59:1607–1616

45. Kelley MS, Jacobs MP, Lowenstein DH, Stewards NEB (2009) The NINDS epilepsy research benchmarks. Epilepsia 50:579–582

46. Kleckner NW, Glazewski JC, Chen CC, Moscrip TD (1999) Subtype-selective antagonism of N-methyl-D-aspartate receptors by felbamate: insights into the mechanism of action. J Pharmacol Exp Ther 289:886–894

47. Klitgaard H, Matagne A, Gobert J, Wulfert E (1998) Evidence for a unique profile of levetiracetam in rodent models of seizures and epilepsy. Eur J Pharmacol 353:191–206

48. Kwan P, Brodie MJ (2000) Epilepsy after the first drug fails: substitution or add-on? Seizure 9:464–468

49. Landmark CJ (2007) Targets for antiepileptic drugs in the synapse. Med Sci Monit 13:RA1–RA7

50. Leite JP, Neder L, Arisi GM, Carlotti CG Jr, Assirati JA, Moreira JE (2005) Plasticity, synaptic strength, and epilepsy: what can we learn from ultrastructural data? Epilepsia 46(Suppl 5):134–141

51. Loscher W (2002) Basic pharmacology of valproate: a review after 35 years of clinical use for the treatment of epilepsy. CNS Drugs 16:669–694

52. Loscher W, Schmidt D (2012) Epilepsy: perampanel-new promise for refractory epilepsy? Nat Rev Neurol 8:661–662

53. Main MJ, Cryan JE, Dupere JR, Cox B, Clare JJ, Burbidge SA (2000) Modulation of KCNQ2/3 potassium channels by the novel anticonvulsant retigabine. Mol Pharmacol 58:253–262

54. Mantegazza M, Curia G, Biagini G, Ragsdale DS, Avoli M (2010) Voltage-gated sodium channels as therapeutic targets in epilepsy and other neurological disorders. Lancet Neurol 9:413–424

55. Marini C, Mantegazza M (2010) Na+ channelopathies and epilepsy: recent advances and new perspectives. Expert Rev Clin Pharmacol 3:371–384
56. McLean MJ, Bukhari AA, Wamil AW (2000) Effects of topiramate on sodium-dependent action-potential firing by mouse spinal cord neurons in cell culture. Epilepsia 41(Suppl 1):S21–S24
57. Mohanraj R, Brodie MJ (2006) Diagnosing refractory epilepsy: response to sequential treatment schedules. Eur J Neurol 13:277–282
58. Ogiwara I, Miyamoto H, Morita N et al (2007) Nav1.1 localizes to axons of parvalbumin-positive inhibitory interneurons: a circuit basis for epileptic seizures in mice carrying an Scn1a gene mutation. J Neurosci 27:5903–5914
59. Pacey LK, Heximer SP, Hampson DR (2009) Increased GABA(B) receptor-mediated signaling reduces the susceptibility of fragile X knockout mice to audiogenic seizures. Mol Pharmacol 76:18–24
60. Pacey LK, Tharmalingam S, Hampson DR (2011) Subchronic administration and combination metabotropic glutamate and GABAB receptor drug therapy in fragile X syndrome. J Pharmacol Exp Ther 338:897–905
61. Patsalos PN (2013) Drug interactions with the newer antiepileptic drugs (AEDs) – part 1: pharmacokinetic and pharmacodynamic interactions between AEDs. Clin Pharmacokinet 52:927–966
62. Perucca E (1997) A pharmacological and clinical review on topiramate, a new antiepileptic drug. Pharmacol Res 35:241–256
63. Perucca E (2006) Clinically relevant drug interactions with antiepileptic drugs. Br J Clin Pharmacol 61:246–255
64. Perucca E, Yasothan U, Clincke G, Kirkpatrick P (2008) Lacosamide. Nat Rev Drug Discov 7: 973–974
65. Petrova LD (2011) Tuberous sclerosis and epilepsy. Am J Electroneurodiagnostic Technol 51:5–15
66. Pfeiffer M, Draguhn A, Meierkord H, Heinemann U (1996) Effects of gamma-aminobutyric acid (GABA) agonists and GABA uptake inhibitors on pharmacosensitive and pharmacoresistant epileptiform activity in vitro. Br J Pharmacol 119:569–577
67. Pisani F, Oteri G, Russo MF, Di Perri R, Perucca E, Richens A (1999) The efficacy of valproate-lamotrigine comedication in refractory complex partial seizures: evidence for a pharmacodynamic interaction. Epilepsia 40:1141–1146
68. Poolos NP, Johnston D (2012) Dendritic ion channelopathy in acquired epilepsy. Epilepsia 53(Suppl 9):32–40
69. Potet F, Chagot B, Anghelescu M, Viswanathan PC, Stepanovic SZ, Kupershmidt S, Chazin WJ, Balser JR (2009) Functional interactions between distinct sodium channel cytoplasmic domains through the action of calmodulin. J Biol Chem 284:8846–8854
70. Reid CA, Berkovic SF, Petrou S (2009) Mechanisms of human inherited epilepsies. Prog Neurobiol 87:41–57

71. Reid CA, Jackson GD, Berkovic SF, Petrou S (2010) New therapeutic opportunities in epilepsy: a genetic perspective. Pharmacol Ther 128:274–280
72. Rheims S, Ryvlin P (2013) Profile of perampanel and its potential in the treatment of partial onset seizures. Neuropsychiatr Dis Treat 9:629–637
73. Rho JM, Donevan SD, Rogawski MA (1994) Mechanism of action of the anticonvulsant felbamate: opposing effects on N-methyl-D-aspartate and gamma-aminobutyric acidA receptors. Ann Neurol 35:229–234
74. Rogawski MA (2013) AMPA receptors as a molecular target in epilepsy therapy. Acta Neurol Scand 127 (Suppl s197):9–18
75. Rogawski MA, Bazil CW (2008) New molecular targets for antiepileptic drugs: alpha(2)delta, SV2A, and K(v)7/KCNQ/M potassium channels. Curr Neurol Neurosci Rep 8:345–352
76. Sake JK, Hebert D, Isojarvi J, Doty P, De Backer M, Davies K, Eggert-Formella A, Zackheim J (2010) A pooled analysis of lacosamide clinical trial data grouped by mechanism of action of concomitant antiepileptic drugs. CNS Drugs 24:1055–1068
77. Sarup A, Larsson OM, Schousboe A (2003) GABA transporters and GABA-transaminase as drug targets. Curr Drug Targets CNS Neurol Disord 2:269–277
78. Sato K, Morimoto K, Okamoto M (1988) Anticonvulsant action of a non-competitive antagonist of NMDA receptors (MK-801) in the kindling model of epilepsy. Brain Res 463:12–20
79. Shandra A, Shandra P, Kaschenko O, Matagne A, Stohr T (2013) Synergism of lacosamide with established antiepileptic drugs in the 6-Hz seizure model in mice. Epilepsia 54:1167–1175
80. Sierra-Paredes G, Sierra-Marcuno G (2007) Extrasynaptic GABA and glutamate receptors in epilepsy. CNS Neurol Disord Drug Targets 6:288–300
81. Sills G, Brodie M (2007) Pharmacokinetics and drug interactions with zonisamide. Epilepsia 48: 435–441
82. Simeone TA, Wilcox KS, White HS (2006) Subunit selectivity of topiramate modulation of heteromeric GABA(A) receptors. Neuropharmacology 50:845–857
83. Simeone TA, Wilcox KS, White HS (2011) Topiramate modulation of beta(1)- and beta(3)-homomeric GABA(A) receptors. Pharmacol Res 64:44–52
84. Sonmezturk HH, Arain AM, Paolicchi JM, Abou-Khalil BW (2012) Similar response to anti-epileptic medications among epileptic siblings. Epilepsy Res 98:187–193
85. Stefani A, Spadoni F, Bernardi G (1997) Voltage-activated calcium channels: targets of antiepileptic drug therapy? Epilepsia 38:959–965
86. Stephen LJ, Brodie MJ (2002) Seizure freedom with more than one antiepileptic drug. Seizure 11:349–351
87. Suzdak PD, Jansen JA (1995) A review of the preclinical pharmacology of tiagabine: a potent and selective anticonvulsant GABA uptake inhibitor. Epilepsia 36:612–626

88. Suzuki S, Kawakami K, Nishimura S, Watanabe Y, Yagi K, Seino M, Miyamoto K (1992) Zonisamide blocks T-type calcium channel in cultured neurons of rat cerebral cortex. Epilepsy Res 12:21–27

89. van Vliet EA, Aronica E, Redeker S, Boer K, Gorter JA (2009) Decreased expression of synaptic vesicle protein 2A, the binding site for levetiracetam, during epileptogenesis and chronic epilepsy. Epilepsia 50:422–433

90. Wada Y, Hasegawa H, Nakamura M, Yamaguchi N (1992) The NMDA receptor antagonist MK-801 has a dissociative effect on seizure activity of hippocampal-kindled cats. Pharmacol Biochem Behav 43:1269–1272

91. Wang SJ, Huang CC, Hsu KS, Tsai JJ, Gean PW (1996) Inhibition of N-type calcium currents by lamotrigine in rat amygdalar neurones. Neuroreport 7:3037–3040

92. White HS, Harmsworth WL, Sofia RD, Wolf HH (1995) Felbamate modulates the strychnine-insensitive glycine receptor. Epilepsy Res 20:41–48

93. White HS, Smith MD, Wilcox KS (2007) Mechanisms of action of antiepileptic drugs. Int Rev Neurobiol 81:85–110

94. Wong M (2010) Mammalian target of rapamycin (mTOR) inhibition as a potential antiepileptogenic therapy: from tuberous sclerosis to common acquired epilepsies. Epilepsia 51:27–36

95. Wrighton SA, Stevens JC (1992) The human hepatic cytochromes P450 involved in drug metabolism. Crit Rev Toxicol 22:1–21

96. Wuttke TV, Seebohm G, Bail S, Maljevic S, Lerche H (2005) The new anticonvulsant retigabine favors voltage-dependent opening of the Kv7.2 (KCNQ2) channel by binding to its activation gate. Mol Pharmacol 67:1009–1017

97. Xu T, Bajjalieh SM (2001) SV2 modulates the size of the readily releasable pool of secretory vesicles. Nat Cell Biol 3:691–698

98. Yue C, Yaari Y (2004) KCNQ/M channels control spike after depolarization and burst generation in hippocampal neurons. J Neurosci 24:4614–4624

99. Yue C, Yaari Y (2006) Axo-somatic and apical dendritic Kv7/M channels differentially regulate the intrinsic excitability of adult rat CA1 pyramidal cells. J Neurophysiol 95:3480–3495

100. Zeng LH, McDaniel S, Rensing NR, Wong M (2010) Regulation of cell death and epileptogenesis by the mammalian target of rapamycin (mTOR): a double-edged sword? Cell Cycle 9:2281–2285

How Can Advances in Epilepsy Genetics Lead to Better Treatments and Cures?

25

Renzo Guerrini and Jeffrey Noebels

Abstract

Advances in genetic analysis are fundamentally changing our understanding of the causes of epilepsy, and promise to add more precision to diagnosis and management of the clinical disorder. Single gene mutations that appear among more complex patterns of genomic variation can now be readily defined. As each mutation is identified, its predicted effects can now be validated in neurons derived from the patient's own stem cells, allowing a more precise understanding of the cellular defect. Parallel breakthroughs in genetic engineering now allow the creation of developmental experimental models bearing mutations identical to the human disorder. These models enable investigators to carry out detailed exploration of the downstream effects of the defective gene on the developing nervous system, and a framework for pursuing new therapeutic target discovery. Once these genetic strategies are combined with interdisciplinary technological advances in bioinformatics, imaging, and drug development, the promise of delivering clinical cures for some genetic epilepsies will be within our reach.

Keywords

Mutation • Phenotype • Gene testing • Comorbidity • Modifier • Complexity

25.1 Introduction

The last decade has witnessed several revolutions in our ability to understand the genetic basis of the epilepsies and its role in diagnosis and treatment. Ten years ago, only a few genes, mostly for ion channels, had been linked to the appearance of epilepsy in large, multigenerational pedigrees. The general belief was that such families were rare, that the numbers of causative genes for epilepsy were few, and that each of the clinically

R. Guerrini (✉)
Neuroscience Department, Children's Hospital A. Meyer-University of Florence, Florence, Italy
e-mail: r.guerrini@meyer.it

J. Noebels
Department of Neurology, Baylor College of Medicine, Houston, TX, USA
e-mail: jnoebels@bcm.edu

H.E. Scharfman and P.S. Buckmaster (eds.), *Issues in Clinical Epileptology: A View from the Bench*, Advances in Experimental Medicine and Biology 813, DOI 10.1007/978-94-017-8914-1_25, © Springer Science+Business Media Dordrecht 2014

defined Mendelian syndromes was the exclusive product of one gene alone. Furthermore, the absence of a positive family history in a patient with epilepsy suggested that a purely genetic etiology was unlikely. Following this logic, it was anticipated that detecting a mutation in one of these genes would be highly predictive of a specific seizure syndrome, and that only a rare, inherited mutation causes disease. Understandably, most investigators concluded that knowledge of the gene defect could lead shortly to dramatically improved treatments, if not a cure, for the disorder.

In fact, none of these pioneering assumptions has proven to be entirely correct. However, as in the field of DNA sequencing, we have entered the 'next generation' of epilepsy genetics, and what began as a search for a few inherited gene errors that could explain why some epilepsies are familial has expanded into a set of powerful research tools and discoveries that have immeasurably accelerated our ability to correctly diagnose and, in some cases, treat the disease. Major strides in clinical phenotyping and classification of epilepsy syndromes have been driven by, and contribute to the identification of, new monogenic epilepsies, both inherited and de novo in origin. Advances in neuroimaging have proven critical to the discovery of genes leading to malformations of cortical development. New methods in molecular genetics and gene sequencing have allowed rapid identification of candidate genes for an increasing number of epilepsy syndromes and potential comorbidities, including sudden unexpected death. Advances in genetic epidemiology, genome-wide association studies and whole exome candidate gene profiling have stimulated the analysis of complex genetic traits. The mathematical aspects of these analyses, as well as the emergence of mutation and polymorphism databases and genotype-phenotype correlations, are now included in the growing new field of epilepsy bioinformatics.

In the neurobiology laboratory, identified genes arising from both human and experimental genetic studies now offer an unparalleled opportunity to examine basic mechanisms of the disease. Genetically engineered models enable the electrophysiological validation of a candidate epilepsy gene using in vivo and in vitro approaches, and are essential to pinpoint the specific brain networks involved. Stem cells derived from patient's fibroblasts can now be reliably transformed into neurons to evaluate the effects of the mutation on cell biology and signaling within the affected nervous system. Contemporary experimental mouse models not only give investigators the ability to selectively express a predefined human gene mutation in the brain at different stages of brain development, but also to reverse its effects with drugs and other genes. High resolution, chronic imaging techniques using fluorescent reporters of gene expression permit the study of the pathophysiology of a genetic lesion over time, tracking the 'downstream' molecular biology of the seizure pathways. Seizures typically arise after prolonged periods of abnormal neural development, and in these cases where the damage is already done, correcting the actual gene defect may come too late to reverse the epileptic condition. However careful analysis of these secondary changes in the physiology and anatomy of the affected neural circuits may offer a second opportunity to discover a novel target for therapy, fulfilling the promise of a cure.

25.2 The Emerging Picture of Epilepsy Genetics

25.2.1 Gene and Mutation Diversity

It is now estimated that genetic factors contribute to at least 40 % of all epilepsies. While there has been considerable progress in identifying genes for Mendelian epilepsies, the extent of genetic susceptibility to more common sporadic epilepsy syndromes remains unknown. Although limited evidence, in both animal models and human disease, has been gathered that susceptibility to epilepsy conferred by specific mutations might be influenced by non-pathogenic alleles at other genetic loci [1, 15] the characterization and validation of susceptibility variants appears particularly complex and requires large-scale collaborative efforts. Moreover, our understanding genetic susceptibility to a major heterogeneous disorder such as epilepsy would likely be incomplete without reference to a specific syndrome. Over 600 entries for pedigrees showing Mendelian phenotypic inheritance patterns can be found in the Online Mendelian Inheritance in Man (OMIM) database, and genetic loci have been identified in over 160 of these cases. These genes arise not only among the >400 members of the ion channel gene family, but across an extraordinarily diverse group of molecular pathways that also regulate membrane excitability, synaptic plasticity, and rhythmic network firing behavior. Causative genes also include those for presynaptic neurotransmitter release, postsynaptic receptors, transporters, cell metabolism, and importantly, many formative steps in early brain development, such as the proliferation and migration of neuronal precursors, dendritogenesis, synaptogenesis, and glial biology. However, inherited mutations in these known epilepsy genes currently only account for a small fraction of patients. Thus, many additional genes causing seizures are likely to be identified. Within each of these genes, the molecular rearrangements themselves are typically novel, or occur with a very low allele frequency within the epilepsy population. Thus monogenic epilepsies are disorders of many, individually rare, errors in an increasingly broad spectrum of biological pathways.

Most of the idiopathic epilepsies arise sporadically among unaffected family members, or do not appear to follow single gene inheritance patterns. From a purely genetic perspective, this finding may be explained by an inadequate family size, or an underlying complex pattern of multigenic inheritance, or even genetic mosaicism, three possibilities which have long bedeviled the analysis of inherited disease. However a new alternative has arisen from an important insight made over the past decade, and promises a steep increase in our ability to isolate genetic risk of epilepsy in individuals, even in small families – namely, the detection of de novo mutation of single genes or copy number variants of even larger chromosomal regions that encompass them. De novo splice site or nonsense mutations that impair function by removing critical portions

of the encoded protein have been identified at convincingly high frequency within specific epilepsy phenotypes, in particular the SCN1A sodium channel linked to the severe myoclonic epilepsy known as Dravet Syndrome. This realization, along with the recent ability to rapidly sequence and assess gene variation in a large list of candidate genes, will greatly contribute to the personal identification of causative genes in the epilepsy clinic.

25.2.2 Phenotype Complexity

Large scale genotype-phenotype studies within monogenic populations have determined that the simple correspondence between genotype and phenotype can break down, resulting in different ages of onset and clinical seizure severity (phenotypic heterogeneity) within those bearing mutations in the same gene. This poor correlation may be due to the many possible structural alterations in the mutant protein leading to either gain or loss of function. However, even in families with single gene inheritance of an identical gene mutation, a degree of complexity remains, as evidenced by 'unaffected carriers' of the 'causative' mutation. In these cases, the phenotypic variability in such families can be attributed to the presence of polymorphisms in modifier genes influencing the phenotypic expression or, alternatively, to environmental factors.

Conversely, identical clinical phenotypes may be due to different underlying genotypes (genetic heterogeneity). Most of the broad phenotypic categories of seizure disorders are now recognized to arise from mutations in more than a single gene. In some cases, the different genes for a clinical epilepsy syndrome all contribute a single functional heteromeric unit, such as the different receptor subunits ($\alpha,\beta,\gamma,\delta$) contributing to a functional GABAa receptor in generalized epilepsy, the pore forming (α) and regulatory ($\beta1$) subunits of the sodium channel in Dravet Syndrome, the different pore forming subunits (KCQ2/3) of the M-current in Benign Neonatal Infantile Epilepsy, or the nicotinic cholinergic receptor subunits ($\alpha2,\beta2,\beta4$) in ADNLFE. In other cases, entirely separate gene pathways may contribute to a very similar phenotype. This property may ultimately explain not only clinical differences in the seizures attached to each gene and their neurological severity in affected patients, but also their pharmacoresistance in clinical subsets of the disorder.

Pharmacoresistance can itself be considered as a phenotypic trait whose intrinsic mechanisms are, at least in part, influenced by genetic variation. Pharmacogenetic studies have attempted to investigate whether drug resistance is influenced by single nucleotide variants in genes for drug targets, or in other genes related to drug uptake and metabolism, which might explain resistance to drugs [16]. These studies, however, are hampered by serious methodological difficulties, since they do not take into account the causal heterogeneity of 'epilepsy' in the populations studied. This oversimplification is reflected in the assumption that a single mechanism would influence drug efficacy in relation to different mechanisms of epileptogenicity, which should be replicated across multiple studies. However results from such studies have not been consistent. For example, a single intronic nucleotide polymorphism in the SCN1A gene was associated with higher prescribed doses of phenytoin and carbamazepine in a UK based study [29], but not in subsequent studies in Austria [34] and Italy [20]. Likewise, studies exploring how gene variants may influence AED penetration into the brain have provided conflicting results [2]. Very large studies on etiologically and ethnically homogeneous populations would be necessary to fully explore the real influence of specific genes on pharmacoresistance.

25.2.3 Discovery of Novel Comorbidity Syndromes

Epilepsy clinicians have long recognized the frequent association of a variety of cognitive and neuropsychiatric symptoms with seizures, and understood that these occur more often than would be predicted by chance. Whether these result as a direct developmental effect of the cause of the epilepsy, the seizures themselves, or

their treatment will always be under debate. Since their co-expression is usually incomplete, other genes may also contribute to the relative penetrance of the co-morbidity.

In the laboratory, mouse models of apparently unrelated disorders, such as Alzheimer's disease, have delivered firm evidence that single genes can produce both epilepsy and cognitive defects unrelated to antiepileptic treatment, and suggest that antiepileptic treatment may be especially neuroprotective in carriers of these genes [24, 25]. Sudden unexpected death (SUDEP) is another important comorbidity, affecting individuals with idiopathic epilepsy. Recent evidence has confirmed the hypothesis that genes underlying cardiac arrhythmias are co-expressed in brain and produce epilepsy [10]. These genes may prove clinically useful in predicting SUDEP risk and exploring treatments to prevent premature lethality in epilepsy patients.

A great deal of interest is now devoted to understanding the developmental causes of epileptic encephalopathies, in relation to autism spectrum disorders (ASD) and intellectual disability. Exome sequencing in a large cohort of individuals with epileptic encephalopathies and subsequent protein-protein interaction analysis revealed a high interconnectivity between genes carrying de novo mutations, with a much greater probability of overlap with ASD and intellectual disability exome sequencing studies [7].

Finally, the ability to probe the full genomic variant profile of unrelated epilepsy patients with and without comorbidities holds enormous promise in understanding the genetic roots of comorbidity. Recently, one such study identified a de novo truncation of the skeletal muscle chloride channel, *CLCN1* in a young woman with a childhood writer's cramp and longstanding pharmacoresistant seizures. This genomic analysis led to the unexpected discovery that CLCN1 is expressed not only in skeletal muscle, but in thalamocortical and cerebellar brain networks, where disruption of chloride-mediated membrane repolarization could lead to hyperexcitability and seizures [5]. This hypothesis-generating study is a harbinger of the kind of novel candidate gene discovery that awaits the widespread use of next generation sequencing (Fig. 25.1).

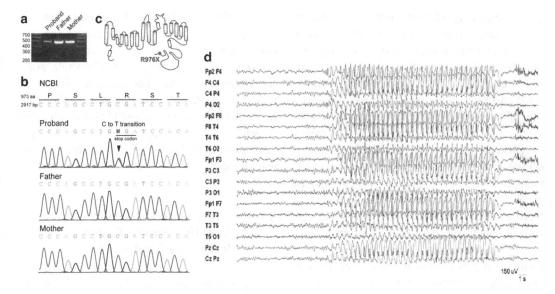

Fig. 25.1 Detection of heterozygous de novo nonsense mutation in *CLCN1*, encoding a premature stop codon in the CLC-1 chloride channel protein in a proband with generalized tonic clonic and absence seizures with a subtle myotonic phenotype. (**a**). PCR amplification of the final coding exon (exon 23) in the trio yielded a 550 bp product. (**b**) Sequence chromatograms for the trio shows the heterozygous base pair substitution encoding a premature stop codon in the proband, but not in either parent. (**c**) Schematic diagram of a single alpha-subunit of the CLC-1 channel protein showing the location of the C-terminal truncation R976X mutation in the proband. (**d**) Typical absence seizure in the proband (From Chen et al. [5])

25.2.4 Gene Testing

Recent identification of causative genes for a number of early-onset severe epilepsies has created the opportunity for diagnostic genetic testing in this population. Some examples include brain malformations and epileptic encephalopathies of infancy. At present, the clinical impact of genetic testing in these syndromes is by itself limited, due to the small percentage of patients in whom a single, causative gene mutation can be identified and the lack of specific, gene-directed, treatment options. However, genetic counseling can certainly be improved by recognizing a specific etiology. It also sets the stage for further research advances in understanding how each of the genes give rise to epileptogenic defects, and discovering which of these may be reversible. It has been claimed that, in some cases, discovery of a single causative gene defect may reduce the need for further diagnostic investigation at the biochemical level. However, from a practical standpoint, since genomic variants require time to analyze, this information typically arrives after

reversible causes have been clinically excluded. It is also essential to understand that recent profiling studies of whole genomes and large sets of candidate exomes such as ion channel genes have determined that patients with sporadic epilepsy often carry more than one potentially causative mutation [17], complicating the interpretation. Furthermore, all individuals carry, on average, 50–100 loss of function variants in disease genes that for the most part produce no apparent clinical effects [30], signifying that the mere presence of a variant does not predict clinical status. This is likely explained by the presence of other 'protective' modifier genes. Thus, as we gain access to a broader view of the genetic landscape in individuals with epilepsy, we expect to routinely encounter patients with a complex genetic basis for their seizure phenotype.

The next steps toward increasing the power of genetic testing in epilepsy include identifying more genes for monogenic epilepsies, and learning to understand the contribution of specific genes in epilepsies with complex inheritance. This will require continued genotype-phenotype correlations, coupled with functional studies of the abnormal proteins to more accurately understand the pathophysiological implications of each new mutation and how they combine to create neural excitability phenotypes. While bioinformatic analysis offers an increasingly powerful way of categorizing the potential damage a gene variant may inflict on protein function, it cannot conclusively predict its actual effect upon a neuron, and indeed, despite being expressed in multiple cell types, it may not affect them all equally. However we are now entering the era of 'personalized' mutation analysis, where the mutant functional defect can be determined directly in the patient's own cells, sometimes finding that it is counter to the expected result. For example, a currently held hypothesis for the mechanism of epilepsy in Dravet Syndrome, as studied in a mouse model of *Scn1a* haploinsufficiency, is based on the failure of interneurons to fire adequately in the face of reduced sodium current through *Scn1a* ion channels [31]. Analysis of membrane excitability in stem-cell derived neurons transformed from a Dravet Syndrome patient showed that the mutation, predicted to reduce the density of functional sodium channels, resulted in increased sodium current and hyperexcitability in cells classified as both excitatory and inhibitory, implying a distinctly different pathogenic mechanism [19]. These studies may have practical implications for diagnosis, genetic counseling and possible treatment, as well as increasing our knowledge of normal brain function and mechanisms of epileptogenesis.

We can now consult lists of epilepsy syndromes in which a chromosomal locus or loci have been mapped, and those in which one or more gene mutations or variants have been identified. These lists are constantly expanding as new loci and genes are identified. Our view on the correlations between phenotype and genotype in genetic epilepsies is also rapidly changing in relation to new findings emerging from exome sequencing and the use of diagnostic panels as unexpected phenotypes become associated with mutations of specific genes and vice-versa. A constantly updated database will be essential to establish all the known gene mutations and

polymorphisms and their clinical correlates, so that genotype-phenotype correlations can be determined. This is the objective of the Human Variome Project [13] and an achievable goal for epilepsy genetics. For example, over 700 mutations in SCN1A have now been reported in the SCN1A Variant Database [14] and offer the possibility of predicting the onset, if not the severity, of the related phenotype with a reasonable likelihood.

25.3 Does Knowledge of a Specific Genetic Cause Influence Treatment?

The appropriateness or inadvisability of a given treatment in a specific condition has in some instances been acquired in clinical practice and then scientifically justified by genetic knowledge. For example, the potential for lamotrigine, a sodium channel blocker, to aggravate seizures in Dravet syndrome was initially reported well before the discovery that loss of function *SCN1A* mutations were the cause of the syndrome [11]. However from a clinical perspective, a number of specific conditions provide evidence that improved understanding of epilepsy genetics, together with enhanced knowledge of molecular pathology and electroclinical characteristics, substantiate more rational and effective treatment choices resulting in better patient management. In cases where genotype-phenotype correlations suggest that the epilepsy may have a benign course, gene testing may support the decision to withhold antiepileptic drug therapy during critical periods of brain maturation. Examples of this are mainly related to benign familial epilepsies starting in the first years of life due to *PRRT2, KCNQ2* and *SCN2A* gene mutations [33].

Only in very rare conditions, however, do the treatment choices specifically target the inherited pathophysiological mechanism. An interesting example is represented by autosomal dominant nocturnal frontal lobe epilepsy (ADNFLE) due to mutation of neuronal nicotinic acid acetylcholine receptor alpha subunit in which the therapeutic effect of nicotine patch treatment on refractory seizures was elegantly albeit anecdotally demonstrated using an N-of-1 trial [32]. Other studies have confirmed that transdermal nicotine administration may be a suitable treatment option for patients with ADNFLE and severe seizures [4]. However, the translational dimension of these observations is limited as nicotine is highly addictive and may cause cardiovascular effects. More specific examples derive from clinical conditions in which a rationale therapeutic approach is prompted by administering a substance that can correct a metabolic defect that causes epilepsy. Pyridoxine-dependent epilepsy, for example, is an autosomal recessive disorder in which seizures manifesting in the neonatal period or in infancy can only be controlled after administration of high doses of pyridoxine [12]. If untreated, the disorder can lead to life threatening status epilepticus. Affected patients require lifelong pyridoxine supplementation but antiepileptic medication is usually unnecessary. While prognosis for seizure control is excellent in most patients, neurodevelopmental impairment is often present and although it has been suggested that children who are treated early have a better outcome, this is not always the case [9]. Pyridoxine-dependent epilepsy is likely underdiagnosed and for this reason in many centers pyridoxine administration is part of a treatment protocol for neonatal seizures. Pyridoxine dependent epilepsy is caused by mutations in the *ALDH7A1* gene, which encodes for an aldehyde dehydrogenase (antiquitin) acting in the cerebral lysine catabolism pathway [21]. Affected individuals have α-aminoadipic semialdehyde (AASA) levels, which cause an intracellular reduction in the active vitamin B6 co-factor pyridoxal-5' -phosphate (PLP) and a concomitant imbalance of glutamic acid and γ-aminobutyric acid (GABA). Folinic acid-responsive seizures are very similar to PDE [8]. Early seizures can also be caused by deficient pyridox(am)ine 5' -phosphate oxidase (PNPO), which respond to pyridoxal-5' -phosphate supplementation [9].

A third important example of a direct link between genetic diagnosis and effective treatment choice is represented by the use of the ketogenic diet in the treatment of the GLUT1 deficiency syndrome, an autosomal dominant disorder due to a mutation in the *SLC2A1* gene. Brain glucose transport occurs by facilitated transport, predominantly via GLUT1, located on the blood–brain barrier endothelium, *SLC2A1* mutations result in insufficient transport of glucose into the brain. Patients with GLUT1 deficiency had originally been described as exhibiting a severe neurological syndrome with early intractable seizures, followed by developmental delay, microcephaly and paroxysmal or continuous dyskinesia. This condition was initially identified and subsequently diagnosed in clinical practice, based on low glucose levels in the CSF (hypoglycorrhachia) in the setting of normal serum glucose or of abnormal CSF/serum glucose ratios [6]. However, a wide range of variants have been described, resulting in variable degrees of impairment of glucose transport [23], complicating the utility of the genetic information in clinical practice [18]. Neurologic consequences of GLUT1 deficiency presumably arise from disordered brain energy metabolism, secondary to reduced transport. D -glucose is the main fuel for the brain, although alternative fuels such as ketone bodies can be used. The treatment of choice for GLUT1 deficiency syndrome is a diet that mimics the metabolic state of fasting and provides ketones as an alternative fuel for the brain, effectively restoring brain energy metabolism. The ketogenic diet is a high-fat, adequate protein, low carbohydrate diet that provides 87–90 % of daily calories as fats and is used in the treatment of drug resistant childhood epilepsy. As the developing brain requires substantially more energy in young children, patients with GLUT1 deficiency syndrome should be started on the diet as early as possible and should remain on the diet at least until adolescence. Although some patients with milder seizure disorders may respond to antiepileptic medication, most do not and seizure response to the ketogenic diet is remarkable. Also, some pharmacological agents such as phenobarbital and diazepam, impair GLUT1 function and should be avoided [3]. In the past few years, the range of clinical epilepsy phenotypes where GLUT1 mutations and a positive response to the ketogenic diet have been identified is expanding

[22, 26–28], raising the possibility that the gene acts as a modifier of other coexisting abnormalities. The usefulness of the ketogenic diet in GLUT1 deficiency syndrome was demonstrated before the syndrome was linked to mutations of the GLUT1 gene, based on the expected pathophysiological consequences of low levels of glycorrachia, however, the possibility of uncovering GLUT1 mutations in patients with atypical clinical presentations of GLUT1 deficiency and even borderline or normal levels of glycorrachia, has brought about invaluable advantages for the diagnosis and treatment of this disorder.

25.4   Conclusion

In the epilepsy clinic, genetic analysis has revealed not only the presence of more monogenic epilepsy syndromes, but, thanks to continually emerging genome-phenome correlations, is also pointing the way to earlier and more accurate diagnosis. Gene-specific classification of patients will aid clinical stratification to tailor more relevant diagnostic testing and better characterization of the natural history of the disease, enabling improved outcome prediction and genetic counseling for family planning. Future clinical treatment trials will almost certainly include genomic characterization to enhance the detection of a drug response signal as well as any gene-linked adverse effects. The relative contributions of major categories of genetic influence, including inherited monogenic epilepsies, de novo mutations, and sporadic individuals with complex multigenic inheritance are under exploration in many different seizure types and are beginning to inform genetic counseling in the neuropediatric setting. The definitions of classical epilepsy syndromes have been enlarged to account for multiple genetic etiologies, and novel comorbidity syndromes.

In the epilepsy neurobiology laboratory, genetics continues to reveal mechanistic insight into the rich biological diversity of gene defects leading to epilepsy phenotypes. Genes linked to epilepsy have opened the door to understanding the neurobiology and pathology of epilepsies and localizing the vulnerable neural pathways at the molecular, cellular, and functional levels. Defects in ion channels and a broad range of other cellular signaling pathways including receptors, transporters, and proteins for exocytotic release of neurotransmitters now constitute primary classes of epileptogenic mechanisms. A second major category of epilepsy genes involves transcription factors regulating the early migration and maturation of interneurons, and a third group controls metabolic functions within the cell and neuron-glia relationships.

Mouse models bearing mutations in each of these gene-delineated pathways allow us to examine the fine details of how they alter developmental plasticity in the epileptic brain, to learn when and where cellular pathology arises, and how it spreads to alter excitability in cortical networks through remodeling of gene expression and synaptic reorganization. Mice engineered to conditionally express gene mutations in specific circuits provide information on which circuits are necessary or sufficient to produce the seizure phenotype. Finally we can learn whether there are critical developmental stages for correcting or reversing the gene defect, exactly what the desired drug effect should be at the cellular level, and which molecular targets are most effective in preventing the epileptic disorder. Taken together, these advances hold great promise for improving the clinical management of seizure disorders.

Acknowledgements We extend our gratitude to Phil Schwartzkroin, a careful scientist, wise Editor-in-Chief, and trusted friend who has inspired us both throughout our careers in epilepsy research.

References

1. Bergren SK, Rutter ED, Kearney JA (2009) Fine mapping of an epilepsy modifier gene on mouse Chromosome 19. Mamm Genome 20:359–366
2. Bournissen FG, Moretti ME, Juurlink DN, Koren G, Walker M, Finkelstein Y (2009) Polymorphism of the MDR1/ABCB1 C3435T drug-transporter and resistance to anticonvulsant drugs: a meta-analysis. Epilepsia 50:898–903
3. Brockmann K (2009) The expanding phenotype of GLUT1-deficiency syndrome. Brain Dev 31:545–552
4. Brodtkorb E, Picard F (2006) Tobacco habits modulate autosomal dominant nocturnal frontal lobe epilepsy. Epilepsy Behav 9:515–520

5. Chen TT, Klassen TL, Goldman AM, Marini C, Guerrini R, Noebels JL (2013) Novel brain expression of ClC-1 chloride channels and enrichment of CLCN1 variants in epilepsy. Neurology 80:1078–1085

6. De Vivo DC, Trifiletti RR, Jacobson RI, Ronen GM, Behmand RA, Harik SI (1991) Defective glucose transport across the bloodbrain barrier as a cause of persistent hypoglycorrhachia, seizures, and developmental delay. N Engl J Med 325:703–709

7. Epi4K Consortium, Epilepsy Phenome/Genome Project, Allen AS, Berkovic SF et al (2013) De novo mutations in epileptic encephalopathies. Nature 501:217–221

8. Gallagher RC, Van Hove JL, Scharer G (2009) Folinic acid-responsive seizures are identical to pyridoxine-dependent epilepsy. Ann Neurol 65:550–556

9. Gospe SM (2011) Pyridoxine-dependent epilepsy. In: Shorvon S, Andermann F, Guerrini R (eds) The causes of epilepsy. Cambridge University Press, Cambridge, pp 237–241

10. Goldman AM, Glasscock E, Yoo J, Chen TT, Klassen TL, Noebels JL (2009) Arrhythmia in heart and brain: KCNQ1 mutations link epilepsy and sudden unexplained death. Sci Transl Med 1:2ra6

11. Guerrini R, Dravet C, Genton P, Belmonte A, Kaminska A, Dulac O (1998) Lamotrigine and seizure aggravation in severe myoclonic epilepsy. Epilepsia 39:508–512

12. Hoffmann GF, Schmitt B, Windfuhr M (2007) Pyridoxal 50-phosphate may be curative in early-onset epileptic encephalopathy. J Inherit Metab Dis 30:96–99

13. The Human Variome Project: http://www.humanvariomeproject.org/

14. SCN1A variant database: http://www.molgen.ua.ac.be/SCN1AMutations/

15. Jorge BS, Campbell CM, Miller AR et al (2011) Voltage-gated potassium channel KCNV2 (Kv8.2) contributes to epilepsy susceptibility. Proc Natl Acad Sci U S A 108:5443–5448

16. Kasperaviciute D, Sisodiya SM (2009) Epilepsy pharmacogenetics. Pharmacogenomics 10:817–836

17. Klassen T, Davis C, Goldman A, Burgess D, Chen T, Wheeler D, McPherson J, Bourquin T, Lewis L, Villasana D, Morgan M, Muzny D, Gibbs R, Noebels J (2011) Exome sequencing of ion channel genes reveals complex profiles confounding personal risk assessment in epilepsy. Cell 145:1036–1048

18. Klepper J (2012) GLUT1 deficiency syndrome in clinical practice. Epilepsy Res 100:272–277

19. Liu Y, Lopez-Santiago LF, Yuan Y, Jones JM, Zhang H, O'Malley HA, Patino GA, O'Brien JE, Rusconi R, Gupta A, Thompson RC, Natowicz MR, Meisler MH, Isom LL, Parent JM (2013) Dravet syndrome patient-derived neurons suggest a novel epilepsy mechanism. Ann Neurol 74:128–133

20. Manna I, Gambardella A, Bianchi A et al (2011) A functional polymorphism in the SCN1A gene does not influence antiepileptic drug responsiveness in Italian patients with focal epilepsy. Epilepsia 52:40–44

21. Mills PB, Struys E, Jakobs C (2006) Mutations in antiquitin in individuals with pyridoxine-dependent seizures. Nat Med 12:307–309

22. Mullen SA, Marini C, Suls A, Mei D, Della Giustina E, Buti D, Arsov T, Damiano J, Lawrence K, De Jonghe P, Berkovic SF, Scheffer IE, Guerrini R (2011) Glucose transporter 1 deficiency as a treatable cause of myoclonic astatic epilepsy. Arch Neurol 68:1152–1155

23. Nickels K, Wirrell E (2010) GLUT1-ous maximus epilepticus: the expanding phenotype of GLUT-1 mutations and epilepsy. Neurology 75:390–391

24. Noebels J (2011) A perfect storm: converging paths of epilepsy and Alzheimer's dementia intersect in the hippocampal formation. Epilepsia 52(Suppl 1):39–46

25. Palop JJ, Chin J, Roberson ED et al (2007) Aberrant excitatory neuronal activity and compensatory remodeling of inhibitory hippocampal circuits in mouse models of Alzheimer's disease. Neuron 55:697–711

26. Striano P, Weber YG, Toliat MR, Schubert J, Leu C, Chaimana R, Baulac S, Guerrero R, LeGuern E, Lehesjoki AE, Polvi A, Robbiano A, Serratosa JM, Guerrini R, Nürnberg P, Sander T, Zara F, Lerche H, Marini C, EPICURE Consortium (2012) GLUT1 mutations are a rare cause of familial idiopathic generalized epilepsy. Neurology 78:557–562

27. Suls A, Dedeken P, Goffin K et al (2008) Paroxysmal exercise induced dyskinesia and epilepsy is due to mutations in SLC2A1, encoding the glucose transporter GLUT1. Brain 131:1831–1844

28. Suls A, Mullen SA, Weber YG, Verhaert K, Ceulemans B, Guerrini R, Wuttke TV, Salvo-Vargas A, Deprez L, Claes LR, Jordanova A, Berkovic SF, Lerche H, De Jonghe P, Scheffer IE (2009) Early-onset absence epilepsy caused by mutations in the glucose transporter GLUT1. Ann Neurol 66:415–419

29. Tate SK, Depondt C, Sisodiya SM et al (2005) Genetic predictors of the maximum doses patients receive during clinical use of the anti-epileptic drugs carbamazepine and phenytoin. Proc Natl Acad Sci U S A 102:5507–5512

30. The 1000 Genomes Project Consortium (2010) A map of human genome variation from population scale sequencing. Nature 467:1061–1073

31. Yu FH, Mantegazza M, Westenbroek RE, Robbins CA, Kalume F, Burton KA, Spain WJ, McKnight GS, Scheuer T, Catterall WA (2006) Reduced sodium current in GABAergic interneurons in a mouse model of severe myoclonic epilepsy in infancy. Nat Neurosci 9:1142–1149

32. Willoughby JO, Pope KJ, Eaton V (2003) Nicotine as an antiepileptic agent in ADNFLE: an N-of-one study. Epilepsia 44:1238–1240

33. Zara F, Specchio N, Striano P et al (2013) Genetic testing in benign familial epilepsies of the first year of life: clinical and diagnostic significance. Epilepsia 54:425–436

34. Zimprich F, Stogmann E, Bonelli S, Baumgartner C, Mueller JC, Meitinger T, Zimprich A, Strom TM (2008) A functional polymorphism in the SCN1A gene is not associated with carbamazepine dosages in Austrian patients with epilepsy. Epilepsia 49:1108–1109

How Might Novel Technologies Such as Optogenetics Lead to Better Treatments in Epilepsy?

Esther Krook-Magnuson, Marco Ledri, Ivan Soltesz, and Merab Kokaia

Abstract

Recent technological advances open exciting avenues for improving the understanding of mechanisms in a broad range of epilepsies. This chapter focuses on the development of optogenetics and on-demand technologies for the study of epilepsy and the control of seizures. Optogenetics is a technique which, through cell-type selective expression of light-sensitive proteins called opsins, allows temporally precise control via light delivery of specific populations of neurons. Therefore, it is now possible not only to record interictal and ictal neuronal activity, but also to test causality and identify potential new therapeutic approaches. We first discuss the benefits and caveats to using optogenetic approaches and recent advances in optogenetics related tools. We then turn to the use of optogenetics, including on-demand optogenetics in the study of epilepsies, which highlights the powerful potential of optogenetics for epilepsy research.

Keywords

On-demand • Responsive • Channelrhodopsin • Halorhodopsin • Arch • AAV • Optrode • Seizure

E. Krook-Magnuson (✉) • I. Soltesz
Department of Anatomy and Neurobiology,
University of California,
192 Irvine Hall, Irvine, CA 92697, USA
e-mail: ekrookma@uci.edu; isoltesz@uci.edu

M. Ledri
Experimental Epilepsy Group, Epilepsy Center,
Department of Clinical Sciences, Lund University
Hospital, Sölvegatan 17, BMC A11,
22184 Lund, Sweden

Laboratory of Molecular Neurobiology, Department
of Molecular and Developmental Neurobiology,
Institute of Experimental Medicine,
Hungarian Academy of Sciences,
Szigony utca, 43, H-1083 Budapest, Hungary
e-mail: marco.ledri@med.lu.se

M. Kokaia
Experimental Epilepsy Group, Epilepsy Center,
Department of Clinical Sciences, Lund University
Hospital, Sölvegatan 17, BMC A11,
22184 Lund, Sweden
e-mail: Merab.kokaia@med.lu.se

H.E. Scharfman and P.S. Buckmaster (eds.), *Issues in Clinical Epileptology: A View from the Bench*,
Advances in Experimental Medicine and Biology 813, DOI 10.1007/978-94-017-8914-1_26,
© Springer Science+Business Media Dordrecht 2014

26.1   Introduction

By enabling unprecedented possibilities for
studying the cell populations and networks
involved in seizure initiation, propagation, and
termination, recent technological advances open
exciting avenues for improving the understanding of mechanisms in a broad range of epilepsies.
Through optogenetics, modulation of select cell
populations is possible at specific times, providing the opportunity to not only record neuronal
activity during seizures, but also to manipulate
neuronal activity. In this way, it is possible to
probe critical networks and circuits, and identify
potential new therapeutic approaches. This chapter
focuses on the development of optogenetics and
on-demand technologies for the study of epilepsy
and the control of seizures.

Optogenetics is a rapidly evolving field
providing powerful tools for neuroscience
[19, 72], including the study of epilepsies [12,
46]. Optogenetics is a technique in which light-
sensitive proteins, called opsins, are introduced
into cells. In this way, it is possible to control
the activity of neuronal populations by shining
light and activating the opsins. Opsins can be
light-sensitive channels, pumps, G-protein-
coupled receptors, or even transcriptional effectors [47]. We focus on light-sensitive channels
and pumps whose activation can inhibit or
excite neurons, and first discuss the benefits
and caveats to using these optogenetic approaches,
as well as recent advances in related tools.
We then turn to the use of optogenetics in the
study of epilepsies specifically.

26.2   Optogenetics: Development
and Technical Advances

Cell-type and temporal precision are two key
strengths to optogenetic approaches. Temporal
precision is achieved by appropriately timed light
delivery (though, of course, this can present its
own challenges, as discussed below for on-
demand approaches). Selective opsin expression
is less straightforward, and is achievable through

distinct methods. In general, expression is often
achieved through the use of viral vectors, (including
adeno-associated virus (AAV) or lentivirus),
electroporation [2], or the use of transgenic animals. Inducible expression [79, 80, 103] and
selective expression of opsins can be achieved for
specified populations of neurons defined by their
neurochemical profile (e.g., expression of parvalbumin), developmental origin [20, 25, 59, 81],
their date of birth (e.g., through the use of retro-
viruses which only infect actively dividing cells
[83]), their location (e.g., by injecting virus in a
restricted region), levels of activity at a specific
time [36], or their long-distance projections (e.g.,
through the use of WGA-Cre, which is retro-
gradely transported transynaptically [34]).

26.2.1 Selective Cell-Type Expression

To achieve selective expression in neurons
defined by their neurochemical profile, two broad
methods are used. The most straight forward
approach is to place the expression of the opsin
under a specific promoter (or even enhancer
[88]). However, this approach has three disadvantages. First, especially when used with
viruses, leaky expression is often noted (that is,
expression in other cell populations). Second,
long promoters do not fit in small vectors (e.g.
adeno-associated viruses (AAV)). Third, in cases
where the promoter is a relatively weak promoter,
the expression of opsins can be insufficient to
achieve strong light-induced currents and alter
the activity of the neurons.

In order to overcome these drawbacks, a second
method was developed: the opsin is instead
placed under a strong promoter, and selectivity is
achieved through the Cre/loxP system. Cre can
mediate either inversion (flipping) or excision
(removal) of DNA, depending on the relative ori-
entations of the loxP sites. For viruses, attempts
at selective expression through the introduction
of a floxed STOP cassette (which would be
excised by Cre) can produce leaky expression
(expression even in cells not expressing Cre).
Additionally, attempting selective expression
through a single inversion (which is then flipped

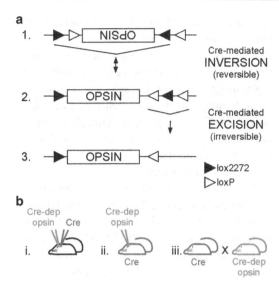

Fig. 26.1 Strategies for selective opsin expression.
(a) The FLEX system makes use of two pairs of loxP sites (*triangles*), including the mutated lox2272 (*dark triangles*). Cre mediates inversion using one set of loxP sites (for simplicity, only the inversion using lox2272 sites are illustrated), flipping the opsin sequence into the correct orientation (stage 2). Cre-mediated excision of one of each loxP sites locks the vector in an active state (stage 3) (Based on Figure 1 from Ref. [9]). (b) Three potential ways to achieve selective opsin expression include (*i*) injecting a Cre-dependent virus (as in **a**) and a Cre-delivering virus (e.g., WGA-Cre, as further discussed in the text [34]), (*ii*) injecting a Cre-dependent virus into a mouse expressing Cre in a subset of neurons, or (*iii*) crossing a mouse line expressing Cre in a subset of neurons with a mouse line expressing opsins in a Cre-dependent manner

by Cre to allow transcription) can produce weak expression, as Cre can continue to mediate flipping, re-inverting the sequence and inhibiting transcription. Therefore, a FLEX system ('flip-excision' [9, 71], also referred to as DIO – double-floxed inverse open reading frame [99]) was implemented (Fig. 26.1a). In this scenario, two sets of loxP sites are used. For one, a mutated sequence is used – lox2272. This sequence is still recognized by Cre, but is only paired with a similarly mutated sequence [51]. Therefore, two distinct sets of loxP pairs can be achieved (one set carrying the mutation, and one set not). One round of Cre-mediated recombination flips the sequence, and another excises one of each type of loxP site, preventing future recombination and

locking the virus in its activated state. This method has proven effective in achieving specific opsin expression, as well as sufficient levels of opsin expression [9]. Cre can be introduced by several methods, including virus injection (note that only low levels of Cre expression are needed). WGA-Cre, mentioned above, can be used to achieve selective expression based on axonal projections [34]. For example, a FLEX-opsin virus can be injected into the hippocampus contralateral to WGA-Cre virus injection, to achieve opsin expression selectively in hippocampal neurons projecting contralaterally, e.g., mossy cells [34]. Alternatively, Cre-dependent virus can be injected into a transgenic mouse (or rat) line expressing Cre in a select population of neurons.

There is a wealth of transgenic mouse lines available, including an ever-growing resource of Cre lines [80], many of which are commercially available (e.g., the Jackson laboratory Cre Repository: cre.jax.org). In addition to being useful in combination with Cre-dependent viral-based opsin expression methods, Cre lines can be crossed with lines expressing opsins in a Cre-dependent fashion [56] (Fig. 26.1b). For example, the Ai32 line developed at the Allen Institute expresses the excitatory opsin channelrhodopsin fused to an enhanced yellow fluorescent reporter protein (ChR2(H134R)-EYFP) from the endogenous *Gt(ROSA)26Sor* locus (a locus active in most cells) with expression enhanced with a CAG promoter [56]. Cre mediates removal of a floxed STOP cassette, and allows expression of the opsin.

An important caveat for Cre-mediated selectivity is that excision of DNA (e.g., removal of the STOP cassette) is permanent, even if Cre-expression itself is transient. This means that opsins can be expressed in cells that are not (currently) expressing Cre. Indeed, even if the cell is simply descended from a cell in which recombination has occurred, opsins will be expressed. This caveat can have significant experimental consequences. For example, following seizures, somatostatin (a neuropeptide whose expression is often used as a biochemical marker for populations of inhibitory interneurons) is transiently expressed in principal cells

[26]. If selective opsin expression in somatostatin-expressing interneurons is being achieved through a Cre-dependent mechanism, selectivity of expression will be (permanently) lost following a seizure.

Another major limitation of available methods for achieving opsin-expression selectivity is the current inability to achieve selectivity in a population defined by multiple characteristics. For example, within a broad neuron population defined by a single neurochemical marker, there are several distinct cell-types. In the hippocampus alone, axo-axonic (also referred to as chandelier cells), dendritically targeting bistratified cells, and a subset of basket cells (which target the perisomatic region of postsynaptic cells) all express the calcium binding protein parvalbumin [8, 30, 41, 45]. Therefore, selective opsin expression in parvalbumin-expressing neurons still results in expression across multiple cell-types. Additionally, there are interneurons that are defined in part by expression of proteins which are also expressed by principal cells. For example, subsets of interneurons express the neuropeptide cholecystokinin (CCK) [30, 45, 53]. However, as principal excitatory cells can also express CCK, selective expression in interneurons cannot be achieved through a Cre-mediated mechanism alone.

Importantly, this is a limitation of current methods which can be overcome through intersectional transgenics [80]. By combining the powerful Cre/loxP system with the Flp/Frt system (an analogous, but distinct, recombination system), it is possible to require expression of two markers for opsin expression. For example, Cre expression could be placed under the CCK promoter (and thus expressed in CCK-expressing cells) and Flp placed under an interneuron-specific marker. Indeed, selective expression of fluorescent proteins has already been achieved in CCK interneurons by using such an approach and a RCE-dual reporter mouse line [80]. However, in order for such an approach to be used for selective expression of opsins, mouse lines or viral vectors requiring both Cre and Flp for opsin expression will need to be generated. Additionally, while there is a vast resource of Cre lines, Flp-lines are markedly scarcer, and the field would certainly benefit from an increase in this resource. Note that beyond allowing access to relatively selective expression in more interneuron types (including neurogliaform and ivy cells, the numerically most dominant interneuron cell type in the hippocampus [7, 16, 31]), intersectional transgenic approaches could also overcome the loss of selectivity for somatostatin interneurons following seizures (described above).

In order to apply optogenetics in humans, a viral-based approach will clearly have to be used. Note that viral vectors have been used in humans, including in the brain [10, 58, 62], and gene-delivery in general is being considered for a range of neurological diseases [10, 87]. Beyond optogenetics, gene-delivery itself may be a new approach in epilepsy [67, 74, 94]. Optogenetic tools to modify gene transcription may also one day be used therapeutically [47]. Note that insertional mutagenesis (and the risk for tumor generation) can be avoided by using vectors which remain extrachromosal (e.g. recombinant AAV).

For animal studies, however, transgenic mouse methods offer several benefits over viral-based approaches. First, injection of virus is an invasive process, which is avoided through a transgenic-only approach. Second, for viral-based expression methods, the level of opsin expression varies depending on the number of copies of viral vector in the cell. Therefore, there can be great cell-to-cell variability in the amount of opsin expression. In some cases expression can be so high that light induces toxic levels of current. Of course, high levels of expression can also be a benefit of viral-based methods, when transgenic lines do not produce strong enough photocurrents. Third, if the site of virus injection and the placement of the optical fiber delivering light to the tissue are improperly aligned, insufficient light may reach the opsin-expressing neurons. In contrast, in the transgenic lines, variability is reduced, even expression is achieved in the select cell population throughout the brain, and spatial selectivity is achieved through the location of light delivery.

In addition to the Cre-dependent opsin expressing mouse lines described above, there are several mouse lines expressing opsins directly under a specific promoter. This avoids the need for crossing strains, and leaves open the door for other Cre-based manipulations. However, a strong promoter must be used to achieve sufficiently high levels of opsin expression. The currently available lines include mice expressing the excitatory opsin channelrhodopsin under the Thy1 promoter [5]. Many of these mice are commercially available. Finally, there have been recent developments in achieving transgenic optogenetic rats [82], further expanding the possibilities for using optogenetics in epilepsy research.

26.2.2 Direction of Modulation of Neuronal Activity

In addition to cell-type specificity, another benefit of optogenetic approaches, over for example electrical stimulation, is the control of direction of modulation of neuronal activity (e.g., excitation versus inhibition).

26.2.2.1 Activation

Two main classes of opsins are available, allowing cell-specific activation or inhibition. Most of the optogenetic tools used for neuronal activation derive from Channelrhodopsin-2 (ChR2), a naturally-occurring, non-selective cation channel expressed by the algae *Clamydomonas reinhardtii*. Upon exposure to blue light (470 nm absorption peak), ChR2 opens and allows passive movement of Na^+, K^+, Ca^{2+} and H^+ following the electrochemical gradient [63], depolarizing cell membranes, and if the cell is depolarized to threshold, generating action potentials. ChR2 possesses fast activation kinetics, and is able to trigger single action potentials in expressing cells following 1–2 ms light exposure, making it a particularly attractive tool for precisely timed stimulation of neuronal populations. ChR2 was the first opsin successfully expressed in mammalian neurons [19]. Since then, researchers have focused on improving its expression levels and its ON/OFF kinetics, and developed different variants with largely diverse properties. Several com-

prehensive reviews about Channelrhodopsin variants are available in the literature (see references [14, 54, 60]), and new variants are constantly being developed. Here we will focus on some important variants most often used in epilepsy research applications.

The first modification to the original ChR2 sequence, an amino acid substitution at position 134, produced a variant with improved expression levels and larger photocurrents in neurons (ChR2-H134R), but presenting slightly lower deactivation kinetics [33]. The lower deactivation kinetics produces lower fidelity of light pulse to action potential generation at high light stimulation frequencies, such that cell firing may not accurately follow the stimulus (missed spikes and/or multiplet spikes per light pulse). Further research therefore then focused on improving kinetics, to allow activation of neuronal populations at higher frequencies (above 40 Hz) with better fidelity of spike generation. The first variant producing higher consistency of high frequency spike generation was developed by a chimeric combination of ChR1 (another channelrhodopsin from *C. reinhardtii*) and ChR2, and was named ChIEF [55]. ChIEF displayed reduced inactivation during persistent light stimulations and improved fidelity at frequencies higher than 25 Hz. Similarly, one amino acid substitution at position 123 of the original ChR2 sequence, led to the development of ChR2(E123T), or ChETA [37], a ChR2 variant displaying dramatically improved activation/deactivation kinetics, allowing consistent and reliable action potential generation at frequencies up to 200 Hz. However, photocurrents generated by ChETA were somewhat smaller than wild-type ChR2, posing a potential drawback for its successful application *in vivo*. To solve this issue, an additional modification of the amino acid sequence at position 159 resulted in the development of ChR2(ET/TC), an improved ChETA variant combining high temporal fidelity with large photocurrent generation [14]. ChR2(ET/TC) still represents to date the channelrhodopsin with the best performance in terms of spike fidelity generation and amplitude of photocurrents.

All of the ChR2 variants described above display an excitation maximum at around 470 nm, and require blue light for their activation. However, the propagation of light in tissue is directly proportional to its wavelength, with blue light presenting high scattering and low penetration compared to higher wavelengths such as red light. Additional penetration through brain tissue is achieved by avoiding wavelengths absorbed by hemoglobin. For experiments requiring coverage of large brain areas, channelrhodopsin variants with red-shifted absorption maxima are therefore preferred, as they allow activation of an increased number of neurons with lower light stimulation intensity. The first attempt towards generating red-shifted activating opsins was made by cloning VChR1, a channelrhodopsin naturally expressed by the spheroidal alga *Volvox carteri*. VChR1 presented an excitation maximum at 550 nm, but significantly lower photocurrents and expression levels in mammalian neurons when compared to ChR2 [100]. To improve VChR1 photocurrents, researchers created a chimera by substituting helices 1 and 2 of VChR1 with their analogs in ChR1, thereby developing C1V1 [97]. Subsequent modification of glutamic acid residues at positions 122 and 162 (resulting in C1V1-T/T) further improved its photocurrents, and resulted in a channelrhodopsin variant with photocurrents comparable to ChR2(H134R) and excitation maximum at 550 nm. C1V1(T/T) also presented vastly increased light sensitivity, allowing its activation with lower light power, making it especially attractive for *in vivo* studies.

The ChR2 variants described above enable fast and precise activation of neuronal populations, but are not optimal for experiments requiring activation of specific neuronal population over longer time windows (minutes). At expression levels typically achieved in neurons, long time activation would require constant delivery of high power to the tissue, with potential and undesirable heating effects. To allow neuronal activation for longer time periods, a separate class of activating opsins was developed, where a single brief pulse of blue light is sufficient to trigger the channel into its active state. Channelrhodopsins with these properties were named Step-Function Opsins (or SFOs), and caused depolarization of cell membranes for periods of 30–60 s after 10 ms blue light exposure [15]. Even slower deactivation kinetics were achieved with a Stabilized SFO [97], displaying dramatically improved light sensitivity and a channel deactivation time constant of about half an hour. A major advantage of SFOs is that they can be used to slightly alter the network contribution of different cell types, as the depolarization they provide following light is generally subthreshold, and therefore does not directly activate expressing cells, but only increases cell sensitivity in responding to physiological network activity.

26.2.2.2 Suppression

The second major class of optogenetic tools available for the study of neuronal networks is constituted by opsins able to hyperpolarize the cell membrane and, if strong enough, silence action potential generation. The first opsin shown to inhibit neuronal activity was halorhodopsin (NpHR), a chloride pump driven by orange light and naturally expressed by the bacterium *Natromonas pharaonis*. When expressed in neurons, exposure to orange light (570 nm absorption maximum) causes active pumping of chloride ions into the cell, thereby hyperpolarizing the membrane potential and inhibiting action potential generation [98]. However, expression of NpHR in neurons was not optimal, and it formed aggregates in the endoplasmic reticulum that could lead to cellular toxicity [33]. Further development of the NpHR sequence focused on decreasing aggregates, improving photocurrents and promoting membrane localization. Several rounds of substantial mutagenesis of the original NpHR sequence allowed researchers to develop a variant (named eNpHR3.0) displaying a threefold increase in photocurrents and twofold increase in membrane hyperpolarization effects, together with a significant red shift of its excitation wavelength, making eNpHR3.0 ideal for a varied range of studies involving neuronal silencing [34].

Although halorhodopsin chloride pumps are able to reduce neuronal activity with high efficiency, actively pumping chloride ions into the neurons could have effects on chloride homeostasis, with potential shifts in the effect of GABAergic inhibition via chloride-permeable $GABA_A$ receptors (i.e., shifting E_{GABA}) [66]. E_{GABA} is already compromised in epileptic tissue [35]. Increasing the intracellular concentration of chloride by its active pumping via NpHR activation could further exacerbate this phenomenon, and cause a shift in E_{GABA} to the point where $GABA_A$ activation becomes depolarizing [66].

Together with halorhodopsins, a separate class of tools to inhibit neuronal activity was developed from naturally-occurring proton pumps derived from different strains of the bacterium *Halorubrum sodomense*. In contrast to NpHR and its variants, proton pumps hyperpolarize cell membranes by actively transporting protons to the extracellular environment, upon exposure to orange/yellow light. The most widely used proton pumps include Archaeorhodopsin-3 (also called Arch [22]) and ArchT [38]. Both have been shown to be able to successfully inhibit neuronal activity *in vitro* and *in vivo*, including when expressed in the brain of non-human primates [38]. Recently, Arch3.0 and ArchT3.0 variants were developed, using modifications similar to those made to the original NpHR sequence, and yielded proton pumps displaying large photocurrents in neurons and increased action potential silencing effects [60]. Due to the fact that these pumps rely on active transport of protons for hyperpolarizing cell membranes (rather than chloride transport), they would not contribute to the disturbance in chloride reversal potential and $GABA_A$-mediated inhibition [66], but may have alternate effects, such as altered pH.

Channelrhodopsins and halorhodopsins or proton pumps can also be expressed simultaneously in the same cells to allow bidirectional control of the cell population of interest [34, 39, 40]. ChR2 and most of its variants are activated by blue light, and are therefore spectrally compatible with NpHR or Arch variants, which are activated by orange/yellow light. Moreover, if particular experimental conditions require simul-

taneous activation of one population and silencing of another, a combination of red-shifted channelrhodopsins could be used together with NpHR or Arch. This could be used, for example, to study the effects of simultaneous pyramidal cell silencing and GABAergic interneuron activation on seizure activity.

Although the expression of opsins can be specific and directed to desired cell populations using the strategies described above, the outcome of neuronal activation and/or silencing in intact networks can be more intricate than perhaps initially expected, due to the extremely complex nature of neuronal circuits. For example, results from *in vivo* experiments using ChR2-mediated light stimulation show some cells being activated (as expected), while others are silenced, likely due to network interactions [38–40]. Similarly, in a study using ArchT activation (expected to inhibit cells), a substantial number of neurons responded to light instead by increasing their firing rate [38]. As epileptic circuits often undergo considerable changes, including axon sprouting and changes in network connections, potentially unexpected network roles should also be considered when using optogenetics with epileptic tissue. Indeed, optogenetics provides a powerful means to explore these changes and their consequences on the functioning of the network in epilepsy. Provided opsin expression remains specific, optogenetics provides the ability to examine the role of specific neuronal populations in health and disease in a manner previously unachievable with techniques such as electrical stimulation.

26.2.3 Light Delivery

In experimental conditions, light is delivered by using a variety of different systems, depending on the needs. Sources able to generate light with suitable wavelength and power include lasers and light emitting diodes (LEDs). Laser sources have the advantage of providing light with narrow wavelengths, and therefore do not require filtering. Additionally, lasers can provide high power, even when coupled to small diameter fibers which are routed through optical commutators. Lasers can be used with mechanical shutters for light on/off switching to avoid delays in reaching maximum power. However, shutters can be expensive, sensitive, and have relatively short life expectancies. A major disadvantage of lasers is their cost. LED light sources are generally more affordable and are becoming increasingly powerful. While LEDs have the disadvantage of delivering light with typical "tails" in excitation spectrum, these can be adequately filtered to ensure proper wavelength excitation. LED sources typically reach maximum power in less than 200 μs even at very high frequency. Therefore, light can be switched on or off by delivering external voltage pulses (rather than via a mechanical shutter) without sacrificing light power.

For *in vitro* preparations light is typically delivered through the lens of the microscope [48, 84, 102], although other methods are also used, including optical fibers positioned in close proximity to the tissue area of interest [50]. For example, small diameter fibers [50] or laser-scanning photostimulation [95] can be used to activate specific regions in the slice, allowing for example circuit mapping and investigation of network alterations occurring after seizures. When light is delivered through the lens of the microscope, filtered light from a mercury or xenon lamp source can also be used, similar to epifluorescence applications.

For *in vivo* situations, light is most commonly delivered through an optical fiber implanted in the region of interest and connected to the light source of choice. Sophisticated light delivery options have also been designed, including multi-waveguides capable of delivering light of different wavelengths to different locations along the guide [104]. Additionally, the optical fiber can be combined with a recording electrode. The combination is termed an optrode, and a number of designs and protocols exist [1, 4, 42, 69, 75, 78, 89, 90, 101], including a recent protocol for simple and relatively low cost optrodes designed for chronic (months long) recordings in rodents [6]. Optical fibers can be directly implanted [6] or guided into the tissue by a cannula previously fixed to the animal's skull [99]. For long-term *in*

vivo applications, an optical commutator is often used to reduce torque on the optical patch cord connecting the animal and the light source. There also exist wireless options for light delivery, including headborne LED devices [90] and injectable μ-LEDs [44].

A major caveat to consider while planning *in vivo* optogenetic experiments is that brain penetrance by light is rather limited, as described above, and progressively reduces with decreasing wavelengths. Therefore, the spatial distance between the light and the cells expressing the opsin can be critical, and will determine the minimal required power for adequate activation of the transgene. The choice of opsin is also important, as some are several fold more sensitive to light than others, or have red-shifted excitation maximum allowing simultaneous activation (or inhibition) of a large number of neurons while maintaining a small diameter optical fiber (reducing tissue damage).

26.3 Optogenetics: Shedding Light on Epilepsy

26.3.1 Review of Recent Studies

The first attempt at using optogenetic approaches for suppressing abnormal hypersynchronized activity involved expression of eNpHR (a slightly improved NpHR protein) in pyramidal cells of the hippocampus [84]. The inhibitory opsin was introduced in excitatory principal cells of organotypic hippocampal slices by using a lentivirus carrying the NpHR transgene under the control of the CaMKII alpha promoter, which is expressed in excitatory cells and absent in inhibitory interneurons. Organotypic hippocampal slices are proposed to represent an *in vitro* model of epileptic tissue, as they exhibit network reorganization, such as cell death, axonal sprouting and synaptic formation, leading to hyperexcitability [3, 11]. The ability of NpHR to inhibit epileptiform activity in such "epileptic" tissue was tested by applying orange light during stimulation train induced bursting (STIB), a stimulation protocol that reliably evokes afterdischarges in

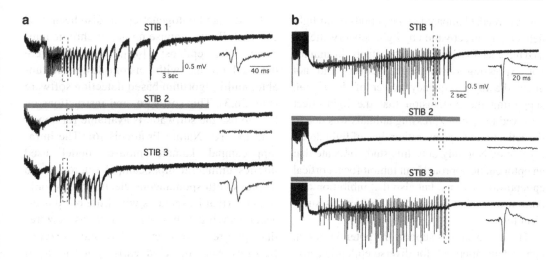

Fig. 26.2 Optogenetic inhibition of epileptiform activity in vitro. NpHR expression in excitatory principal cells of organotypic hippocampal slices is efficient in inhibiting stimulation train induced bursting (STIB) when activated by *orange light*, in both CA3 (**a**) and CA1 (**b**) areas. Stimulation with *blue light* failed to alter STIB-induced bursting (**b**, *bottom*) (Reproduced with permission from Ref. [84])

the CA1 and CA3 area. Orange light application effectively suppressed STIB-induced activity, while blue light application was ineffective, indicating the specificity of the approach used (Fig. 26.2) [84].

Optogenetic approaches have also had success in inhibiting seizures *in vivo* across a range of epilepsies, including induced (acute) seizures [76], focal cortical seizures [94], temporal lobe seizures during the chronic (spontaneous seizures) phase of the disease [48], and thalamocortical epilepsy in a model of cortical stroke [65].

Using the rat pilocarpine model of acute induced seizures in awake behaving male rats, Sukhotinsky and colleagues examined the ability to inhibit seizures using optogenetic inhibition of the hippocampus [76]. The inhibitory opsin halorhodopsin (eNpHR3.0 [34]) was expressed in principal excitatory cells in the hippocampus using adeno-associated virus (AAV) and a CamKIIα promoter. Animals receiving light and expressing the opsin showed an increase in time to seizure onset from the time of pilocarpine injection compared to controls (time to seizure onset with opsin activation: 21 ± 1.8 min versus 15.2 ± 1.1 min in controls). Controls included animals not injected with virus and not receiving

light, animals injected with virus but not receiving light, and animals receiving light but not expressing the opsin. Therefore, the activation of opsins (and inhibition of hippocampal principal excitatory cells) delayed the time to seizure onset. This study supports the notion that optogenetics can be used to inhibit seizures. Moreover, it indicates that targeted inhibition of principal cells in the hippocampus can delay the onset of pilocarpine induced seizures.

Wkyes and colleagues demonstrated the successful use of an optogenetic approach to inhibit focal cortical seizures [94]. Neocortical epilepsy is frequently drug-resistant, and new therapeutic approaches are being actively sought. Focal cortical epilepsy was induced in rats by focal injection of tetanus toxin into the motor cortex. Lentivirus was co-injected with the tetanus toxin, in order to transduce excitatory pyramidal neurons in the epileptic focus with the inhibitory opsin halorhodopsin (NpHR2.0, under a CamKIIα promoter). Seven to 10 days after the injection of tetanus toxin, the ability of an optogenetic approach to inhibit seizures was investigated. EEG was recorded for a 1,000 s baseline period, then intermittent light (20 s on, 20 s off) was delivered for 1,000 s, and then a final 1,000 s of post-light EEG was recorded. Compared to the periods of no-light delivery, opsin activation by light delivery attenuated recorded EEG epileptiform activity. In animals not expressing opsins, light delivery did not affect the high-frequency power of the signal, supporting the conclusion that the light-effect observed in opsin-expressing animals was due to the activation of opsins, rather than of light delivery *per se*. Not only does this study indicate that an optogenetic approach can inhibit focal cortical epileptiform activity, but also that inhibition of a portion of excitatory cells at the focus is sufficient to do so.

These two studies support the potential for an optogenetic approach for diverse epileptic activity, and make use of the power of optogenetics to selectively target specific populations of cells. An additional major benefit of optogenetics is the temporal precision which it can provide. That is, an optogenetic approach could be employed in an on-demand or responsive fashion, such that intervention only occurred either immediately before a seizure would occur (seizure prediction) or early during a seizure onset (seizure detection). In addition to the experimental benefits of an on-demand approach, by limiting intervention to only those times when it is needed, an on-demand approach may reduce negative side effects associated with chronic treatments.

On-demand optogenetics have been used in two models of epilepsy – thalamocortical and temporal lobe epilepsy. Using a cortical stroke model of thalamocortical epilepsy, and line-length threshold crossing for automated seizure detection, Paz and colleagues demonstrated the successful inhibition of seizures [65]. The inhibitory opsin halorhodopsin (eNpHR3.0 [34]) was expressed under a CamKIIα promoter in the ventrobasal thalamus ipsilateral to the site of induced cortical stroke. On-demand light activation of opsins interrupted seizures. In addition to illustrating the potential for on-demand optogenetics to stop seizures, these findings supported the theory that the cortical strokes produced thalamocortical seizures; that is, optogenetics can provide insight into the mechanisms of seizures, including critical brain regions and networks.

On-demand optogenetics has also been used successfully in a mouse model of chronic temporal lobe epilepsy [48]. Seizures were detected on-line with custom-designed, tunable, multi-algorithm based detection software (Fig. 26.3). This software, and instructions on how to use the software, is available for download through Nature Protocols [6]. The intrahippocampal kainate mouse model used mimics unilateral hippocampal sclerosis, and displays both spontaneous electrographic-only seizures (that is, seizures with little or no overt accompanying behavior) as well as seizures that progress to overt behavioral seizures. Seizures were detected early, prior to overt behavior. Selective expression of the inhibitory opsin halorhodopsin (eNpHR3.0) was achieved by crossing mice expressing halorhodopsin in a Cre-dependent fashion with mice expressing Cre under the CamKIIα promoter [56]. On-demand light delivery to the hippocampus, inhibiting excitatory cells, dramatically truncated seizures (Fig. 26.4).

Krook-Magnuson et al. [48] then went on to try a second approach. Rather than inhibiting excitatory cells directly through optogenetics, the authors instead used optogenetics to excite a subpopulation of inhibitory neurons. Selective expression of the excitatory opsin channelrhodopsin (ChR2) was achieved by crossing mice expressing ChR2 in a Cre-dependent manner with mice expressing Cre selectively in parvalbumin-expressing neurons. In the hippocampus, parvalbumin-expressing interneurons represent less than 5 % of the total neuronal population [16, 30, 92]. Remarkably, seizures were significantly inhibited through this approach. Seizures were also significantly inhibited when light was delivered to the contralateral hippocampus. Finally, light delivery reduced the number of seizures progressing to overt behavioral seizures. These data indicate that focal light delivery can have a significant effect on temporal lobe seizures, that an on-demand approach can work in temporal lobe epilepsy, and that a strategy directly targeting only a small fraction of cells (that is, parvalbumin-expressing interneurons) can significantly inhibit temporal lobe seizures.

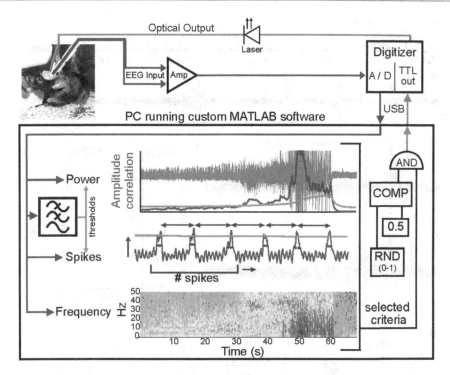

Fig. 26.3 Schematic of online seizure detection for on-demand optogenetics. EEG input (*blue*) recorded from the animal is amplified (Amp), digitized (A/D), and relayed to a PC running real-time seizure detection software. This software is tuned for each animal, with user-defined thresholds (*green*). Seizure detection algorithms utilize features of signal power (*top*), spikes (*middle*), or frequency (*bottom*). Once a seizure has been detected using the selected criteria, the software can activate, via a TTL signal from the digitizer to the laser, the optical output (*orange*) delivered to the animal. Signal power related calculations (*purple*, during an example seizure shown in *grey*), spike characteristics (e.g., amplitude, rate, regularity, and spike width, shown in *red*), and frequency characteristics (shown for the same seizure, with warmer colors representing higher energy) are illustrated. COMP: digital comparator. This on-line seizure detection software is available for download through reference [6] (Figure reproduced with permission from Ref. [48])

26.3.2 Obstacles and Future Potential in Epilepsy Research

While the results from these studies are promising, a number of hurdles need to be overcome before optogenetics, and hopefully on-demand optogenetics, can be realized in the clinical setting. These include demonstration of safe and stable opsin expression in humans, as well as a safe implantable device for on-line seizure detection and light delivery. However, on-demand optogenetics, with its cell-type, spatial, and temporal-specificity, may one day aid patients currently suffering from uncontrolled seizures and the negative side-effects of systemic treatment options. An example patient population that could benefit from the clinical realization of an on-demand optogenetic therapeutic is patients with refractory bilateral temporal lobe epilepsy for whom surgical resection is not an option.

Optogenetics additionally presents a powerful tool for expanding our understanding of mechanisms of epilepsy. While the studies discussed here have demonstrated a wide potential for optogenetics in the field of epilepsy, there is much more to be gained from fully harnessing the power of optogenetics. Through optogenetics it is possible to test hypotheses regarding critical cell-types and networks involved in the initiation, continuation, propagation, and (natural or induced) cessation of seizures. The studies described above inhibited seizures using optogenetic techniques, but it is also possible to use optogenetic approaches to study mechanisms

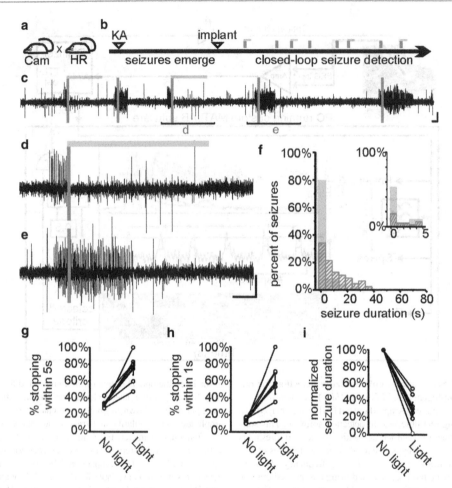

Fig. 26.4 Seizure control in vivo in mice expressing HR in principal cells in a model of temporal lobe epilepsy. (a) Crossing CamK-Cre and Cre-dependent halorhodopsin (HR) mouse lines generated mice expressing the inhibitory opsin HR in excitatory cells (Cam-HR mice). (b) Experimental timeline. (c–e) Example electrographic seizures detected (*vertical green bars*), activating amber light (589 nm) randomly for 50 % of events (light: *amber line*, example in **d**; no-light example in **e**). (f) Typical example distribution of post-detection seizure durations (5 s bin size) during light (*solid amber*) and no-light inter-

nal control conditions (*hashed gray*). Inset: first 5 s bin expanded, 1 s bin size. Note that most seizures stop within 1 s of light delivery. (g–i) Group CamHR data showing the percent of seizures stopping within 5 s of detection (**g**), within 1 s of detection (**h**), and the average post-detection seizure duration (normalized to average no-light post-detection duration for each animal) (**i**). Note that in one animal (shown in **c–e**), all seizures were stopped within 1 s of light delivery. Averaged data: filled circles. Error bars represent s.e.m. Scale bars in **c–e**: 100 µV, 5 s (Reproduced with permission from Ref. [48])

of epilepsy through the induction (rather than inhibition) of seizures [64]. The information gained from optogenetic experiments can in turn open the door for new therapeutic approaches beyond optogenetics, including new drugs targeting key cell types or electrical stimulation targeting key brain regions.

While the field is benefiting greatly from recent technological advances, there is a continuing need

for additional developments. A reliable and inexpensive long-term EEG monitoring system, with fully computerized analysis of EEG and video for automated detection and analysis of electrographic and behavioral seizures, would push the field forward dramatically. For example, this would increase the feasibility (and statistical power by allowing more animals to be monitored and analyzed) of studies with mild or moderate head injury for which only a small subset of animals go on to develop epilepsy. Recent advances in wireless devices, including those capable of delivering light [44, 90], and improvements in seizure detection [6, 91] are paving the way for such future advances. These advances will additionally improve the utility of optogenetics for epilepsy research by allowing chronic on-demand light delivery to freely moving, untethered, animals.

26.4 Other Technical Advances: New Avenues, New Insights

This chapter has focused on optogenetics. Clearly, however, the field takes advantage of a large range of new technological advances, several of which are being rapidly integrated with optogenetics. For example, on-demand approaches (which as described above can be successfully integrated with optogenetics) have the potential to provide both experimental and therapeutic benefits. While electrical stimulation lacks the cell-type specificity of optogenetics, it can provide temporal precision, and thus can also be used in an on-demand fashion. Previously, on-demand electrical stimulation was found to provide superior seizure control in rats [32]. More recently, on-demand transcranial electrical stimulation (TES) was used to reduce spike-and-wave episodes in absence seizures in rats [13]. There is also intense clinical interest in an on-demand therapeutic option, and clinical trials have shown promise (reviewed in reference [93]).

A step beyond early seizure detection is seizure prediction. A recent study in patients with drug-resistant partial-onset epilepsy was able to predict for a subset of patients periods of high seizure risk and periods where the chances of having a seizure were relatively low, based on an analysis of the frequency bands recorded from intracranial EEG [24]. Further supporting the possibility of seizure prediction, changes in multi-unit activity are reported in human patients prior to seizure onset [18, 85]. Unfortunately, there is considerable variability in this early activity from seizure to seizure [17, 18], which may limit the ability to have accurate seizure pre-

diction. However, detecting seizures early (prior to overt behavioral manifestations) and intervening (optogenetically or otherwise) to truncate seizures to this pre-clinical stage could have a large impact on patient quality of life.

Imaging techniques are an additional example of the wide-range of expanding techniques that are being increasingly applied to the study of epilepsy, and include diffusion tensor imaging (DTI, reviewed in reference [28]), magnetic resonance imaging (MRI, which can be combined with optogenetics [52]), positron emission tomo-graphy (PET), single-photon emission computed tomography (SPECT, for a review see reference [57]), the new clarity brain [23], calcium imaging and voltage sensitive dyes (for recent reviews see references [28, 77]). Anatomical imaging techniques of neuronal projections in intact brains allow examination of network connections between brain regions in health and disease. Appreciating long-distance network connections, and how these shape local network connections [49, 86], will undoubtedly provide crucial information on seizure propagation mechanisms, as well as potentially mechanisms behind seizure initiation and termination. Functional imaging can reveal local as well as long-distance network dynamics, and is contributing substantially to our understanding of mechanisms in epilepsy. For example, a recent study using calcium imaging of epileptic tissue found not only variability in firing between neurons during epileptiform events, but also variability between epileptiform events, with each event comprised of different patterns of co-activated clusters of neurons [29].

Advances are certainly not limited to seizure detection or imaging techniques. Whole-genome sequencing, which is providing ever-expanding information on the genetics of epilepsies (reviewed in reference [61]), is an excellent example of the driving force that new technological advances can provide to the field. Additional diverse technological advances, including uncaging of GABA [96] and devices allowing focal cooling [68], are introducing unique new opportunities for studying and treating epilepsy. Advances in recording techniques are providing unprecedented information regarding the activity of neurons during epileptiform events. It is now possible to record from hundreds of units in human epileptic patients (for a discussion of the spike sorting techniques involved see reference [27]), providing a wealth of information on the involvement of neurons in seizures [18, 43, 85]. The novel information gained from these new techniques can aid in seizure detection and prediction discussed above. Importantly, this data can also be incorporated into "big data"-driven large-scale computational models [16, 70]. Hypotheses can then be tested *in silico*, and new hypotheses in turn generated to be tested *in vitro* or *in vivo* (for reviews of computational neuroscience in epilepsy, see references [21, 73]).

From the genetics, to the proteins, to the cell-types and networks critical in epilepsy, advances are being made and insights gained. Optogenetics, together with a vast array of novel technological developments, is expected to continue to light new avenues for studying the mechanisms of the epilepsies.

Acknowledgements This chapter on optogenetic approaches to epilepsy highlights the fundamental veracity of Phil's overarching conceptual framework that placed a major emphasis on the critical importance of rigorous, quantitative mechanistic understanding of epileptic neuronal circuits in order to develop new generations of temporally and spatially selective, more effective seizure control strategies.

Other Acknowledgements This work was supported by US National Institutes of Health grant NS74432 and the Swedish Brain Foundation.

References

1. Abaya TVF, Blair S, Tathireddy P, Rieth L, Solzbacher F (2012) A 3D glass optrode array for optical neural stimulation. Biomed Opt Express 3:3087–3104
2. Adesnik H, Bruns W, Taniguchi H, Huang ZJ, Scanziani M (2012) A neural circuit for spatial summation in visual cortex. Nature 490:226–231
3. Albus K, Wahab A, Heinemann U (2008) Standard antiepileptic drugs fail to block epileptiform activity in rat organotypic hippocampal slice cultures. Br J Pharmacol 154:709–724
4. Anikeeva P, Andalman AS, Witten I, Warden M, Goshen I, Grosenick L, Gunaydin LA, Frank LM, Deisseroth K (2012) Optetrode: a multichannel readout for optogenetic control in freely moving mice. Nat Neurosci 15:163–170
5. Arenkiel BR, Peca J, Davison IG, Feliciano C, Deisseroth K, Augustine GJ, Ehlers MD, Feng G (2007) In vivo light-induced activation of neural circuitry in transgenic mice expressing channelrhodopsin-2. Neuron 54:205–218
6. Armstrong C, Krook-Magnuson E, Oijala M, Soltesz I (2013) Closed-loop optogenetic intervention in mice. Nat Protoc 8:1475–1493
7. Armstrong C, Krook-Magnuson E, Soltesz I (2012) Neurogliaform and ivy cells: a major family of nNOS expressing GABAergic neurons. Front Neural Circuits 6:23
8. Armstrong C, Soltesz I (2012) Basket cell dichotomy in microcircuit function. J Physiol 590:683–694
9. Atasoy D, Aponte Y, Su HH, Sternson SM (2008) A FLEX switch targets channelrhodopsin-2 to multiple cell types for imaging and long-range circuit mapping. J Neurosci 28:7025–7030
10. Bartus RT, Baumann TL, Siffert J, Herzog CD, Alterman R, Boulis N, Turner DA, Stacy M, Lang AE, Lozano AM, Olanow CW (2013) Safety/feasibility of targeting the substantia nigra with AAV2-neurturin in Parkinson patients. Neurology 80:1698–1701
11. Bausch SB, McNamara JO (2000) Synaptic connections from multiple subfields contribute to granule cell hyperexcitability in hippocampal slice cultures. J Neurophysiol 84:2918–2932
12. Bentley JN, Chestek C, Stacey WC, Patil PG (2013) Optogenetics in epilepsy. Neurosurg Focus 34:E4
13. Berenyi A, Belluscio M, Mao D, Buzsaki G (2012) Closed-loop control of epilepsy by transcranial electrical stimulation. Science 337:735–737
14. Berndt A, Schoenenberger P, Mattis J, Tye KM, Deisseroth K, Hegemann P, Oertner TG (2011) High-efficiency channelrhodopsins for fast neuronal stimulation at low light levels. Proc Natl Acad Sci U S A 108:7595–7600
15. Berndt A, Yizhar O, Gunaydin LA, Hegemann P, Deisseroth K (2009) Bi-stable neural state switches. Nat Neurosci 12:229–234
16. Bezaire MJ, Soltesz I (2013) Quantitative assessment of CA1 local circuits: knowledge base for interneuron-pyramidal cell connectivity. Hippocampus 23:7595–7600
17. Bower MR, Buckmaster PS (2008) Changes in granule cell firing rates precede locally recorded spontaneous seizures by minutes in an animal model of temporal lobe epilepsy. J Neurophysiol 99: 2431–2442
18. Bower MR, Stead M, Meyer FB, Marsh WR, Worrell GA (2012) Spatiotemporal neuronal correlates of seizure generation in focal epilepsy. Epilepsia 53:807–816

19. Boyden ES, Zhang F, Bamberg E, Nagel G, Deisseroth K (2005) Millisecond-timescale, genetically targeted optical control of neural activity. Nat Neurosci 8:1263–1268

20. Butt SJ, Fuccillo M, Nery S, Noctor S, Kriegstein A, Corbin JG, Fishell G (2005) The temporal and spatial origins of cortical interneurons predict their physiological subtype. Neuron 48:591–604

21. Case MJ, Morgan RJ, Schneider CJ, Soltesz I (2012) Computer modeling of epilepsy. In: Jasper's basic mechanisms of the epilepsies, 4th edn. Oxford, New York, pp 298–311

22. Chow BY, Han X, Dobry AS, Qian X, Chuong AS, Li M, Henninger MA, Belfort GM, Lin Y, Monahan PE, Boyden ES (2010) High-performance genetically targetable optical neural silencing by light-driven proton pumps. Nature 463:98–102

23. Chung K, Wallace J, Kim SY, Kalyanasundaram S, Andalman AS, Davidson TJ, Mirzabekov JJ, Zalocusky KA, Mattis J, Denisin AK, Pak S, Bernstein H, Ramakrishnan C, Grosenick L, Gradinaru V, Deisseroth K (2013) Structural and molecular interrogation of intact biological systems. Nature 497:332–337

24. Cook MJ, O'Brien TJ, Berkovic SF, Murphy M, Morokoff A, Fabinyi G, D'Souza W, Yerra R, Archer J, Litewka L, Hosking S, Lightfoot P, Ruedebusch V, Sheffield WD, Snyder D, Leyde K, Himes D (2013) Prediction of seizure likelihood with a long-term, implanted seizure advisory system in patients with drug-resistant epilepsy: a first-in-man study. Lancet Neurol 12:563–571

25. Corbin JG, Butt SJ (2011) Developmental mechanisms for the generation of telencephalic interneurons. Dev Neurobiol 71:710–732

26. Drexel M, Kirchmair E, Wieselthaler-Holzl A, Preidt AP, Sperk G (2012) Somatostatin and neuropeptide Y neurons undergo different plasticity in parahippocampal regions in kainic acid-induced epilepsy. J Neuropathol Exp Neurol 71:312–329

27. Einevoll GT, Franke F, Hagen E, Pouzat C, Harris KD (2012) Towards reliable spike-train recordings from thousands of neurons with multielectrodes. Curr Opin Neurobiol 22:11–17

28. Engel J Jr, Thompson PM, Stern JM, Staba RJ, Bragin A, Mody I (2013) Connectomics and epilepsy. Curr Opin Neurol 26:186–194

29. Feldt Muldoon S, Soltesz I, Cossart R (2013) Spatially clustered neuronal assemblies comprise the microstructure of synchrony in chronically epileptic networks. Proc Natl Acad Sci U S A 110:3567–3572

30. Freund TF, Buzsaki G (1996) Interneurons of the hippocampus. Hippocampus 6:347–470

31. Fuentealba P, Begum R, Capogna M, Jinno S, Marton LF, Csicsvari J, Thomson A, Somogyi P, Klausberger T (2008) Ivy cells: a population of nitric-oxide-producing, slow-spiking GABAergic neurons and their involvement in hippocampal network activity. Neuron 57:917–929

32. Good LB, Sabesan S, Marsh ST, Tsakalis K, Treiman D, Iasemidis L (2009) Control of synchronization of brain dynamics leads to control of epileptic seizures in rodents. Int J Neural Syst 19:173–196

33. Gradinaru V, Thompson KR, Zhang F, Mogri M, Kay K, Schneider MB, Deisseroth K (2007) Targeting and readout strategies for fast optical neural control in vitro and in vivo. J Neurosci 27:14231–14238

34. Gradinaru V, Zhang F, Ramakrishnan C, Mattis J, Prakash R, Diester I, Goshen I, Thompson KR, Deisseroth K (2010) Molecular and cellular approaches for diversifying and extending optogenetics. Cell 141:154–165

35. Graves TD (2006) Ion channels and epilepsy. QJM 99:201–217

36. Guenthner CJ, Miyamichi K, Yang HH, Heller HC, Luo L (2013) Permanent genetic access to transiently active neurons via TRAP: targeted recombination in active populations. Neuron 78: 773–784

37. Gunaydin LA, Yizhar O, Berndt A, Sohal VS, Deisseroth K, Hegemann P (2010) Ultrafast optogenetic control. Nat Neurosci 13:387–392

38. Han X, Chow BY, Zhou H, Klapoetke NC, Chuong A, Rajimehr R, Yang A, Baratta MV, Winkle J, Desimone R, Boyden ES (2011) A high-light sensitivity optical neural silencer: development and application to optogenetic control of non-human primate cortex. Front Syst Neurosci 5:18

39. Han X, Qian X, Bernstein JG, Zhou HH, Franzesi GT, Stern P, Bronson RT, Graybiel AM, Desimone R, Boyden ES (2009) Millisecond-timescale optical control of neural dynamics in the nonhuman primate brain. Neuron 62:191–198

40. Han X, Qian X, Stern P, Chuong AS, Boyden ES (2009) Informational lesions: optical perturbation of spike timing and neural synchrony via microbial opsin gene fusions. Front Mol Neurosci 2:12

41. Howard A, Tamas G, Soltesz I (2005) Lighting the chandelier: new vistas for axo-axonic cells. Trends Neurosci 28:310–316

42. Hung C, Ling G, Mohanty SK, Chiao JJ (2013) An integrated μLED optrode for optogenetic stimulation and electrical recording. IEEE Trans Biomed Eng 60:225–229

43. Keller CJ, Truccolo W, Gale JT, Eskandar E, Thesen T, Carlson C, Devinsky O, Kuzniecky R, Doyle WK, Madsen JR, Schomer DL, Mehta AD, Brown EN, Hochberg LR, Ulbert I, Halgren E, Cash SS (2010) Heterogeneous neuronal firing patterns during interictal epileptiform discharges in the human cortex. Brain 133:1668–1681

44. Kim TI, McCall JG, Jung YH, Huang X, Siuda ER, Li Y, Song J, Song YM, Pao HA, Kim RH, Lu C, Lee SD, Song IS, Shin G, Al-Hasani R, Kim S, Tan MP, Huang Y, Omenetto FG, Rogers JA, Bruchas MR (2013) Injectable, cellular-scale optoelectronics with applications for wireless optogenetics. Science 340:211–216

45. Klausberger T, Somogyi P (2008) Neuronal diversity and temporal dynamics: the unity of hippocampal circuit operations. Science 321:53–57

46. Kokaia M, Andersson M, Ledri M (2013) An optogenetic approach in epilepsy. Neuropharmacology 69:89–95

47. Konermann S, Brigham MD, Trevino AE, Hsu PD, Heidenreich M, Cong L, Platt RJ, Scott DA, Church GM, Zhang F (2013) Optical control of mammalian endogenous transcription and epigenetic states. Nature 500:472–476

48. Krook-Magnuson E, Armstrong C, Oijala M, Soltesz I (2013) On-demand optogenetic control of spontaneous seizures in temporal lobe epilepsy. Nat Commun 4:1376

49. Krook-Magnuson E, Varga C, Lee SH, Soltesz I (2012) New dimensions of interneuronal specialization unmasked by principal cell heterogeneity. Trends Neurosci 35:175–184

50. Ledri M, Nikitidou L, Erdelyi F, Szabo G, Kirik D, Deisseroth K, Kokaia M (2012) Altered profile of basket cell afferent synapses in hyper-excitable dentate gyrus revealed by optogenetic and two-pathway stimulations. Eur J Neurosci 36:1971–1983

51. Lee G, Saito I (1998) Role of nucleotide sequences of loxP spacer region in Cre-mediated recombination. Gene 216:55–65

52. Lee JH, Durand R, Gradinaru V, Zhang F, Goshen I, Kim DS, Fenno LE, Ramakrishnan C, Deisseroth K (2010) Global and local fMRI signals driven by neurons defined optogenetically by type and wiring. Nature 465:788–792

53. Lee SY, Soltesz I (2011) Cholecystokinin: a multifunctional molecular switch of neuronal circuits. Dev Neurobiol 71:83–91

54. Lin JY (2011) A user's guide to channelrhodopsin variants: features, limitations and future developments. Exp Physiol 96:19–25

55. Lin JY, Lin MZ, Steinbach P, Tsien RY (2009) Characterization of engineered channelrhodopsin variants with improved properties and kinetics. Biophys J 96:1803–1814

56. Madisen L, Mao T, Koch H, Zhuo JM, Berenyi A, Fujisawa S, Hsu YW, Garcia AJ 3rd, Gu X, Zanella S, Kidney J, Gu H, Mao Y, Hooks BM, Boyden ES, Buzsaki G, Ramirez JM, Jones AR, Svoboda K, Han X, Turner EE, Zeng H (2012) A toolbox of Cre-dependent optogenetic transgenic mice for light-induced activation and silencing. Nat Neurosci 15:793–802

57. Maehara T (2007) Neuroimaging of epilepsy. Neuropathology 27:585–593

58. Markert JM, Medlock MD, Rabkin SD, Gillespie GY, Todo T, Hunter WD, Palmer CA, Feigenbaum F, Tornatore C, Tufaro F, Martuza RL (2000) Conditionally replicating herpes simplex virus mutant, G207 for the treatment of malignant glioma: results of a phase I trial. Gene Ther 7:867–874

59. Matta JA, Pelkey KA, Craig MT, Chittajallu R, Jeffries BW, McBain CJ (2013) Developmental origin dictates interneuron AMPA and NMDA receptor subunit composition and plasticity. Nat Neurosci 16:1032–1041

60. Mattis J, Tye KM, Ferenczi EA, Ramakrishnan C, O'Shea DJ, Prakash R, Gunaydin LA, Hyun M, Fenno LE, Gradinaru V, Yizhar O, Deisseroth K (2012) Principles for applying optogenetic tools derived from direct comparative analysis of microbial opsins. Nat Methods 9:159–172

61. Merwick A, O'Brien M, Delanty N (2012) Complex single gene disorders and epilepsy. Epilepsia 53(Suppl 4):81–91

62. Murphy AM, Rabkin SD (2013) Current status of gene therapy for brain tumors. Transl Res 161: 339–354

63. Nagel G, Szellas T, Huhn W, Kateriya S, Adeishvili N, Berthold P, Ollig D, Hegemann P, Bamberg E (2003) Channelrhodopsin-2, a directly light-gated cation-selective membrane channel. Proc Natl Acad Sci U S A 100:13940–13945

64. Osawa S, Iwasaki M, Hosaka R, Matsuzaka Y, Tomita H, Ishizuka T, Sugano E, Okumura E, Yawo H, Nakasato N, Tominaga T, Mushiake H (2013) Optogenetically induced seizure and the longitudinal hippocampal network dynamics. PLoS One 8:e60928

65. Paz JT, Davidson TJ, Frechette ES, Delord B, Parada I, Peng K, Deisseroth K, Huguenard JR (2013) Closed-loop optogenetic control of thalamus as a tool for interrupting seizures after cortical injury. Nat Neurosci 16:64–70

66. Raimondo JV, Kay L, Ellender TJ, Akerman CJ (2012) Optogenetic silencing strategies differ in their effects on inhibitory synaptic transmission. Nat Neurosci 15:1102–1104

67. Richichi C, Lin EJ, Stefanin D, Colella D, Ravizza T, Grignaschi G, Veglianese P, Sperk G, During MJ, Vezzani A (2004) Anticonvulsant and antiepileptogenic effects mediated by adeno-associated virus vector neuropeptide Y expression in the rat hippocampus. J Neurosci 24:3051–3059

68. Rothman SM (2009) The therapeutic potential of focal cooling for neocortical epilepsy. Neurotherapeutics 6:251–257

69. Royer S, Zemelman BV, Barbic M, Losonczy A, Buzsaki G, Magee JC (2010) Multi-array silicon probes with integrated optical fibers: light-assisted perturbation and recording of local neural circuits in the behaving animal. Eur J Neurosci 31:2279–2291

70. Schneider CJ, Bezaire M, Soltesz I (2012) Toward a full-scale computational model of the rat dentate gyrus. Front Neural Circuits 6:83

71. Schnutgen F, Doerflinger N, Calleja C, Wendling O, Chambon P, Ghyselinck NB (2003) A directional strategy for monitoring Cre-mediated recombination at the cellular level in the mouse. Nat Biotechnol 21:562–565

72. Smedemark-Margulies N, Trapani JG (2013) Tools, methods, and applications for optophysiology in neuroscience. Front Mol Neurosci 6:18

73. Soltesz I, Staley K (eds) (2008) Computational neuroscience in epilepsy. London/San Diego/Burlington
74. Sorensen AT, Nikitidou L, Ledri M, Lin EJ, During MJ, Kanter-Schlifke I, Kokaia M (2009) Hippocampal NPY gene transfer attenuates seizures without affecting epilepsy-induced impairment of LTP. Exp Neurol 215:328–333
75. Stark E, Koos T, Buzsaki G (2012) Diode probes for spatiotemporal optical control of multiple neurons in freely moving animals. J Neurophysiol 108:349–363
76. Sukhotinsky I, Chan AM, Ahmed OJ, Rao VR, Gradinaru V, Ramakrishnan C, Deisseroth K, Majewska AK, Cash SS (2013) Optogenetic delay of status epilepticus onset in an in vivo rodent epilepsy model. PLoS One 8:e62013
77. Takano H, Coulter DA (2012) Imaging of hippocampal circuits in epilepsy. In: Noebels JL, Avoli M, Rogawski MA et al (eds) Jasper's basic mechanisms of the epilepsies, 4th edn. Oxford, New York, pp 190–201
78. Tamura K, Ohashi Y, Tsubota T, Takeuchi D, Hirabayashi T, Yaguchi M, Matsuyama M, Sekine T, Miyashita Y (2012) A glass-coated tungsten microelectrode enclosing optical fibers for optogenetic exploration in primate deep brain structures. J Neurosci Methods 211:49–57
79. Tanaka KF, Matsui K, Sasaki T, Sano H, Sugio S, Fan K, Hen R, Nakai J, Yanagawa Y, Hasuwa H, Okabe M, Deisseroth K, Ikenaka K, Yamanaka A (2012) Expanding the repertoire of optogenetically targeted cells with an enhanced gene expression system. Cell Rep 2:397–406
80. Taniguchi H, He M, Wu P, Kim S, Paik R, Sugino K, Kvitsiani D, Fu Y, Lu J, Lin Y, Miyoshi G, Shima Y, Fishell G, Nelson SB, Huang ZJ (2011) A resource of Cre driver lines for genetic targeting of GABAergic neurons in cerebral cortex. Neuron 71:995–1013
81. Taniguchi H, Lu J, Huang ZJ (2013) The spatial and temporal origin of chandelier cells in mouse neocortex. Science 339:70–74
82. Tomita H, Sugano E, Fukazawa Y, Isago H, Sugiyama Y, Hiroi T, Ishizuka T, Mushiake H, Kato M, Hirabayashi M, Shigemoto R, Yawo H, Tamai M (2009) Visual properties of transgenic rats harboring the channelrhodopsin-2 gene regulated by the thy-1.2 promoter. PLoS One 4:e7679
83. Toni N, Laplagne DA, Zhao C, Lombardi G, Ribak CE, Gage FH, Schinder AF (2008) Neurons born in the adult dentate gyrus form functional synapses with target cells. Nat Neurosci 11:901–907
84. Tonnesen J, Sorensen AT, Deisseroth K, Lundberg C, Kokaia M (2009) Optogenetic control of epileptiform activity. Proc Natl Acad Sci U S A 106:12162–12167
85. Truccolo W, Donoghue JA, Hochberg LR, Eskandar EN, Madsen JR, Anderson WS, Brown EN, Halgren E, Cash SS (2011) Single-neuron dynamics in human focal epilepsy. Nat Neurosci 14:635–641
86. Varga C, Lee SY, Soltesz I (2010) Target-selective GABAergic control of entorhinal cortex output. Nat Neurosci 13:822–824
87. Vezzani A (2007) The promise of gene therapy for the treatment of epilepsy. Expert Rev Neurother 7:1685–1692
88. Visel A, Taher L, Girgis H, May D, Golonzhka O, Hoch RV, McKinsey GL, Pattabiraman K, Silberberg SN, Blow MJ, Hansen DV, Nord AS, Akiyama JA, Holt A, Hosseini R, Phouanenavong S, Plajzer-Frick I, Shoukry M, Afzal V, Kaplan T, Kriegstein AR, Rubin EM, Ovcharenko I, Pennacchio LA, Rubenstein JL (2013) A high-resolution enhancer atlas of the developing telencephalon. Cell 152:895–908
89. Wang J, Wagner F, Borton DA, Zhang J, Ozden I, Burwell RD, Nurmikko AV, Wagenen R, Diester I, Deisseroth K (2012) Integrated device for combined optical neuromodulation and electrical recording for chronic in vivo applications. J Neural Eng 9:016001
90. Wentz CT, Bernstein JG, Monahan P, Guerra A, Rodriguez A, Boyden ES (2011) A wirelessly powered and controlled device for optical neural control of freely-behaving animals. J Neural Eng 8:046021
91. White AM, Williams PA, Ferraro DJ, Clark S, Kadam SD, Dudek FE, Staley KJ (2006) Efficient unsupervised algorithms for the detection of seizures in continuous EEG recordings from rats after brain injury. J Neurosci Methods 152:255–266
92. Woodson W, Nitecka L, Ben-Ari Y (1989) Organization of the GABAergic system in the rat hippocampal formation: a quantitative immunocytochemical study. J Comp Neurol 280:254–271
93. Wu C, Sharan AD (2013) Neurostimulation for the treatment of epilepsy: a review of current surgical interventions. Neuromodulation 16:10–24, discussion 24
94. Wykes RC, Heeroma JH, Mantoan L, Zheng K, Macdonald DC, Deisseroth K, Hashemi KS, Walker MC, Schorge S, Kullmann DM (2012) Optogenetic and potassium channel gene therapy in a rodent model of focal neocortical epilepsy. Sci Transl Med 4:161ra152
95. Xu X, Olivas ND, Levi R, Ikrar T, Nenadic Z (2010) High precision and fast functional mapping of cortical circuitry through a novel combination of voltage sensitive dye imaging and laser scanning photostimulation. J Neurophysiol 103:2301–2312
96. Yang X, Rode DL, Peterka DS, Yuste R, Rothman SM (2012) Optical control of focal epilepsy in vivo with caged gamma-aminobutyric acid. Ann Neurol 71:68–75
97. Yizhar O, Fenno LE, Prigge M, Schneider F, Davidson TJ, O'Shea DJ, Sohal VS, Goshen I, Finkelstein J, Paz JT, Stehfest K, Fudim R, Ramakrishnan C, Huguenard JR, Hegemann P, Deisseroth K (2011) Neocortical excitation/inhibition balance in information processing and social dysfunction. Nature 477:171–178
98. Zhang F, Aravanis AM, Adamantidis A, de Lecea L, Deisseroth K (2007) Circuit-breakers: optical tech-

nologies for probing neural signals and systems. Nat Rev Neurosci 8:577–581

99. Zhang F, Gradinaru V, Adamantidis AR, Durand R, Airan RD, de Lecea L, Deisseroth K (2010) Optogenetic interrogation of neural circuits: technology for probing mammalian brain structures. Nat Protoc 5:439–456

100. Zhang F, Prigge M, Beyriere F, Tsunoda SP, Mattis J, Yizhar O, Hegemann P, Deisseroth K (2008) Redshifted optogenetic excitation: a tool for fast neural control derived from Volvox carteri. Nat Neurosci 11:631–633

101. Zhang J, Laiwalla F, Kim JA, Urabe H, Wagenen RV, Song Y-K, Connors BW, Zhang F, Deisseroth K, Nurmikko AV (2009) Integrated device for optical stimulation and spatiotemporal electrical recording of neural activity in light-sensitized brain tissue. J Neural Eng 6:055007

102. Zhang YP, Oertner TG (2007) Optical induction of synaptic plasticity using a light-sensitive channel. Nat Methods 4:139–141

103. Zhu P, Narita Y, Bundschuh ST, Fajardo O, Scharer YP, Chattopadhyaya B, Bouldoires EA, Stepien AE, Deisseroth K, Arber S, Sprengel R, Rijli FM, Friedrich RW (2009) Optogenetic dissection of neuronal circuits in zebrafish using viral gene transfer and the Tet system. Front Neural Circuits 3:21

104. Zorzos AN, Boyden ES, Fonstad CG (2010) Multiwaveguide implantable probe for light delivery to sets of distributed brain targets. Opt Lett 35: 4133–4135

Index

A

AAV. *See* Adeno-associated virus (AAV)
Aberrant neurogenesis, 198, 204
Absence seizures, 6, 16, 59, 68, 82–86, 88, 110, 134, 172, 212–215, 245, 279, 313, 331
Acetylcholine, 278, 314
Acquired epilepsy, 26, 111, 112, 117, 126, 152–156, 216, 221, 292
Adeno-associated virus (AAV), 157, 320, 322, 327
Adult neurogenesis, 139, 143, 144
Animal models, 14–16, 19, 26, 31, 48, 56, 58–60, 64, 66–68, 83, 84, 98, 99, 110, 112, 118, 126, 127, 134, 139–144, 152, 154, 155, 162, 172, 174, 176, 177, 196, 200–202, 204, 212, 215–217, 221, 224, 234, 235, 238, 243–250, 256, 266, 270, 273–279, 284, 288, 291, 292, 299, 301, 302, 310
Antiepileptic drugs (AEDs), 175, 223, 234, 235, 238, 254, 256, 257, 276, 284, 286, 287, 290, 312, 314
Anti-seizure drugs (ASDs), 51, 296–304, 312
Apoptosis, 113–115, 177
Arch, 325
Arousal, 66–69, 87
α1 subunit, 128, 137, 138, 142, 143, 221
Autophagy, 113–114

B

Behavioral arrest, 16–20, 67, 68
Brain endothelium, 315
Brainstem, 7, 66, 68, 69, 84, 88, 96, 99, 101, 102, 278

C

Cellular electrophysiology, 185, 190
Cerebrovasculature, 254, 288
Channelrhodopsin, 321, 323–325, 328
Chloride channel, 135, 137, 214, 312, 313
Chloride cotransporter, 200, 291
Chronic models, 186–191, 288, 291
Comorbidity, 26, 51, 265–270, 273–276, 278, 312–313, 316
Complexity, 27, 112, 128, 144, 284, 311–312
Complex partial seizure, 57–59, 65, 66, 110
Consciousness, 4, 5, 31, 57, 59, 66, 67, 69
Convulsion, 7, 16, 155, 300

Convulsive, 86–88, 102, 110, 126, 288
Cortex, 4, 5, 7, 9, 13, 17, 18, 29, 30, 32, 34, 35, 46, 48, 49, 56–60, 64–69, 73, 83–87, 96, 102, 112, 125, 154, 155, 176, 190, 200, 202, 213–216, 219–221, 224, 236, 254, 327

D

Dentate gyrus (DG), 17, 32, 35, 47, 134–144, 152–158, 162, 163, 165, 174, 188, 196–198, 200–202, 216–224, 248
Depression, 13–16, 30, 33, 35, 51, 64, 144, 267–269, 273–279
Disease models,
Drug-refractory epilepsy, 296
Drug resistance, 25, 97, 235, 254, 256, 257, 312
Dysplasia, 11, 13, 74, 162, 171, 177, 196, 200, 201

E

Ectopic granule cell, 143, 162
Electrocorticogram, 75
Electroencephalogram (EEG), 5–20, 28, 29, 43–46, 48–52, 56–60, 65, 66, 72, 78, 82–84, 87, 88, 97, 113, 126, 189, 190, 212, 287, 327–331
Epilepsy
  surgery, 7, 11, 34, 35, 98, 212
  syndromes, 134, 196, 212, 235, 247, 275, 310, 311, 314, 316
Epileptic
  dogs, 288–290
  rodents, 17, 36, 301
  seizures, 4, 5, 44, 57, 97, 99, 188, 190, 212, 248, 253–262, 268, 289
Epileptiform, 4–10, 15, 25–37, 44, 48, 50, 51, 56, 57, 73, 74, 76, 78, 96, 128–130, 134, 141, 175, 177, 187, 196, 235, 237–239, 259, 300, 326–328, 331, 332
Epileptogenesis, 19, 26–30, 33, 35, 37, 44, 56, 58, 60, 72, 96–99, 102, 110, 115–118, 126–127, 129, 130, 133–145, 162, 165, 166, 171, 175, 176, 178, 187, 188, 198, 202, 204, 216, 222, 224, 235–239, 248, 249, 254, 255, 258–262, 274, 275, 279, 284, 292, 314
Epileptogenic zone, 10, 12, 14, 56–60, 189–191
Epileptogenisis, 26, 29, 36–37

H.E. Scharfman and P.S. Buckmaster (eds.), *Issues in Clinical Epileptology: A View from the Bench*,
Advances in Experimental Medicine and Biology 813, DOI 10.1007/978-94-017-8914-1,
© Springer Science+Business Media Dordrecht 2014

Extracellular calcium, 34
Extracellular matrix (ECM), 215, 254, 255, 257,
    260–261

F
Febrile seizures, 77, 97, 98, 134, 144, 172, 199, 248
Focal seizure, 5–7, 12–14, 56, 63–69, 110
Fragile X, 244–247, 304

G
GABA, 18, 31, 32, 134, 135, 137–139, 141–144,
    151–158, 163, 172–174, 201, 204, 213, 257, 258,
    287, 290, 291, 298, 315, 331
GABA$_A$ receptor, 8, 76, 77, 87, 117, 125, 127, 128, 134,
    138, 143, 144, 155, 156, 173, 196, 198–203, 214,
    221–223, 298, 311, 325
GABA receptor, 32, 134, 137–138, 141–143, 171, 172,
    212, 214, 216, 217, 276
Gap junctions, 74, 76, 77, 174, 176, 257
Generalized tonic clonic seizures, 6, 82, 87–88
Gene testing, 313–314
Genetic absence models, 87, 88
Genetic epilepsy, 134, 303
Glia-neuronal interactions, 254, 255, 257–260
Granule cell (GC), 33, 100, 115, 117, 134–136, 138, 142,
    152, 153, 155, 157, 158, 162–165, 196, 197,
    200–202, 216–219, 221–223

H
Halorhodopsin, 324, 325, 327, 328, 330
HCN channel. *See* Hyperpolarization-activated cyclic
    nucleotide-gated (HCN) channel
High frequency oscillation (HFO), 7, 47, 48, 72, 76,
    187, 189–191
Hilus, 135, 136, 141, 143, 144, 152–154, 156–158, 162,
    188, 217, 218, 222–224, 248, 249
Homeostasis, 28, 47, 114, 117, 124–130, 188, 257, 325
Hyperpolarization-activated cyclic nucleotide-gated
    (HCN) channel, 56, 188, 212, 213

I
Ictal, 3–20, 32, 35, 36, 50, 56–61, 64–69, 82, 84, 88, 96,
    97, 102, 203, 254, 259, 267
Idiopathic generalized epilepsies, 86–88
IL-6, 171, 172, 174, 178
IL-1β, 115–117, 170–178
Inflammation, 29, 58, 98, 114–117, 169–178, 215,
    235–237, 256, 257
Interictal, 3–20, 28–30, 34–36, 44, 47–52, 56, 58, 65, 74,
    75, 96, 134, 188, 190, 200, 238, 254, 268, 276, 284
Interictal spike, 8–11, 15, 17, 18, 29, 30, 34, 56, 58
Interneuron, 15, 18, 31, 46, 50, 72, 73, 77, 78, 83, 110,
    112, 115, 118, 125, 135–137, 140, 141, 152–156,
    158, 162, 166, 187, 188, 222, 223, 258, 260, 303,
    314, 316, 321, 322, 325, 326, 328
In vitro models, 27, 29, 31, 176, 201, 326

L
Latent period, 97, 98, 162, 176, 187, 238, 248,
    249, 288
Learning and memory, 101, 127–129, 278
Ligand-gated ion channel, 214, 221–224, 303
Local field potential oscillations, 45, 65, 85, 259

M
Maladaptive, 100, 102, 115, 129, 204, 237, 238
Mammalian target of rapamycin (mTOR), 37, 171, 203,
    279, 304
Mechanism of action, 290, 296–298
Modifier, 311, 313, 316
Monotherapy, 276, 296, 297, 299, 303, 304
mTOR. *See* Mammalian target of rapamycin (mTOR)
Mutation, 19, 37, 134, 155, 187, 212, 213, 215,
    244–247, 254, 261, 262, 286, 292, 300,
    303, 304, 310–316, 321
Myoclonic juvenile seizures, 87, 97

N
Necroptosis, 114
Necrosis, 114
Neural network, 169
Neural plasticity, 96
Neurodegeneration, 100, 114–117, 175, 215, 216,
    221–224, 286
Neurogenesis, 98, 99, 139, 143, 144, 175, 196–199,
    204, 215, 259

O
On-demand, 320, 328, 329, 331
Optrode, 326
Oscillation, 7, 16–18, 20, 44–49, 52, 72–78, 83, 84, 134,
    187, 189–191, 213, 224

P
Parvalbumin, 137, 140, 153, 188, 260, 320, 322, 328
Pathophysiology, 27, 36, 83, 87, 186–189, 191, 213,
    234–236, 239, 255, 275, 279, 299, 310
Pedunculopontine tegmental nucleus, 68
pH, 72, 77, 124, 325
Phagoptosis, 114
Pharmacoresistant epilepsy, 31, 34, 171, 284, 286,
    288, 296
Phenotype, 14, 116, 117, 154, 196, 212, 215, 236,
    244–247, 261, 278, 279, 303, 304, 310–316
Pilocarpine, 33, 50, 67, 98, 99, 116, 118, 128, 139,
    142–144, 153, 155–158, 162–165, 177, 187, 236,
    249, 256, 277, 278, 284, 327
Plasticity, 47, 50–52, 96, 98–102, 124–130, 134, 140,
    144, 156, 165, 172–174, 203, 205, 222, 223, 236,
    258–261, 298, 304, 311, 316
Polytherapy, 296, 297, 301–304
Posttraumatic, 99, 237
Potentiation, 51, 124, 173, 259

Pre-ictal, 11, 14, 15, 48
Primary generalized seizure, 59–60, 87–88
Prophylaxis, 234–235, 238
Psychiatric comorbidity, 267–269
Pyroptosis, 114–115

R
Reactive astrogliosis, 176
Responsive, 8, 174, 235, 245, 270, 286, 315, 328

S
Seizure-like, 14–16, 28, 30–32, 34, 96, 102, 128, 140,
    155, 284
Sharp wave, 5, 6, 17–18, 46, 47, 51, 72, 76
Simple (elementary) partial seizure, 57, 66, 67
Sleep, 8, 16–19, 28, 44–48, 50–52, 64–69, 83, 101, 124,
    238, 267, 273, 277
Slow waves, 5, 18, 44, 46, 50–52, 64–68
Somatic comorbidity, 267
Somatostatin, 135, 137, 152, 153, 157, 187, 188, 321, 322
Spike-wave discharge, 8, 19–20, 59, 82, 85, 279
Sprouting, 33, 59, 98, 112, 115, 117, 126, 140, 143, 144,
    156, 158, 162–166, 188, 221, 235, 236, 238, 248,
    325, 326
Status epilepticus, 16, 27–29, 31, 32, 35, 50, 52, 59, 67,
    97, 110, 111, 115, 116, 118, 128, 129, 134, 136,
    152, 153, 155, 157, 162–165, 175, 187–189, 199,
    215, 216, 235–238, 248, 258, 261, 278, 289, 315

Synapses, 72–74, 76, 96, 101, 115, 117, 124–129,
    136–138, 141, 142, 144, 155, 157, 162–164, 173,
    176, 178, 188, 199, 214, 215, 221, 235, 246,
    258–260, 298, 300
Synaptic plasticity, 47, 50, 51, 124, 126, 129, 174, 223,
    258–261, 298, 304, 311
Synaptic transmission, 47, 77, 125, 128, 141, 172–174,
    189, 215, 259, 261, 298

T
Temporal lobe epilepsy, 13, 14, 21, 48, 51, 57, 58, 61,
    64, 67, 97, 110, 111, 134, 152, 162, 164–166,
    171, 187–189, 196, 217, 238, 244, 247, 256, 268,
    276–278, 291, 299, 328–330
Thalamus, 20, 59, 65, 66, 68, 69, 83, 84, 86, 87, 101,
    112, 115, 213–215, 328
Theta, 6, 16–19, 44–47, 65
TNF-α, 114, 115, 117, 171–174, 178
Transition to seizure, 13, 15, 18, 32, 72, 77, 78
Translation, 25, 27, 29, 234, 239, 246, 247,
    288, 303

V
Voltage-gated ion channel, 156, 170, 171, 174, 177, 187,
    188, 212–217, 224, 297, 300

Z
Zebrafish, 244, 284–286, 288, 291–292

CPSIA information can be obtained at www.ICGtesting.com
Printed in the USA
LVOW02*1209290714

396543LV00003B/29/P